T0387901

Elastomer Blends and Composites

Elastomer Blends and Composites

Principles, Characterization, Advances, and Applications

Edited by

Sanjay Mavinkere Rangappa

Senior Research Scientist, King Mongkut's University of Technology North Bangkok (KMUTNB), Bangkok, Thailand

Jyotishkumar Parameswaranpillai

Associate Professor, Department of Science, Faculty of Science & Technology, Alliance University, Bengaluru, Karnataka, India

Suchart Siengchin

President, King Mongkut's University of Technology, North Bangkok (KMUTNB), Thailand

Togay Ozbakkaloglu

Professor, Ingram School of Engineering, Texas State University, San Marcos, TX, United States

Elsevier
Radarweg 29, PO Box 211, 1000 AE Amsterdam, Netherlands
The Boulevard, Langford Lane, Kidlington, Oxford OX5 1GB, United Kingdom
50 Hampshire Street, 5th Floor, Cambridge, MA 02139, United States

Library of Congress Cataloging-in-Publication Data
A catalog record for this book is available from the Library of Congress

British Library Cataloguing-in-Publication Data
A catalogue record for this book is available from the British Library

ISBN: 978-0-323-85832-8

For information on all Elsevier publications visit our website at
https://www.elsevier.com/books-and-journals

Publisher: Matthew Deans
Acquisitions Editor: Edward Payne
Editorial Project Manager: John Leonard
Production Project Manager: Prem Kumar Kaliamoorthi
Cover Designer: Greg Harris

Typeset by TNQ Technologies

Contents

Contributors xiii

Preface xvii

1. Introduction to elastomers 1

Jyotishkumar Parameswaranpillai, C.D. Midhun Dominic, Sanjay Mavinkere Rangappa, Suchart Siengchin and Togay Ozbakkaloglu

1.1 Introduction 1

1.2 Vulcanization/cross-linking in elastomers 2

1.3 Elastomeric composites and blends 2

1.4 Recent developments in elastomeric blends and composites 3

1.5 Conclusion 6

Acknowledgments 7

References 7

2. Manufacturing methods of elastomer blends and composites 11

M. Ramesh, D. Balaji, L. Rajeshkumar and V. Bhuvaneswari

2.1 Introduction 11

2.2 Preparation techniques 13

2.3 Conclusion 29

References 29

3. Elastomer-based blends 33

Aswathy Jayakumar, Sabarish Radoor, Jyotishkumar Parameswaranpillai, E.K. Radhakrishnan, Indu C. Nair and Suchart Siengchin

3.1 Introduction 33

3.2 Compatibilization of elastomer-based blends 34

3.3 Impact of nanofillers on elastomer-based blends 34
3.4 Fabrication methods of elastomers 35
3.5 Processing and characterization methods of elastomers-based blends 35
3.6 Properties of elastomers-based blends 36
3.7 Applications of elastomer-based blends 36
3.8 Conclusion 41
References 41

4. Elastomer-based filler composites 45

S.N. Vasantha Kumar and P.C. Sharath

4.1 Introduction 45
4.2 Preparation and properties of fillers 46
4.3 Conclusions and perspectives 52
References 53

5. Engineering applications of elastomer blends and composites 57

Naga Srilatha Cheekuramelli, Dattatraya Late, S. Kiran and Baijayantimala Garnaik

5.1 Introduction 57
5.2 Elastomer blends and composites processing methods 58
5.3 Elastomer blends and composites engineering applications 62
5.4 Conclusion 76
Acknowledgments 77
References 77

6. Rheology of elastomer blends and composites 83

Mariacristina Gagliardi

6.1 Introduction 83
6.2 Basic aspects of rheology 84
6.3 Basic key terms 85

6.4 Rheological models 86
6.5 Newtonian fluids (viscous liquids) 86
6.6 Non-Newtonian fluids 88
6.7 Conditions affecting the rheological properties of materials 89
6.8 Effect of temperature 90
6.9 Effect of the system structure at the micro-/nano-scale 90
6.10 Applied rheology in elastomers, blends, and composites thereof 91
6.11 Static versus dynamic rheological tests 92
6.12 Laboratory tests and instrumentations 96
6.13 Cone-and-plate rheometer 97
6.14 Capillary viscometer 97
6.15 Mooney viscometer 97
6.16 Constitutive rheological models 98
6.17 Uncured rubber melts 98
6.18 Elastomer blends 100
6.19 Elastomer composites 101
6.20 Conclusions 101
References 102

7. Morphological characteristics of elastomer blends and composites 103

A.V. Kiruthika

List of abbreviations 103
7.1 Introduction 103
7.2 Morphology 104
7.3 Effect of plant fiber-reinforced elastomer composites 117
7.4 Effect of synthetic fiber-reinforced elastomer composites 120
7.5 Conclusions 121
References 122

8. Mechanical behavior of elastomer blends and composites 127

G. Rajeshkumar and S. Arvindh Seshadri

8.1 Introduction 127
8.2 Mechanical behavior of elastomer blends 128
8.3 SMP of elastomer blends 128
8.4 DMP of elastomer blends 133
8.5 Mechanical behavior of elastomer composites 136
8.6 SMP of elastomer composites 136
8.7 DMP of elastomer composites 141
8.8 Conclusions 144
References 144

9. Thermal behavior of elastomer blends and composites 149

Atul Kumar Maurya, Rupam Gogoi and Gaurav Manik

9.1 Introduction 149
9.2 Thermodynamics of the rubber–rubber and rubber–polymer blends 150
9.3 Thermal behavior of blends 152
9.4 Thermal behavior of elastomeric composites 161
9.5 Conclusion 164
References 165

10. Viscoelastic behavior of elastomer blends and composites 171

Rupam Gogoi, Gaurav Manik and Sushanta K. Sahoo

10.1 Introduction 171
10.2 Viscoelasticity of elastomer blends 180
10.3 Viscoelasticity of elastomer composites 185
10.4 Conclusion 190
References 191

11. Spectroscopy of elastomer blends and composites 195

Sabarish Radoor, Jasila Karayil, Amritha Bemplassery, Aswathy Jayakumar, Jyotishkumar Parameswaranpillai and Suchart Siengchin

11.1 Introduction 195
11.2 FT-IR and Raman spectroscopy 196
11.3 Fluorescence spectroscopy 199
11.4 NMR spectroscopy 202
11.5 Conclusion 204
Acknowledgments 204
References 204

12. Wide-angle X-ray diffraction and small-angle X-ray scattering studies of elastomer blends and composites 209

Angel Romo-Uribe

12.1 Focus 209
12.2 X-ray diffraction 210
12.3 Methods in X-ray scattering 215
12.4 Wide-angle X-ray diffraction, WAXD 217
12.5 Small-angle X-ray scattering (SAXS) 224
12.6 Applications 234
12.7 Synchrotron scattering 237
12.8 Conclusions 239
References 240
Further reading 241

13. Theoretical modeling and simulation of elastomer blends and nanocomposites 243

Jitha S. Jayan, B.D.S. Deeraj, Appukuttan Saritha and Kuruvilla Joseph

13.1 Introduction 243
13.2 Simulations of elastomers 244

13.3 Modeling study of elastomer blends and composites 250
13.4 Major concern/challenges 256
13.5 Conclusion and future scope 257
References 257

14. Recycling of elastomer blends and composites 269
Jitha S. Jayan, A.S. Sethulekshmi, Gopika Venu, B.D.S. Deeraj, Appukuttan Saritha and Kuruvilla Joseph
14.1 Introduction 269
14.2 Devulcanization methods 271
14.3 Value-added products from revulcanized elastomeric blends and composites 286
14.4 Conclusion 296
14.5 Future perspectives 297
References 297
Further reading 304

15. Applications of elastomer blends and composites 305
Sudheer Kumar, Sukhila Krishnan and Smita Mohanty
15.1 Introduction 305
15.2 Polyurethane-based elastomer blends and composites 307
15.3 Silicone-based elastomer blends and composites 314
15.4 Ethylene-propylene-diene monomer (EPDM)-based elastomer 317
15.5 Other elastomers 321
15.6 Conclusions 324
References 324

16. Properties of elastomer—biological phenolic resin composites 331

Kushairi Mohd Salleh, Marhaini Mostapha, Kam Sheng Lau and Sarani Zakaria

16.1 Introduction 331
16.2 Biological phenolic resin 334
16.3 Properties of blended composite 340
16.4 Conclusion 346
16.5 Future trend 346
Acknowledgments 347
References 347

17. Advances in stimuli-responsive and functional thermoplastic elastomers 353

Jiaqi Yan and Richard J. Spontak

17.1 Overview of thermoplastic elastomers and their applications 353
17.2 Introduction to model block copolymers as TPEs 356
17.3 Physical modification of nonpolar TPEs and their applications 366
17.4 Chemical modification of nonpolar TPEs and their applications 381
17.5 Morphological development and applications of charged TPEs 385
17.6 Concluding remarks 391
Acknowledgments 392
References 392

Index 405

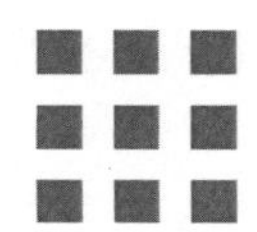

Contributors

D. Balaji Department of Mechanical Engineering, KPR Institute of Engineering and Technology, Coimbatore, Tamil Nadu, India

Amritha Bemplassery Department of Chemistry, National Institute of Technology, Calicut, Kerala, India

V. Bhuvaneswari Department of Mechanical Engineering, KPR Institute of Engineering and Technology, Coimbatore, Tamil Nadu, India

Naga Srilatha Cheekuramelli Polymer Science and Engineering Division, CSIR-National Chemical Laboratory, Pune, MH, India; Academy of Scientific and Innovative Research, ACSIR Headquarters, CSIR-HRDC Campus, Ghaziabad, UP, India

B.D.S. Deeraj Department of Chemistry, Indian Institute of Space Science and Technology, Valiamala, Thiruvananthapuram, Kerala, India

Mariacristina Gagliardi NEST, Istituto Nanoscienze-CNR and Scuola Normale Superiore, Pisa, Italy

Baijayantimala Garnaik Polymer Science and Engineering Division, CSIR-National Chemical Laboratory, Pune, MH, India; Academy of Scientific and Innovative Research, ACSIR Headquarters, CSIR-HRDC Campus, Ghaziabad, UP, India

Rupam Gogoi Department of Polymer and Process Engineering, Indian Institute of Technology Roorkee, Saharanpur, UP, India

Aswathy Jayakumar School of Biosciences, Mahatma Gandhi University, Kottayam, Kerala, India; Department of Mechanical and Process Engineering, The Sirindhorn International Thai-German Graduate School of Engineering (TGGS), King Mongkut's University of Technology North Bangkok, Bangkok, Thailand; Materials and Production Engineering, The Sirindhorn International Thai-German Graduate School of Engineering (TGGS), King Mongkut's University of Technology North Bangkok, Bangkok, Thailand

Jitha S. Jayan Department of Chemistry, School of Arts and Sciences, Amrita Vishwa Vidyapeetham, Amritapuri, Kollam, Kerala, India

Kuruvilla Joseph Department of Chemistry, Indian Institute of Space Science and Technology, Valiamala, Thiruvananthapuram, Kerala, India

Jasila Karayil Government Women's Polytechnic College, Calicut, Kerala, India

S. Kiran Polymer Science and Engineering Division, CSIR-National Chemical Laboratory, Pune, MH, India; Academy of Scientific and Innovative Research, ACSIR Headquarters, CSIR-HRDC Campus, Ghaziabad, UP, India

A.V. Kiruthika Department of Physics, Seethalakshmi Achi College for Women, Karaikudi, Tamil Nadu, India

Sukhila Krishnan Sahrdaya College of Engineering and Technology, Department of Applied Science and Humanities, Kodakara, Kerala, India

Sudheer Kumar School for Advanced Research in Petrochemicals (SARP), Laboratory for Advanced Research in Polymeric Materials (LARPM), Central Institute of Petrochemicals Engineering and Technology (CIPET), Bhubaneswar, Odisha, India

Dattatraya Late Academy of Scientific and Innovative Research, ACSIR Headquarters, CSIR-HRDC Campus, Ghaziabad, UP, India; Physical and Materials Chemistry Division, CSIR-National Chemical Laboratory, Pune, MH, India

Kam Sheng Lau Bioresource & Biorefinery Research Group, Department of Applied Physics, Faculty of Science and Technology, Universiti Kebangsaan Malaysia, Bangi, Selangor, Malaysia

Gaurav Manik Department of Polymer and Process Engineering, Indian Institute of Technology Roorkee, Saharanpur, UP, India

Atul Kumar Maurya Department of Polymer and Process Engineering, Indian Institute of Technology Roorkee, Saharanpur, UP, India

Sanjay Mavinkere Rangappa Materials and Production Engineering, The Sirindhorn International Thai-German Graduate School of Engineering (TGGS), King Mongkut's University of Technology North Bangkok, Bangkok, Thailand

C.D. Midhun Dominic Department of Chemistry, Sacred Heart College (Autonomous), Kochi, Kerala, India

Smita Mohanty School for Advanced Research in Petrochemicals (SARP), Laboratory for Advanced Research in Polymeric Materials (LARPM), Central Institute of Petrochemicals Engineering and Technology (CIPET), Bhubaneswar, Odisha, India

Marhaini Mostapha Bioresource & Biorefinery Research Group, Department of Applied Physics, Faculty of Science and Technology, Universiti Kebangsaan Malaysia, Bangi, Selangor, Malaysia; HICOE-Centre for Biofuels and Biochemical Research (CBBR), Institute of Self-Sustainable Building (ISB), Universiti Teknologi Petronas, Seri Iskandar, Perak, Malaysia

Indu C. Nair Department of Biotechnology, Sahodaran Ayyappan Smaraka Sree Narayana Dharma Paripalana Yogam (SAS SNDPYOGAM), College, Konni, Pathanamthitta, Kerala, India

Togay Ozbakkaloglu Department of Civil Engineering, Texas State University, San Marcos, TX, United States

Jyotishkumar Parameswaranpillai Department of Science, Faculty of Science & Technology, Alliance University, Bengaluru, Karnataka, India

E.K. Radhakrishnan School of Biosciences, Mahatma Gandhi University, Kottayam, Kerala, India

Sabarish Radoor Materials and Production Engineering, The Sirindhorn International Thai-German Graduate School of Engineering (TGGS), King Mongkut's University of Technology North Bangkok, Bangkok, Thailand; Department of Mechanical and Process Engineering, The Sirindhorn International Thai-German Graduate School of Engineering (TGGS), King Mongkut's University of Technology North Bangkok, Bangkok, Thailand

G. Rajeshkumar Department of Mechanical Engineering, PSG Institute of Technology and Applied Research, Coimbatore, TN, India

L. Rajeshkumar Department of Mechanical Engineering, KPR Institute of Engineering and Technology, Coimbatore, Tamil Nadu, India

M. Ramesh Department of Mechanical Engineering, KIT-Kalaignarkarunanidhi Institute of Technology, Coimbatore, Tamil Nadu, India

Angel Romo-Uribe Research & Development, Advanced Science & Technology Division, Johnson & Johnson Vision Care Inc., FL, Jacksonville, United States

Sushanta K. Sahoo Materials Science and Technology Division, CSIR – National Institute for Interdisciplinary Science and Technology, Thiruvananthapuram, Kerala, India

Kushairi Mohd Salleh Bioresource & Biorefinery Research Group, Department of Applied Physics, Faculty of Science and Technology, Universiti Kebangsaan Malaysia, Bangi, Selangor, Malaysia

Appukuttan Saritha Department of Chemistry, School of Arts and Sciences, Amrita Vishwa Vidyapeetham, Amritapuri, Kollam, Kerala, India

S. Arvindh Seshadri Department of Mechanical Engineering, PSG Institute of Technology and Applied Research, Coimbatore, TN, India

A.S. Sethulekshmi Department of Chemistry, School of Arts and Sciences, Amrita Vishwa Vidyapeetham, Amritapuri, Kollam, Kerala, India

P.C. Sharath Metallurgical Engineering Department, Jain University, Bangalore, Karnataka, India

Suchart Siengchin Materials and Production Engineering, The Sirindhorn International Thai-German Graduate School of Engineering (TGGS), King Mongkut's University of Technology North Bangkok, Bangkok, Thailand; Department of Mechanical and Process Engineering, The Sirindhorn International Thai-German Graduate School of Engineering (TGGS), King Mongkut's University of Technology North Bangkok, Bangkok, Thailand

Richard J. Spontak Department of Chemical & Biomolecular Engineering, North Carolina State University, Raleigh, NC, United States; Department of Materials Science & Engineering, North Carolina State University, Raleigh, NC, United States

S.N. Vasantha Kumar Mechanical Engineering Department, Canara Engineering College, Mangalore, Karnataka, India

Gopika Venu Department of Chemistry, School of Arts and Sciences, Amrita Vishwa Vidyapeetham, Amritapuri, Kollam, Kerala, India

Jiaqi Yan Department of Chemical & Biomolecular Engineering, North Carolina State University, Raleigh, NC, United States

Sarani Zakaria Bioresource & Biorefinery Research Group, Department of Applied Physics, Faculty of Science and Technology, Universiti Kebangsaan Malaysia, Bangi, Selangor, Malaysia

Preface

Elastomers are also called rubbers and have many specific features such as elasticity, good strength, and high toughness. Some of the examples of elastomers are natural rubber, polyurethane, ethylene–propylene rubber, silicone rubbers, etc. The performance of the elastomer depends on the cross-link density, molecular weight, and type of elastomer. The elastomer as such possesses poor thermomechanical properties. Therefore, cross-linking of elastomers is essential for improving their performance. Sulfur or no-sulfur compounds are used for cross-linking elastomers. The cross-link density or vulcanization can be controlled by changing the cross-linking agents, varying the amount of fillers, etc. Generally cross-linking improves the performance of the elastomer.

In recent years, blends and composites of elastomers are formulated for improved performance. Some of the potential applications of elastomeric blends and composites are in the oil and gas industry, automobile, conveyer belts, adhesive, and sealants, etc. This book comprises all the important areas of elastomeric blends and composites and their characteristic properties such as structural, thermomechanical, morphology, rheology, and modeling. The recycling and potential applications of elastomeric blends and composites are also discussed. We hope this important book will benefit students, scientists, industrialists, and those who are interested in rubber-based materials. The editors thank all the authors for their contributions.

Acknowledgments

This work was supported by King Mongkut's University of Technology North Bangkok has received funding support from the National Science, Research and Innovation Fund (NSRF) with Grant No of KMUTNB-FF-65-19.

Editors
Sanjay Mavinkere Rangappa (Thailand)
Jyotishkumar Parameswaranpillai (India)
Suchart Siengchin (Thailand)
Togay Ozbakkaloglu (USA)

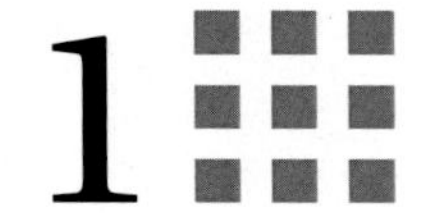

Introduction to elastomers

Jyotishkumar Parameswaranpillai[1], C.D. Midhun Dominic[2], Sanjay Mavinkere Rangappa[3], Suchart Siengchin[3], Togay Ozbakkaloglu[4]

[1]DEPARTMENT OF SCIENCE, FACULTY OF SCIENCE & TECHNOLOGY, ALLIANCE UNIVERSITY, BENGALURU, KARNATAKA, INDIA; [2]DEPARTMENT OF CHEMISTRY, SACRED HEART COLLEGE (AUTONOMOUS), KOCHI, KERALA, INDIA; [3]MATERIALS AND PRODUCTION ENGINEERING, THE SIRINDHORN INTERNATIONAL THAI-GERMAN GRADUATE SCHOOL OF ENGINEERING (TGGS), KING MONGKUT'S UNIVERSITY OF TECHNOLOGY NORTH BANGKOK, BANGKOK, THAILAND; [4]DEPARTMENT OF CIVIL ENGINEERING, TEXAS STATE UNIVERSITY, SAN MARCOS, TX, UNITED STATES

1.1 Introduction

Elastomers are viscoelastic polymers, showing both viscous and elastic properties. The elastomer possesses long polymer chains held together with weak intermolecular forces. These weak forces make them flexible and sticky, with high elongation upon the application of stress. Once the stress is released, it could come back to the original shape and hence called elastomer. Most of these polymers are having 100% recoverability, thanks to their low degree of cross-links. The low degree of cross-links in the elastomer makes it amorphous; however, once it is stretched, crystalline sites can be observed. The elastomer retains its original shape because of these cross-links. In order words, cross-linking within the elastomer provides a better shape and stiffness. The other features of elastomers are low glass transition temperature, hydrophobicity, good adhesion, good breaking resistance, tear strength, abrasion resistance, resistant to gas, water, and steam, and are excellent insulators. It finds application in adhesives, sealing applications, and insulations [1]. Elastomers are of natural and synthetic origin. The synthetic elastomers are petroleum-based, while natural rubbers are of plant origin. Examples for elastomers are natural rubber, EPDM rubber, silicone rubber, olefin thermoplastic elastomer, etc. The properties of a pure elastomer such as elasticity, strength, hardness, and resilience are poor, which restricts its use in advanced applications. Therefore, cross-linking is necessary. The properties and performance of the elastomer can be dramatically enhanced with vulcanization/cross-linking [2].

Elastomer Blends and Composites. https://doi.org/10.1016/B978-0-323-85832-8.00002-X

1.2 Vulcanization/cross-linking in elastomers

Vulcanization or cross-linking in elastomers can be achieved by sulfur [3] and nonsulfur [4] compounds. The cross-linking results in a three-dimensional network structure. The cross-linking of elastomer with sulfur is a slow process and may take a long time for complete curing. However, a shorter cure time is preferred, and the process can speed up by including other ingredients along with sulfur. But, the cross-linking reactions in elastomers should be controlled for their usefulness and success for specific applications. Premature cross-linking results in poor performance, and the products formed are not useable. Effective control over the cross-linking is therefore essential for achieving the right physical property. The following agents are used along with sulfur for the optimum cross-linking (i) accelerators, (ii) activators (iv) retarders, (v) prevulcanization inhibitor [5].

The accelerators (examples: N-t-butylbenzothiazole-2-sulfenamide, 2-mercaptobenzothiazole, 2,2′ -dithiobisbenzothiazole, dithiocarbamates, tetramethyl thiuram disulfide, tetramethyl thiuram monosulfide, etc) are used for fast cross-linking; it also enables the cure reaction to take place at a lower temperature and reduced the consumption of sulfur [6,7]. The activators (zinc oxide) and coactivators (fatty acids) are widely employed to improve the vulcanization efficacy of elastomers [8]. The retarders (examples: phthalic anhydride, salicylic and benzoic acids, and N-nitrosodiphenylamine) are used to retard the vulcanization of elastomer while the prevulcanization inhibitor (N-cyclohexylthiophthalimide) prevents early vulcanization [2]. Many studies had been carried out for the understanding of the mechanism of vulcanization. It is proposed that in the first step, the ZnO reacts with stearic acid forming a zinc/stearate complex. In the second step, the zinc/stearate complex reacts with the accelerator forming an active accelerator complex. In the third step, the active accelerator complex reacts with sulfur forming an active sulfurating agent. The active sulfurating agent reacts with elastomer resulting in a cross-linked intermediate, followed by polysulfide cross-links, and finally fully vulcanized network [8]. Further, the performance of the elastomer can be enhanced by the addition of improving agents such as antioxidants (dithiocarbamate) [9], plasticizers (oil, resins, waxes) [10,11], fillers (carbon black, nanomaterials, etc), and dyes (TiO_2, ZnO, ZnS, etc.) [12].

1.3 Elastomeric composites and blends

In recent years, there has been a boom in the application of nanotechnology in elastomers. This caused the researchers to attempt the applications of different nanomaterials in different elastomers. The nanoparticles such as graphene, CNTs, CNF, silica, metal nanoparticles, $CaCO_3$, cellulose, plant fibers, etc., have been widely used for the modification of elastomers [13,14]. It was observed that the mechanical, thermal stability, viscoelastic properties, flame retardancy, and barrier properties can be improved with the modification with nanofillers. However, the properties of the nanocomposites depend on several factors such as size, shape, l/d ratio, method of manufacturing, and distribution of nanoparticles. The methods used for the preparation of elastomer-based composites are solution mixing, latex compounding, melt mixing, and in-situ polymerization [14]. Similarly, elastomeric blend systems have also been reported. Blending is an effective method to improve performance. Developing a new elastomer takes time and is not cost-effective. However, blending with appropriate elastomer will bring attractive properties [15]. Examples of elastomeric blends are PP/SEBS, polyamide 6 (PA6)/fluoroelastomer, natural rubber-polypropylene (NR/PP), etc. [16–18]. Blending depends on several factors such as compatibility, blend ratio, morphology, etc. Various characterization techniques such as universal testing machine, SEM, TEM, XRD, FTIR, TGA, DSC, DMA, etc, may be utilized for the successful characterization and evaluation of the blends and composites. These high-performance elastomeric blends and composites find applications in lightweight composites, space shuttles, tire inner liners, cable industry, shape memory polymers, soft robotic applications, and flame-retardant applications [19–24].

1.4 Recent developments in elastomeric blends and composites

1.4.1 NR-based elastomers

Bayat et al. [25] developed composite foams based on natural rubber and silica nanoparticles. Azodicarbonamide is used as the forming agent. One-step (140 °C) and two-step forming cycles (90 and 140 °C) were used, and the resulted foams were compared. The presence of silica reduced the cell size, however, the increased the cell wall thickness and foam density. The cell size and cell wall thickness are more or less in agreement for composites prepared by one-step and two-step forming cycles. However, the two-step process has a lower foam density compared to the single-step process. Yang et al. [26] fabricated conducting NR composites having 10 wt% PEDOT: PSS (organic conductor). The composites showed excellent conductivity (87 S/cm) and elongation at break (490%); furthermore, showed good temperature and strain sensing properties. Alam et al. [27] reported good mechanical properties for anisotropic NBR/carbonyl iron particles composites in the presence of an applied magnetic field. The authors recommend the NBR composites for magnetorheological applications. Li et al. [28] reviewed the magnetorheological elastomers and discussed their potential application in sensors, isolators, and absorbers. The magnetorheological elastomers are having many potential functionalities when compared with MR fluids, such as particle settling, sealing problems, environmental contamination, etc., and are potential candidates for vibration control applications. In a recent study, natural rubber latex was modified with graphene oxide and zwitterionic chitin nanocrystals isolated from crab shell powder [4]. The reinforcement caused an enhancement in tensile strength and elongation at break. The increase in tensile properties was due to the hydrogen bonding between the natural rubber latex and zwitterionic chitin nanocrystals and zwitterionic chitin nanocrystals and graphene oxide. This approach paves a new method for the development of high-performance elastomeric composites without sulfur vulcanization for advanced applications.

High-performance PP/NR/MWCNTs composites with superior elongation at break, tensile toughness, and impact strength had been developed [29]. It was observed that the T g of the NR phase increased by approximately 10°C with MWCNTs loading due to the selective localization of nanofillers in the NR phase. The dielectric properties of NR/millable polyurethane composites with nonblack and black fillers reported improved dielectric properties [30]. The improvement in properties was due to the filler network formed in the composites. Zhang et al. [31] developed a new method for the fabrication of NR/CNTs composites with superior conductivity and mechanical properties. The method of fabrication of the composites involves the following steps. In the first step highly cross-linked NR was first prepared and made into granules using a two-roll mill. The second step involves mixing NR + CNT masterbatch, NR gum, NR granules, and curing additives using a two-roll mill followed by the hot pressing. The incorporation of the NR granules caused selective dispersion of the CNTs in NR continuous phase resulting in an increase in the conductivity. Also, the disulfides and polysulfides present in the NR granules interact with NR continuous phase enabling strong interfacial adhesion resulting in good mechanical properties.

1.4.2 EPDM-based elastomers

The reuse and recycling of cross-linked rubber are a major environmental problem, and devulcanization is one of the green solutions for it. Pirityi and Pölöskei [32] devulcanized sulfur cured EPDM using a two-roll mill at various temperatures. It was observed that at low temperatures of milling, the degradation can be reduced. Later the authors successfully reused the devulcanized EPDM by (1) adding an extra curing agent, (2) mixing it with virgin rubber. The addition of recycled EPDM (25 and 50 wt%) in virgin rubber showed reasonably good mechanical properties. Liu et al. [33] studied and compared the aging behavior of EPDM cross-linked with sulfur and EPDM cross-linked with peroxide in acid solution (fuel cell environment). The EPDM cross-linked with peroxide undergoes chain breakage during aging in acidic solution (after 10 weeks) resulting in deteriorated mechanical properties after aging. On the other hand, the EPDM cross-linked with sulfur was more stable in an acidic medium, and no such molecular backbone breakage was observed.

The variations in relaxation modulus and creep of EPDM at different temperatures and strain conditions were studied and modeled using the time–temperature superposition principle by Wang et al. [34]. According to the time–temperature superposition principle, the relaxation modulus may reduce to ca. 45%–10% strain, while creep increases by 191% at 1 MPa stress after 100 years of service. Shafranska et al. [35] studied the plasticizing effect of saturated and unsaturated soybean oil on EPDM/carbon black composites. It was observed that the oil with the highest degree of unsaturation gives the highest plasticizing effect; however, the cross-link density and mechanical properties were reduced. On the other hand, oil with all-saturated fatty acid chains shows a poor plasticizing effect, but the cross-link density and mechanical properties were not affected. Su et al. [36] studied the tribology of EPDM/carbon black composites. The authors observed a good improvement in tribological properties of EPDM with the incorporation of carbon black. Salimi et al. [37] observed that the incorporation of reclaimed EPDM in virgin EPDM reduced the mechanical and viscoelastic properties due to the breakage of the EPDM chains during the reclaiming process.

1.4.3 Silicone rubber

The silicone rubber is mostly prepared by the hydrolysis of dimethyldichlorosilane [38]. A significant development in the commercial manufacturing of silicone has been reported over the last few decades. The siloxane bonds between the polymer units permit flexibility and viscoelasticity. The silicone rubber as such has very low cross-linked networks; therefore, the reinforcement with nanoparticles is recommended to enhance the strength, stiffness, and other properties [39]. Demand for flexible, thermally conductive, and insulative material is growing especially in the electrical and electronic industry. Silicone rubber is a flexible polymer with excellent weather resistance. In a recent study [40], Al_2O_3 was grafted with boron nitride sheets to form heterostructured filler with excellent thermal conductivity. The heterostructured filler is then incorporated in room-temperature vulcanized silicone rubber to improve its functionality. The composites showed enhanced thermomechanical properties, electrical insulation property, and thermal conductivity and were an ideal composite material for electrical and electronic applications. In a similar study, Kumar et al. [41] modified room-temperature vulcanized silicone rubber with carbon nanotube (CNT), nanographene, and CNT-GR hybrid fillers. The incorporation of filler improved the modulus of the composites and actuation displacement in the order CNTs $>$ CNT-GR $>$ GR. The CNTs and CNT-GR-based composites showed good stretchability and durability. Thus, these composites are a good choice for sensors and actuators.

The degradation of high-temperature silicone rubber modified with nano/micro silica fillers under long-term environmental aging conditions was studied by Rashid et al. [42]. The composites with 2 wt.% nano/20 wt.% micro silica fillers showed the best resistance to accelerated environmental aging conditions. The flame retardancy of silicon rubber foam modified with expanded graphite (EG) and modified HNTs was studied by Pang et al. [43]. The limiting oxygen index, peak heat release rate, smoke production rate, CO and CO_2 production, etc., of silicon rubber foam, were reduced with the incorporation of expanded graphite and modified HNTs. That means the flame retardancy of silicon rubber foam can be significantly improved by the synergistic effect of expanded graphite and HNTs. In recent work, Guo et al. [44] developed self-healing PDMS by incorporating hydrogen and disulfide bonds. The polymer showed high stretchability and autonomous healing under universal temperature conditions.

1.4.4 Olefin thermoplastic elastomer

Thermoplastic elastomer consists of polyolefin thermoplastics such as PP and PE and olefin copolymers such as EPDM, SEBS, and NR. The continuous thermoplastic part provided the strength and modulus, while the olefin copolymer part provided flexibility and toughness [45]. Dynamic vulcanization of ethylene-octene copolymer (POE) and polypropylene (PP) was carried out for formulating olefin thermoplastic elastomer [46]. The vulcanizing agent bis(1-(tert-butylperoxy)-1-methylethyl)-benzene (BIPB) was used. The olefin thermoplastic elastomer with 0.6 phr BIPB and 60/40 PP/POE composition showed the highest tensile strength and modulus. Further, the silicone powder was introduced in 60/40 olefin thermoplastic elastomer containing bromine-phosphorus flame-retardant material. The incorporation of silicone powder (3 phr) improved the flow properties, limiting oxygen index, and aging resistance.

Kiziltas et al. [47] utilized recycled PP and guayule latex for the preparation of thermoplastic elastomer. The PP-g-MA was used as the coupling agent. Graphene nanoplatelets (2.5–10 wt.%) were used as a reinforcement. It was observed that the thermal and mechanical properties of thermoplastic elastomer improved with the incorporation of graphene nanoplatelets; 10 wt% of filler gives the best results. The mechanical properties of the composites were compared with Ford's materials specification "WSS-M4D954-A and CA 387A (used in automotive applications). Compared to the materials used in automotive applications, the graphene nanoplatelets-based composites showed better mechanical properties. Wang et al. [48] modified the properties of SBS by modifying it with 3,6-di(2-pyridyl)-1,2,4,5-tetrazine (DPT) and $CuSO_4$. The SBS-DPT/$CuSO_4$ composite presented good tensile strength, toughness, and shape memory effect with a shape fixity ratio of 78%–82%. In another work [49], a two-way shape memory effect was observed in olefin block copolymer/silicone elastomer with 70/30 composition. DHBP was used as the flexibilizer.

1.4.5 Biodegradable elastomers

Wang et al. [50] synthesized poly(glycerol-sebacate), a biodegradable elastomer, and observed good mechanical properties and biocompatibility by in vitro and in vivo studies. Several other biodegradable elastomers such as poly(glycerol dodecanoate) [51] poly (3-hydroxybutyrate-co-3-hydroxyvalerate) [52], polyurethanes [53], and their potential applications have been reported.

1.5 Conclusion

Elastomers an important polymer material with interesting properties such as flexibility, high elongation at break, and elasticity. However, the elastomer in its pure form possesses poor elasticity, strength, and resilience. Therefore, one must cross-link elastomer with sulfur or nonsulfur components to have desirable properties. The blending allows to development of new systems with interesting properties with little effort. Blends of elastomer with other plastics such as PP, PE are well known. Recently several nanofillers and other modifiers are used widely along with neat elastomer or with elastomeric blends to enhance the properties of the elastomer-based composites.

Acknowledgments

This work was supported by King Mongkut's University of Technology North Bangkok has received funding support from the National Science, Research and Innovation Fund (NSRF) with Grant No of KMUTNB-FF-65-19.

References

[1] T. Özdemir, Elastomeric micro-and nanocomposites for neutron shielding, in: Micro and Nanostructured Composite Materials for Neutron Shielding Applications, Woodhead Publishing, 2020, pp. 125–137.

[2] M.A. Akiba, A.S. Hashim, Vulcanization and crosslinking in elastomers, Prog. Polym. Sci. 22 (3) (1997) 475–521.

[3] M. Maciejewska, A. Sowińska, A. Grocholewicz, Zinc complexes with 1, 3-diketones as activators for sulfur vulcanization of styrene-butadiene elastomer filled with carbon black, Materials 14 (2021) 3804.

[4] C. Liu, S. Huang, J. Hou, W. Zhang, J. Wang, H. Yang, J. Zhang, Natural rubber latex reinforced by graphene oxide/zwitterionic chitin nanocrystal hybrids for high-performance elastomers without sulfur vulcanization, ACS Sustain. Chem. Eng. 9 (18) (2021) 6470–6478.

[5] J. Kruželák, R. Sýkora, I. Hudec, Sulphur and peroxide vulcanisation of rubber compounds—overview, Chem. Pap. 70 (12) (2016) 1533–1555.

[6] P. Ghosh, S. Katare, P. Patkar, J.M. Caruthers, V. Venkatasubramanian, K.A. Walker, Sulfur vulcanization of natural rubber for benzothiazole accelerated formulations: from reaction mechanisms to a rational kinetic model, Rubber Chem. Technol. 76 (3) (2003) 592–693.

[7] N.P. Cheremisinoff, Condensed Encyclopedia of Polymer Engineering Terms, Butterworth-Heinemann, 2001.

[8] S. Mostoni, P. Milana, B. Di Credico, M. D'Arienzo, R. Scotti, Zinc-based curing activators: new trends for reducing zinc content in rubber vulcanization process, Catalysts 9 (8) (2019) 664.

[9] Y. Zhang, K.L. Wade, T. Prestera, P. Talalay, Quantitative determination of isothiocyanates, dithiocarbamates, carbon disulfide, and related thiocarbonyl compounds by cyclocondensation with 1, 2-benzenedithiol, Anal. Biochem. 239 (2) (1996) 160–167.

[10] R. Koley, R. Kasilingam, S. Sahoo, S. Chattopadhyay, A.K. Bhowmick, Synthesis and characterization of epoxidized neem oil: a bio-derived natural processing aid for elastomer, J. Appl. Polym. Sci. 138 (20) (2021) 50440.

[11] M. Włoch, U. Ostaszewska, J. Datta, The effect of polyurethane glycolysate on the structure and properties of natural rubber/carbon black composites, J. Polym. Environ. 27 (2019) 1367–1378.

[12] A. Marzec, M. Zaborski, Pigment and dye modified fillers as elastomeric additives, in: Anna Boczkowska (Ed.), Advanced Elastomers–Technology, Properties and Applications, IntechOpen, 2012. https://doi.org/10.5772/50735.

[13] S.K. Thomas, J. Parameswaranpillai, S. Krishnasamy, P.S. Begam, D. Nandi, S. Siengchin, J.J. George, N. Hameed, N.V. Salim, N. Sienkiewicz, A comprehensive review on cellulose, chitin, and starch as fillers in natural rubber biocomposites, Carbohydr. Polym. Technol. Appl. 2 (2021) 100095.

[14] M. Maiti, M. Bhattacharya, A.K. Bhowmick, Elastomer nanocomposites, Rubber Chem. Technol. 81 (3) (2008) 384–469.

[15] A. Nihmath, M.T. Ramesan, Development of novel elastomeric blends derived from chlorinated nitrile rubber and chlorinated ethylene propylene diene rubber, Polym. Test. 89 (2020) 106728.

[16] S.S. Banerjee, S. Burbine, N. Kodihalli Shivaprakash, J. Mead, 3D-printable PP/SEBS thermoplastic elastomeric blends: preparation and properties, Polymers 11 (2) (2019) 347.

[17] S.S. Banerjee, A.K. Bhowmick, Novel nanostructured polyamide 6/fluoroelastomer thermoplastic elastomeric blends: influence of interaction and morphology on physical properties, Polymer 54 (24) (2013) 6561–6571.

[18] N.R. Choudhury, T.K. Chaki, A.K. Bhowmick, Thermal characterization of thermoplastic elastomeric natural rubber-polypropylene blends, Thermochim. Acta 176 (1991) 149–161.

[19] Ankit, N. Tiwari, F. Ho, F. Krisnadi, M.R. Kulkarni, L.L. Nguyen, S.J.A. Koh, N. Mathews, High-k, ultrastretchable self-enclosed ionic liquid-elastomer composites for soft robotics and flexible electronics, ACS Appl. Mater. Interfaces 12 (33) (2020) 37561–37570.

[20] Q. Ma, S. Liao, Y. Ma, Y. Chu, Y. Wang, An ultra-low-temperature elastomer with excellent mechanical performance and solvent resistance, Adv. Mater. 33 (36) (2021) 2102096.

[21] K. Banerjee, J. Chanda, S. Ghosh Chowdhury, K. Pal, R. Mukhopadhyay, S.K. Bhattacharyya, A. Bandyopadhyay, Effect of pre-mastication on dispersion of nanoclay in presence of carbon black in an inner liner compound: studies on physicomechanical and functional properties, Polym. Eng. Sci. 61 (3) (2021) 906–917.

[22] L.C. Wang, Q. Sun, C.C. Zhang, The charring effect and flame retardant properties of thermoplastic elastomers composites applied for cable, Fibers Polym. 21 (11) (2020) 2599–2606.

[23] Y. Gao, W. Liu, S. Zhu, Thermoplastic polyolefin elastomer blends for multiple and reversible shape memory polymers, Ind. Eng. Chem. Res. 58 (42) (2019) 19495–19502.

[24] D. Zhalmuratova, H.J. Chung, Reinforced gels and elastomers for biomedical and soft robotics applications, ACS Appl. Polym. Mater. 2 (3) (2020) 1073–1091.

[25] H. Bayat, M. Fasihi, Y. Zare, K.Y. Rhee, An experimental study on one-step and two-step foaming of natural rubber/silica nanocomposites, Nanotechnol. Rev. 9 (1) (2020) 427–435.

[26] Y. Yang, G. Zhao, X. Cheng, H. Deng, Q. Fu, Stretchable and healable conductive elastomer based on PEDOT: PSS/natural rubber for self-powered temperature and strain sensing, ACS Appl. Mater. Interfaces 13 (12) (2021) 14599–14611.

[27] M.N. Alam, V. Kumar, S.R. Ryu, J. Choi, D.J. Lee, Anisotropic magnetorheological elastomers with carbonyl iron particles in natural rubber and acrylonitrile butadiene rubber: a comparative study, J. Intell. Mater. Syst. Struct. (2021). https://doi.org/10.1177/1045389X20986995.

[28] Y. Li, J. Li, W. Li, H. Du, A state-of-the-art review on magnetorheological elastomer devices, Smart Mat. Struct. 23 (12) (2014) 123001.

[29] S.T. Nair, S.C. George, N. Kalarikkal, S. Thomas, Enhanced mechanical and thermal performance of multiwalled carbon nanotubes-filled polypropylene/natural rubber thermoplastic elastomers, New J. Chem. 45 (11) (2021) 4963–4976.

[30] R. Jose, L.A. Varghese, U.G. Panicker, Tailoring dielectric properties of natural rubber/millable polyurethane elastomer blends by filler embedding, Polym. Bull. (2021). https://doi.org/10.1007/s00289-021-03595-z.

[31] C. Zhang, Z. Tang, X. An, S. Fang, S. Wu, B. Guo, Generic method to create segregated structures toward Robust, flexible, highly conductive elastomer composites, ACS Appl. Mater. Interfaces 13 (2021) 24154–24163.

[32] D.Z. Pirityi, K. Pölöskei, Thermomechanical devulcanisation of ethylene propylene diene monomer (EPDM) rubber and its subsequent reintegration into virgin rubber, Polymers 13 (7) (2021) 1116.

[33] Q. Liu, J. Li, Y. Jiang, C. Cong, L. Xu, Y. Zhang, X. Meng, Q. Zhou, Effect of crosslinked structure on the chemical degradation of EPDM rubber in an acidic environment, Polym. Degrad. Stabil. 185 (2021) 109475.

[34] Z.N. Wang, S.L. Shen, A. Zhou, H.M. Lyu, Investigation of time-dependent characteristics of EPDM rubber gasket used for shield tunnels, J. Mater. Civ. Eng. 33 (9) (2021) 04021251.

[35] O. Shafranska, A. Jones, A. Perkins, J. Dahlgren, J. Tardiff, D.C. Webster, Low-unsaturated soybean oils in EPDM rubber compounds, J. Appl. Polym. Sci. (2021). https://doi.org/10.1002/app.51499.

[36] Q. Sun, S. Wang, X. Lv, Research on tribological properties and viscoelasticity of EPDM rubber in the main drive sealing system of shield machine, J. Tribol. 144 (5) (2021) 051201.

[37] A. Salimi, F. Abbassi-Sourki, M. Karrabi, M.H.R. Ghoreishy, Investigation on viscoelastic behavior of virgin EPDM/reclaimed rubber blends using Generalized Maxwell Model (GMM), Polym. Test. 93 (2021) 106989.

[38] S. Setiadji, E. Sumiyanto, N. Syakir, A.R. Noviyanti, I. Rahayu, Synthesis of polydimethylsiloxane and its monomer from hydrolysis of dichlorodimethylsilane, in: Key Engineering Materials, vol. 860, Trans Tech Publications Ltd, 2020, pp. 234–238.

[39] E.L. Warrick, O.R. Pierce, K.E. Polmanteer, J.C. Saam, Silicone elastomer developments 1967–1977, Rubber Chem. Technol. 52 (3) (1979) 437–525.

[40] H. Yan, X. Dai, K. Ruan, S. Zhang, X. Shi, Y. Guo, H. Cai, J. Gu, Flexible thermally conductive and electrically insulating silicone rubber composite films with BNNS@ Al_2O_3 fillers, Adv. Comp. Hybrid Mater. 4 (1) (2021) 36–50.

[41] V. Kumar, M.N. Alam, A. Manikkavel, J. Choi, D.J. Lee, Investigation of silicone rubber composites reinforced with carbon nanotube, nanographite, their hybrid, and applications for flexible devices, J. Vinyl Addit. Technol. 27 (2) (2021) 254–263.

[42] A. Rashid, J. Saleem, M. Amin, S.M. Ali, A.A. Khan, M.B. Qureshi, S. Ali, D. Dancey, R. Nawaz, Investigation of 9000 hours multi-stress aging effects on High-Temperature Vulcanized Silicone Rubber with silica (nano/micro) filler hybrid composite insulator, PLoS One 16 (7) (2021) e0253372.

[43] Q. Pang, F. Kang, J. Deng, L. Lei, J. Lu, S. Shao, Flame retardancy effects between expandable graphite and halloysite nanotubes in silicone rubber foam, RSC Adv. 11 (2021) 13821–13831.

[44] H. Guo, Y. Han, W. Zhao, J. Yang, L. Zhang, Universally autonomous self-healing elastomer with high stretchability, Nat. Commun. 11 (2020) 2037.

[45] Ç. Girişken, S.A. Seven, O.G. Ersoy, Y.Z. Menceloğlu, Investigation of structure-morphology-function relationship of plastomers used to produce low mold shrinkage thermoplastic olefins, Eur. Polym. J. 159 (2021) 110758.

[46] M. Li, Y. Wang, C. Shen, S. Gao, PP/POE thermoplastic elastomer prepared by dynamic vulcanization and its flame retardant modification, J. Elastomers Plastics (2021). https://doi.org/10.1177/00952443211029039.

[47] A. Kiziltas, S. Tamrakar, J. Rizzo, D. Mielewski, Characterization of graphene nanoplatelets reinforced sustainable thermoplastic elastomers, Composites Part C 6 (2021) 100172.

[48] Q. Wang, Y. He, Q. Li, C. Wu, SBS thermoplastic elastomer based on dynamic metal-ligand bond: structure, mechanical properties, and shape memory behavior, Macromol. Mater. Eng. 306 (2021) 2000737.

[49] S.M. Lai, Y.J. Chen, B.Y. Yu, Preparation and characterization of two-way shape memory olefin block copolymer/silicone elastomeric blends, J. Appl. Polym. Sci. 138 (2021) 51238.

[50] Y. Wang, G.A. Ameer, B.J. Sheppard, R. Langer, A tough biodegradable elastomer, Nat. Biotechnol. 20 (6) (2002) 602–606.

[51] H. Ramaraju, L.D. Solorio, M.L. Bocks, S.J. Hollister, Degradation properties of a biodegradable shape memory elastomer, poly (glycerol dodecanoate), for soft tissue repair, PLoS One 15 (2) (2020) e0229112. https://doi.org/10.1371/journal.pone.0229112.

[52] H.S. Kim, J. Chen, L.P. Wu, J. Wu, H. Xiang, K.W. Leong, J. Han, Prevention of excessive scar formation using nanofibrous meshes made of biodegradable elastomer poly (3-hydroxybutyrate-co-3-hydroxyvalerate), J. Tissue Eng. (2020). https://doi.org/10.1177/2041731420949332.

[53] Q. Luo, J. Chen, P. Gnanasekar, X. Ma, D. Qin, H. Na, J. Zhu, N. Yan, A facile preparation strategy of polycaprolactone (PCL)-based biodegradable polyurethane elastomer with a highly efficient shape memory effect, New J. Chem. 44 (3) (2020) 658–662.

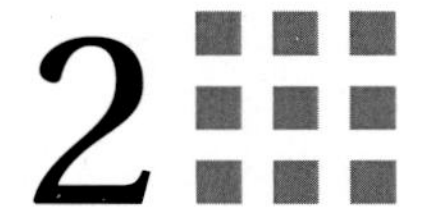

Manufacturing methods of elastomer blends and composites

M. Ramesh[1], D. Balaji[2], L. Rajeshkumar[2], V. Bhuvaneswari[2]

[1]DEPARTMENT OF MECHANICAL ENGINEERING, KIT-KALAIGNARKARUNANIDHI INSTITUTE OF TECHNOLOGY, COIMBATORE, TAMIL NADU, INDIA; [2]DEPARTMENT OF MECHANICAL ENGINEERING, KPR INSTITUTE OF ENGINEERING AND TECHNOLOGY, COIMBATORE, TAMIL NADU, INDIA

2.1 Introduction

"Composite" is a word that is often heard in the world of materials. It is a part of each and every new product being manufactured in the world every day. The material possesses biodegradability along with plenty of availability, which are the characteristics of today's functional materials. Elastomers are commonly used due to their relatively changeable internal to external characteristics that depend on curing. Even though initial modulus and elastomeric applications require and use the nonlinear structural flexibility and adaptability of the elastomer, the initial ductility and reliability of the elastomers are quite low [1]. Additives such as metallic nanoparticles, carbon black (CB), silica particles, and carbon nanotubes (CNTs) are typically used to help reinforce a design because of the variety of ways in which they are able to provide assistance. Fillers such as stiffening agents are added to elastomers, causing them to stiffen while keeping significant characteristics like huge strain insulin resistant and release shows. Elastomers' elevated and modifiable deformation is critical in industry. They are popular for a variety of reasons, along with everyone's especially harmful and longer product existence, great shrinkage, pressure soothing resistance, and low price. In motor vehicles and vibration mounts, fillers are used to improve the rigidity and toughness of the rubber. Additionally, these grow and improve the production process, making the materials more resilient, hard, and fracture resistant. Padding reinforcement has been shown to enhance the manufacture of stronger, lighter, and therefore less costly composites. Fillers like silica, $CaCO_3$, CB, and so forth, have been used to increase the mechanical and physical properties of rubber since the beginning. Elastomers strengthened with CB molecules, which range in size from 100 to 1000 Å, is by far the most easily found elastomeric composite [2–4]. Attributed to the prevalence of a stiff pollutants

Elastomer Blends and Composites. https://doi.org/10.1016/B978-0-323-85832-8.00011-0

phase and the elastic chains' interactions with it, CB-filled elastic composites have a distinct behavior. However, the major downside of these additives is that only increased fiber loadings provide appreciable improvements in properties. Inventive strengthening with superior characteristics must be created whenever there is a requirement for new information implementation for outstanding thermoplastic composites [5]. Nanoparticles have the necessary characteristics to serve as fillers because of their tiny population and high surface area. You can significantly enhance properties by only using a minuscule bit of nanoparticles. It is expected that the use of unique CNT fillers will enhance the properties of any and all varieties of elastomers, especially rubbers. In the situation of CNTs, the enhanced interfacial bonding between a greater filler matrix and a softer leather shell leads to an immobilized leather shell [6]. Also important is the shape factor of the padding aggregates, which is known as the anisometry. Increasing the anisometry of substances with the same particle size and surface chemistry but different patterns will enhance the crystallite size. CNTs feature a larger shape factor than CB, and this explains why they are more reinforced. When inserting multiwalled carbon nanotubes (MWCNTs) into thermoplastics, the alignment and bonding problems encountered may be used to the benefit their incorporation. When nanotubes are added to silicone rubber (SR), a significant improvement in the final product can be noticed [2,7]. At limited filler loading, CNTs act as an efficient group in extremely durable rubber [8]. In the early stages of deformation, the actual impact of the physicochemical covalent bonds of the chain stores contribute to the overall stiffness of the material. Including generated inorganic polymers, all could allow conducted lightweight structures because they are electrical and thermal insulators, with improved thermal permeability behavior of the cellular proteins. The specialized crystal structures of the CNTs could be demonstrated by the improvement of CNT-based thermoplastic composites.

Improvements in the material characteristics of composite material have been found to be possible through the use of very low margin concentrations of CNTs [8]. CNT-filled aerogel coatings could be used in various industries like metal rods, wheel elements, and sensors, but they could also be used for sheltering electric circuit and insulating electronic wiring. Aerogel materials rely on lateral load reliability between the matrixes and nanomaterials to be efficient. In the event that the proportion of reinforced concrete is extremely low or has been spread over a wide area, there are much more good implementations that hold back the crack's progress, which makes it suitable in aircraft materials that have a light weight. Additional problems can arise, including a rise in frequency due to the nanotubes' particle size, and a loss of responsibility over the coordination or position of the nanotubes. A further challenge is that the nanotube–elastomer bond is weak. The force exerted by CNTs on the elastomer is absent because the CNT and the epoxy coating do not share an electric attraction because the elastomer's matrixes hydrogen embrittlement coefficients are misaligned. A great deal of these problems can be remedied by altering CNTs, which may need to be done when working with the coatings [9–11]. All the features of these remarkable composites are discussed in detail in this review.

2.2 Preparation techniques

The most important elastomer based composite fabrication approaches generally in existence are elucidated in subsequent sections in detail.

2.2.1 Solvent casting

Mixing between both fluid layers is performed in different additives and the result is diffusion of both fluid layers. To obtain dry observations of the composite, the solvent has to evaporate. By applying this methodology, the team of researchers from the aforementioned study produced CNT/styrene butadiene rubber composites and increased the modulus and tensile strength by 45% and 70%, respectively, while using just 1 phr (per hundred rubber) of multiwalled CNT. Sustainable tourism has been the most practical method to use in the scenario of silicone matrixes, as no increased solvent is required regardless of the type of filler. If the water-based elastomer matrix is incompatible only with filler, both the filler and elastomer suspensions are prepared in many other solid support mixtures. They distributed CNTs in nonaqueous solvent-casting toluene had been previously done by few experimenters. The use of solvents such as toluene cannot be supported, regardless of this method's effectiveness [12].

Thick layers with an external thickness of between 0.15 and 0.20 mm in diameter and a radius of 31.5 mm were created with solvent-casting polymer samples. The thickness of the bonnet accommodation the dehumidifier varied because of a slight altitude great cost in the hood, but this angle did not influence the dispersion of particles in the layers. Care would have to be taken to prevent that both the bleaching ground and the survey were category while the resin evaporates. Due to this, mechanized mixture and examining just before appropriate distribution has posed a problem for both the worker doing the merging and the machine using the mixture. The presence of dust particles made the mixture more soluble, thus increasing the severity of this problem. Adding more fluid to led to a decrease in viscosity because of the relation of piezoelectric particulate. Particles in the air were wet and added with polyurethane solution, and thus resulted in superior simplicity of blending and drawing compared to the dry method described [13,14].

2.2.2 Freeze drying

The technique of freeze rinsing involves placing solutions in a cryogenic chamber where they are frozen, and afterward the frozen alternatives are discarded via diffusion under void, resulting in porous material. Defrost heat, contact time, essence of substance, and freeze control all influence the pore size, width, volume, and morphology. Permeable walls are commonly created using aqueous, natural, colloidal, and supercooled CO_2 solutions [15]. Filler is dispensed into the matrix material with basic stirring and the polyurethane is stored in refrigerator temperature and then pressed. CNTs and elastomer composites have not, until now, been synthesized using this technique. The authors synthesized polystyrene CNT latex composites using a freezing and compression molding process, as reported by many authors [16]. Elastomer acrylic composites can be prepared using that technique.

This technique was used to produce a hard hydrogel by adding nanoclay to an oxidative gel in situ. Formulated membrane was soaked in nitric acid for some time, after which it was exposed to UV light for a brief period. Utilizing simple solvents eliminates the need for use of any reduction process. The hydrogel's porosity was created by ball milling, which improved the hydrogel's surface area. It is similar to an elastomer because of the mechanical behavior of the nanoparticles (elongation at break around 700%). There is a possibility that the stream mortgage of the hydrogel may cause light to appear during segmentation where a high level of biocompatible content and a high level of laponite content help increase the strength of the gel. Ag^+ ions that were trapped in the polymers were the principal reduction centers. In addition, the particular order rate values are connected to the permeability and cross-linking agent concentration through an inverse effect. At least five major periods of reuse were seen with the hydrogel. Lastly, the hydrogel exhibited bactericidal activity due to the confinement of nanocrystals within the gel matrix. Thus, it can be indicated that the "soak and irradiate" method is an accurate nonenergy rigorous technique to prepare high catalytic systems and have such strong matrix supported components [17,18].

2.2.3 Spray drying

In order to turn CNT suspensions into CNT/elastomer powder composites, the suspension of CNTs can be spray-dried. Pigments of composites are prepared by powder drying the revocation of CNTs, which is mixed homogeneously, atomized, and dried using elastomer and water reducing reagents. Spray drying, which disperses the CNTs well in polymeric globules, made unified particles of composites (diameter less than 10 μm) in which the CNTs are well dispersed. SBR powders with same size of 5–10 μm were synthesized by spray drying the powders with uniformly sized CNTs. Compared to those filled with true SBR composites, the 60 phr CNT composites have an increased toughness and wear resistance by 70% and an enhanced tensile strength by 250%. On the other hand, the flexural modulus of the coatings filled with 50 phr CNTs was significantly improved by 600% [19].

2.2.4 Latex stage compounding

Leather resin could be mixed with CNTs, alongside elements used in the curing process. There are a number of cross-linking methods, the most common being the use of sulfur in the existence of zinc oxide, stearic acid, etc. The cross-linking among polymer matrix in the coatings is improved as a result of this process. When starting the curing process, the curing agents are initially blended with the latex using the milling method, which is called intensifying. After that, the exacerbated resin is mixed with the CNTs [20,21]. In the few experiments this method has been successfully employed, a synthesis of MWCNT/natural rubber latex (NRL) coatings was successfully achieved. Roughly two-thirds of the NRL used in this study was prepared via ball milling, with subsequent composites produced via continuous methods and observed noteworthy improvements in electrical, rheological, and thermal conductivity.

The addition of additives together with curing agents, such as sulfur in the appearance of zinc oxide, stearic acid, etc., is also done with NRL. The treating agents are blended (or compounded) with 2-octyl 4-N-lauroyl-aminocaproic acid (2-octyl 4-N-lauroyl-L-carnitine) accompanied by filler dispersion. In this process, cross-linking occurs in the polymer chains of the nanocomposites, and thus, greater properties can be obtained. Although in situ polymerization methods, which are another efficient method for significantly improving the distribution of fillers in polymers and achieving stable filler-polymer interaction, are limited in number, reports on this are lacking for NR. This is because there is a high concentration of NR in its polymer form, which makes it easier to control the crystallization of styrene into cis polyisoprene. A variety of methods have been discovered to synthesize NR/CNT polymers [22].

2.2.5 Heterocoagulation approach

The traditional heteroinflammation process involves the use of a mixture of smaller charged particles, which are placed on top of larger, oppositely charged particles. This algorithm depends on chemical precipitation among oxidized nanoparticles of cation and anion [23]. The electronic structure among all the padding and polymer is thought to increase the bonding ability between the filler and polymer [24,25]. Furthermore, homogeneously dispersed polymers (coagulation) were also effectively often used to prepare clay-filled composite materials [26,27]. In addition, this method is applied by few experimenters for the synthesis of semiconducting MWCNT/NR composites. In addition to this, the communication between some of the magnetic fields keeps them well-distributed, resulting in a fine mixture of composites [28].

2.2.6 In situ polymerization

The use of in-precipitation method to manufacture CNT-based polymers utilizes the in-precipitation method. The technique offered here is very simple and easy to implement and allows for the distribution of CNTs in the epoxy resin evenly. Many factors influence the rate of polymerization reaction used, including the nature of the so-called disturbances, the reaction conditions, and the type of carbon fiber required. Treatment of polymers can be based on the direct addition of coating materials as well as in situ. The polymers hydrolyze in the appearance of CNTs, resulting in the formation of chemical bonds in between monomers and also with the nanostructured materials. Such robust matrix-filler bonds enhance interfacial interaction, which in turn results in exceptional properties like enhanced shear fracture toughness [29–31]. These techniques are popularly used with insoluble polymers, and in the case of melt mixing, insoluble polymers could be equipped by solution processing (heat-based unbalanced polymers).

2.2.7 Melt blending/extrusion

Since the title clearly identifies the nature of melt processing, it requires great heat-stress mixture for reinforced composites. This process includes heating exotic alloys in inner blenders or twin-screw mills, followed by the addition of CNTs. In comparison to other methods of composite fabrication, this may be the most commercially preferred. Because of this, there are no solvents needed. The machine used for this purpose is modified in a way that the rivets within this can relocate at great velocities, which then in turn leads in effective distribution of CNT bunches within the elastic matrix. While this commercial-scale process is the better placed process for the task, it does suffer from two downsides. The first issue to overcome is the issue of efficient scattering of CNTs only occurring at greater quantities within the elastic matrix [32,33] and the main factor is the struggle to append trying to cure operatives to polymers at huge computational temperatures. As you can see, this method is very focused on polymer composite materials. However, there are a few academic papers on the process of fabricating elastic composites as well. In the study done by few researchers, for instance, the diffusion of CNTs in SBR and BR blend was successfully carried out by melting mixing. MWCNT/CR composites were fabricated by attaching the CNT dispersion to mill mixed CR, then the ethanol dispersion was used to disperse the CNT dispersion. The ethanol dispersion then made this solution ideal for dissolving the CNT dispersion, resulting in improved mechanical properties [34].

2.2.8 Solid-state shear pulverization

CNTs are blended with composites and ground into a fine powder by griddle mill or rotary drum extruder, after which plastics are transplanted on to the CNTs. The significant benefit of this method is that it is quick to scale up. This is a less aggressive solvent free method. Many experimenters were able to prepare CNT/polymer composites by using solid-state shear pulverization and demonstrated that this method significantly enhances dissipation percentage of the nanofillers in the matrices and also ends up in a substantial characteristic's development by the process of fractional distillation, pigments of the composite materials are produced. Both these methodologies of composite manufacturing comprise increasing a prealigned assortment of CNTs in a membrane by CVD and then in situ polymerization of a monomer into these arrays and a subsequent CVD deposition of the resulting polymer. Through the consequent composite, the nanotubes were dispersed evenly throughout the polymer, while the crystalline structure was also enhanced. Additionally, capillary electrophoresis deposition can be gotten by performing a solution containing magnetic fields displaced in a common liquid and directing the solution to an anode in an alternating magnetic field [35]. CNTs' functional groups allow them to dissolve in water and cause them to have a negative charge, thus making them transfer to the anode (CNTs have binding sites that enable them to dissolve in water and result in them having a negative charge, causing it to move toward the anode). This results in several benefits, such as lower cost, a consistent coating that can be applied to irregular shape substrates, and the ability to deposit coatings on irregular ones. While CNT diffusion methods are discussed above, there are a number of other methodologies that can be used. Several of the recent researches employ a combination of ultrasonication (either on its own or with other techniques) such as ultrasonic treatment coupled with ball milling [36,37], ultrasonication coupled with deformation, etc. Many authors found that a combination of the silicone phase mitigating to self-assembly technique allowed them to synthesize MWCNT/NR latex composites, and that the resulting composites exhibited adhesion behavior with only a 3% CNT content. To explain the enhanced dispersion, they proposed the enhanced distribution centered on skeletal anatomy and temperature analyses, as a consequence of increasing techniques [38].

2.2.9 Liquid crystal elastomer

An advanced form of 3D printing called digital light processing (DLP) can create hierarchical, complicated shapes with microeconomic and mesoscopic structures that are otherwise not possible to produce using conventional manufacturing techniques. In order to meet the increased demand for thermoelectric energy-dispersing devices, we have lengthened the structure to the macroscopic wavelength. It was reported that a photopolymer such as DLP-based liquid crystal elastomer (LCE) was used to manufacture a variety of complex shaped, high-resolution energy dissipating electronic devices. A set of 3D LCE lattice structures printed utilizing compressive mechanical testing have been shown to have 12-fold higher price and up to 27 times greater strain-energy dissipation compared to available commercially photocurable elastomer resin 3D LCE lattice structures were printed and a compressive mechanical test was used to examine the stress—strain responses. It is also thought that the reported behavior of such structures provides another important insight on the energy properties of LCEs, and that may encourage the benefits of new electronic devices [39,40].

The materials used for the processing were ethylenedioxy diethanethiol (EDDET), phenylbis 2, 4, 6 trimethylbenzoyl phosphine oxide (PPO), 2, 6-ditert butylphenol (BHT), triethylamine (TEA), and ethyl acetate. C10 has also been used along with 1,4-Bis-4-3-acryloyloxy propyloxy benzoyloxy-2-methylbenzene (RM257), Tango Black, polyvinyl chloride (PVC), polytetrafluoroethylene (PTFE), and silicone were also used to prepare an experimental prototype as in Fig. 2.1. For an experimental prototype, two distinct polymers were used. Using 1 wt.% of RM257 and toluene along with 2 wt.% of BHT, printed parts were made by heating the raw materials to 100°C. RM257 and toluene was used in equal proportions. To introduce an acrylate-capped oligomer EDDET was introduced in an equal mole ratio of mesogenic to dithiol binding sites, which served to accomplish an acrylate-capped oligomer. Sodium polyphosphate at 2 wt.% were then added to the above solution. For the spinal cage, utilization of EDDET binding sites actively boosted the polyethylene crystalline nature in the printed part. The remainder of the technique was similar to the one that was used to create the exploratory printed parts. Custom DLP 3D

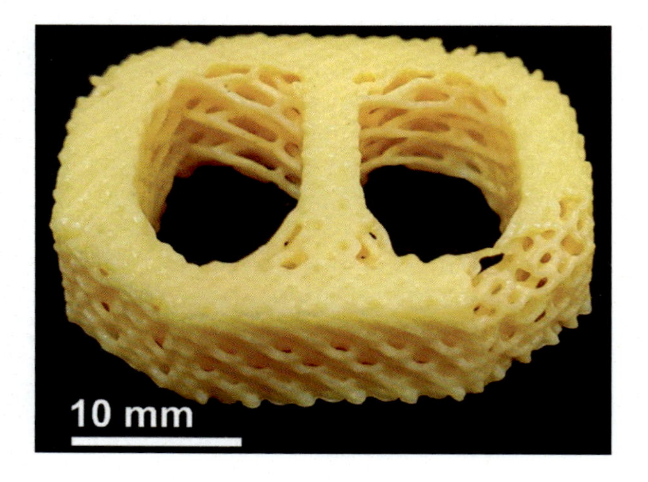

FIGURE 2.1 Digital light processed spinal ring with porous architecture [41].

printer used during past projects general engineering research team was used to print LCE and TangoBlack resins at room temperature, with the experimental parts printed in an experimental fashion using LCE and TangoBlack resin [41].

The LCE resin was printed in an SL1 DLP 3D printer at ambient temperature. Compressive stress tests were conducted using a TA Instruments ElectroForce 3200, a device that uses uniaxial compression to perform stress–strain measurements. A $9 \times 9 \times 9$ mm^3 lattice LCE and a $9 \times 9 \times 9$ mm^3 TangoBlack were built and tested. This machine is able to check if all structures have dimensions which fit properly and that the print direction corresponds to the z-axis. In addition, the constructions were evaluated at strain rates of 0.21%, 2.2%, 22%, and 220% and were quantified to compare the strain rate dependency behavior of the LCE with TangoBlack printed materials. In addition, the stress developed was also utilized to specify the amount of energy and to analyze orthotropic reactions in the nanostructures as well as orthotropic reactions that result from publish orientation [42].

2.2.10 Soft and biostable elastomer

New progress in 3D printers have revolutionized environmental science by making it possible to create complicated and usable devices at a lower price, with modifiable features, and in a small-batch manner. While gentle elastomers are extremely significant for biological devices because they are able provide the characteristics similar to cells with enhanced biocompatibility, they are generally unsuitable for applications in the general manufacturing sector. On the other hand, there are few polymeric thermoplastics with 3D printing, yet nothing is recognized more about process parameters of polymeric digital 3D elastomers. Our new methodology can produce a gentle, excellent biocompatible, and biostable PCU-Sil from polycarbonate, and this is the first study to publish an improved PCU-Sil creation methodology. We use a systematic method to characterize the thermophysical and temperature characteristics of the resulting, as well as establish a variety of process conditions, to help direct the 3D printing process. Parametric studies aimed at minimizing porosity while maximizing the sculptural accurateness of the 3D-printed samples are performed through parametric studies and studies involving parametric variables to arrive at ideal renowned definition like tool rotational speed, heat, and layer height. We describe the surface roughness of the 3D-printed structures, their degradation behavior, and biocompatibility, under static and cyclic loading, as well as biodegradation. When exposed to the heat treatment, the 3D-printed polymers yield a Young's modulus of 6.9 MPa and a failure strain of 457% while maintaining excellent cell viability. The models are printed to show the wide range of applications for the 3D-printed material: Patient-specific heart models and subdividing principal structures are created to highlight the characteristics of the 3D-printed material. This work forecasts that the 3D printing framework developed here will usher in a plethora of new possibilities not only for PCU-Sil, but also for other soft, biocompatible, and thermoplastic polymers used in various biomedical applications requiring strength and flexibility coupled with bioactivity, such as vascular implants, heart valves, and catheters [43]. Fig. 2.2 depicts the

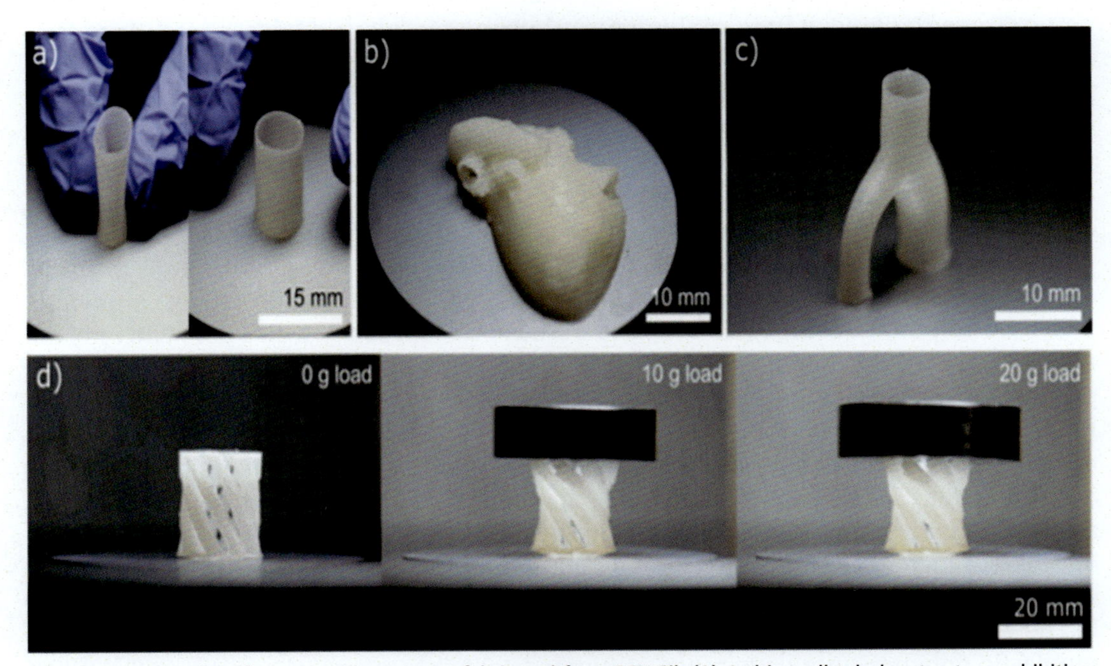

FIGURE 2.2 Examples of 3D-printed structures fabricated from PCU-Sil. (A) A thin-wall tubular structure exhibiting compliance, (B) a patient-specific heart model, (C) an arterial bifurcation, and (D) a progressively loaded twisting arm mechanism [44].

manufactured PU-Sil elastomer composites with which various real-time models were fabricated.

In this study, PCU-Sil is 3D-printed to make a soft and biodegradable elastomer, and 3D printing methods and the structural and cytocompatibility of published samples are reported. In order to guide the 3D printing process, the heat and organoleptic spectral lines were examined. It was also discovered that by systematically researching 3D printing parameters, an optimal printing set of parameters for PCU-Sil was found, likely to result in an optimal print variable of 260°C nozzle heat, 9 mm/s scan speed, and 0.15 mm wall thickness. The print values were measured utilizing micro-CT to create objects with a minimum of porosity and a high level of geometric fidelity. The walls of structures that are this thin were printed using this method, illustrating their capacity for publishing structures with truce properties. In order to quantify the mechanical properties of the 3D-printed samples, three-dimensional tensile tests were performed at 5% strain and at 45.67% strain. In this test, the Young's modulus increased from 6.9 MPa at 5% strain to 6.97 MPa at 45.67% strain, which is evidence of good extensibility. 3D-printed dog bone samples did not show any dramatic drop in material characteristics from a larger samples procedure in PBS, implying biostability and making it a candidate for medical implants that last for the long term. Additionally, three-dimensional printed data were tested to discomfort testing to see if they exhibited cyclic behavior. To determine the wear resistance of the compatible structures, the result will be used to approximate the fatigue lives of the project integration to avoid additional fatigue failures. While the in vitro biocompatibility testing for cytotoxicity, NO production, and endothelial activation revealed that high cell viability with no negative impacts on cell function or cytoskeletal structure were maintained throughout exposure to PCU-Sil, a wide range of possible outcomes were observed. While these results are promising, a limitation to these findings is the lack of a biobath in the experiment to assess material characteristics. For a better understanding of how PCU-Sil might perform during the process of being implemented as a medical implant, such as how it could possibly change in properties if it is submerged in body samples and at body temp, a biobath would be helpful. Decomposition studies were conducted solely with PBS, which meant that PCU-Sil would only be exposed to hydrophilic nature, and all other degradation mechanisms, like oxidative degradation, were ignored. In spite of this, the results for the PCU-Sil characterization indicate that it has great potential for printing dental components in three dimensions. Based on the findings from this research, it appears that FFF 3D printing of PCU-Sil (nonbiodegradable, bio-compliant, and structurally compliant) is an appropriate fabrication approach for the production of additively manufactured biostable, biocompliant, and structurally compliant medical devices. As shown in this work, we believe that the 3D FFF printing framework presented here will pave the way for new possibilities, not just for PCU-Sil but for other soft, biocompatible, and thermoplastic polymers in a wide range of biomedical applications where strong flexibility and strength are required, as well as biocompatibility like vascular implants, heart valves, and catheters [44–46].

2.2.11 Short fiber reinforced elastomer composite

Several layers of polydimethylsiloxane (PDMS) film strengthened with simple nickel-coated carbon fibers (NiCF) were experimentally verified, with the filaments aligned along different directions due to the use of an external magnetic field. Fig. 2.3 illustrates the scanning electron microscopy image of the nickel coated carbon fibers. Before they were cured for 48 h in the magnetic field, the fibers were distributed in the solid solution utilizing sonication and structural mixing, after which they were cured in the low impact field (<0.2 T) for another 48 h. As a result, the fibers were oriented in the nickel functionalized matrix with the fiber length being limited by the presence of intermingled material. The aim of this research was to find out the acceleration capability of this thermal polymer. With this information, motion was accomplished, and deceleration was shown to be large. The results are in agreement with the model, as predicted, and the results support the hypothesis that the instability is caused by the thermal torque. During the flexing deformed configuration, the composite films go through a transformation from a bending-only setup for the 0 degree fiber sample, to a wiggling configuration for the fibers at 90 degree. Even so, all transitional dimensions demonstrate respectively bending and twisting. This kind of actions has also been observed in dolphins to help them pick their way through complex waters [47]. Fig. 2.4 denotes the PDMS-NiCF specimen after loading.

To fabricate metal oxide specimens of a novel NiC-PDMS composite with reduced electromagnetic waves to reorient the filaments and create isotropic material samples, this study utilized samples made with reduced electromagnetic waves to orientate the fibers. Homogenization and computer-controlled mixing were used to distribute the fibers throughout the host material. For the sake of consistency, it was found that

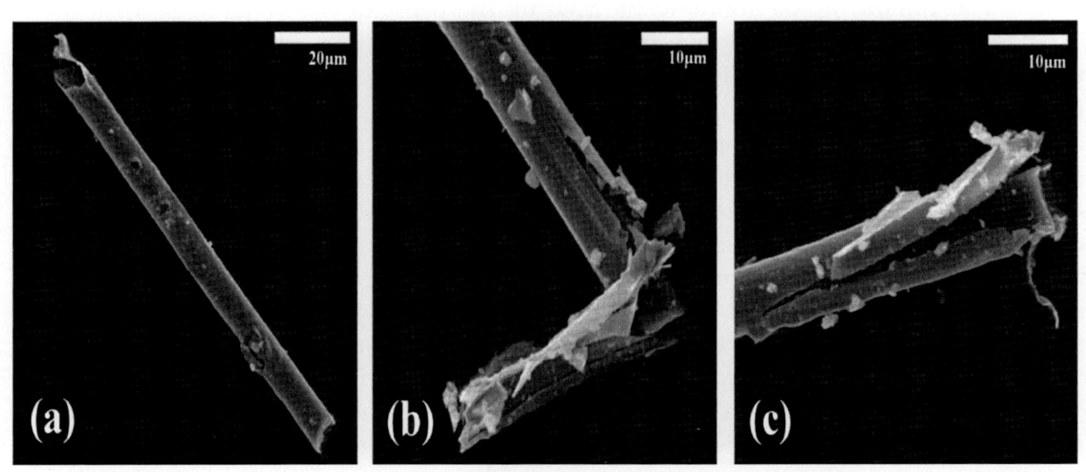

FIGURE 2.3 SEM morphology of NiC fibers. (A) Undamaged NiCF; (B) damaged NiCF fiber; (C) inner view of damaged NiCF [48].

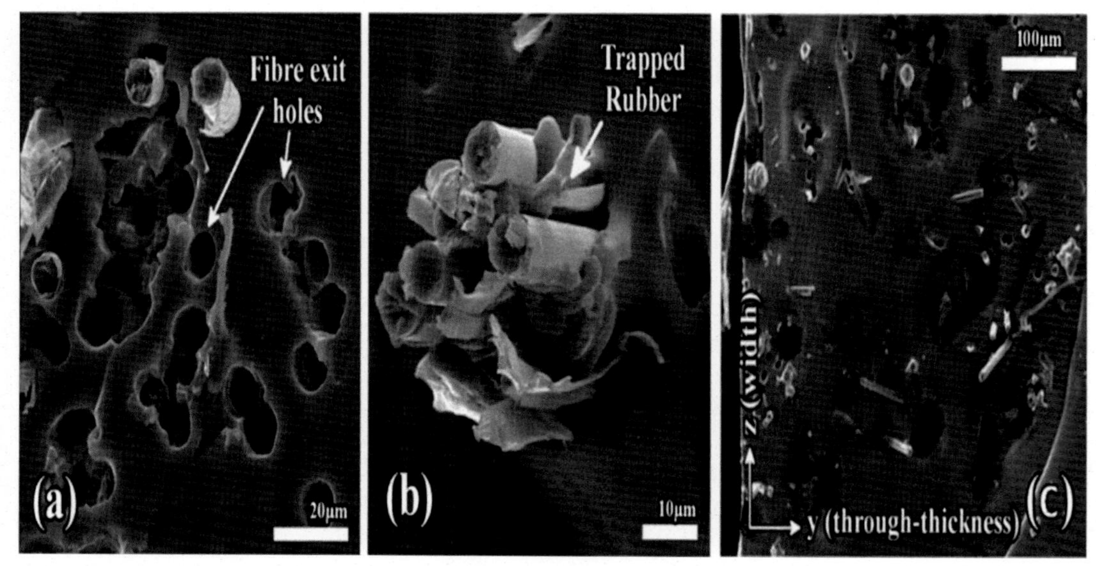

FIGURE 2.4 Fiber–matrix interface of PDMS-NiCF composites. (A) Fiber pullout from PDMS matrix. (B) Rubber trapped within fiber clusters. (C) Fiber dispersion in PDMS matrix [48].

consistent dispersion and orientation was possible for all carbon fibers, since a tiny proportion of conglomerations and malformed particles were observed. The existence of unaffiliated fibers was the result of the fibers eventually losing their nickel coating, as can be seen by SEM images. When mixing the catalysts and metals, the polymer composites had tensile strength that were governed by the resilience and gravitational force and during trying to cure phase. To test its strength, this novel composite was acted upon magnetically, allowing its ability to be used in microactuator systems to be measured. Thanks to the nickel functionalization of the fibers, a low gravitational force was employed (a field strength of no more than 0.2 T). It was found that the approximate orthotropic predisposition of the fibers recognized from the experiments was $Xa = 2.20 \times 10 - 3$, which is approximately 30 times larger than the value reported in the literature for neat carbon fibers, and thus indicates the beneficial impact of the nickel coating on the magnetic properties. The polymer impeller has a highly nonlinear capability, and the effect of the thermal torque on the fibers gives rise to an online dating behavior that changes with an increasing gravitational flux: the samples did not change shape up to a particular field value, but after the critical point, the curvature increased suddenly. It was found that wiggling and flexing behavior was discovered at various step identities of the fiber, and it was explained by the simplistic formula that was introduced.

By regulating the direction of the electric field, the mixture of flexing and contorting in the sensors could be needed to construct complicated navigation mechanisms that are suitable for a wide range of nano systems, like nano and links back. Further development of the device will be required in order to verify whether this material can operate at Reynold's numbers flow and, if this can be done, a prototype design will have to be built and tested in order to discover how quickly this system can travel. For instance, by using the percussive power and speed, it is possible to ascertain whether the newly developed precision scheme is of similar capability to time limited devices. The preliminary research described in partial substitute has implied that the yield stress reliance of the substance is restricted, and that the management information could be applied to a segments and subdevice if the device operated at a moderate pace. It has been shown that cyclic deformation has only a very small number of liquid deformations, supporting its use as a tensile stress actuating device. Though future strain requirements will also necessitate enhancements to the fiber matrix interface; however, envisioning strain material applications also calls for enhancements to the fiber matrix interaction [48,49].

2.2.12 Surface-modified flax elastomer composites

Coatings demonstrate a greater degree of particular structural rigidity, exceptional acoustical soundproofing, and incredible chemical stability when related to organic fibers. Fig. 2.5 denotes the treated and untreated flax organic fiber and their suspended liquid. Alongside, Fig. 2.5 also shows the SEM micrograph of flax fiber depicting its diameter. The traits of these candidates led them to be future material options for the magenta, affordable, and cost-effective heavy industry of next laminated composites. The subject of this research dealt with the issue of material extrusion in 3D printing of natural fibers utilizing direct ink writing (DIW) method. Many studies have been

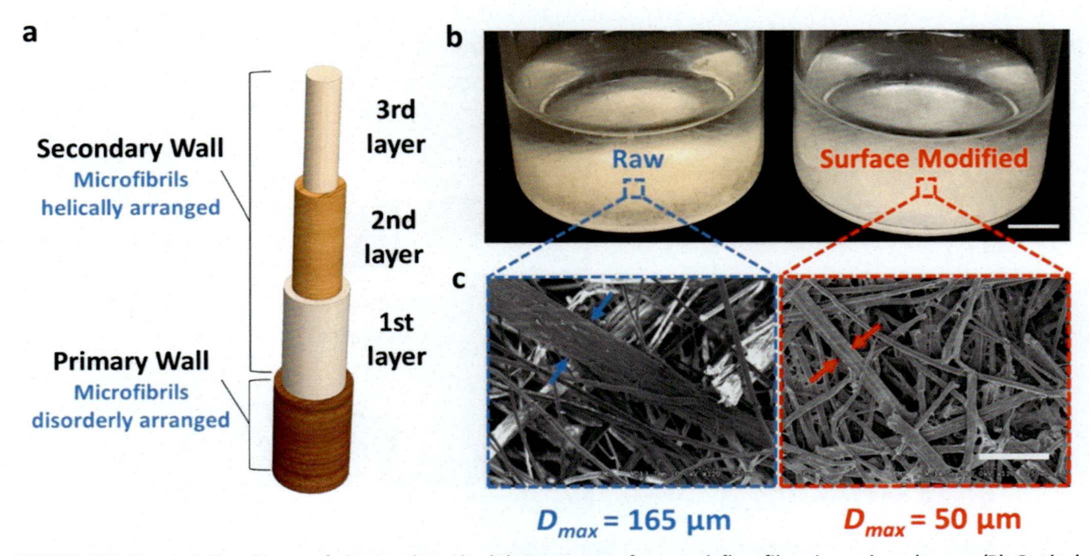

FIGURE 2.5 Treated flax fibers of shorter length. (A) Structure of treated flax fiber in various layers. (B) Optical images of untreated and treated flax fiber suspended in solutions. (C) SEM micrographs of treated fibers of various diameters [51].

conducted on flax-fiber, a frequent plant fiber. An examination is being done on the impact of the bulk material of the flax-fiber on the printability in the DIW process. Different approaches may be utilized for achieving a variety of results. It has been found that the external coating of the fiber can be governed and removed by the sonication process, and the processed fiber can be distributed universally in the polymer to obtain a solid ink (as shown in Fig. 2.6), that also helps to overcome the suggestion cluttering problem in extrusion-based 3D printing techniques. A complete understanding and quantification of the effect can be obtained by identifying and quantifying the thermophysical behavior. Desktop and mobile devices are also investigated, especially with regards to how readable the layout of the document is. Furthermore, 3D printed flax-fiber elastomer composites are examined to see how they respond to mechanical stress. This investigation has shown that only 0.2 wt% ground flax fibers, while contributing only 0.2 wt% to the overall composite weight, can yield tensile properties that are almost as great as or even greater than those of synthetic fiber composites with far more higher fiber loading fractions. It appears that the research team found a book and feasible technique to engineer composite materials through using system. A greater level of house tunability and content biocompatibility both indicate great promise of the complex 3d carbon fiber in a wide range of fields, such as flexible electronics, medical imaging, and wearable electronics [50]. Fig. 2.7 denotes the experimental setup used for DIW.

A brief report on linseed elastic composites was written with a bio-friendly ink directly onto the skin, which could be adjusted to produce a wide range of dynamic material characteristics. The thermal conductivity of pigment was designed to ensure that ink is extruded smoothly. This investigation focuses on the printing of the composites, which are provided in various dispensing tips. We discovered that a small amount of short flax fibers can have a major impact on the mechanical behavior. Our

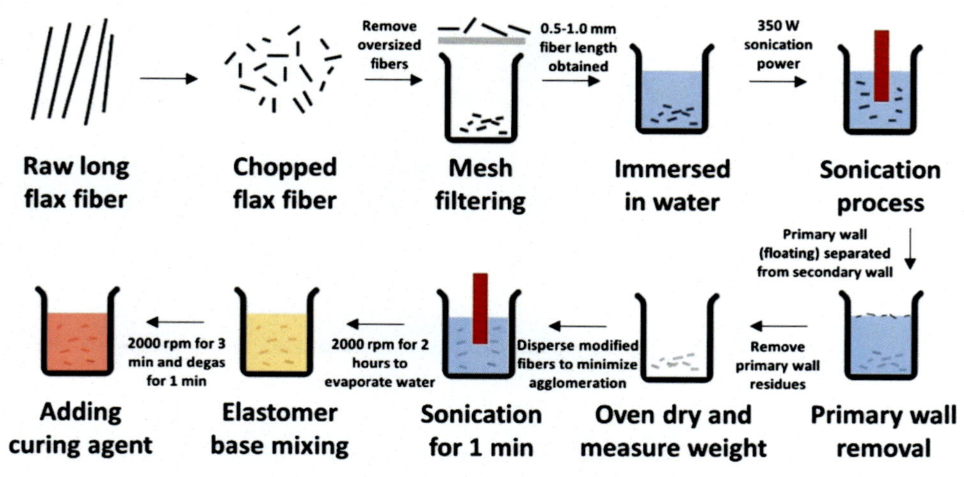

FIGURE 2.6 Ink preparation process [51].

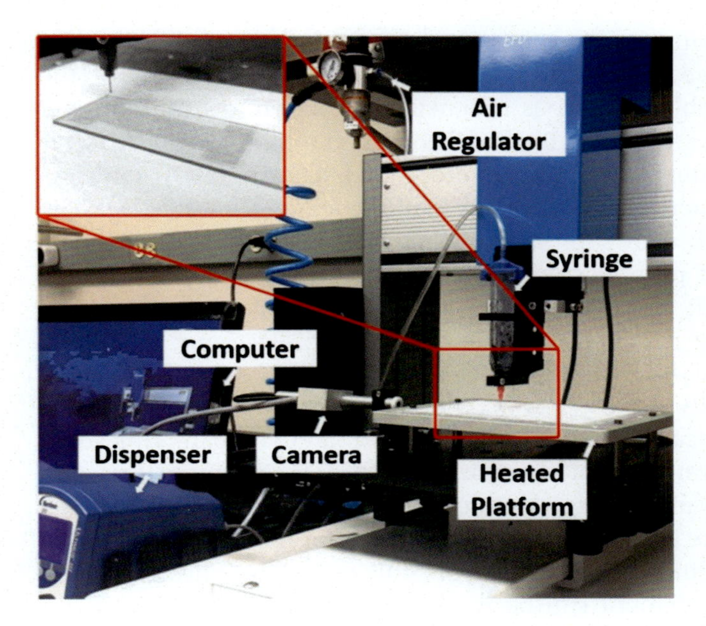

FIGURE 2.7 DIW system setup [51].

findings show that the short flax-fiber-embedded composite has a substantially better mechanical property capability than the many composite materials reported in literature. As a simple and inexpensive way to tune mechanical properties, it is possible to fine-tune elastomer composite materials' mechanical properties with only a slight adjustment to the filler concentration. This offers new possibilities in numerous fields, such as flexible electronics and dynamic wearable electronics. While the softness, flexibility, and bioactivity that is correlated with the surface-modified flax-fiber aerogel composites is promising, its conductivity is a potential drawback. Using a wide array of engineering strategies, we will significantly broaden the architectural and material intricacy of natural light as possible fiber elastic polymers [51,52].

2.2.13 Modeling for randomly oriented multimaterial

An advanced production process, called Rapid Prototyping, previously existed to only serve as a visual demonstration of the procedure and is now also utilized to produce structural improvements. Since additive manufacturing technologies have evolved to be able to their manufacturing capable, this component should be used to make more modern methods and metal components, it is later on in the process. These multimaterial systems can be of a lot of different structures, such as polymer composites, homogeneous or structure-generated heterogeneous materials, and gelled/augmented materials, and porosity materials are hugely beneficial in the advancement of AM. As applied to the aerospace, automobile, medical, and pharmaceutical industries, these frameworks have a significant practical impact. In metallurgy, for some of these constructions, there are issues with uniformity and geometric control of size and distribution, and highest percentages of the primary materials; however, a few of the traditional approaches cause problems such as irregular distribution, distribution and roundness control of the volume, and shape. Randomly oriented multimaterial (ROM) is fabricated on a Polyjet 3D printer utilizing computer-aided design (CAD) software. CATIA has been used for product design development modeling with ROMM. The Universal Testing Machine (UTM) allows for stress–strain testing of Polyjet 3DP materials (for those that are made of pure elastomer and for those which the reinforcing elastomer is randomly deposited). Fig. 2.8 denotes the pure elastomer material after being loaded in universal testing machine. Using elastomer (an elastomer combined with PVC has been shown to be stiffer than the elastomer alone) has been shown to greatly boost ROMM. Compatibility also extends to different Polyjet 3DP modules, which were taken at slightly different locations (horizontal, inclined, and vertical). This study shows that reinforcements are given to all of equal size. The ROMM technique is used to check for the uniformity and randomness of volumetric dispersion in elastomeric components in order to make sure they meet certain standards for the typical range and shape allocation required. According to the experiments, this technique's designing and production abilities, the ROMM element can be made to have increased stiffness for structural parts, along with improved form and fit [53,54]. Fig. 2.9 showcases the ROM material with respect to its failure surface morphology with fibers in various orientations.

Even though it is possible to fabricate complicated products with dimensional parts using AM processes, these are less powerful when compared to the others. The application compatibility issue mentioned above prohibits them from use in usable settings. According to the existing optimization, it will be beneficial to the AM process, helping to increase its overall strength. The experiments were conducted, and the final conclusion was derived. In this research, the ROMM CAD modeling expands the possible length and location of components along the x- and y-axes of a gridlines, as well as along the x-, y-, and z-axes. The initial extension grows 22% when the test is applied to a specimen that is 0% percent of its final length. The stress-strain curve for the ROMM component is less symmetrical, than that of pure elastomers. Based on this there is a difference in the elongation

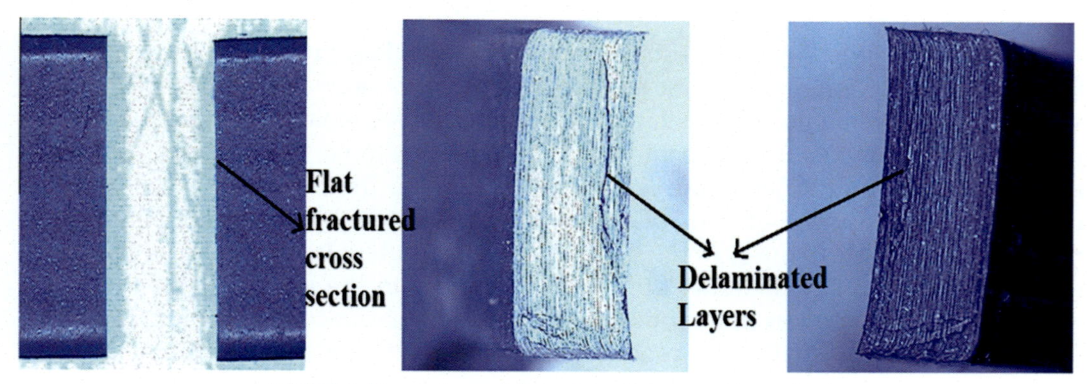

FIGURE 2.8 Morphology of delaminated elastomer layers [55].

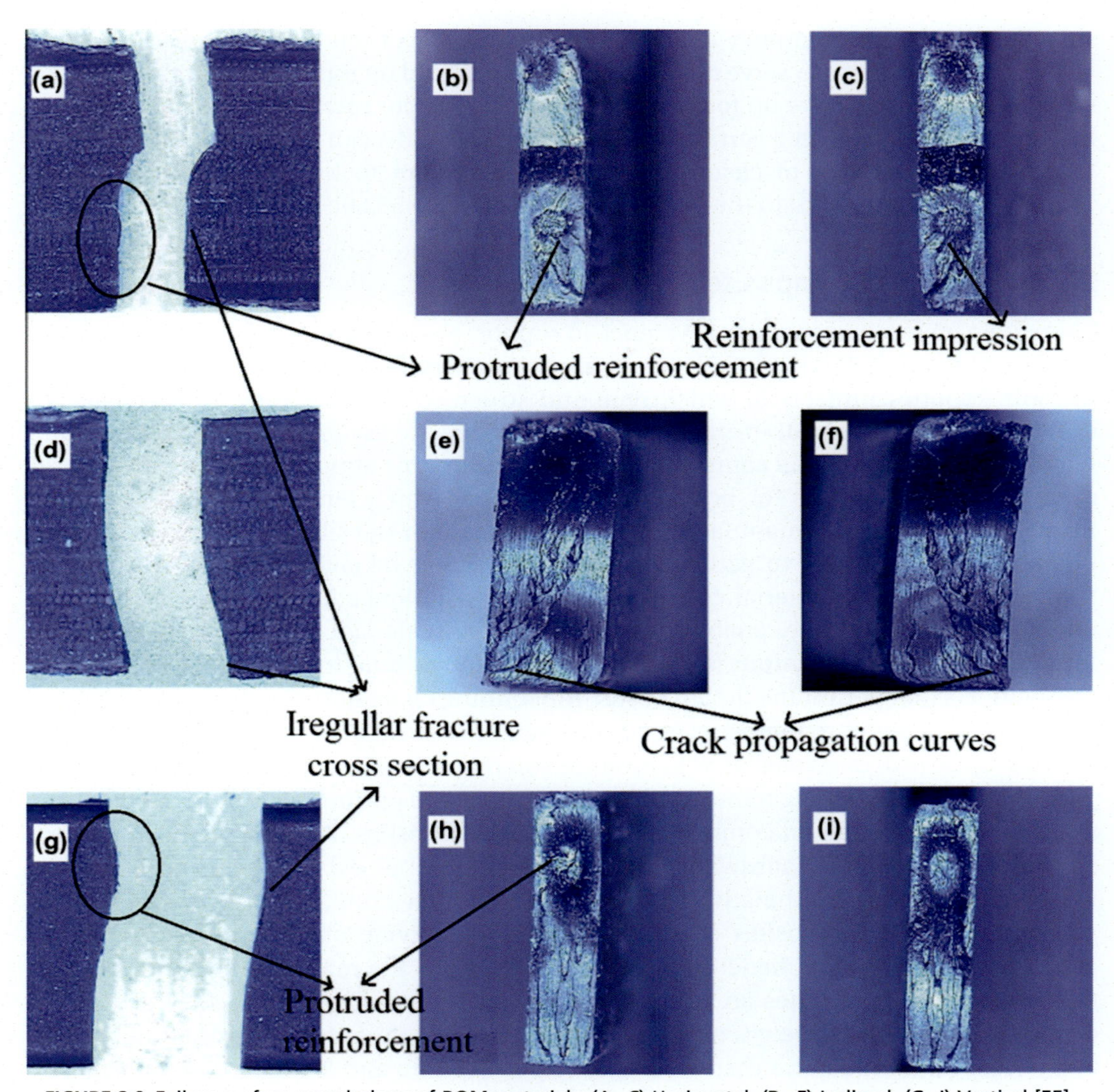

FIGURE 2.9 Failure surface morphology of ROM materials: (A–C) Horizontal, (D–F) Inclined, (G–I) Vertical [55].

ratio of elastomer to plastic reinforcement: the preference for elongation at transverse planes of different magnitudes and an inverse preference for plastic orientation at the plane of fracture. Due to the above fractured reinforcement analysis, there is not enough uninterrupted reinforcement to transfer the load from the matrix to the bar and to the bar attachment. Fracturing of the ROMM happens primarily at the interface among matrix and reinforcement, which ends with failings via the matrix. When running the fractograms on all three independent hemispheres, the widely held analyses are carried to be the same seen among two immobility regimes HO and IO, but different from VO. These polymeric substances have an elastomer-like property: they can be stretched, deformation, along with gaskets and gets can serve as going to live hinges, and on top of the fact, they have an elastomericity, they make up for hinges are elastic. Without sacrificing some of the flexibility, it is essential to get a certain amount of stiffness. This can be accomplished by utilizing various amounts of clear plastic as reinforcement to bring up the elastomeric properties of the material to the desired level of stiffness [55,56].

2.2.14 Silicone composites

There is an increasing amount of evidence to support elastic composite materials as viable actuation materials. It is practically necessary to have the ability to produce these materials for Soft Robotics projects in which type and structure are intricately intertwined. We describe a multimaterial 3D printer and ase printer that are used for in situ manufacturing of polymeric actuators and composites. We will be speaking about polymer synthesis and encapsulation technique for both silicones, both of which employ adjustable passive mixing, then review the manufacturing methods used for the silicone skin print and the different types of prints to gather information on their behavior. A study of rheology showed the range of material conditions that allow prolonged 3D printing of silicone/ethanol-based composite, enabling small and large-scale silicone to be continuously formed. A curvilinear printout is used for the creation of functional designs by selecting print models that permit the design of soft-functioning characteristics. We demonstrate a single-job functional manufacturing of multimaterial structures for soft actuation, and propose these processes enable us to produce soft robots with many actuation modes [57].

In some of the investigations, a silicone composite 3D printer was fabricated, which was constructed based on composite material characteristics. It is being looked at from both a software and a hardware perspective. Isothermal expansion using empirical measurements controls the advanced 3D printer, and custom 2-part (two-component) platinum silicone composites can be expanded for varying object complexity of use. properties to ensure that inkjet printing will not disturb the materials in the work area, the 3D printer incorporates an adjustable passive mixer which was specially developed depends on the test of the viscosity properties. The 3D printer has a workstation of 40.6 in 6 × 40.6 in (a little more than a square feet) × 76.2 cm, and is able to use up to three different types of silicone composites in the same job. Material expansion through three-dimensional printing was done by an automated mechanism which allowed complex geometries such as silicone and a wide range of ethanol to work in an enveloping rubber skin to be presented in several shapes. Print parameters were tested under varying printing conditions (vertical, horizontal, and crossed, and along the surface of the specimen) to find out if the print materials produced, changed, degraded, or held the shape of the design correctly. In addition, anisotropy in mechanical properties was introduced for the tailoring of the finished part at the design stage, which allowed for a direction to be chosen that enabled the fine-tuning of the property parameters of the design [58–60].

2.3 Conclusion

The processing techniques of elastomer blends, which provide the pathway to choose the appropriate technique to manufacture the elastomer blends and composites, were discussed in this chapter. The processing of manufacturing is consolidated to have a clear overview about the manufacturing process of elastomer blends and composites. All the manufacturing methods were discussed with a specific emphasis on 3D printing techniques where all the methods were directly or indirectly relying on additive manufacturing technologies. The specific 3D printing technique is also discussed in a few of the manufacturing methods, but the technology varies for each and every elastomer material production. According to the common understanding from the aforementioned discussion, it could be stated that DIW and shape-deposition modeling are the preferred techniques in additive manufacturing for elastomer blends and composites. The 3D-printing elastomer blends and composites accommodate the materials at various phases including the surface-modified elastomer composites, ROM, silicone composites, and other variants too. The 3D printing intrusion is one of the many remarkable manufacturing techniques for elastomer blends and composites. It could be concluded that the manufacturing of elastomer blends and composites could go a long way when they are integrated with additive manufacturing technologies, but a few limitations like time of printing and material heterogeneity need to be addressed to have a wider scope of applications.

References

[1] L.D. Perez, M.A. Zuluaga, T. Kyu, J.E. Mark, B.L. Lopez, Preparation, characterization, and physical properties of multiwall carbon nanotube/elastomer composites, Polym. Eng. Sci. 49 (5) (2009) 866–874.

[2] M.D. Frogley, D. Ravich, H.D. Wagner, Mechanical properties of carbon nanoparticle-reinforced elastomers, Compos. Sci. Technol. 63 (11) (2003) 1647–1654.

[3] A. Das, K.W. Stöckelhuber, R. Jurk, J. Fritzsche, M. Klüppel, G. Heinrich, Coupling activity of ionic liquids between diene elastomers and multi-walled carbon nanotubes, Carbon 47 (14) (2009) 3313–3321.

[4] M. Khalid, A.F. Ismail, C.T. Ratnam, Y. Faridah, W. Rashmi, M.F. Al Khatib, Effect of radiation dose on the properties of natural rubber nanocomposite, Radiat. Phys. Chem. 79 (12) (2010) 1279–1285.

[5] X.W. Zhou, Y.F. Zhu, J. Liang, Preparation and properties of powder styrene–butadiene rubber composites filled with carbon black and carbon nanotubes, Mater. Res. Bull. 42 (3) (2007) 456–464.

[6] M.J. Jiang, Z.M. Dang, S.H. Yao, J. Bai, Effects of surface modification of carbon nanotubes on the microstructure and electrical properties of carbon nanotubes/rubber nanocomposites, Chem. Phys. Lett. 457 (4–6) (2008) 352–356.

[7] K.Q. Xiao, L.C. Zhang, The stress transfer efficiency of a single-walled carbon nanotube in epoxy matrix, J. Mater. Sci. 39 (14) (2004) 4481–4486.

[8] D. Baskaran, J.W. Mays, M.S. Bratcher, Polymer-grafted multiwalled carbon nanotubes through surface-initiated polymerization, Angew. Chem. 116 (16) (2004) 2190–2194.

[9] A. Saravana Kumar, P. Maivizhi Selvi, L. Rajeshkumar, Delamination in drilling of sisal/banana reinforced composites produced by hand lay-up process, in: Applied Mechanics and Materials, vol. 867, Trans Tech Publications Ltd., 2017, pp. 29–33.

[10] K. Subramaniam, A. Das, L. Häubler, C. Harnisch, K.W. Stöckelhuber, G. Heinrich, Enhanced thermal stability of polychloroprene rubber composites with ionic liquid modified MWCNTs, Polym. Degrad. Stabil. 97 (5) (2012) 776–785.

[11] L. Liu, Z. Zheng, C. Gu, X. Wang, The poly (urethane-ionic liquid)/multi-walled carbon nanotubes composites, Compos. Sci. Technol. 70 (12) (2010) 1697–1703.

[12] J. Liu, A.G. Rinzler, H. Dai, J.H. Hafner, R.K. Bradley, P.J. Boul, A. Lu, T. Iverson, K. Shelimov, C.B. Huffman, F. Rodriguez-Macias, Fullerene pipes, Science 280 (5367) (1998) 1253–1256.

[13] M. Ramesh, L. Rajeshkumar, Wood flour filled thermoset composites. Thermoset composites: preparation, properties and applications, Mater. Res. Forum 38 (2018) 33–65. https://doi.org/10.21741/9781945291876-2.

[14] M.J. O'connell, S.M. Bachilo, C.B. Huffman, V.C. Moore, M.S. Strano, E.H. Haroz, K.L. Rialon, P.J. Boul, W.H. Noon, C. Kittrell, J. Ma, Band gap fluorescence from individual single-walled carbon nanotubes, Science 297 (5581) (2002) 593–596.

[15] H. Hyung, J.D. Fortner, J.B. Hughes, J.H. Kim, Natural organic matter stabilizes carbon nanotubes in the aqueous phase, Environ. Sci. Technol. 41 (1) (2007) 179–184.

[16] C.A. Dyke, J.M. Tour, Covalent functionalization of single-walled carbon nanotubes for materials applications, J. Phys. Chem. A 108 (51) (2004) 11151–11159.

[17] L. Bokobza, C. Belin, Effect of strain on the properties of a styrene–butadiene rubber filled with multiwall carbon nanotubes, J. Appl. Polym. Sci. 105 (4) (2007) 2054–2061.

[18] T.J. Lee, A.H. Morgenstern, T. Höft, B.B. Nelson-Cheeseman, Dispersion of particulate in solvent cast magnetic thermoplastic polyurethane elastomer composites, AIMS Mater. Sci. 6 (2019) 354–362.

[19] L. Qian, H. Zhang, Controlled freezing and freeze drying: a versatile route for porous and micro-/nano-structured materials, J. Chem. Technol. Biotechnol. 86 (2) (2011) 172–184.

[20] J. Yu, K. Lu, E. Sourty, N. Grossiord, C.E. Koning, J. Loos, Characterization of conductive multiwall carbon nanotube/polystyrene composites prepared by latex technology, Carbon 45 (15) (2007) 2897–2903.

[21] M. Ramesh, L. Rajesh Kumar, A. Khan, A.M. Asiri, Self-healing polymer composites and its chemistry, in: Self-Healing Composite Materials, Woodhead Publishing, 2020, pp. 415–427.

[22] S. Ganguly, P. Das, T.K. Das, S. Ghosh, S. Das, M. Bose, M. Mondal, A.K. Das, N.C. Das, Acoustic cavitation assisted destratified clay tactoid reinforced in situ elastomer-mimetic semi-IPN hydrogel for catalytic and bactericidal application, Ultrason. Sonochem. 60 (2020) 104797.

[23] X. Zhou, Y. Zhu, Q. Gong, J. Liang, Preparation and properties of the powder SBR composites filled with CNTs by spray drying process, Mater. Lett. 60 (29–30) (2006) 3769–3775.

[24] P.M. Visakh, S. Thomas, A.K. Chandra, A.P. Mathew, Advances in Elastomers II, Springer, Berlin, 2013.

[25] M. Ramesh, L. RajeshKumar, V. Bhuvaneshwari, Bamboo fiber reinforced composites, in: Bamboo Fiber Composites, Springer, Singapore, 2021, pp. 1–13.

[26] A. Anand K, S. Jose T, R. Alex, R. Joseph, Natural rubber-carbon nanotube composites through latex compounding, Int. J. Polym. Mater. 59 (1) (2009) 33–44.

[27] D. Ponnamma, K.K. Sadasivuni, K.T. Varughese, S. Thomas, M.A.A. AlMa'adeed, Natural polyisoprene composites and their electronic applications, in: Flexible and Stretchable Electronic Composites, Springer, Cham, 2016, pp. 1–35.

[28] M. Ramesh, L. Rajesh Kumar, Bioadhesives, in: Green Adhesives: Preparation, Properties and Applications, 2020, pp. 145–164.

[29] Y. Xu, B. Higgins, W.J. Brittain, Bottom-up synthesis of PS–CNF nanocomposites, Polymer 46 (3) (2005) 799–810.

[30] E.J. Park, S. Hong, D.W. Park, S.E. Shim, Preparation of conductive PTFE nanocomposite containing multiwalled carbon nanotube via latex heterocoagulation approach, Colloid Polym. Sci. 288 (1) (2010) 47–53.

[31] X. Huang, W.J. Brittain, Synthesis and characterization of PMMA nanocomposites by suspension and emulsion polymerization, Macromolecules 34 (10) (2001) 3255–3260.

[32] M. Ramesh, C. Deepa, L. Rajesh Kumar, M.R. Sanjay, S. Siengchin, Life-cycle and environmental impact assessments on processing of plant fibres and its bio-composites: a critical review, J. Ind. Textil. (2020). https://doi.org/10.1177/1528083720924730.

[33] Y. Xu, W.J. Brittain, C. Xue, R.K. Eby, Effect of clay type on morphology and thermal stability of PMMA–clay nanocomposites prepared by heterocoagulation method, Polymer 45 (11) (2004) 3735–3746.

[34] Z. Peng, C. Feng, Y. Luo, Y. Li, L.X. Kong, Self-assembled natural rubber/multi-walled carbon nanotube composites using latex compounding techniques, Carbon 48 (15) (2010) 4497–4503.

[35] B.Z. Tang, H. Xu, Preparation, alignment, and optical properties of soluble poly (phenylacetylene)-wrapped carbon nanotubes, Macromolecules 32 (8) (1999) 2569–2576.

[36] J. Fan, M. Wan, D. Zhu, B. Chang, Z. Pan, S. Xie, Synthesis, characterizations, and physical properties of carbon nanotubes coated by conducting polypyrrole, J. Appl. Polym. Sci. 74 (11) (1999) 2605–2610.

[37] D. Balaji, M. Ramesh, T. Kannan, S. Deepan, V. Bhuvaneswari, L. Rajeshkumar, Experimental investigation on mechanical properties of banana/snake grass fiber reinforced hybrid composites, Mater. Today Proc. (2020). https://doi.org/10.1016/j.matpr.2020.09.548.

[38] A. Star, J.F. Stoddart, D. Steuerman, M. Diehl, A. Boukai, E.W. Wong, X. Yang, S.W. Chung, H. Choi, J. R. Heath, Preparation and properties of polymer-wrapped single-walled carbon nanotubes, Angew. Chem. 113 (9) (2001) 1771–1775.

[39] A. Das, K.W. Stöckelhuber, R. Jurk, M. Saphiannikova, J. Fritzsche, H. Lorenz, M. Klüppel, G. Heinrich, Modified and unmodified multiwalled carbon nanotubes in high performance solution-styrene–butadiene and butadiene rubber blends, Polymer 49 (24) (2008) 5276–5283.

[40] M. Ramesh, C. Deepa, M. Tamil Selvan, L. Rajeshkumar, D. Balaji, V. Bhuvaneswari, Mechanical and water absorption properties of Calotropis gigantea plant fibers reinforced polymer composites, Mater. Today Proc. (2020). https://doi.org/10.1016/j.matpr.2020.11.480.

[41] N.A. Traugutt, D. Mistry, C. Luo, K. Yu, Q. Ge, C.M. Yakacki, Liquid-crystal-elastomer-based dissipative structures by digital light processing 3D printing, Adv. Mater. 32 (28) (2020) 2000797.

[42] J. Fritzsche, H. Lorenz, M. Klüppel, A. Das, R. Jurk, K.W. Stöckelhuber, G. Heinrich, Elastomer–carbon nanotube composites, in: Polymer–Carbon Nanotube Composites, Woodhead Publishing, 2011, pp. 193–229.

[43] H. Xia, Q. Wang, K. Li, G.H. Hu, Preparation of polypropylene/carbon nanotube composite powder with a solid-state mechanochemical pulverization process, J. Appl. Polym. Sci. 93 (1) (2004) 378–386.

[44] E.O. Bachtiar, O. Erol, M. Millrod, R. Tao, D.H. Gracias, L.H. Romer, S.H. Kang, 3D printing and characterization of a soft and biostable elastomer with high flexibility and strength for biomedical applications, J. Mech. Behav. Biomed. Mater. 104 (2020) 103649.

[45] J.I. Masuda, J.M. Torkelson, Dispersion and major property enhancements in polymer/multiwall carbon nanotube nanocomposites via solid-state shear pulverization followed by melt mixing, Macromolecules 41 (16) (2008) 5974–5977.

[46] M. Moniruzzaman, K.I. Winey, Polymer nanocomposites containing carbon nanotubes, Macromolecules 39 (16) (2006) 5194–5205.

[47] V. Bhuvaneswari, M. Priyadharshini, C. Deepa, D. Balaji, L. Rajeshkumar, M. Ramesh, Deep learning for material synthesis and manufacturing systems: a review, Mater. Today Proc. (2021). https://doi.org/10.1016/j.matpr.2020.11.351.

[48] D.C. Stanier, J. Ciambella, S.S. Rahatekar, Fabrication and characterisation of short fiber reinforced elastomer composites for bending and twisting magnetic actuation, Compos. Appl. Sci. Manuf. 91 (2016) 168–176.

[49] T. Villmow, P. Pötschke, S. Pegel, L. Häussler, B. Kretzschmar, Influence of twin-screw extrusion conditions on the dispersion of multi-walled carbon nanotubes in a poly (lactic acid) matrix, Polymer 49 (16) (2008) 3500–3509.

[50] P.C. Ma, S.Q. Wang, J.K. Kim, B.Z. Tang, In-situ amino functionalization of carbon nanotubes using ball milling, J. Nanosci. Nanotechnol. 9 (2) (2009) 749–753.

[51] Y. Jiang, J. Plog, A.L. Yarin, Y. Pan, Direct ink writing of surface-modified flax elastomer composites, Compos. B Eng. 194 (2020) 108061.

[52] M. Ramesh, J. Maniraj, L. Rajesh Kumar, Biocomposites for energy storage, in: Biobased Composites: Processing, Characterization, Properties, and Applications, 2021, pp. 123–142.

[53] M. Moniruzzaman, F. Du, N. Romero, K.I. Winey, Increased flexural modulus and strength in SWNT/epoxy composites by a new fabrication method, Polymer 47 (1) (2006) 293–298.

[54] Z. Peng, C. Feng, Y. Luo, Y. Li, Z. Yi, L.X. Kong, Natural rubber/multiwalled carbon nanotube composites developed with a combined self-assembly and latex compounding technique, J. Appl. Polym. Sci. 125 (5) (2012) 3920–3928.

[55] M. Sugavaneswaran, G. Arumaikkannu, Modelling for randomly oriented multi material additive manufacturing component and its fabrication, Mater. Des. 54 (2014) 779–785.

[56] M. Ramesh, L. Rajeshkumar, D. Balaji, V. Bhuvaneswari, Green composite using agricultural waste reinforcement, in: Green Composites, Springer, Singapore, 2021, pp. 21–34.

[57] M. Ramesh, Kenaf (*Hibiscus cannabinus* L.) fibre based bio-materials: a review on processing and properties, Prog. Mater. Sci. 78 (2016) 1–92.

[58] A. Miriyev, B. Xia, J.C. Joseph, H. Lipson, Additive manufacturing of silicone composites for soft actuation, 3D Print. Addit. Manuf. 6 (6) (2019) 309–318.

[59] M. Ramesh, C. Deepa, M. Tamil Selvan, K.H. Reddy, Effect of alkalization on characterization of ripe bulrush (*Typha domingensis*) grass fiber reinforced epoxy composites, J. Nat. Fibers (2020) 1–12.

[60] M. Ramesh, R. Vimal, K.H. Hara Subramaniyan, C. Aswin, B. Ganesh, C. Deepa, Study of mechanical properties of jute-banana-glass fiber reinforced epoxy composites under various post curing temperature, in: Applied Mechanics and Materials, vol. 766, Trans Tech Publications Ltd., 2015, pp. 211–215.

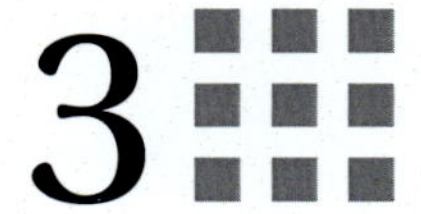

Elastomer-based blends

Aswathy Jayakumar[1,2], Sabarish Radoor[2], Jyotishkumar Parameswaranpillai[3], E.K. Radhakrishnan[1], Indu C. Nair[4], Suchart Siengchin[2]

[1]SCHOOL OF BIOSCIENCES, MAHATMA GANDHI UNIVERSITY, KOTTAYAM, KERALA, INDIA; [2]DEPARTMENT OF MECHANICAL AND PROCESS ENGINEERING, THE SIRINDHORN INTERNATIONAL THAI-GERMAN GRADUATE SCHOOL OF ENGINEERING (TGGS), KING MONGKUT'S UNIVERSITY OF TECHNOLOGY NORTH BANGKOK, BANGKOK, THAILAND; [3]DEPARTMENT OF SCIENCE, FACULTY OF SCIENCE & TECHNOLOGY, ALLIANCE UNIVERSITY, BENGALURU, KARNATAKA, INDIA; [4]DEPARTMENT OF BIOTECHNOLOGY, SAHODARAN AYYAPPAN SMARAKA SREE NARAYANA DHARMA PARIPALANA YOGAM (SAS SNDPYOGAM), COLLEGE, KONNI, PATHANAMTHITTA, KERALA, INDIA

3.1 Introduction

Elastomers are polymeric materials having low intermolecular forces with flexible and elastic nature (Fig. 3.1) [1]. Due to their recyclability, renewability, and cost-effectiveness, they found application in several sectors. Examples of elastomers include natural rubber, synthetic polyisoprene, butyl rubber, polybutadiene, styrene-butadiene, nitrile rubber, chloroprene rubber, ethylene propylene rubber, polyacrylic rubber, silicone, fluoro silicone, fluoro elastomers, perfluoro elastomers, polyether block amides, chloro sulfonated polyethylene, ethylene-vinyl acetate, thermoplastic elastomers, thermoplastic polyurethane, thermoplastic olefins, resilin, elastin, and polysulfide. Elastomers are blended to improve the physical, mechanical and for better processing of compounds [2].

Elastomeric blends are of two types depending upon the interaction between the polymers and can be miscible and partially miscible. They can be immiscible rubber

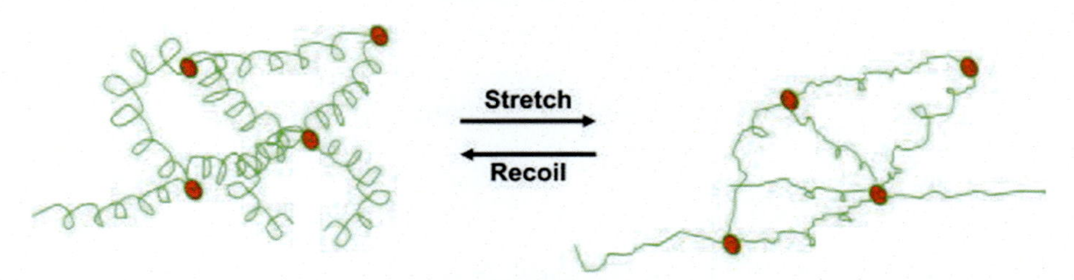

FIGURE 3.1 Stretching and recoiling of elastomers. *Reproduced with permission from Elsevier, License Number-5080840325667.*

Elastomer Blends and Composites. https://doi.org/10.1016/B978-0-323-85832-8.00017-1

blends, rubber/thermoplastic blends, and thermoset blends. The immiscible blends are found as continuous phase having one phase dispersed within the continuous phase [3]. During the mechanical mixing of the same, the material with lower viscosity deforms and encapsulates the compounds having higher viscosity to produce a globular shape. But the morphology of this miscible phase is not at all in equilibrium condition. Thermoplastic polymer/rubber blends are obtained by the blending of rubber-rich mixtures, which forms soft thermoplastic elastomers, whereas plastic-rich blends form rubber toughened thermoplastic [3].

The use of conventional rubber in several applications such as automotive, medical, engineering, and other appliances are replaced by these thermoplastic elastomers. In the thermoplastic elastomers, block polymers consist of soft and rubbery blocks. The thermoset plastics can be toughened by the addition of rubber particles [3]. Epoxy resins are the most prominent class of thermosetting polymers. The brittle nature of the thermoplastics can be improved by the blending with proper liquid rubbers.

The elastomers are often blended with other materials in order to enhance its properties [2]. But these blending processes not always led to the improvement in properties due to the immiscibility faced by these elastomers. Hence, several researchers utilized block polymers, reactive copolymers, nonreactive copolymers, and nanofillers for the compatibilization. The higher performance of elastomer-based blends can be accomplished by compatibilization and/or by enhancing the interaction of polymers (mechanically or chemically) [2,4]. The current chapter is designated to provide an overview on elastomeric polymer blends, fabrication process, mechanical and thermal performances along with its application in various sectors.

3.2 Compatibilization of elastomer-based blends

The compatibilization with materials of interest has of great significance in the development of elastomer-based blends as it improves the performance of blends. The compatibilizer reduces the interfacial energy, thereby improving the adhesion of the polymer phases [5]. This will result in the fine distribution of components and reduces the agglomeration. The process of compatibilization can be of reactive or nonreactive. There are several types of fillers that can act as compatibilizing agents. Silica, carbon black, aluminum oxide, magnesium silicate, aluminum trihydrate, mica, talc, calcium carbonate, clay, zinc oxide, and calcium are the commonly used compatibilizing agents [3]. There are several research studies that suggest the improvement in the performance of elastomer-based blends with the use of these compatibilizers.

3.3 Impact of nanofillers on elastomer-based blends

Recently, nanofillers are used as compatibilizing agent for the development of elastomer-based blends. The compatibilization effect of nanoparticles makes them suitable for its application in developing elastomer-based blends. The effect of compatibilization depends on the distribution of particles between the two polymer phases. There are several studies that suggest the compatibilizing effect of nanoparticles. This might depends upon the distribution of nanoparticles, restriction of coalescence, and the alteration in viscosity. There are organic and inorganic nanomaterials that are used as compatibilizing agent in elastomeric blends such as nanoclay, titanium and zinc oxide nanoparticles, carbon nanotubes, graphene and carbon black fillers [6].

3.4 Fabrication methods of elastomers

The fabrication methods of elastomeric blends are highly dependent on the composition of monomer, its ability to cross-link, and the type of application. The reported fabrication methods involve polycondensation via step addition polymerization, photo cross-linking, and alternative network formation [7]. The commonly employed method for the development of polymer-based blends is polycondensation via step addition polymerization. Polyesters, polyamides, and polyurethanes are the synthetic polymers produced by this technique. The other method employed is photocross-linking, which utilizes photoactive diene-based modifications. Alternative network formation utilizes cross-linking agents for the better functionality of the systems.

3.5 Processing and characterization methods of elastomers-based blends

Depending upon the applications, the synthesized elastomers are developed into films or complex structures. The processing method can be categorized into two types depending upon the type of elastomers (thermoplastics or thermoset). Basically, thermoplastic elastomers are processed using extrusion, injection molding, compression molding, transfer molding, whereas thermosetting elastomers involve compression molding and transfer molding. The morphological, mechanical, thermal, rheological, and conductivity studies of elastomer-based blends are characterized by several techniques. Most commonly used techniques include wide-angle X-ray diffraction (WAXD), small-angle X-ray scattering (SAXS), atomic force microscopy (AFM), thermal gravimetric analysis (TGA), differential scanning calorimetry (DSC), universal testing machine (UTM), scanning electron microscopy (SEM), transmission electron microscopy (TEM), Fourier transform infrared spectroscopy (FT-IR) analysis, nuclear-magnetic resonance (NMR), and dynamic mechanical analysis (DMA). WAXD is used to analyze the dispersion of components, morphology, whereas the phase behavior and structural evolution can be obtained by SAXS. The crystallization behavior, surface roughness, and distribution of particle size can be measured by using AFM. The interaction between components and the functional groups can be analyzed using FT-IR. The thermal behavior, melting point, dynamics, and crystallization behavior of elastomer-based blends can be observed using TGA. The mechanical parameters such as tensile strength, Young's modulus, and elongation at break percentage can be analyzed by UTM.

3.6 Properties of elastomers-based blends

The main challenges associated with elastomer-based polymers are its high stretchability, good mechanical strength, good tear resistance, and self-healing and adhesive properties (Song and his coworkers [8]. The emergence of thermoplastic elastomers is considered as one of the significant development in the area of polymer science and technology [9].

Jalalifar and his coworkers studied the reinforcing effect of halloysite nanotube (HNT) particles and polyolefin elastomer–grafted–maleic anhydride (POE-g-MA) compatibilizer on polylactic acid/polyolefin elastomer (PLA/POE) blends [10]. Their study revealed the enhancement in viscoelastic and thermal properties of PLA/POE blends and could be due to the better interactions and interfacial adhesions between them. Previously, the blending of elastomer with other polymers has been reported to have enhancement in mechanical properties. For e.g., Varsavas and his colleague studied the effect of blending E-glass fibers (GF) with thermoplastic polyurethane elastomer (TPU) on the mechanical properties of PLA [11]. Their study suggested that the addition of TPU has resulted in the significant increase in fracture toughness and the ductility of the blend while GF incorporation led to the enhancement in strength and elastic modulus of the blends.

The disadvantages of carbon-based or metal-based sensors can be overcome by the use of elastomer-based sensors due to its better compatibility, flexibility, bendability, and gauge factor. The studies conducted by Panwar and his colleague reported the development of piezoresistive strain sensor by the incorporation of Poly(vinylidene fluoride-trifluoroethylene- chlorotrifluoroethylene) P(VDF-TrFE-CTFE) (TER)/poly(3,4-ethylene dioxythiophene) (PEDOT) and poly(styrene sulfonate) (PSS). The developed sensor has efficient gauge factor and high ductility with the application in electronics [12]. There are several studies regarding the studies on properties of elastomer-based blends as indicated in Table 3.1.

3.7 Applications of elastomer-based blends

3.7.1 Self-healable elastomer blends

Imato and his coworkers utilized charge transfer interaction mechanism to develop self-healing elastomer based on polyurethane, electron-rich pyrene and electron-deficient naphthalene diimide. Under air and water condition at 30–100°C, the blends undergo self-healing, and the mechanical properties of the same have reported to been restored after the healing process (air, 100°C) [13].

The blending of polyol and zwitterionic polyol has been reported to result in the development of self-healable elastomers [14]. The developed elastomers have good

Table 3.1 Properties of elastomer-based blends.

Elastomeric blends	Properties	Applications	Reference
Polyolefin elastomer—grafted—maleic anhydride (POE-g-MA) polylactic acid/ polyolefin elastomer (PLA/POE) blends	Improvement in viscoelastic and thermal properties		[10]
E-glass fibers (GF), (TPU)	Enhanced fracture toughness and the ductility		[11]
Poly(vinylidene fluoride-trifluoroethylene-chlorotrifluoroethylene) P(VDF-TrFE-CTFE) (TER)/poly(3,4-ethylene dioxythiophene) (PEDOT) and poly(styrene sulfonate) (PSS)	Efficient gauge factor and high ductility	Electronics	[12]
Polyurethane and naphthalene diimide	Self-healable and improved mechanical performances		[13]
Polyol and zwitterionic polyol	Improved optical, mechanical, self-healable, high transmittance and low yellow index.		[14]
Bromobutyl rubber (BIIR) and terpene resin (pαp)	Enhanced mechanical performances, improved stretchability percentage, good healability and tear resistance		[8]
Poly(lactide) with biobased polyamide elastomer	Enhancement in toughness	Packaging, medical and automotive	[25]
Polylactide *graft*-ESA	Enhancement in melt memory effect		[18]
Polycaprolactone (PCL) and polystyrene-*block*-poly(ethylene-*co*-butylene)-*block*-polystyrene (SEBS)	Enhanced mechanical performances	Soft actuators, deployable devices and electronics.	[15]
Maleinized linseed oil, polylactide and polystyrene-*b*-(ethylene-*ran*-butylene)-*b*-styrene	Enhanced impact strength, toughness, elongation at break, and ultraviolet light protection	Food industry	[19]
Polyurethane/polylactide shape-memory blends	Enhanced optical transparence with balanced mechanical properties	Shape memory polymeric materials	[17]
Poly (styrene-butadiene-styrene) and polyurethane	Improved mechanical, thermal, and abrasion resistance		[20]
Poly(ether-block-amide) (PEBA), glycidyl methacrylate (GMA) and poly (lactic acid)	Improved tensile toughness and impact strength		[21]
Poly (butylene-sebacate—*co*—terephthalate) (PBSET), and hexamethylene diisocyanate (HDI), poly (lactic acid) (PLA)	Enhance the tensile strength and elongation at break		[23]

optical, mechanical, and self-healable properties with high transmittance and low yellow index (Fig. 3.2). The self-healing percentage of developed material shows 94%—100% (zwitterionic clusters), ~69% (nonionic elastomers), and 80%—87% (ionic elastomers). The zwitterionic clusters eliminate the trade-off between mechanical and self-healing properties. This could be due to the strong intermolecular interactions between the zwitterionic clusters.

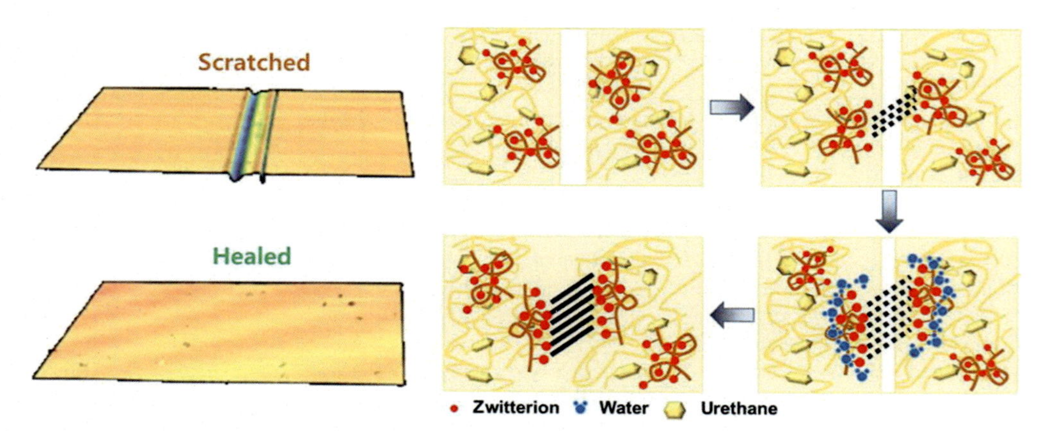

FIGURE 3.2 Self-healing mechanism of zwitterionic elastomer blend. *From J. Kang, J. Kim, K. Choi, P.H. Hong, H.J. Park, K. Kim, Y.K. Kim, G. Moon, H. Jeon, S. Lee, M.J. Ko, S.W. Hong, A water-triggered highly self-healable elastomer with enhanced mechanical properties achieved using localized zwitterionic assemblies, Chemical Engineering Journal (2020). https://doi.org/10.1016/j.cej.2020.127636. Reproduced with permission from Elsevier, License Number-5015321101387.*

Song and his coworkers developed a novel polymeric elastomer by mixing bromobutyl rubber (BIIR) and terpene resin (PαP). The developed elastomer has been reported to have sufficient mechanical strength, good healability, and tear resistance along with adhesive nature. Their study revealed the improved stretchability percentage (strain ~1720%), tensile strength (11.5 MPa), adhesion strength (35 kPa), and tearing energy (2300 J/m^2). The strong covalent interaction between bromobutyl rubber and terpene resin has reported to result in the improvement of properties [8].

Peng and his coworkers employed 4D printing technology to obtain self-healable and shape memory elastomer based on polycaprolactone (PCL) and polystyrene-*block*-poly(ethylene-co-butylene)-*block*-polystyrene (SEBS) [15]. The developed material has been reported to have enhanced mechanical performances along with multifunctional in nature having application as soft actuators, deployable devices, and electronics. Ding and his coworkers have successfully developed a novel poly (epichlorohydrin-*co*-ethylene oxide)-*g*-poly (methyl methacrylate) copolymer (ECO-*g*-PMMA) and PLA-based blends by atom transfer radical polymerization (ATRP) method [16]. Here, improvement in mechanical properties along with the synergistic toughening effect has been noticed.

By utilizing the bilayer structure design, Ji et al. fabricated polyurethane/polylactide shape memory blends with adjustable optical and mechanical properties. The better phase continuity and the improved shearing effect between the two layers optimized the temporary shape fixation and the permanent shape recovery. Similarly, high optical transparency and balanced mechanical properties are the other properties of the developed shape memory polymeric materials [17]. The poor melting processability and low fracture toughness of stereo complex-type polylactide (SC-PLA) can be overcome by the blending of the same with poly (ethylene-*co*-vinyl acetate) (EVA) [18]. They revealed the generation

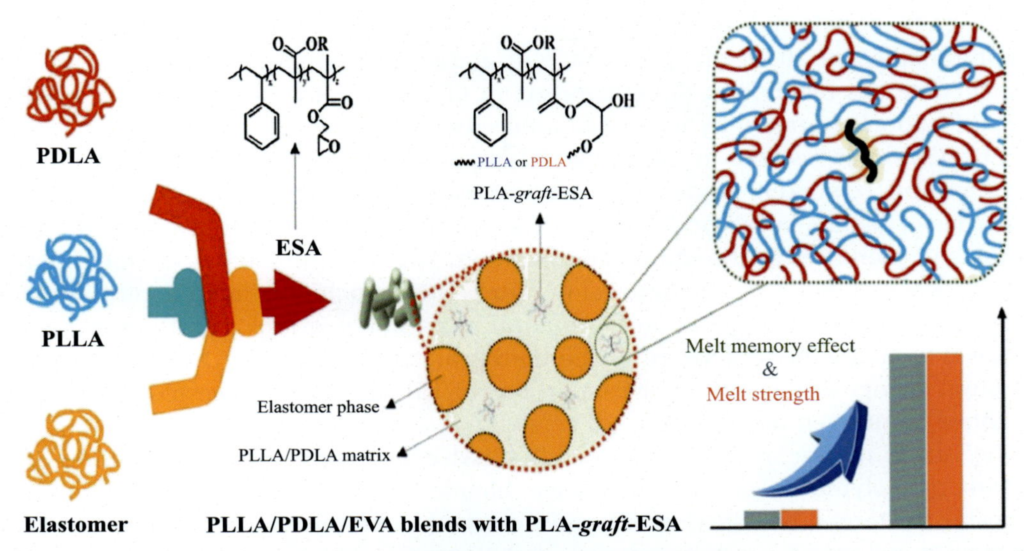

FIGURE 3.3 PLA-graft-ESA copolymer with melt memory effect and strength. *From S. Deng, J. Yao, H. Bai, H. Xiu, Q. Zhang, Q. Fu, A generalizable strategy toward highly tough and heat-resistant stereocomplex-type polylactide/elastomer blends with substantially enhanced melt processability, Polymer, 224 (2021). https://doi.org/10.1016/j.polymer.2021.123736. Reproduced with permission from Elsevier, License Number-5074801263294.*

of PLA-*graft*-ESA copolymer with the enhancement in melt memory effect of the matrix (Fig. 3.3). The combatibilizing effect of this copolymer might have stabilized the chain clusters in the blends and further stimulated the formation of stereo complex crystals.

3.7.2 Food packaging application of elastomer-based blends

A cost-effective and sustainable elastomer-based blend with maleinized linseed oil, polylactide, and thermoplastic elastomer (polystyrene-*b*-(ethylene-*ran*-butylene)-*b*-styrene) has been reported for the application in food industry [19]. They studied the effect of uncompatibilized and compatibilized polylactide/polystyrene-*b*-(ethylene-*ran*-butylene)-*b*-styrene blends developed by extrusion method. The compatibilization with maleinized linseed oil has reported to result in the improvement in impact strength, toughness, elongation at break, and ultraviolet light protection, which is significant for packaging application.

3.7.3 Mechanical performance of elastomer-based blends

Tsai and his coworkers developed thermoplastic elastomer based on poly (styrene-butadiene-styrene) and polyurethane [20]. Their studies revealed the improvement in mechanical, thermal, and abrasion resistance of the blends by the addition of thermoplastic polyurethane.

Biobased polylactic acid (PLA) blends developed by reactive blending of poly(ether-block-amide) (PEBA) elastomer grafted glycidyl methacrylate (GMA) and poly (lactic acid) have been reported to extent the potential of polylactic acid. Their study revealed that the loading of 20 weight percentage of grafted poly(ether-block-amide)- glycidyl methacrylate has resulted in the enhancement of tensile toughness. Here, the reactive compatilization promoted by interfacial adhesion between PLA with the epoxide group of PEBA-GMA has resulted in the enhanced impact strength of PLA blends [21].

Xu and his coworkers studied the effect of water on the mechanical and structural properties of blends based on TPU [22]. They observed a tremendous decrease in the modulus of blends after the immersion in water, whereas the increased modulus was found in wet samples after the removal of water. They concluded that, the plasticization effect of water, new hydrogen bond formation by carbonyl groups, and the breakage of hydrogen bond are the reasons for the hardening and softening of TPU blends. Here, the structural change of the soft and hard portions of TFU has been suggested to involve in the above mentioned mechanisms.

The brittle nature of Poly (lactic acid) (PLA) can be overcome by blending with elastomeric polymers. For e.g., PLA when blended with biopolymer poly (butylene-sebacate—*co*—terephthalate) (PBSeT) and hexamethylene diisocyanate (HDI) has been reported to enhance the tensile strength and elongation at break than the control. The blending also resulted in the enhanced degradation potential than the neat PLA [23].

Wang and his coworkers developed supertoughened polylactide (PLA) blend by the incorporation of PLA, ethylene—acrylic ester-glycidyl methacrylate random terpolymer (EGMA) and polyamide-11 (PA11). The notched Izod impact strength of the blends has been reported to get increased for PLA/EGMA/PA11 (mass ratio 80/20/1 have 31 kJ/m^2 and mass ratio 80/20/5 have 61 kJ/m^2) than PLA/EGMA (10 kJ/m^2). The reported impact strength of neat PLA was 1.9 kJ/m^2 [24] (Fig. 3.4).

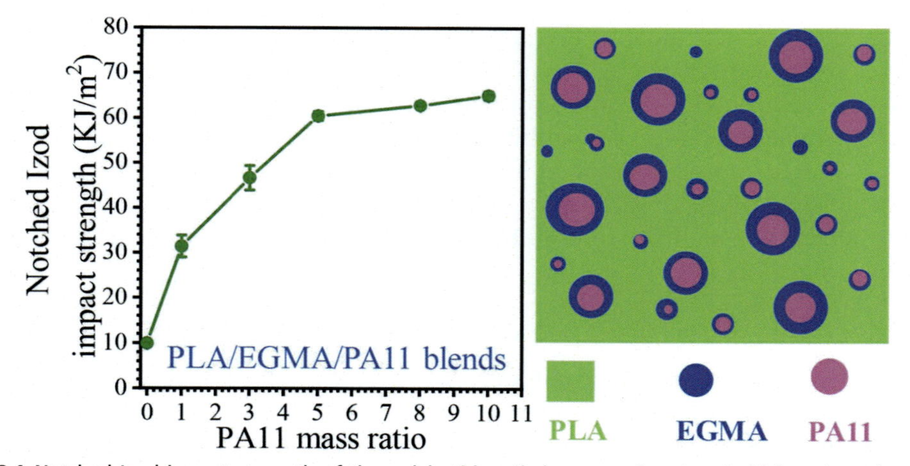

FIGURE 3.4 Notched Izod impact strength of the polylactide, ethylene—acrylic ester-glycidyl methacrylate random terpolymer, and polyamide-11 blends (PLA/EGMA/PA11). *From Q. Wang, J. Zhang, X. Wang, Z. Wang, Significant enhancement of notched Izod impact strength of PLA-based blends through encapsulating PA11 particles of low amounts by EGMA elastomer, Applied Surface Science, 526 (2020) 146657. https://doi.org/10.1016/j.apsusc.2020.146657. Reproduced with permission from Elsevier, License Number-5015720823423.*

Zhang and his coworkers have reported the blending of Poly(lactide) with biobased polyamide elastomer. The fabricated blend showed an enhancement in toughness due to the excellent hydrogen bonding between the Poly(lactide) and polyamide copolymer elastomer. Their studies revealed its potential application in packaging, medical, and automotive purposes [25].

3.8 Conclusion

Elastomers are polymers having viscoelastic nature with weak intermolecular forces. Compared with other materials, they have low Young's modulus and high failure strain. Hence, they are often combined with other materials of interest for enhancing their desirable properties. Polycondensation, photo cross-linking, and alternative network formation are the commonly used methods to develop elastomeric blends. Several compatibilizing agents such as silica, carbon black, aluminum oxide, magnesium silicate, aluminum trihydrate, mica, talc, calcium carbonate, clay, zinc oxide, nanoparticles, and calcium are used to improve the performance of elastomer-based blends. There are several studies regarding the synthesis, characterization and applications studies of elastomer-based blends. The blending of elastomer with material of interest has immense promises as it improves the morphological, mechanical, thermal, stretchability percentage, tensile strength, adhesion strength, tearing energy, rheological, viscoelastic, optical, self-healable properties, and elastic modulus of the blends. Hence, blends with desirable properties can have application in industrial as well as medical sectors.

References

[1] T. Özdemir, Elastomeric Micro- and Nanocomposites for Neutron Shielding, 2020, pp. 125–137, https://doi.org/10.1016/b978-0-12-819459-1.00005-2.

[2] D. Mangaraj, Elastomer blends, Rubber Chem. Technol. 75 (3) (2002) 365–427, https://doi.org/10.5254/1.3547677.

[3] P. Deepalekshmi, P.M. Visakh, A.P. Mathew, A.K. Chandra, S. Thomas, Advances in Elastomers: Their Blends and Interpenetrating Networks-State of Art, New Challenges and Opportunities 11 (2013) 1–9, https://doi.org/10.1007/978-3-642-20925-3_1.

[4] P.J. Corish, Elastomer Blends, 1994, pp. 545–599, https://doi.org/10.1016/b978-0-08-051667-7.50017-2.

[5] L.A. Utracki, Compatibilization of polymer blends, Can. J. Chem. Eng. 80 (6) (2002) 1008–1016, https://doi.org/10.1002/cjce.5450800601.

[6] J. Abraham, L. Somasekharan, T. S., L.R. Pillai, H.J. Maria, S. Thomas, Elastomer Blends: The Role of Nanoparticles on Properties, 2014, pp. 1–9, https://doi.org/10.1007/978-3-642-36199-9_290-1.

[7] V. Kanyanta, A. Ivankovic, N. Murphy, Bio-medical applications of elastomeric blends, Composites 12 (2013) 227–252, https://doi.org/10.1007/978-3-642-20928-4_8.

[8] S. Song, H. Hou, J. Wang, P. Rao, Y. Zhang, A self-healable, stretchable, tear-resistant and sticky elastomer enabled by a facile polymer blends strategy, J. Mater. Chem. 9 (7) (2021) 3931–3939, https://doi.org/10.1039/d0ta11497a.

[9] R. Asaletha, M.G. Kumaran, S. Thomas, Thermoplastic elastomers from blends of polystyrene and natural rubber: morphology and mechanical properties, Eur. Polym. J. 35 (2) (1999) 253–271, https://doi.org/10.1016/s0014-3057(98)00115-3.

[10] N. Jalalifar, B. Kaffashi, S. Ahmadi, The synergistic reinforcing effects of halloysite nanotube particles and polyolefin elastomer-grafted-maleic anhydride compatibilizer on melt and solid viscoelastic properties of polylactic acid/polyolefin elastomer blends, Polym. Test. 91 (2020) 106757. https://doi.org/10.1016/j.polymertesting.2020.106757.

[11] S.D. Varsavas, C. Kaynak, Effects of glass fiber reinforcement and thermoplastic elastomer blending on the mechanical performance of polylactide, Compos. Commun. 8 (2018) 24–30. https://doi.org/10.1016/j.coco.2018.03.003.

[12] V. Panwar, G. Anoop, Flexible piezoresistive strain sensor based on optimized elastomer-electronic polymer blend, Measurement 168 (2021) 108406. https://doi.org/10.1016/j.measurement.2020.108406.

[13] K. Imato, H. Nakajima, R. Yamanaka, N. Takeda, Self-healing polyurethane elastomers based on charge-transfer interactions for biomedical applications, Polym. J. 53 (2) (2020) 355–362. https://doi.org/10.1038/s41428-020-00432-4.

[14] J. Kang, J. Kim, K. Choi, P.H. Hong, H.J. Park, K. Kim, Y.K. Kim, G. Moon, H. Jeon, S. Lee, M.J. Ko, S. W. Hong, A water-triggered highly self-healable elastomer with enhanced mechanical properties achieved using localized zwitterionic assemblies, Chem. Eng. J. (2020) 127636. https://doi.org/10.1016/j.cej.2020.127636.

[15] B. Peng, Y. Yang, T. Ju, K.A. Cavicchi, Fused filament fabrication 4D printing of a highly extensible, self-healing, shape memory elastomer based on thermoplastic polymer blends, ACS Appl. Mater. Interfaces (2020). https://doi.org/10.1021/acsami.0c18618.

[16] Y. Ding, X. Chen, D. Huang, B. Fan, L. Pan, K. Zhang, Y. Li, Post-chemical grafting poly(methyl methacrylate) to commercially renewable elastomer as effective modifiers for polylactide blends, Int. J. Biol. Macromol. 181 (2021) 718–733. https://doi.org/10.1016/j.ijbiomac.2021.03.139.

[17] X. Ji, F. Gao, Z. Geng, D. Li, Fabrication of thermoplastic polyurethane/polylactide shape-memory blends with tunable optical and mechanical properties via a bilayer structure design, Polym. Test. 97 (2021).

[18] S. Deng, J. Yao, H. Bai, H. Xiu, Q. Zhang, Q. Fu, A generalizable strategy toward highly tough and heat-resistant stereocomplex-type polylactide/elastomer blends with substantially enhanced melt processability, Polymer (2021). https://doi.org/10.1016/j.polymer.2021.123736.

[19] R. Tejada-Oliveros, R. Balart, J. Ivorra-Martinez, J. Gomez-Caturla, N. Montanes, L. Quiles-Carrillo, Improvement of impact strength of polylactide blends with a thermoplastic elastomer compatibilized with biobased maleinized linseed oil for applications in rigid packaging, Molecules 26 (1) (2021) 240. https://doi.org/10.3390/molecules26010240.

[20] Y. Tsai, M.-C. Kuo, J.-H. Wu, Fabrication and characterization of poly(styrene-butadiene-styrene)/thermoplastic polyurethane blends, J. Mater. Sci. Chem. Eng. 09 (02) (2021) 1–10. https://doi.org/10.4236/msce.2021.92001.

[21] Y. Xia, G. Wang, Y. Feng, Y. Hu, G. Zhao, W. Jiang, Highly toughened poly(lactic acid) blends prepared by reactive blending with a renewable poly(ether-block-amide) elastomer, J. Appl. Polym. Sci. 138 (13) (2020) 50097. https://doi.org/10.1002/app.50097.

[22] D.-H. Xu, F. Liu, G. Pan, Z.-G. Zhao, X. Yang, H.-C. Shi, S.-F. Luan, Softening and hardening of thermal plastic polyurethane blends by water absorbed, Polymer 218 (2021) 123498. https://doi.org/10.1016/j.polymer.2021.123498.

[23] S.J. Kim, H.W. Kwak, S. Kwon, H. Jang, S.-i. Park, Characterization of PLA/PBSeT blends prepared with various hexamethylene diisocyanate contents, Materials 14 (1) (2021) 197. https://doi.org/10.3390/ma14010197.

[24] Q. Wang, J. Zhang, X. Wang, Z. Wang, Significant enhancement of notched Izod impact strength of PLA-based blends through encapsulating PA11 particles of low amounts by EGMA elastomer, Appl. Surf. Sci. 526 (2020) 146657. https://doi.org/10.1016/j.apsusc.2020.146657.

[25] Y. Zhang, J. Chen, Q. Peng, L. Song, Z. Wang, Z. Wang, Hydrogen bonding assisted toughness enhancement of poly(lactide) blended with a bio-based polyamide elastomer of extremely low amounts, Appl. Surf. Sci. 506 (2020). https://doi.org/10.1016/j.apsusc.2019.144684.

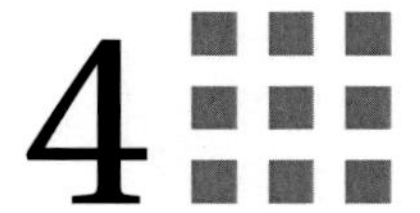

4 Elastomer-based filler composites

S.N. Vasantha Kumar[1], P.C. Sharath[2]

[1]*MECHANICAL ENGINEERING DEPARTMENT, CANARA ENGINEERING COLLEGE, MANGALORE, KARNATAKA, INDIA;* [2]*METALLURGICAL ENGINEERING DEPARTMENT, JAIN UNIVERSITY, BANGALORE, KARNATAKA, INDIA*

4.1 Introduction

In recent studies, understanding the polymer materials has gained much importance due to their high specific strength and modulus, which suits industrial applications. However, neat polymers possess low mechanical performance as compared to the compounding polymer with fillers or hybrid polymer composites. Many studies are carried out to develop conductive polymer composites that enhance the range of applications in electronics to sensors applications [1]. Elastomers also referred to as rubbers, are one type of polymers among many polymers that may exhibit superplasticity under some conditions. Their features are limited to kinked and coiled molecular chains that have higher-molecular-weight amorphous property, molecular bonds are free to rotate or stretch when supplied with external load or force, have viscoelastic behavior, and a low degree of cross-link. Raw elastomers are subject to vulcanization processes to produce robust and applicable elastomers. Suitable catalysts are used under varying operating pressure and temperature to produce the required property, based on the form of raw material. The stiffness of the elastomers is highly dependent on cross-link density [2].

As projected by the Global News Network, the annual use of elastomers is currently over 18 million tonnes, with production increasing at a steady rate each year by 5.1% from 2018 to 2026 [3]. When these components are commonly used in tires and consumer goods such as conveyor belts, hoses, and automotive mounts, the significance of the elastomer industry is illustrated [4]. The majority of the highway products are prepared by elastomers filled with plastic blends. Low-density polyethylene (LDPE) can be used for road applications as a filling material in bitumen [5]. The new term "hybrid magnetic elastomer" is coined by few researchers who enhanced the applications of elastomers filled with complex fillers in the field of magneto-responsive materials and rheological space applications [6,7].

The inclusion of carbon black and silica (traditional fillers) has affected by improving their unique mechanical properties [8], although, with certain drawbacks such as the use of large-sized particles and huge quantities. Nowadays, discovery of nanoscale particles

Elastomer Blends and Composites. https://doi.org/10.1016/B978-0-323-85832-8.00004-3

fled the use of fillers to nanofillers, which increased the surface area and diminished the size of elastomer composite.

While considering the elastomer-based filler material for suitable applications, its processing route is a most essential thing to consider. As a material analyst/specialist and design engineer should have a broad knowledge on selection of suitable material to intended applications. Even though statistical analysis tools are available in today's market, one should understand the basics of material behavior in conjunction to its properties when subjected to ambient atmosphere. Hence, this chapter focuses on the preparation and properties of fillers used in the elastomers with recent advancements in the elastomer field. Conventional fillers such as carbon black and silica are replaced with nano and complex fillers, which find applications of elastomer composites in actuators, sensors, and magneto-responsive to the rheological field.

4.2 Preparation and properties of fillers

4.2.1 Carbon black

The carbon black (inks, pigments) and elastomer interactions were made by chemical bonding, which dated back to the 1960s [9]. Further growth leads to the enhancement of carbon black surface oxidation by passing oxygen at high temperatures or using nitric acid to determine the amount and form of functions produced. Comprehension of the molecular slippage process in 1975 by Danneberg made a significant breakthrough in the posttreatment of carbon, converting polymeric chain chemical grafting into a black surface [10]. Subsequent chemical modifications of carbon black by functionalization, silica, and alumina surface coating created high-reactivity compounds [11–13].

4.2.2 Silica

Silica-reinforced elastomers exhibit lower modulus and abrasion resistance. The use of silica as fillers in elastomers can be of two types. They are (i) precipitated silica and (ii) fumed silica.

Precipitated silica: These forms of silica are produced by combining pure sand and alkaline salt with silica glass. At moderate pressure and temperature, it is later solubilized in water. Reinforced silica is obtained by filtering and washed in an acid medium with different drying conditions [11,13,14].

Fumed silica: It is prepared by oxide composition of SiH_4 or methyl hydride precursors at high temperature: $n\ SiH_4 + 2n\ O_2 \rightarrow n\ SiO_2 + 2n\ H_2O$

In contrast to precipitation silica, fumed silica is typically in a fluffy shape and very stable with morphology and few surface silanols.

4.2.3 Different fillers

4.2.3.1 Magnetic fillers

Aloui et al. [15] studied the effect of nano-sized Magnesium-Silica with micro-sized Carbonyl-Iron particle (CIP). They observed symbiotic interaction between the filler when an external magnetic field was used for curing. Relative modulus increases to 140% by using coupling agent Silane in the polymer mixture. The high increase in the property was due to the mobility of the particles along the magnetic field line and the organization of particles. It finds use in the field of magneto-rheology, where shear stiffness and a magnetically conductive filler with higher sensitivity are needed. Also, it is observed that if micro-sized CIPs were replaced with nano MgSi, tensile strength increases with reduction insensitivity. It was concluded that micro-sized particle fillers achieved a higher magnetic moment and the usage of Silane decreases the switching effect, but better mechanical efficiency was observed.

4.2.3.2 Copper nanowire

Copper Nanowires (CuNWs) were used as fillers in the ultrasoft silicone elastomer [16]. The CuNWs were prepared by using the chemical wet synthesis method. A schematic representation of a composite made from a silicone elastomer with a Shore hardness of 30 is shown. A detailed procedure for fabrication was given by Ref. [16].

The manufactured composites show 3.1±0.2 W/mK of thermal conductivity, which is 19 times greater in contrast with matrix in the temperature range 20–80 °C. It is done by the freeze-drying process that produces a nanowire continuous network. Additionally, microwave irradiation seizes the nanowire junction and interfacial thermal resistance, which makes these materials find applications in thermal interface materials.

4.2.3.3 Hybrid fillers (TiO_2-Graphene)

Chen et al. [17] used titanium dioxide functionalized graphene as fillers to study the properties of polyurethane dielectric elastomers. The solvothermal approach was adopted to reduce the graphene oxide to reduced graphene oxide (RGO) and rearrangements of TiO_2 nanoparticles onto RGO. Fig. 4.1 shows the dispersion of

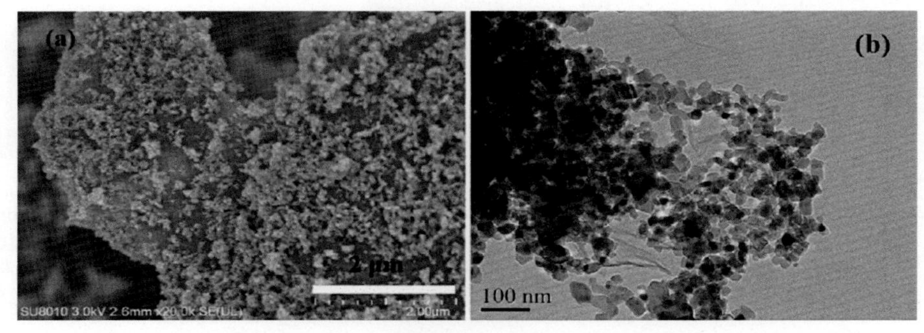

FIGURE 4.1 (A) SEM and (B) TEM image of TiO_2-RGO nanocomposite. *From T. Chen, J. Qiu, K. Zhu, J. Li, Electromechanical performance of polyurethane dielectric elastomer flexible micro-actuator composite modified with titanium dioxide-graphene hybrid fillers, Materials and Design 90 (2016) 1069–1076.*

nanoparticles of TiO_2 in the RGO nanosheet. They observed that wettability and hydrophilicity can be improved by proper arrangements with nano-TiO_2 onto graphene resulting in enhanced interface interactions. They concluded that modified TiO_2 in RGO nanofillers finds applications in microactuators, sensors, and microelectronic devices.

4.2.3.4 Piezoelectric (PZT) and silver-coated glass microsphere fillers

Research on the production of high-performance stretchable piezoelectric nanogenerator (SPENG) was performed by Chou et al. [18] where PZT particles and Ag-coated glass microspheres fillers were used in silicone rubber matrixes. SPENG was used as the stretchable piezoelectric composite (SPC) layer, with $PbZr_{0.52}Ti_{0.48}O_3$ (Lead Zirconate Titanate) (65 vt%)/rubbery matrix, and Ag-coated glass microspheres (72 wt%) rubbery matrix as the stretchable electrode composite (SEC) layer. Its fabrication process is shown in Fig. 4.2. Because of its superior stretchability, more viscous molecular network, and low-cost, solid-state silicon rubber was chosen as a matrix. Fillers were blended into the matrix by mixing, then cured and molded at a suitable temperature. Scanning electron images show that PZT particles were tightly embedded in Silicon rubber. A strong interface adhesion between the electrode layer and the piezoelectric layer is found when the Ag-coated glass microsphere is integrated into the matrix.

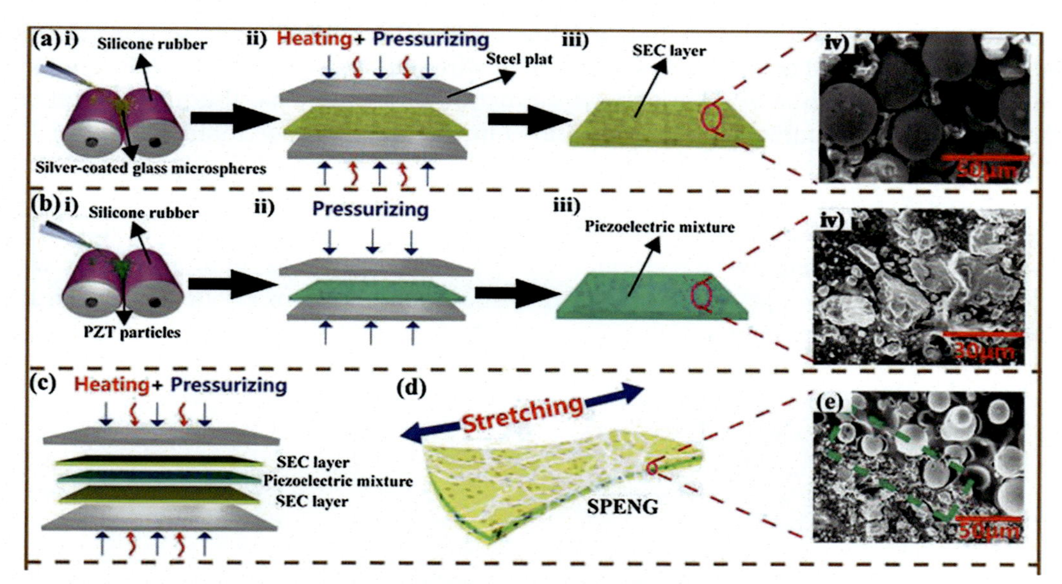

FIGURE 4.2 The fabrication process and characterizations of the SPENG. The schematic diagrams of fabrication processes of (A) SEC layer, (B) SPC mixture, and (C) SPENG. (D) The SPENG produces electricity when it is stretched. (E) The SEM image of the SPENG cross-section view. *From X. Chou, J. Zhu, S. Qian, X. Niu, J. Qian, X. Hou, J. Mu, W. Geng, J. Cho, J. He, C. Xue, All-in-one filler-elastomer-based high-performance stretchable piezoelectric nanogenerator for kinetic energy harvesting and self-powered motion monitoring, Nano Energy 53 (2018), 550–558. https://doi.org/10.1016/j.nanoen.2018.09.006.*

4.2.3.5 SBS (styrene–butadiene–styrene/multiwall) carbon nanotubes fillers

Costa et al. have studied the SBS (styrene-butadiene styrene) block with multiwalled carbon nanotube (MWCNT) filler and its effect [19]. It was prepared by solution casting route. Electrical conductivity and percolation threshold were more influenced by the filler content and elastomer matrix. Morphology and orientation of the SBS block much affected by deformation [20]. Stress level increases linearly with an increase in the volume fraction of styrene material in the copolymers. The morphology and styrene material of Nanocomposites affect their electrical conductivity.

4.2.3.6 Carbon nanotubes and hybrid fillers

Melt blending produced thermoplastic elastomers (TPE) loaded with carbon nanotubes (CNTs), carbon black (CB), and hybrid CNT with CB studied by Dang et al. [21]. Its effect on the percolation threshold and electrical properties was also studied with different volume fractions of fillers. Electrical conductivity for all the types of composites is much affected near the percolation threshold. It was due to the network structures and provides passage for current to flow. The addition of CB concentration levels above 8% by volume showed a small improvement in composite conductivity. It implies that above the percolation threshold, substantial conducting networks evolved with little or no impact on conductivity. The percolation threshold was marginally higher in the case of CNT filler. For all three forms of fillers, Fig. 4.3 indicates the conductivity trends. It was

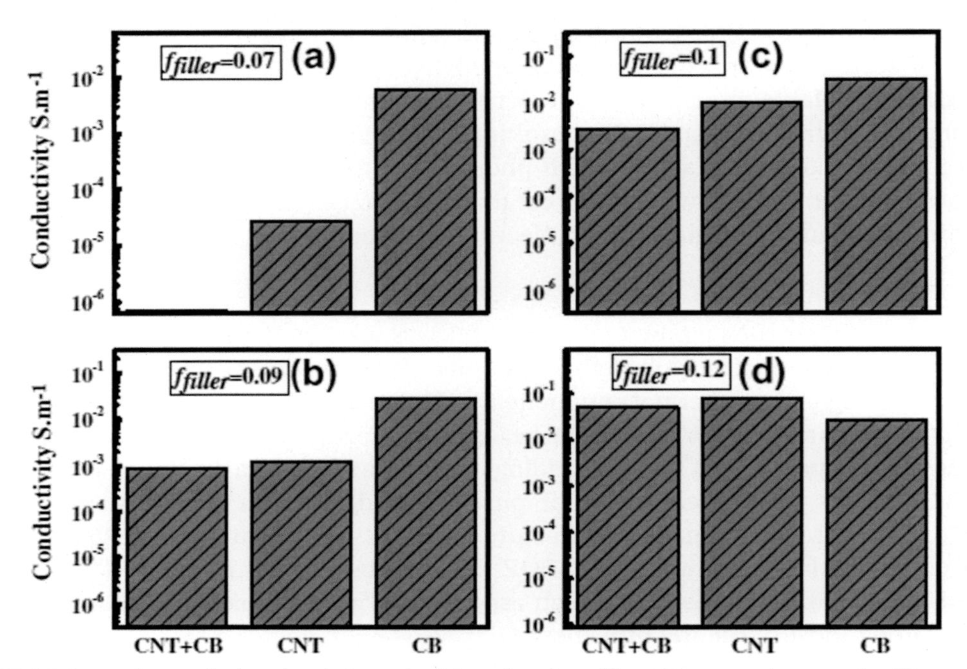

FIGURE 4.3 Dependence of the electrical conductivity of carbon fillers/TPE composites on the filler volume fraction: (A) filler = 0.07, (B) filler = 0.09, (C) filler = 0.1, and (D) filler = 0.12. *From Z.M. Dang, K. Shehzad, J.W. Zha, A. Mujahid, T. Hussain, J. Nie, C.Y. Shi, Complementary percolation characteristics of carbon fillers based electrically percolative thermoplastic elastomer composites, Composites Science and Technology 72 (1) (2011) 28–35. https://doi.org/10.1016/j.compscitech.2011.08.020.*

proved that the percolation threshold depends on filler dispersion along with its aspect ratio. The filler conductivity (percolation threshold) was in the order of CB > CNT > hybrid (CNT-CB).

4.2.3.7 Graphene nanoplatelets (GnPs), expanded graphite (EG), and multiwalled carbon nanotubes (MWCNTs)

Das et al. studied the solution of styrene-butadiene rubber (S-SBR) composites filled with graphene nanoplatelets (GnPs), expanded graphite (EG), and MWCNTs [22]. The use of EG improves Young's modulus, with the maximum modulus being found in the case of 5% CNT filler. It was shown that by disentanglement of the bundle or delamination of the EG, but it is greater in the dispersion of MWCNT. MWCNT's high aspect ratio and higher exposed surface area contribute to a larger degree of physical contact with the matrix and network creation, leading to a higher value of storage modules. They found that adding a small amount of MWCNT greatly enhanced the properties.

4.2.3.8 3D graphene foam filler

Fang et al. [23] studied elastomers filled with graphene foam (GF) and polydimethylsiloxane (PDMS). They successfully produced three-dimensional GF coated with polydopamine (PDA) and made a reaction with a 3-aminopropyltriethoxysilane (APTS) composite. Thermal conductivity of such produced composites shows 28.77 W/mK in-plane and 1.62 W/mK out of plane at 11.62wt% GF loading condition. Compression of the sample connecting the GF arm to lay down along the plane was the cause for seeing an improvement in conductivity. Also, the interface between GF and matrix had a notable impact. The van Der Waals force was weaker than the chemical bonding between PDMS and PDA, resulting in a decrease in conductivity. However, APTS molecules form a chain in which phonons can pass smoothly through vibration, improving heat transfer. It was found that this material can be used in the cooling of the ceramic heater and managing heat in electronic devices.

4.2.3.9 Boron nitride filled in polyolefin elastomer

When boron nitride (BN) content of 43.75 vol % was used by Feng et al. [24], they stated that thermally conductive polyolefin elastomer (POE) loaded with BN demonstrated excellent through-plane thermal conductivity with outstanding electrical insulation. The manufacturing process was carried out using melt mixing on a two-roll mill and compressed at a temperature of 90 °C by adding a pressure of 10 MPa for a period of 15 min. Fig. 4.4 displays SEM micrographs of pure BN and different BN mixtures. Tensile strength of 16.2 MPa, elongation to fracture of 80.2%, and modulus of 1326.6 MPa are some of the mechanical properties of POE/BN composites. It was due to flake slipping and well-oriented BN flakes with a high aspect ratio, which avoided crack propagation and absorbed most of the fracture energy.

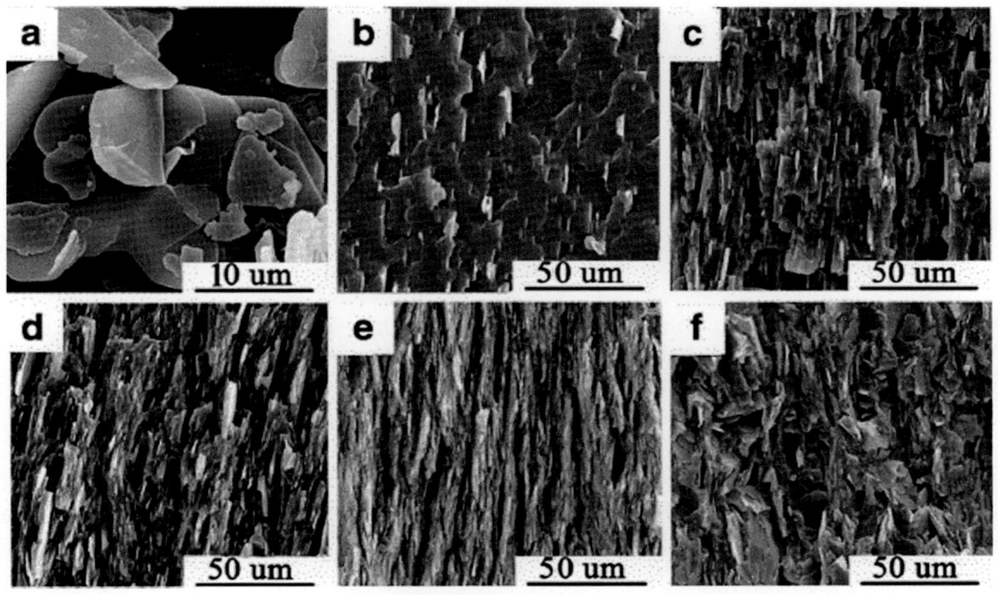

FIGURE 4.4 SEM images of pure BN and POE/BN composites with different content of BN. (A) Pure BN, (B) 3.73BN, (C) 16.29BN, (D) 25.93BN, (E) 43.75BN, and (F) 25.93BN-comparison (note that 25.93BN and 25.93BN comparison were fabricated by the two-roll milling method and the torque rheometer method, respectively). *From C.P. Feng, L. Bai, R.Y. Bao, Z.Y. Liu, M.B. Yang, J. Chen, W. Yang, Electrically insulating POE/BN elastomeric composites with high through-plane thermal conductivity fabricated by two-roll milling and hot compression, Advanced Composites and Hybrid Materials 1 (1) (2018) 160–167. https://doi.org/10.1007/s42114-017-0013-2.*

4.2.3.10 Expanded graphite filled with styrene isoprene styrene block copolymer

Krupa et al. [25] developed styrene-isoprene-styrene block copolymer-based conductive polymer composites loaded with extended graphite for oil sensor applications. The sensor response was checked using motor oil after the composite was cast from a toluene solvent solution. The response was measured by geometrical changes such as stretching and specific electrical conductivity. Because of the minimal oil permeability, increasing the filler content increases electrical conductivity but decreases response time. Hence, they used the new method for measuring the response of oil by deformation. Adhesion between polymer and filler is the basic understanding of deformation behavior, with low results in microcrack formation and voids at the interface. It allows mechanical properties to deteriorate. The voids created around the particles of the filler improve oil permeability and increase the rate of response. A deeper investigation is needed in this context for a better understanding of oil absorption sensitivity. They concluded that stretched sample response rate was 14.3 times higher than the unstretched one. The degree to which materials are stretched has a substantial effect on their basic electrical conductivity.

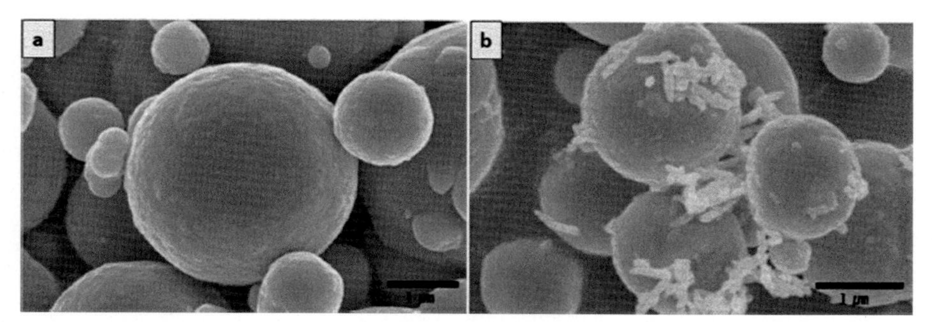

FIGURE 4.5 SEM images of pure CI (A) and (B) CI with γ-Fe_2O_3 particles. *From C.J. Lee, S.H. Kwon, H.J. Choi, K.H. Chung, J.H. Jung, Enhanced magnetorheological performance of carbonyl iron/natural rubber composite elastomer with gamma-ferrite additive, Colloid and Polymer Science 296 (9) (2018) 1609–1613. https://doi.org/10.1007/s00396-018-4373-0.*

4.2.3.11 Gamma-ferrite additive to carbonyl iron (CI) natural rubber (NR) composite

Magnetorheological (MR) elastomers have gained much attention in recent years. The efficiency of carbonyl iron (CI)/natural rubber (NR) reinforced with gamma ferrite (γ -Fe_2O_3) additive was investigated by Lee et al. [26]. By considering pure CI, the composite was prepared, and γ-Fe_2O_3 rod-shaped was later mixed to which NR and carbon black (CN) were applied and retained under predetermined heating conditions. The micrographs of pure CI and CI with γ-Fe_2O_3 particles are seen in Fig. 4.5. It can be observed from Fig. 4.5 that rod-shaped γ-Fe_2O_3 particles were surrounded on spherical shape CI microspheres and also on the free space. They found that CI with γ-Fe_2O_3 particles had a higher modulus than pure CI and that the modulus increased as the magnetic field intensity increased. Fig. 4.5.

4.2.4 Glycerol filler

Glycerol droplets that can be used as a dielectric actuator have been loaded with commercially available PDMS by Mazurek et al. [27]. The justification for the use of droplets of glycerol is that it serves as a softener that reduces the composite's elastic modulus and raises the dielectric constant. The dielectric losses of the glycerol–PDMS composites were found to be very low. The high-voltage dielectric spectroscopy measurements show that with rising voltages and filler loadings, the conductivity continues to increase.

4.3 Conclusions and perspectives

The synthesis and properties of various elastomer fillers have been reviewed. It has been shown that, with their specific assets, fillers such as graphene, boron nitride, and carbon nanotubes can be used for a variety of applications and display excellent promise as filler agents in a different elastomer class.

In the performance of the processed elastomers, many types of particles such as pyrolysis product, water-based product, sol–gel product, layered fillers, electromagnetic phases, nanotubes, and porous type fillers play a major role. However, the preparation process must be selected based on the material's intended use.

The properties of the different fillers with elastomer are in direct relation with the preparation method along with how the dispersion of the filler takes place and its physicochemical reactions. Microscopy is the evasion required for the analysis of matrix filler dispersion, the measurement of electrical/thermal conductivity, permeability, and other mechanical properties using sufficient instruments. It is primarily dependent on the contact with the elastomer used between the structure/surface of the filler.

The integration of different fillers in the elastomeric matrix has been shown an adverse and diversified effect on mechanical and thermal properties. By using the casting method, one can find that higher tensile property over the sol–gel method can be observed. It all depends on the interaction between filler–filler and dispersion of the fillers. In the future, one can use Nanocomposites as the best choice of filler material for usage ranging from tires to automotive components, biomedical to sensing devices.

References

[1] R. Ou, S. Gupta, C.A. Parker, R.A. Gerhardt, Fabrication and electrical conductivity of poly(methyl methacrylate) (PMMA)/Carbon black (CB) composites: comparison between an ordered carbon black nanowire-like segregated structure and a randomly dispersed carbon black nanostructure, J. Phys. Chem. B 110 (45) (2006) 22365–22373, https://doi.org/10.1021/jp064498o.

[2] Y. Wang, Y. Wang, M. Tian, L. Zhang, J. Ma, Influence of prolonging vulcanization on the structure and properties of hard rubber, J. Appl. Polym. Sci. 107 (1) (2008) 444–454, https://doi.org/10.1002/app.27086.

[3] Research and Market, Global Elastomers Market Size, Market Share, Application Analysis, Regional Outlook, Growth Trends, Key Players, Competitive Strategies and Forecasts, 2018 to 2026, 2018. https://www.globenewswire.com/news-release/2018/11/30/1659969/0/en/Global-Elastomers-Market-Size-Market-Share-Application-Analysis-Regional-Outlook-Growth-Trends-Key-Players-Competitive-Strategies-and-Forecasts-2018-To-2026.html.

[4] M. Porter, Rubber technology handbook werner hofmann, carl hanser verlag, munich, 1989. pp. xv + 611, price dm 86.00. isbn 3–446–14895–7, Br. Polym. J. 23 (4) (1990) 359–359, https://doi.org/10.1002/pi.1990.4980230405.

[5] A. Nag, S. Koppula, J. Jose, S. Satapathy, G.B. Nando, A Novel Technique for the Utilization of Waste Plastic in Road Making, CIPET Bulletin, 2006, pp. 30–35.

[6] G.V. Stepanov, D.Y. Borin, A.V. Bakhtiiarov, P.A. Storozhenko, Magnetic properties of hybrid elastomers with magnetically hard fillers: rotation of particles, Smart Mater. Struct. 26 (3) (2017) 035060, https://doi.org/10.1088/1361-665x/aa5d3c.

[7] G.V. Stepanov, D.Y. Borin, E.Y. Kramarenko, V.V. Bogdanov, D.A. Semerenko, P.A. Storozhenko, Magnetoactive elastomer based on magnetically hard filler: synthesis and study of viscoelastic and damping properties, Polym. Sci. 56 (5) (2014) 603–613, https://doi.org/10.1134/S0965545X14050149.

[8] A. Ciesielski, An Introduction to Rubber Technology, Rapra Technology Ltd, 1999. http://library.lol/main/47F057FEBC3902C0D349EBD9580BA49B.

[9] F. Bueche, Molecular basis for the mullins effect, J. Appl. Polym. Sci. 4 (10) (1960) 107–114, https://doi.org/10.1002/app.1960.070041017.

[10] E. Dannenberg, Molecular slippage mechanism of reinforcement, Trans. Inst. Rubber Ind. 42 (1966) 26–32.

[11] E. Custodero, L. Simonot, J.-C. Tardivat, Carbon Black Coated With an Aluminous Layer and Method for Obtaining Same, 1997. https://patentscope.wipo.int/search/en/detail.jsf?docId=WO1999028391.

[12] W.L. Hergenrother, J.C. Oziomek, M. William, Addition of Salts to Improve the Interaction of Silica With Rubber, 1997. https://patents.google.com/patent/US5872176A/en.

[13] N. Tsubokawa, M. Hosoya, Reactive carbon black having acyl imidazole or acid anhydride groups: preparation and reaction with functional polymers having hydroxyl or amino groups, React. Polym. 14 (1) (1991) 33–40.

[14] M.J. Wang, T.A. Brown, W.J. Patterson, R.A. Francis, in: International Rubber Conference, 1997, pp. 6–9. Kuala Lumpur.

[15] S. Aloui, M. Klüppel, Magneto-rheological response of elastomer composites with hybrid-magnetic fillers, Smart Mater. Struct. 24 (2) (2014) 025016.

[16] S. Bhanushali, P.C. Ghosh, G.P. Simon, W. Cheng, Copper nanowire-filled soft elastomer composites for applications as thermal interface materials, Adv. Mater. Interf. 4 (17) (2017). https://doi.org/10.1002/admi.201700387.

[17] T. Chen, J. Qiu, K. Zhu, J. Li, Electro-mechanical performance of polyurethane dielectric elastomer flexible micro-actuator composite modified with titanium dioxide-graphene hybrid fillers, Mater. Des. 90 (2016) 1069–1076.

[18] X. Chou, J. Zhu, S. Qian, X. Niu, J. Qian, X. Hou, J. Mu, W. Geng, J. Cho, J. He, C. Xue, All-in-one filler-elastomer-based high-performance stretchable piezoelectric nanogenerator for kinetic energy harvesting and self-powered motion monitoring, Nano Energy 53 (2018) 550–558. https://doi.org/10.1016/j.nanoen.2018.09.006.

[19] P. Costa, J. Silva, V. Sencadas, R. Simoes, J.C. Viana, S. Lanceros-Méndez, Mechanical, electrical and electro-mechanical properties of thermoplastic elastomer styrene–butadiene–styrene/multiwall carbon nanotubes composites, J. Mater. Sci. 48 (3) (2013) 1172–1179. https://doi.org/10.1007/s10853-012-6855-7.

[20] T.A. Huy, R. Adhikari, G.H. Michler, Deformation behavior of styrene-block-butadiene-block-styrene triblock copolymers having different morphologies, Polymer 44 (4) (2003) 1247–1257. https://doi.org/10.1016/S0032-3861(02)00548-7.

[21] Z.-M. Dang, K. Shehzad, J.-W. Zha, A. Mujahid, T. Hussain, J. Nie, C.-Y. Shi, Complementary percolation characteristics of carbon fillers based electrically percolative thermoplastic elastomer composites, Compos. Sci. Technol. 72 (1) (2011) 28–35. https://doi.org/10.1016/j.compscitech.2011.08.020.

[22] A. Das, G.R. Kasaliwal, R. Jurk, R. Boldt, D. Fischer, K.W. Stöckelhuber, G. Heinrich, Rubber composites based on graphene nanoplatelets, expanded graphite, carbon nanotubes and their combination: a comparative study, Compos. Sci. Technol. 72 (16) (2012) 1961–1967. https://doi.org/10.1016/j.compscitech.2012.09.005.

[23] H. Fang, Y. Zhao, Y. Zhang, Y. Ren, S.-L. Bai, Three-dimensional graphene foam-filled elastomer composites with high thermal and mechanical properties, ACS Appl. Mater. Interfaces 9 (31) (2017) 26447–26459. https://doi.org/10.1021/acsami.7b07650.

[24] C.-P. Feng, L. Bai, R.-Y. Bao, Z.-Y. Liu, M.-B. Yang, J. Chen, W. Yang, Electrically insulating POE/BN elastomeric composites with high through-plane thermal conductivity fabricated by two-roll milling and hot compression, Adv. Compos. Hybrid Mater. 1 (1) (2018) 160–167. https://doi.org/10.1007/s42114-017-0013-2.

[25] I. Krupa, M. Prostredný, Z. Špitalský, J. Krajči, M.A.S. AlMaadeed, Electrically conductive composites based on an elastomeric matrix filled with expanded graphite as a potential oil sensing material, Smart Mater. Struct. 23 (12) (2014) 125020. https://doi.org/10.1088/0964-1726/23/12/125020.

[26] C.J. Lee, S.H. Kwon, H.J. Choi, K.H. Chung, J.H. Jung, Enhanced magnetorheological performance of carbonyl iron/natural rubber composite elastomer with gamma-ferrite additive, Colloid Polym. Sci. 296 (9) (2018) 1609–1613. https://doi.org/10.1007/s00396-018-4373-0.

[27] P. Mazurek, L. Yu, R. Gerhard, W. Wirges, A.L. Skov, Glycerol as high-permittivity liquid filler in dielectric silicone elastomers, J. Appl. Polym. Sci. 133 (43) (2016). https://doi.org/10.1002/app.44153.

Engineering applications of elastomer blends and composites

Naga Srilatha Cheekuramelli[1,2], Dattatraya Late[2,3], S. Kiran[1,2], Baijayantimala Garnaik[1,2]

[1]POLYMER SCIENCE AND ENGINEERING DIVISION, CSIR-NATIONAL CHEMICAL LABORATORY, PUNE, MH, INDIA; [2]ACADEMY OF SCIENTIFIC AND INNOVATIVE RESEARCH, ACSIR HEADQUARTERS, CSIR-HRDC CAMPUS, GHAZIABAD, UP, INDIA; [3]PHYSICAL AND MATERIALS CHEMISTRY DIVISION, CSIR-NATIONAL CHEMICAL LABORATORY, PUNE, MH, INDIA

5.1 Introduction

Elastomer blend and composite have progressed as advanced material systems for the agriculture, biomedical, and ocean engineering applications [1–3]. The main advantages of these material systems are biocompatibility and degradability, which have promoted the use of these materials [4]. These materials afford good mechanical properties and other physical properties over ceramics and metals, which are important in biomedical field, ocean, and agricultural environment [4].

These materials consist of electrical-conductive polymers for example, polypyrrole and polythiophene, etc., and biopolymers such as gums, gelatin, chitosan, and cellulose, etc. [5] The ability to control the micro- or macroscale functionality, and produce its influence on other functionality with the usage of the biopolymers or conductive polymers within elastomeric blends and/or composites, has possibly been required for various applications [6]. Applications facilitating by these systems are due to their functional properties in addition to the biocompatibility and degradability, responsiveness, eco-friendly behavior toward specific biocells or tissues [7]. Thus elastomer blends and composites are widely recommended for biomedical, ocean, agricultural, and environmental applications.

In addition, functional properties, the mechanical strength, flexural properties, high-level flexibility and curing behavior, and other physical properties are relevant to the application, which make these class of systems more significant. To enhance functional and physical properties, fillers are incorporated in the matrix material or with the addition of new functionalities via grafting or blending of matrix material with the other essential materials. In the past, the attractive property is the mechanical strength of these systems, improvement of mechanical strength has already reached significantly.

Elastomer Blends and Composites. https://doi.org/10.1016/B978-0-323-85832-8.00015-8

The main aim is to tune the properties of mechanical and application-driven functional properties to achieve the optimum level and get success in particular applications in prospective area of synthetic elastomer blends and composites. Properties and name will be varied for the composites obtained according to their properties of filler size and the dimensions. The fabrication of micro- or nanocomposites, micro- or nanofillers is added to rubbers, which significantly facilitate the application.

The elastomer blends and composites are used in medical field such as cardiovascular applications. These systems are also used to cover many reconstructive medical and dental applications (biomedical applications with the desired mechanical property), which include hip joints and knee replacements, etc. [8,9] Wide progress has been achieved in the development of elastomeric blends and composites systems in all application fields; however, the applications still need to be explored particularly in ocean and agriculture engineering. As per worldwide demand for using these systems in 2000, the demand has estimated at 200 billion US dollars per year. This type of demand significantly raises annually at the rate about 10% [10].

In this context, the current advancement of elastomer blends and/or composite systems for the biomedical, ocean, and agricultural engineering applications has been described. In addition to that, the biodegradable and other biofunctionalities relevant to the application are highlighted in this chapter. Moreover, this chapter illustrates the elastomer blend and/or composite systems' properties and their functionalities obtained via processing methods. The wide functionalities are obtained via synthetic routes, and their use in elastomeric blends and/or composites is highlighted.

5.2 Elastomer blends and composites processing methods

The advantages of elastomer blends and composites exhibit in many ways and can be processed for functional uses such as carpeting, fabrics in fibers, and varieties in molded shapes. The processing methods of most important elastomer system are elaborated as follows:

Internal mixing method: This method is widely used and preferred processing method for elastomeric blends and composites. Elastomeric blends and composites are predefined material, processing temperatures become compatible for blending of two or more materials or dispersing nanofillers within the elastomeric matrix material. Then, it can be extruded into different shapes or fibers or films with the help of molds. This kind of processing method instigated the process that involves molecular level of mixing or compositing than the simple mixing of materials to form a different physical shape and has an ultimate role in the performance of the end product.

5.2.1 Extrusion (twin or single screw)

For the process of elastomeric blends and composites films or sheets, most favorable convenient process is melting extrusion, which is both environmentally and economically. In Screw (twin/single) extruder as shown in Fig. 5.1A and B, the materials such as elastomeric blends and/or composites are fed through feed hopper and subsequently transported

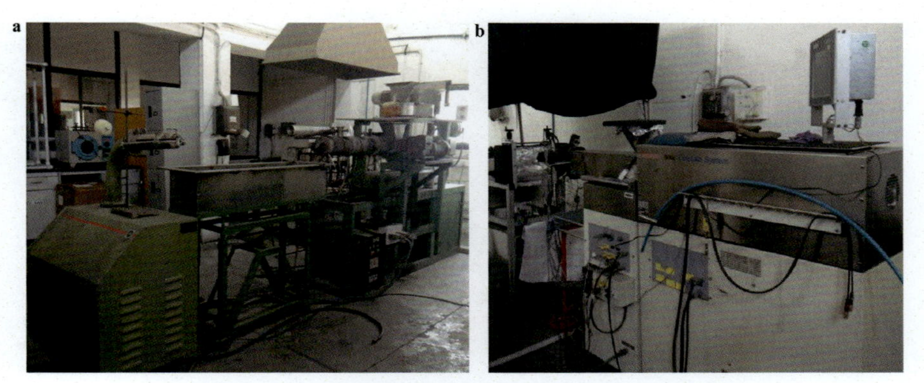

FIGURE 5.1 (A) Twin screw extruder (B) Single screw extruder.

through rotating screw and heating chambers and then finally shape-forming die. The extruder screw design is compatible with computer-assisted modeling. The compound fed strip width less with respect to the feed hopper width lightly. The flight of the screw is equal to or a little beneath to strip thickness. The barrel has a gap for steam or cold-water flow and more rottenly is made up of stainless hard. Initially, barrel needs to heat and at that stage, cold water is circulated to maintain the optimal temperature. Frequently mixing, compounding, and devolatilization are also involved in process, the formulations include special additives, such as plasticizers, flame retardants, lubricants, pigments, fillers, and other polymers. The elastomeric blends and/or composites coming out from the die are usually carried to the next stage of processing through the conveyor system. The cooling of the extrudate is carried out by immersion in water or by a spray of cold water. Films or sheets are formed through extruded thin-walled molds. Optimization of the process requires a fundamental understanding of material (elastomeric blends and/or composites) properties and processing characteristics.

5.2.2 Brabender

These mixers are interchangeable measuring hands in connection with a lab station to serve for mixing the possibility of elastomeric materials and composites and many other materials that can be processed under accurate conditions with a minimum amount of required material. The advantages include real-time transfer of results and actual values, consistent and meaningful results even with small sample quantities, no interruption of production process and modular design allow application flexibility. In principle, the gelation process is carried out during the mixing of compound in the chamber between two blades. The run of torque recorded and graphically illustrated as a planogram in the lab station allows researchers to follow time-dependent changes as gelation process progressing. There are two important characteristic points of the Brabender plastogram as shown in Fig. 5.2; one is minimum of torque curve and the inflection point of torque curve. The maximum torque curve M_X, which occurs at point X on the torque curve, and

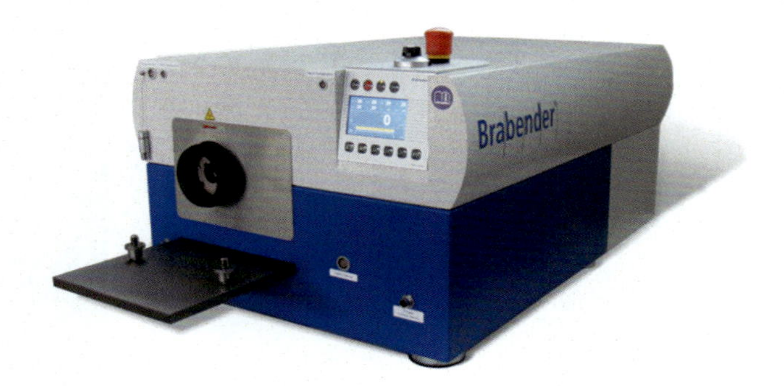

FIGURE 5.2 Plastograph of Brabender.

the time to reach point X is t_X, these values M_X and t_X are defined as fusion torque and fusion time, respectively. Researchers claimed that at the peak material, for example, elastomeric blends and/or composites influence an effective void/bubble-free state and start to melt at the interface between the compacted material and the hot metal blade surface. Mixing materials, for instance, elastomeric blends and/or composites emerge from the blade, are usually carried out to the next stage of processing through the mold or die system. Films or sheets are formed through the thin-walled mold or die system. Optimization of process requires a fundamental understanding of material (elastomeric blends and/or composites) properties and processing characteristics.

5.2.3 Two roll mills

Fig. 5.3 depicts two roll mills, which consist of two horizontal parallel heavy metal rolls with temperature control. The individual motor is directly connected to these rolls through gears to adjust the speed. These two rolls turn toward each other with a preset adjustable gap to allow elastomer/rubber/composite to pass through to achieve high shear mixing. When an elastic rubber is fed into the mixer for blending or compositing, it converts to a state in which it will accept particulate additives/blend material. Subsequently, incorporation of blend material/additives and enclosing them with a rubber matrix material are allowed. Better incorporation and distribution can be obtained with the help of a cutting knife as a helper. However, during the distribution of filler within the rubber matrix, filler particle agglomeration is a major drawback for the product, which probably reduces the production efficiency and reliability. However, the compressibility of the mixture with the help of two rolls allows high forces, which are applied to the particle agglomerates causing them to fracture or the blended material to cross-link. It serves the purpose of separating the agglomerated particulates once they have been fractured. Further, the addition of plasticizer facilitates easy incorporation of fillers. After thorough incorporation of all the ingredients, the mixture is homogenized and the batch is then shaped. Films or sheets are formed through the thin-walled mold/die system. Optimization of the process requires a fundamental understanding of material (elastomeric blends and/or composites) properties and processing characteristics.

FIGURE 5.3 Two roll mills.

In summary of above three methods for processing of elastomeric blends and composites, the properties of outcome product strongly depend on the processing parameters of particular processing machine that include uni/biaxial orientation, degree of crystallinity, the morphology of amorphous and/or crystalline regions, internal stress and dimensional control, etc. The complete system from material design, synthesis, and formulation to product design and processing must be addressed to achieve optimal material selection for a specific application. All these elements of the system are important. Material scientists and engineers who realize the structure–property relationship and can manipulate the molecular design work closely with process engineers and product designers in a systematic approach to meet the growing demand for the outcome of elastomeric blends and/or composites that have specific characteristics.

5.2.4 Radiation method

High energy chemistry is the branch of chemistry that applies a physical source of energy to achieve reactions of chemical compounds, i.e., within the chemical compound or between the compounds, for example, between the elastomers or between the elastomer and a polymer to form a blend or between the matrix and fillers to form a functionalized composite [11,12]. High energy chemistry involvement in processing of elastomer blends and/or composites by supplying high energy, i.e., specific energy to specific material chemistry, which is contrast to thermal energy, effects and produces a change in position of electrons in the outer shell of electron from the outer orbital of system leading to ionization and specific chemical reactions [11,12].

Radiation chemistry deals with the interaction of ionizing radiations with the material (elastomer blends and/or composites); for instance, electromagnetic radiation energy higher than 24eV ionizes any atom or molecule. Electromagnetic quanta of lower energy than 24eV, for example, infrared energy, and microwave energy cannot influence ionization of the components of elastomer blends and/or composites. However, if it is absorbed, it can lead to internal thermal effects. On the other side of UV-vis spectrum, higher quanta with a shorter wave and higher energy possibly ionize the elastomer blends and/or composites at the molecular level. Moving toward the shorter end of wave in the UV range on elastomers systems, they are connected with the elastomeric system's resistance parameter to influence. With increasing the energy ionization starts the sequence of secondary and next generations of ionization by degraded quanta and secondary electrons. The ionization capability described by the energy of quanta also applies to the charged particles, e.g., electrons, protons, and neutrons either directly or indirectly [11,12].

Dealing with the radiation chemistry, firstly, we need to consider what kind of source of radiation, then the nature of material/irradiated matters, and the distribution of radiation energy on material/matter. Recognition of mechanisms of chemical reactions assists to find general rules and make predictions concerning the final results, sometimes feasibility is not achieved via experimental demonstrations [11,12]. That is the way the radiation energy induces the chemical reactions via increasing thermal effects to form the desired elastomeric blend and/or composite system.

5.3 Elastomer blends and composites engineering applications

5.3.1 Biomedical engineering applications

Biomedical elastomers are divided into three categories: first one is commodity elastomers, which are useful in the biomedical field; the second one is medical-grade elastomers, these are certified for food contact or for short-term physiological contact; and the third category consists of a small group of elastomers that are well suitable for long-term physiological contact or implantation. Although a wide range of commodity elastomer blends and composites have been developed, current developments and uses of commodity elastomers in the biomedical field are discussed in this chapter.

A wide range of biodegradable polymeric materials have been proposed for biomedical applications and environmental applications, which include poly (lactic acid), polycaprolactone, and poly(3-hydroxybutyrate). However, these polymers exhibit limited biomedical applications due to their inferior mechanical properties and less biocompatibility with the cells and tissues. Also, low elongation and triggering inflammatory reactions upon long time interaction with the cells and tissues limit its potential biomedical uses. Recently, the researchers have developed poly(glycerol sebacate) (PGS) elastomer, which significantly shows its applicability as a biodegradable elastomer for biomedical applications and may address the above concerns [13]. This poly(glycerol sebacate) elastomer is synthesized by polycondensation reaction method of biobased chemical sebacic acid and glycerol, other similar biochemicals such as itaconic acid, succinic acid, 1,3-propanediol, and 1,4-butanediol were thought as agents for producing new elastomeric systems. Glycerol is nontoxic, colorless, orderless, and used as a hygroscopic compound in food and approved by Food and Drug administration (FDA). Due to the existence of hydroxyl groups, it is water-soluble and hygroscopic. On the other hand, sebacic acid is also FDA-approved and nontoxic biomedical compound in the ω-oxidation of medium to long-chain fatty acids [13].

A family of poly (glycerol sebacate) and other polyester-based biodegradable elastomers have been developed with the help of biochemical reactants, which are suitable to the biological cells and tissues [14]. These polymers exhibit reasonable chemical and physical properties, where biodegradable functionality rate can be easily altered by changing the reactant. The biodegradability rate of poly (polyol sebacate) (PPS) polymer is comparable with the poly(lactide-co-glycolide), which is easily miscible with the PPS-prepolymers to create new biocompatible and degradable blends and/or composites. Moreover, poly(-xylitol sebacate) has been prepared from the biochemical sebacic acid reactant and xylitol, this polymer has superior physical properties, applicable in the biomedical field without having immunological threatening [14]. The polymer can be combined with the glutamate to form poly(xylitol glutamate sebacate) by condensation reaction, which provides excellent hydrophilicity, increased degradation rate, and elongation rate [15]. The polymer blends and/or composites with various processing systems with optimum chemical interactions/bonding can provide excellent structural compatibility, biocompatibility, and other functional properties, which are appropriate for biomedical engineering applications. Also, some studies showed that the modifications of PGS has predominated the limitation in terms of mechanical strength. Some other studies reveal that the cellulose-based, carbon-based, and silicon-based nanofillers incorporation in PGS has modified physical characteristics. For instance, the solution cast method has been implemented for the preparation of PGS/cellulose nanocrystals [16]. These nanocrystals showed an increment in tensile strength and modulus with an improved cross-linking density as well as hydrophobicity. Carbon nanotubes (CNTs) are incorporated into the PGS to enhance stiffness, electrical conductivity, and thermal stability. Other reports suggest that the multiwalled CNTs embedded with the poly (glycerol sebacate citrate) (PGS/CNT) elastomer have been used for biomedical applications [17,18]. Various literature reports divulge that the mechanical properties increase and the biomedical properties decrease in case of carbon-based materials. However, bio carbon-based materials or green synthesis of carbon materials may overcome the limitations presented by these traditional carbon-based composites for biomedical applications. One of the challenges confronted by PGS applications where cytotoxicity was caused by acidic biodegradation and rapid biodegradation in vivo. Chen et al. developed PGS-based elastomeric nanocomposites blended with nanoparticles of bioglass [19]. The alkaline bioglass neutralized the acidity caused by the biodegradation of PGS without affecting the physical characteristics. The newly developed PGS nanobioglass composites demonstrated improved biocompatibility as compared to PGS. The mechanical properties of nanocomposites also improved with regard to PGS and with other composites, which showed compatible with the cardiac muscle. Moreover, the bioactive silica glass phase dispersed in the PGS matrix at a molecular level, which greatly enhances the physical properties, hydrophilicity, biomineralization activity, and osteoblast compatibility. These hybrid elastomers showed characteristic elastic behavior and adjustable physical properties (tensile strength and modulus) superior to PGS elastomers [20–22].

Conductive biopolymer-based materials have been widely used in a variety of biomedical fields. For example, these materials are used for regulating cellular behavior with the electrical stimulus response in tissue engineering applications [23–27]. These materials are also used for bio actuating and as a sensor in the biomedical field. However, these conductive families have some limitations, for example, for soft tissue engineering, conductive material with softness, elasticity, and full degradation cannot be achieved except the elastomeric conductive blends and/or composites. It creates difficulties in processing and also results in high mechanical stiffness, which negatively influences the mechanobiological interaction between cell and polymer in soft tissue engineering applications [28,29]. Blending a soft elastomer with intrinsically conductive polymer is simple and effective to improve its flexibility and stretch ability [30–33]. Poly(glycerol sebacate), or its derivative blended with polyaniline or poly(1,6-heptadiyne), is used as a soft conductive scaffold for cardiac or skeletal tissue repair [34,35]. Polyurethane elastomer with the polythiophene or polypyrrole is prepared to produce an electroactive composite with enhanced mechanical resilience [36,37]. Poly(1,6-heptadiyne) is a conductive polymer, is similar to that of polyacetylene. In recent years, it is possible to control the conducting characteristics ranging from an insulating polymers to the metals by various synthetic methods where substantial efforts have been carried out to produce conducting polymeric materials for the electronic industry [38–41]. Among them, ring-closing metathesis (RCM) is a kind of cyclopolymerization technique, which is more attractive and efficient method to synthesize functionalized macrocycles instigated conducting polymers [40,41] and is less explored. The 1,6-heptadiyne's ring-closing polymer (RCP) has been synthesized via cyclopolymerization technique [41]. The application has not been explored due to the probability of polyenes cyclopolymerization, which provides a highly π-conjugated system in the polymer backbone [38–41].

The electrical conductivity of these nondegradable electrically conductive polymers is generally high for the desired application. If this kind of system is introduced within the body, it can cause several issues. For example, nonbiodegradable polymer will still remain in the body after degradation of the matrix material, which may induce chronic inflammation and infection and subsequently cause implant failure. Direct conjugation of appropriate conductive oligomers into the polymer backbone can be achieved with a conductive elastomer with the desirable biodegradation, electrical, and mechanical property. This may address the above concerns relevant to the conductive polymer. Moreover, biodegradable 1,6-heptadiynes may lead to the development of biomedical engineering applications [36,37].

Nanostructured particles have shown the diverse applications ranging from medical devices to drug delivery systems. The antibacterial properties of copper and its oxides have been known for centuries. The current advanced textile technology where material composite applications demonstrated their potential as antimicrobial at the micro- or nanoscale. Similarly, polyurethane (PU) has proved its implementation ability in several biomedical applications such as biomedical coatings, blood bags, catheters, heart valves, dental fillers, protective clothing, and even tissue engineering constructs. The bio-based PU materials are also extensively developed by the researchers, maintaining mechanical and elastic properties, which can be used as futuristic material for biomedical engineering applications. Recently, researchers have developed composites of copper oxide nanoparticles and PU matrix material. The antibacterial activity of these composites has been performed; the composite solutions are electro-spun with different compositions of CuO in PU matrix to form thin films. These thin films have been tested for their antibacterial activity against methicillin-resistant staphylococcus aureus (MRSA). Significant reduction of populations is demonstrated with 10 wt.% of CuO for 4 h period. Finally, they demonstrated the potential of generating tailored antimicrobial structures for a host application, such as filters' design, patterned coatings, breathable fabrics, adhesive films, and mechanically supporting structures [42]. The process of formation electrospun fiber mat via electrospinning technique is most important for biomedical applications, for example, water filtering and toxic ion extracting, etc. This method of fabricating nonwoven fibrous mats with tailored antimicrobial properties can be used as a mask for reducing virus attachment such as currently facing Covid-19 pandemic situations.

The progress of implantable biomedical devices is the combination of various fields such as materials science, microelectronics, and biomaterials. Modern implantable medical devices have been evolved beyond serving as physical substitutes and also play an active role in restoring body functions. For instance, neural synaptic responses can be artificially created as depicted in Fig. 5.4A. The use of materials in biomedical implantable applications is highly attractive especially in soft tissue applications as they can cause appropriate host tissue response. Moreover, organic transistors for generating synaptic responses are shown in Fig. 5.4A. The homopolymers do not often exhibit multifunctionalities such as biodegradability, biocompatibility, biofunctionality, and mechanical properties, etc. Therefore, the concepts of blends and composites were introduced to facilitate multiple advantages rather than using homo polymer in cardiovascular applications. The polymer blending and/or polymer scaffolds made up of two or more polymers can produce scaffold, which exhibits stiffness, closer to the human myocardium, and improves hydrophilicity for better cell interaction and biodegradability. Polymer composites have demonstrated the improvement of compliance of the vascular graft similar to human native vessels [43]. A nonleaching antibacterial coating is shown in Fig. 5.4B, the infection resistant polymer brush coating on a biodegradable substrate, which has been explored recently. It can be utilized in migrating serious biomedical implant-related complications arising from generation of biocide-resistant bacterial strains, losing antibacterial property over time without significantly compromising the cytocompatibility of biomaterials. The kind of stress such as less infection, less polymeric materials could be a significant material for biomedical implantable applications [44].

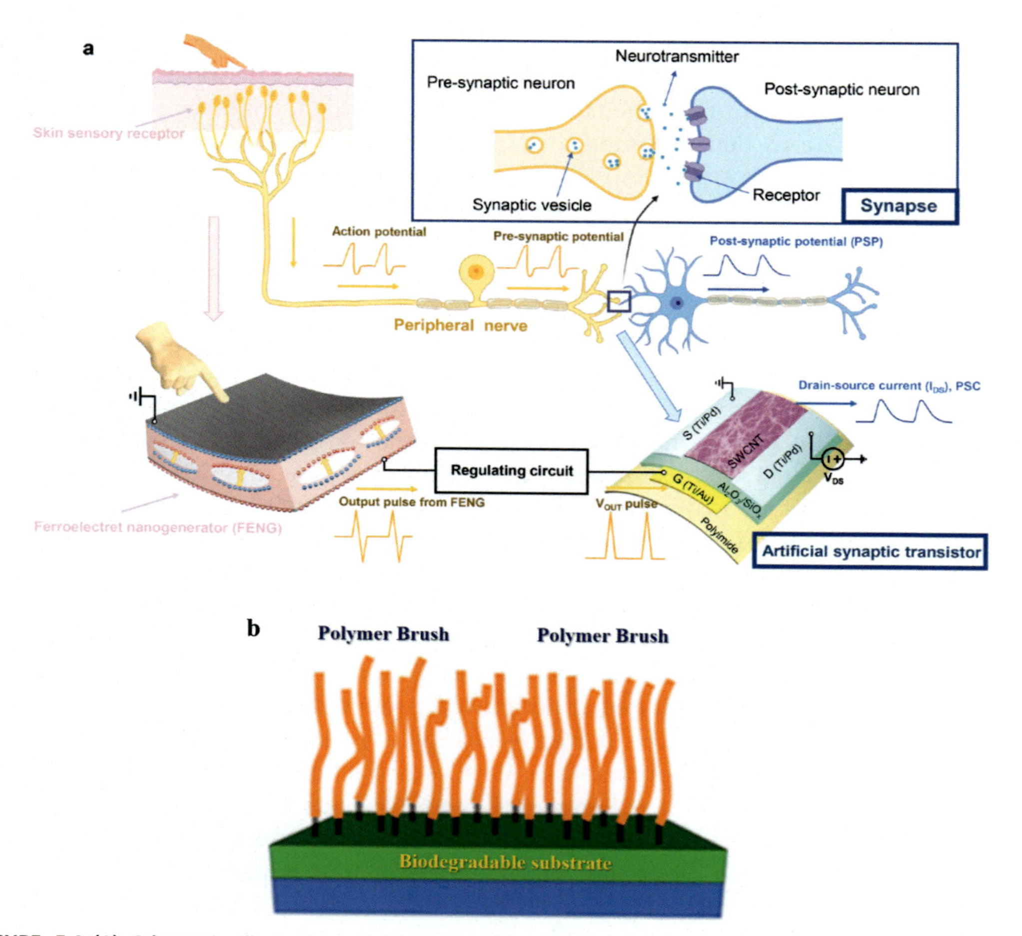

FIGURE 5.4 (A) Schematic illustration of biological skin, its peripheral nerve and synapse (upper part), in comparison with its artificial electrical counterpart (lower part). The schematic illustration is partially created with BioRender.com (B) Infection-resistant polymer brush coating on the surface of biodegradable polyester. *Reproduced with permission H.W. Toh, D.T.W. Yee, J.C.K. Ng, V.Ow, S. Lu, L.P. Tan, P.E.H. Wong, S. Venkatraman, Y. Huang, H.Y. Ang, Polymer blends and polymer composites for cardiovascular implants, Eur. Polym. J. (2021) 110249. Copyright 2021, S. Dhingra, A. Joshi, N. Singh, S. Saha, Infection resistant polymer brush coating on the surface of biodegradable polyester, Mater. Sci. Eng. C 118 (2021) 111465, American Chemical Society: ACS.*

The silicone elastomers show excellent biocompatibility and reasonable mechanical properties. These elastomers find wide medical applications especially as breast implants and urinary catheters. Silicone plastics or composites are also widely used for finger prosthesis implants as shown in Fig. 5.5 [45].

The artificial heart valves are constructed from various material combinations such as metal and polymers, which mainly consist of a mechanical type (Fig. 5.6, bottom left) [46]. The unidirectional blood flow is allowed through mechanical closure of a tilting disc valve. In recent years, prosthetic heart valves are being fabricated from polyurethane elastomers, which are biochemically inert (Fig. 5.6, bottom right). It helps to manufacture these devices, which overcome a variety of problems, for instance, material fatigue,

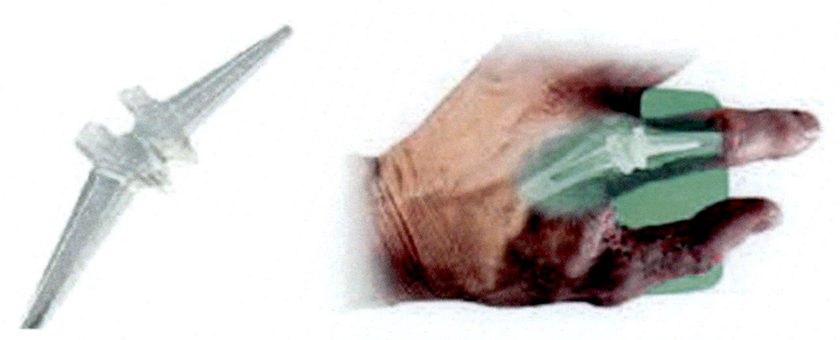

FIGURE 5.5 Figure showing finger prosthesis implant (left) and when implanted (right) [45]. Picture from http://www.fingerreplacement.com.

while maintaining natural hemodynamic and functional characteristics of heart valves simultaneously. The heart valves closely simulate natural body hemodynamic due to their soft texture and flexible nature, which simulate the lubricity as shown by natural heart valves and contract and expand freely as the blood flows through them. The high resistance to tear and abrasion and fatigue of polyurethane elastomers make them more attractive application in prosthetic heart valves [46].

5.3.2 Ocean engineering applications

The marine ecosystem is mainly regulated by many biological and biogeochemical processes and between many subcategories of them. Electronic technology and electroanalytical chemistry have been explored and implemented in marine environments to understand how marine chemistry and life interact with each other. Various processes are involved in the marine ecosystem, these processes and changes that are occurring in the processes are needed to be monitored to maintain sustainable marine ecosystem services. Traditionally, these marine system requirements were realized manually via sampling procedures and then the tested in the laboratory. However, these traditional measurements only analyze local environmental parameters but also, these fixed interval measurements. This may not be sufficient for dynamic measurement of a wide variety of parameters and their limitations.

Automated in-situ measurements such as wireless/remote sensing of the marine ecosystem are the other possible measurement technologies, which facilitate on time measurement and overcome the limitations of offered traditional technologies. Sensors and systems are the main purposes of development in ocean engineering applications because these systems provide a significant measurement of complementary processes that include physical and biological processes and variable over long time intervals and large volumes. Also, submillimeter-scale molecular processes of the marine ecosystem to climate change with a large volume of data measurements reveal via a wide variety of sensor network systems. Such kind of sensor developments evaluate ranging from ocean research, which starts from natural bio geochemical cycles to marine pollution monitoring. For example, in the case of marine pollution reduction strategies, detection of pollutants such as volatile organic compounds is most important, even in outer marine environment pollution. Therefore,

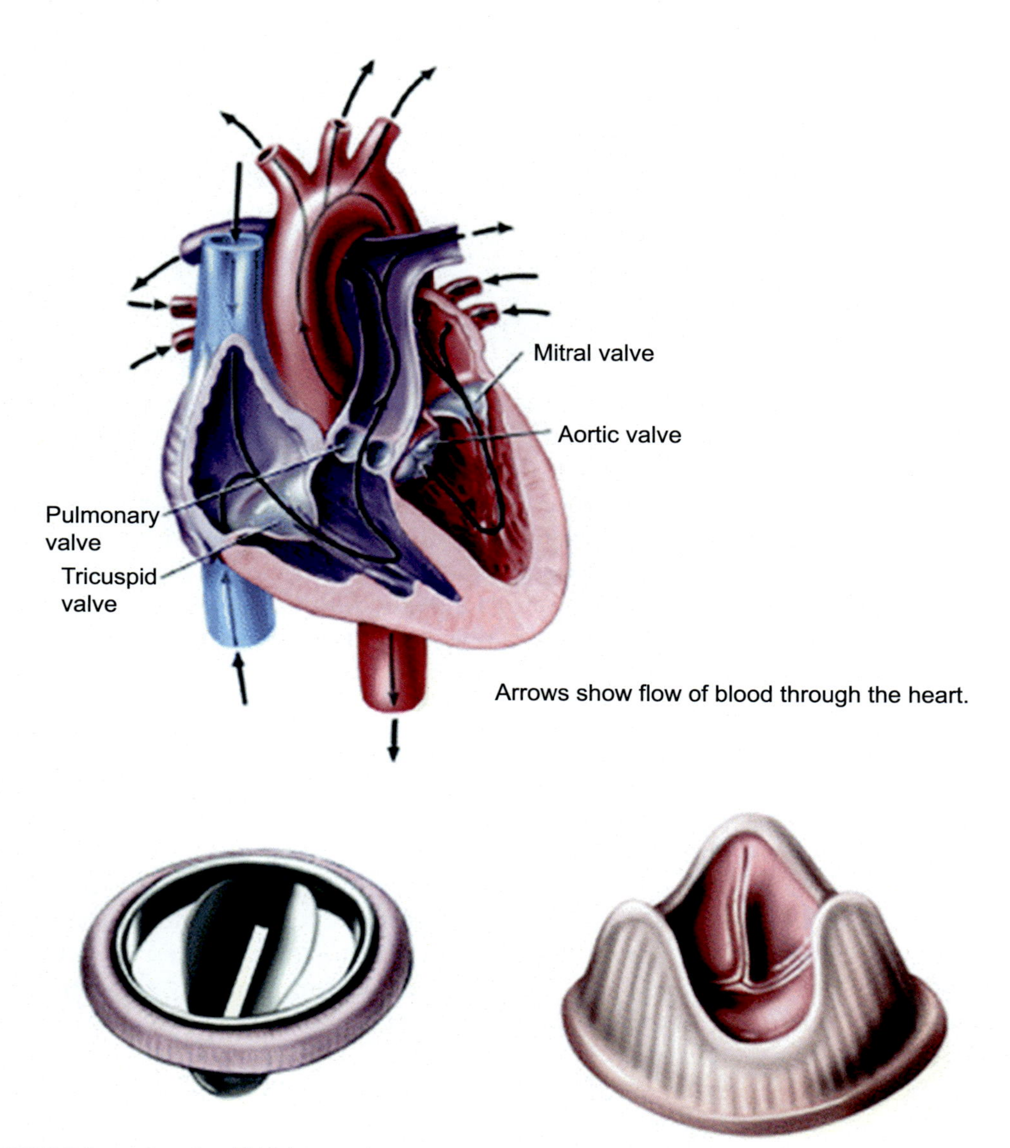

FIGURE 5.6 Examples of artificial heart valves, top: heart showing heart valves positions, bottom left: mechanical heart valve, and bottom right: polymeric heart valve. *Images based on Relay Health, 2009, V. Kanyanta, A. Ivankovic, N. Murphy, Bio-medical applications of elastomeric blends, composites, in: Advances in Elastomers II, 2013, pp. 227–252.*

researchers have developed a chemical sensor, which consists of a glass substrate with a 3D colloidal crystal and poly(dimethylsiloxane) elastomer. The solvents such as benzene, toluene, and xylene, capable of swelling the elastomer matrix, are applied to the surface of the crystal, the lattice constant alters the wavelength of the Bragg diffraction. Based on the mechanism, they demonstrated the chemical sensor for the detection of pollutants in the environment. In addition to the above lattice constant, the optical constant of the demonstrated chemical sensor changes with the influence of specific pollutants, which can be visually observed with the naked eye. With these significant characteristics, the chemical sensor is a useful candidate for monitoring of environment on site [47]. These kinds of elastomeric composites are good choice for measuring pollutants in the marine ecosystems. Other researchers have explored electrically percolative thermoplastic elastomer composites filled with CNTs, carbon black, and combination of both. These composites were developed by the melt blending method. The effect of filler concentration on electrical properties of composites was investigated, the result showed that after a certain threshold, the composites exhibited a transition from insulator to conductive with variation of the filler composition. They also explained filler morphology and dispersion effect on the improvement of composites' electrical conductivity characteristics. Such kind of elastomer-based composites with high electrical conductivity are useful for detecting biogeochemical processes or physical parameters of the marine environment. Though very few numbers of elastomeric systems such as blends and composites have been developed for ocean engineering application, the functional elastomeric compound interactions and their effect influencing capabilities fundamentally have been confirmed. Such kind of compounds have been developed for biomedical field applications and other fields. The utilization of elastomeric systems experiences nontoxic behavior and biodegradable nature via bio-oil-based synthetic compounds or blending with the other biocompounds or with a specific functionalizing agent, which lead to an excellent candidate for ocean engineering applications [48].

Apart from biological and chemical measurements, physical measurements such as temperature, salination content, pressure, etc., are also important in marine metrology. Each of these physical parameters individually influences all marine/ocean relevant biogeochemical processes. These physical parameters' measurement data are also useful for realizing the tsunami strategic conditions. Moreover, surface vessel movements of ships and boats, as well as submarine and subsea-mining systems detection, are most important these days.

To overcome all the difficulties based on traditional technologies and to satisfy the requirements of biogeochemical processes from the marine environment, high resolution and advanced measurement technologies such as carbon-based multiparameter assessment lab on a chip (LoC) or lab on box (LoB) devices are required. To satisfy less power consumption, low production cost and size, multienvironmental stability, used as disposal devices, LoCs/LoBs, and nanotechnology carbon-based materials, will be considered as the technologies of the future. However, the deficiency of advanced sensor development is primarily due to the demand of the system, which influences environmental conditions. The reliable challenges are to achieve long-term stability, which leads to advanced sensor technologies and suitable for ocean engineering applications. The working environment and challenges of ocean environment include electrolysis, corrosion, biofouling (packaging for operation in ocean environment), sensor stability parameters, and device miniaturization capability. These drawbacks can be overcome to achieve reliable and economical device [49–51].

In marine environment, circumventing the pollution due to plastics is the most challenging task. Recently, the impact of micro plastics on marine life, biodiversity, and potential toxicity and their consequences have been studied by various researchers. It has been observed that most of the marine life is affected by the marine-poly-waste and their nondegradability character. Hence, prevention strategies are needed to implement and prevent unnecessary polymeric waste in order to provide a better life to the marine organisms and animals. Though the degradable polymeric systems have been developed widely, the degradability in marine environment depends on different factors that affect the degradation of polyesters in different environments, which are represented in Table 5.1. Hence, specific marine environment degradability of polymeric waste is very vital. More recently, researchers have come up with an idea of degradability of polymeric materials within the seawater at a given period of time as shown in Fig. 5.7. Newly designed polymers that degrade in seawater are potential alternative to the commodity polymers in certain applications. Seawater degradable polymeric materials such as PLA, poly (butylene adipate-co-terephthalate) (PBAT), poly (butylene succinate) (PBS), poly (hydroxyalkanoate) (PHA), and poly(ε-caprolactone) (PCL), etc. These polymers have been explored as seawater degradable polymers. However, the degradability of plastics in seawater is complex and requires an in-depth investigation and evaluation to provide an accurate basis for the practical application of materials that further facilitate marine life protection (Fig. 5.7) [50,51,53,54].

The processing of elastomer blends and composites is most important for improving its functionalities for applications. Among all processes, reactive extrusion is cost-effective and environmentally friendly method to produce new materials with enhanced functional properties. It allows in-situ polymerization, modification, functionalization of polymers, or chemical bonding of two or more immiscible systems,

Table 5.1 Factors that affect the degradation of polyesters in different environments [50].

Key factors	Pure water	River water	Seawater	Compost
Moisture	+++++	+++++	+++++	+++++
Temperature	Adjustable +++	0–25°C +++	0–17°C +	48–65°C +++++
Salinity	0 –	0%–1% –	~35% –	<5% –
Microorganisms	0 –	10^3–10^6 +++	1–10^5 ++	10^9 +++++
pH	Neutral ++++	Neutral ++++	8.6 (alkaline) ++++	Neutral ++++
Degradability	++	++++	+++	+++++

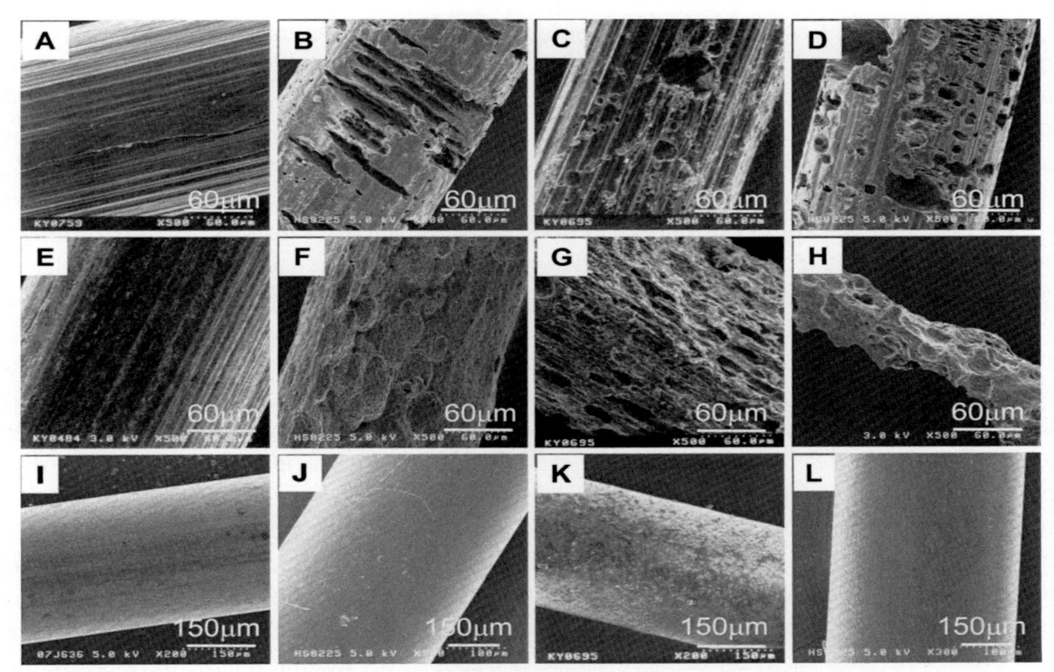

FIGURE 5.7 Scanning electron micrographs of PCL (AeD), PHB/V (EeH), and PBS (IeL) monofilament fibers before and after soaking in deep seawaters for 12 months. A, E, I: not soaked; B, F, J: soaked in Rausu water; C, G, K: soaked in Toyama water; D, H, L: soaked in Kume water. White bars show scales. *Reproduced with permission G.X. Wang, D. Huang, J.H. Ji, C. Völker, F.R. Wurm, Seawater-degradable polymers: seawater-degradable polymers—fighting the marine plastic pollution, Adv. Sci., 8 (1) (2021) 2170004. Copyright 2021, Elsevier.*

which can be carried out on commonly used extrusion lines. Although reactive extrusion has been known for many years, its applications for processing bio-based polymer blends and composites are relatively new in scientific research. The conventional and dynamic cross-linking of bio-based polymer blends and composites are prepared using an effective way in order to prepare new materials with unique properties such as biodegradable thermoplastic elastomers or shape-memory materials possessing the environmental stability and other physical properties, etc. The current trends confirmed that the reactive extrusion can be considered as a fast solvent-free low cost and pro-ecologic alternative for conventional modification methods of natural filler/fibers [54].

5.3.3 Agriculture engineering applications

The field of agriculture mainly contributes for the production of food, feed, and fiber. The demand is continuously increasing due to increase in population day by day; consequently, the crop production cost is increasing at an alarming rate while natural resources are being depleted. To overcome these limitations, advanced technology instigated precise farming and is required while maximizing agricultural production with a less price [55].

In recent years, the advancement of the blends and composites has been increasing in the field of agriculture, various technologies such as smart delivery of fertilizers and pesticides, detection of contaminants and pathogens, which are available for improving the practices of precise farming [55]. More recently, carbon-based nanomaterials have received contemplation due to their distinctive physiochemical properties. The carbon-based nanomaterials include fullerene, graphene, nano diamonds, carbon nanodots, and CNTs [56]. Carbon and its allotropic forms exhibited functional characteristics that are significantly demanded in the agricultural industry. Carbon-nanotubes-based materials are used as controlled release fertilizers, which show an excellent impact on seed germination. The researchers reveal a twofold increase in germination rate due to the nanotube penetration into the seed, which interns ameliorating effects on germination and growth [57]. The germination rate of tomato seeds is improved by using CNTs. The improved rate is due to pores/channels of CNT, which allow better penetration of water into the seeds [57]. Moreover, the growth of crops for better yield is suggested with the help of carbon-based nanoparticles. The translocation of nutrients and storage of compounds are also considered [58]. Carbon-coated gold nanoparticles are utilized as a gene gun for transmitting DNA into the *Oryza sativa, Leucaena leucocephala*, and Nicotiana tabacum [59]. Similarly, researchers have studied CNT, CNT-bounded biomolecule, which intern prevention of degradation and provide stabilized cellular bio metabolism as compared to the free molecules [60]. They confirmed that plant injuries are most significant and recovery of plant tissue is easier and valuable for the agriculture industry and ecosystem [61].

The combination of carbon-based materials with the elastomer either in the form of blends or composites exhibits significant advantages in agricultural industry. For example, recently researchers have developed polyester elastomer of biobased diacids and dialcohols via condensation polymerization [62]. In general, these polyester materials possess a higher degree crystallinity. To suppress high crystallization, five biobased chemicals such as sebacic acid, itaconic acid, succinic acid, 1,3-propanediol, and 1,4-butanediol have been introduced in order to form elastomeric compounds. Here, itaconic acid has been used to introduce double bonds for further cross-linking and sebacic acid has been used for high flexibility. The developed elastomeric system is shown in Scheme 5.1. Though the developed polyester-based system's molecular segments are in less number than the traditional one, additional cross-linking of said system leads to higher-molecular-weight systems. Further, the crystallization degrees, glass transition temperature, and other physical characteristics of these systems have been appreciated by reaction conditions and monomer ratio [62]. Such kind of tuned properties with the combination of carbon-based material are highly recommended as an eco-friendly system for real-time agricultural engineering applications.

However, these elastomeric systems are demanding more specific functionality, i.e., biodegradability, which has been explored in response to the plastic waste generated problems in the agricultural field environment. Researchers have prepared biodegradable elastomers in the elastomer family, such as polyhydroxyalkanoates, i.e., polyesters are feasible for plant production. This kind of elastomeric system accumulates as

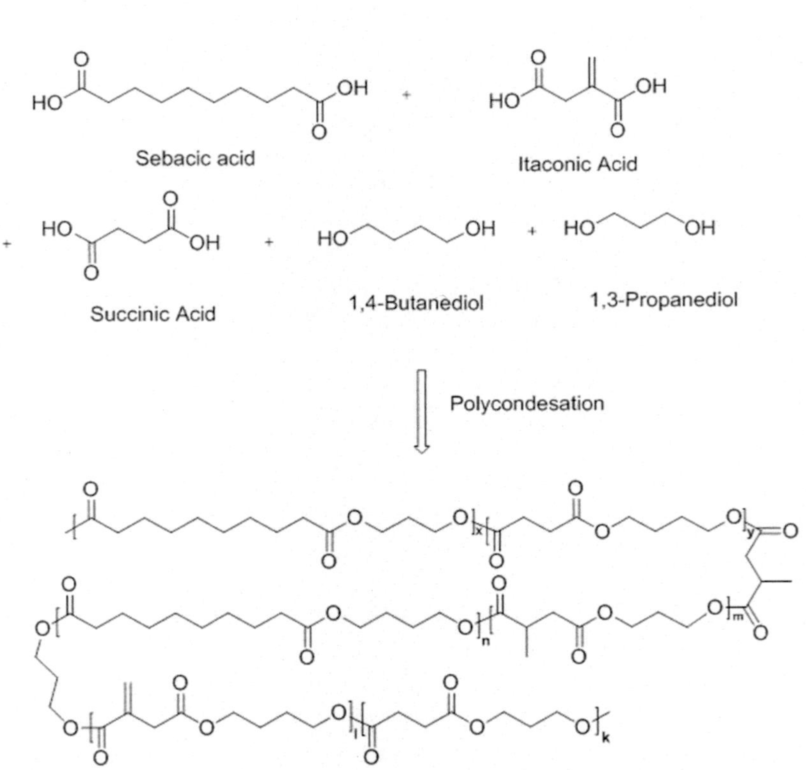

Sebacic acid

Itaconic Acid

Succinic Acid

1,4-Butanediol

1,3-Propanediol

Polycondensation

SCHEME 5.1 Reaction formula of bio-based polyester elastomer. *Reproduced with the permission of copyright from https://doi.org/10.1016/j.compscitech.2016.07.019).*

inclusions in a wide variety of bacteria to its biodegradability. Moreover, the possibility of producing PHAs has arisen from the demonstration of the plastic accumulation in transgenic Arabidopsis plants expressing the bacterial PHA biosynthetic genes. Synergism between knowledge of the enzymes and genes contributing to PHA synthesis in bacteria and engineering of plant metabolic pathways will be necessary for the development of crop plants that produce biodegradable plastics [63,64]. Because of the biodegradable nature along with the other physical and chemical properties, and environmental stability, its use has been attracting agricultural engineering applications, speciality, and commodity products.

More importantly, bio-oil-based elastomeric systems have been suggested for yielding crop production. Among them, PU elastomer has been developed by researchers and widely applied in the biomedical field, which is attributed to the biological and mechanical characteristics [65]. Similarly, Soybean oil and castor oil-based PU materials have been developed [66,67]; however, among the bio-oils, the development of bio-based polyol to prepare PU is one of the most important among the production of PU systems. Currently, vegetable oil-based PU materials, where vegetable oil is a soft segment to copolymerize with various isocyanates. These kinds of PU systems provide environmental requirements along with the physical and chemical functionalities. In many aspects, these materials are superior to petroleum-based PU materials.

With the realization of functional properties of the abovementioned elastomeric systems either in the form of blends or composites and with the addition of carbon-based material, the reliable system is a suitable candidate for real-time agricultural applications, which fulfills the traditional requirements such as economic feasibility for applications, environmental stability, and other requirements, as well as chemical and physical properties, and suitable processing technologies. Moreover, these elastomeric blends and/or composites facilitate smart delivery of fertilizers and pesticides, detection of contaminants and pathogens and are available in the agricultural field for improving the practices of precise farming. Further, these systems provide improved germination and stabilized cellular biometabolism as compared to the free molecules for the growth and development of better yield crops.

Since the last decade, the extensive research on carbon-based materials has been carried out. These materials with the combination of bio-based elastomeric systems (blends and/or composites) exhibit as unique materials for the development of devices for detection, monitoring, and diagnosing in the agricultural field. However, few literature reports are available based on these devices for understanding the development and formation of tissues, design of highly organized biomaterial scaffolds to provide controlled cell adhesion and spatial organization, the regulation capability of a biomolecule, surface modification of biological molecule to maintain biological compatibility and activity, as well as, controlled and targetable drug delivery. Although the researchers have explored innovative developments, which may develop elastomers and composites at the nanoscale level, device technologies could facilitate a path for quick advancement in the agricultural and tissue engineering field. For instance, recently, a nondiagnostic kit has been developed for measuring biological aspects of plants [68]. The developed small device can help to detect pathogens and fungal attachment on crops such as wheat, barley, and corn to prevent plant-based diseases [69,70]. This device is also useful for other purposes such as detection of antigens, antibodies and nucleotide sequence, gene target, isolation, and purification of specific genes. They suggested that this device could be prominent in the agriculture sector for the required crop production. They also suggested that if this device enables the improved multifunctionalities and is tested under field trials, it can lead to a functional device for the applied agriculture sector.

Recycling of polymeric materials is currently one of the most significant industrial challenges in Agriculture area, particularly in agriculture farm. It is essential to look at potential methods to utilize agricultural polymeric waste (agri-poly-waste). Recently, the agricultural polymeric waste's utilization has been explored after recycling these materials considering their structure, properties, and recycling line, which are shown in Figs. 5.8 and 5.9. The agricultural polymeric waste with less weight percent of commercially available compatibilizers has been blended. It was observed that the use of small quantity of modifiers could overcome the drawbacks caused by the presence

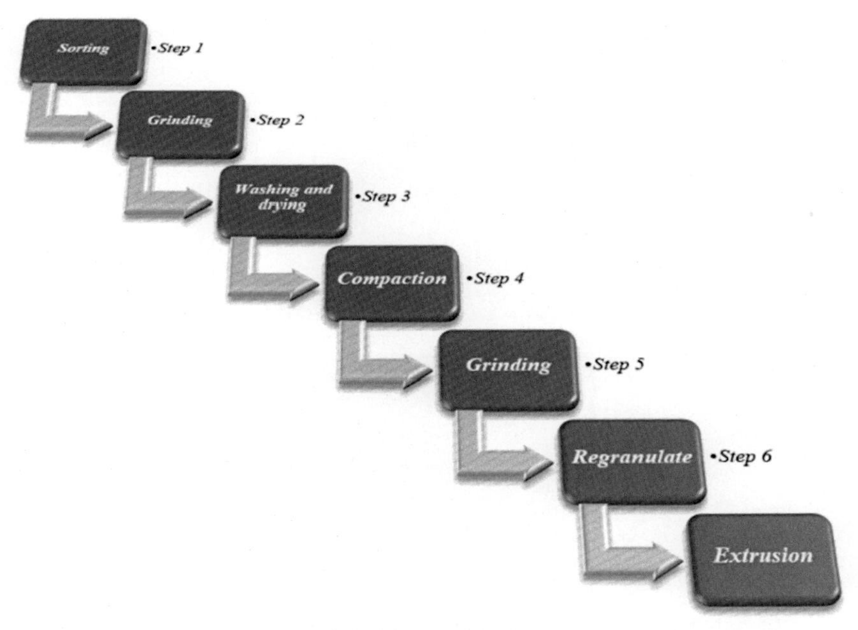

FIGURE 5.8 The recycling line from which agri-poly-waste is collected [71].

of impurities. For example, incorporation of maleic anhydride-grafted compounds enhanced the tensile strength of waste by 13%–25%. The use of more ductile automates such as ethylene vinyl acetate and paraffin increased the elongation at break by 55%–64%. However, some of modified/recycled agri-poly-waste leads to reduce performance. For example, the presence of compatibilizer such as styrene-ethylene-butadiene-styrene and ethylene vinyl acetate copolymers, which cause up to 20% less of young modulus. Thermal performance such as slight reduction in melting temperature and crystallinity were also explored recently. The above properties are estimated under complete recyclization, depending on desired application and properties. These materials need not require the whole recycling line that further causes stabilized properties to that of desired application [71,72].

In summary, there have been research and development of various high-performance synthetic bio-based elastomers for agriculture engineering applications; however, they are quite few. Advancement in these elastomeric systems with applications-driven capability leads to an excellent choice of these materials for agricultural applications. Moreover, carbon-based material and its functionalized hybrids (elastomer blends or composites) with wide functionality were extensively explored. The researchers confirmed that these kinds of significant materials with the elastomeric systems as composite enabled electronic devices, which will be developed in the field of agriculture for detection, monitoring, and diagnosing the quality crop production with reduced economic burden.

FIGURE 5.9 The appearance of M1(cycle-1), M2(cycle-2), M3(cycle-3) wastes and impurities removed during washing and retained on sieves with (A) 4 mm, (B) 2 mm, (C) 1 mm, (D) 0.5 mm, (E) 0.125 mm, and (F) residue [73].

5.4 Conclusion

Elastomer blends and/or composite materials have been increasing rapidly due to their potential applications. Recently, biodegradability and other biofunctionality characteristics of these elastomeric systems have been further recommended in biomedical, ocean, and agricultural engineering applications. Elastomer blends and composites processes are most significant for improving its functionalities for applications. Among all processes, reactive extrusion is a cost-effective and environmentally friendly process to produce new materials with improved functional properties. The in-situ polymerization, modification, functionalization of polymers, or chemical bonding of two or more immiscible systems can be carried out on commonly used extrusion lines. Though, the research and development in various high-performance synthetic bio-based elastomers have been carried out, their blends and composites, in agriculture, ocean engineering applications, and even in biomedical applications are quite few. The advancement in these elastomeric systems with applications-driven capability leads to an excellent choice of these materials for the suggested applications. Moreover, carbon-based materials and its functionalized hybrids (elastomer blends or composites) with wide functionality mention that these significant materials with the elastomeric systems allow advanced electronic devices in the field of biomedical, ocean, and agriculture for detection, monitoring, and diagnosing with less economic burden.

Acknowledgments

Naga Srilatha Cheekuramelli gratefully acknowledges the Department of Science and Technology, New Delhi, India, for the award of Women Scientist Scheme-A (WOS-A) (GAP-330026) for her doctoral research. The authors also gratefully acknowledge LIA MATSUCAT, France, for the collaboration.

References

[1] N. Dubey, C.S. Kushwaha, S.K. Shukla, A review on electrically conducting polymer bionanocomposites for biomedical and other applications, Int. J. Polym. Mater. Polym. Biomater. 69 (11) (2020) 709–727.

[2] M. Mucha, K. Wańkowicz, J. Balcerzak, Analysis of water adsorption on chitosan and its blends with hydroxypropylcellulose, e-Polymers 7 (1) (2007) 1–10.

[3] J.R. Aggas, J. Lutkenhaus, A. Guiseppi-Elie, Chemiresistive and chemicapacitive devices formed via morphology control of electroconductive bio-nanocomposites, Adv. Electr. Mater. 4 (2) (2018) 1700495, https://doi.org/10.1002/aelm.201700495.

[4] V. Kanyanta, A. Ivankovic, N. Murphy, Bio-medical applications of elastomeric blends, composites, in: Advances in Elastomers II, Springer, Berlin, Heidelberg, 2013, pp. 227–252, https://doi.org/10.1007/978-3-642-20928-4_8.

[5] A. Tiwari, A.P. Mishra, S.R. Dhakate, R. Khan, S.K. Shukla, Synthesis of electrically active biopolymer–SiO2 nanocomposite aerogel, Mater. Lett. 61 (23–24) (2007) 4587–4590, https://doi.org/10.1016/j.matlet.2007.02.076.

[6] X.H. Xu, G.L. Ren, J. Cheng, Q. Liu, D.G. Li, Q. Chen, Self-assembly of polyaniline-grafted chitosan/glucose oxidase nanolayered films for electrochemical biosensor applications, J. Mater. Sci. 41 (15) (2006) 4974–4977, https://doi.org/10.1007/s10853-006-0118-4.

[7] S.K. Shukla, A. Tiwari, Synthesis of chemical responsive chitosan–grafted-polyaniline bio-composite, in: Advanced Materials Research, vol. 306, Trans Tech Publications Ltd, 2011, pp. 82–86, https://doi.org/10.4028/www.scientific.net/AMR.306-307.82.

[8] M.T. Khorasani, M. Zaghiyan, H. Mirzadeh, Ultra high molecular weight polyethylene and polydimethylsiloxane blend as acetabular cup material, Colloids Surf. B Biointerfaces 41 (2–3) (2005) 169–174.

[9] J.I. Onate, M. Comin, I. Braceras, A. Garcia, J.L. Viviente, M. Brizuela, N. Garagorri, J.L. Peris, J.I. Alava, Wear reduction effect on ultra-high-molecular-weight polyethylene by application of hard coatings and ion implantation on cobalt chromium alloy, as measured in a knee wear simulation machine, Surf. Coating. Technol. 142 (2001) 1056–1062.

[10] M.J. Lysaght, J.A. O'Loughlin, Demographic scope and economic magnitude of contemporary organ replacement therapies, Am. Soc. Artif. Intern. Organs J. 46 (5) (2000) 515–521.

[11] Z.P. Zagórski, Radiation chemistry of spurs in polymers, in: Advances in Radiation Chemistry of Polymers, 2004, pp. 21–31. IAEA-TECDOC-1420, Vienna.

[12] J. Bik, W. Głuszewski, W.M. Rzymski, Z.P. Zagórski, EB radiation crosslinking of elastomers, Radiat. Phys. Chem. 67 (3–4) (2003) 421–423.

[13] X.J. Loh, A.A. Karim, C. Owh, Poly (glycerol sebacate) biomaterial: synthesis and biomedical applications, J. Mater. Chem. B 3 (39) (2015) 7641–7652.

[14] J.P. Bruggeman, B.J. de Bruin, C.J. Bettinger, R. Langer, Biodegradable poly (polyolsebacate) polymers, Biomaterials 29 (36) (2008) 4726–4735.

[15] W. Dong, T. Li, S. Xiang, P. Ma, M. Chen, Influence of glutamic acid on the properties of poly (xylitol glutamate sebacate) bioelastomer, Polymers 5 (4) (2013) 1339–1351.

[16] L. Zhou, H. He, C. Jiang, S. He, Preparation and characterization of poly (glycerol sebacate)/cellulose nanocrystals elastomeric composites, J. Appl. Polym. Sci. 132 (27) (2015).

[17] Q. Liu, J. Wu, T. Tan, L. Zhang, D. Chen, W. Tian, Preparation, properties and cytotoxicity evaluation of a biodegradable polyester elastomer composite, Polym. Degrad. Stabil. 94 (9) (2009) 1427–1435.

[18] A.K. Gaharwar, A. Patel, A. Dolatshahi-Pirouz, H. Zhang, K. Rangarajan, G. Iviglia, S.R. Shin, M.A. Hussain, A. Khademhosseini, Elastomeric nanocomposite scaffolds made from poly (glycerol sebacate) chemically crosslinked with carbon nanotubes, Biomater. Sci. 3 (1) (2015) 46–58.

[19] Q.Z. Chen, S.L. Liang, J. Wang, G.P. Simon, Manipulation of mechanical compliance of elastomeric PGS by incorporation of halloysite nanotubes for soft tissue engineering applications, J. Mechan. Behav. Biomed. Mater. 4 (8) (2011) 1805–1818.

[20] X. Zhao, Y. Wu, Y. Du, X. Chen, B. Lei, Y. Xue, P.X. Ma, A highly bioactive and biodegradable poly (glycerol sebacate)–silica glass hybrid elastomer with tailored mechanical properties for bone tissue regeneration, J. Mater. Chem. B 3 (16) (2015) 3222–3233.

[21] S.L. Liang, W.D. Cook, G.A. Thouas, Q.Z. Chen, The mechanical characteristics and in vitro biocompatibility of poly (glycerol sebacate)-Bioglass® elastomeric composites, Biomaterials 31 (33) (2010) 8516–8529.

[22] S. Liang, W.D. Cook, Q. Chen, Physical characterization of poly (glycerol sebacate)/Bioglass® composites, Polym. Int. 61 (1) (2012) 17–22.

[23] N.K. Guimard, N. Gomez, C.E. Schmidt, Conducting polymers in biomedical engineering, Prog. Polym. Sci. 32 (8–9) (2007) 876–921.

[24] M. Xie, L. Wang, B. Guo, Z. Wang, Y.E. Chen, P.X. Ma, Ductile electroactive biodegradable hyperbranchedpolylactide copolymers enhancing myoblast differentiation, Biomaterials 71 (2015) 158–167.

[25] A. Kotwal, C.E. Schmidt, Electrical stimulation alters protein adsorption and nerve cell interactions with electrically conducting biomaterials, Biomaterials 22 (10) (2001) 1055–1064.

[26] S. Meng, M. Rouabhia, Z. Zhang, Electrical stimulation modulates osteoblast proliferation and bone protein production through heparin-bioactivated conductive scaffolds, Bioelectromagnetics 34 (3) (2013) 189–199.

[27] J.G. Hardy, D.J. Mouser, N. Arroyo-Currás, S. Geissler, J.K. Chow, L. Nguy, J.M. Kim, C.E. Schmidt, Biodegradable electroactive polymers for electrochemically-triggered drug delivery, J. Mater. Chem. B 2 (39) (2014) 6809–6822.

[28] F. Guilak, D.L. Butler, S.A. Goldstein, F.P. Baaijens, Biomechanics and mechanobiology in functional tissue engineering, J. Biomech. 47 (9) (2014) 1933–1940.

[29] D.L. Butler, S.A. Goldstein, R.E. Guldberg, X.E. Guo, R. Kamm, C.T. Laurencin, L.V. McIntire, V.C. Mow, R.M. Nerem, R.L. Sah, L.J. Soslowsky, The impact of biomechanics in tissue engineering and regenerative medicine, Tissue Eng. B Rev. 15 (4) (2009) 477–484.

[30] C.L. Kolarcik, S.D. Luebben, S.A. Sapp, J. Hanner, N. Snyder, T.D. Kozai, E. Chang, J.A. Nabity, S.T. Nabity, C.F. Lagenaur, X.T. Cui, Elastomeric and soft conducting microwires for implantable neural interfaces, Soft Matter 11 (24) (2015) 4847–4861.

[31] S. Seyedin, J.M. Razal, P.C. Innis, A. Jeiranikhameneh, S. Beirne, G.G. Wallace, Knitted strain sensor textiles of highly conductive all-polymeric fibers, ACS Appl. Mater. Interfaces 7 (38) (2015) 21150–21158.

[32] B.U. Hwang, J.H. Lee, T.Q. Trung, E. Roh, D.I. Kim, S.W. Kim, N.E. Lee, Transparent stretchable self-powered patchable sensor platform with ultrasensitive recognition of human activities, ACS Nano 9 (9) (2015) 8801–8810.

[33] M. Li, H. Li, W. Zhong, Q. Zhao, D. Wang, Stretchable conductive polypyrrole/polyurethane (PPy/PU) strain sensor with netlike microcracks for human breath detection, ACS Appl. Mater. Interfaces 6 (2) (2014) 1313–1319.

[34] T.H. Qazi, R. Rai, D. Dippold, J.E. Roether, D.W. Schubert, E. Rosellini, N. Barbani, A.R. Boccaccini, Development and characterization of novel electrically conductive PANI–PGS composites for cardiac tissue engineering applications, Acta Biomater. 10 (6) (2014) 2434–2445.

[35] I. Jun, S. Jeong, H. Shin, The stimulation of myoblast differentiation by electrically conductive submicron fibers, Biomaterials 30 (11) (2009) 2038–2047.

[36] C.R. Broda, J.Y. Lee, S. Sirivisoot, C.E. Schmidt, B.S. Harrison, A chemically polymerized electrically conducting composite of polypyrrole nanoparticles and polyurethane for tissue engineering, J. Biomed. Mater. Res. 98 (4) (2011) 509–516.

[37] M.M.P. Madrigal, M.I. Giannotti, G. Oncins, L. Franco, E. Armelin, J. Puiggalí, F. Sanz, L.J. del Valle, C. Alemán, Bioactive nanomembranes of semiconductor polythiophene and thermoplastic polyurethane: thermal, nanostructural and nanomechanical properties, Polym. Chem. 4 (3) (2013) 568–583.

[38] S.K. Podiyanachari, S. Moncho, E.N. Brothers, S. Al-Meer, M. Al-Hashimi, H.S. Bazzi, One-pot tandem ring-opening and ring-closing metathesis polymerization of Disubstituted Cyclopentenes featuring a terminal alkyne functionality, Macromolecules 53 (11) (2020) 4330–4337.

[39] X. Liu, F. Liu, W. Liu, H. Gu, ROMP and MCP as versatile and forceful tools to fabricate dendronized polymers for functional applications, Polym. Rev. (2020) 1–53.

[40] D. Pasini, A. Nitti, Free radical cyclopolymerization: a tool towards sequence control in functional polymers, Eur. Polym. J. 122 (2020) 109378.

[41] E.H. Kang, I.S. Lee, T.L. Choi, Ultrafast cyclopolymerization for polyene synthesis: living polymerization to dendronized polymers, J. Am. Chem. Soc. 133 (31) (2011) 11904–11907.

[42] Z. Ahmad, M.A. Vargas-Reus, R. Bakhshi, F. Ryan, G.G. Ren, F. Oktar, R.P. Allaker, Antimicrobial properties of electrically formed elastomeric polyurethane–copper oxide nanocomposites for medical and dental applications, in: Methods in Enzymology, vol. 509, Academic Press, 2012, pp. 87–99.

[43] H.W. Toh, D.T.W. Yee, J.C.K. Ng, V. Ow, S. Lu, L.P. Tan, P.E.H. Wong, S. Venkatraman, Y. Huang, H. Y. Ang, Polymer blends and polymer composites for cardiovascular implants, Eur. Polym. J. (2021) 110249.

[44] S. Dhingra, A. Joshi, N. Singh, S. Saha, Infection resistant polymer brush coating on the surface of biodegradable polyester, Mater. Sci. Eng. C 118 (2021) 111465.

[45] H. Wan, Y. Cao, L.W. Lo, J. Zhao, N. Sepulveda, C. Wang, Flexible carbon nanotube synaptic transistor for neurological electronic skin applications, ACS Nano 14 (8) (2020) 10402–10412.

[46] V. Kanyanta, A. Ivankovic, N. Murphy, Bio-medical applications of elastomeric blends, composites, in: Advances in Elastomers II, 2013, pp. 227–252.

[47] T. Endo, Y. Yanagida, T. Hatsuzawa, Colorimetric detection of volatile organic compounds using a colloidal crystal-based chemical sensor for environmental applications, Sensor. Actuator. B Chem. 125 (2) (2007) 589–595.

[48] Z.M. Dang, K. Shehzad, J.W. Zha, A. Mujahid, T. Hussain, J. Nie, C.Y. Shi, Complementary percolation characteristics of carbon fillers based electrically percolative thermoplastic elastomer composites, Compos. Sci. Technol. 72 (1) (2011) 28–35.

[49] D. Sun, M. Tweedie, D.R. Gajula, B. Ward, P.D. Maguire, High-strength thermoplastic bonding for multi-channel, multi-layer lab-on-chip devices for ocean and environmental applications, Microfluid. Nanofluidics 19 (4) (2015) 913–922.

[50] F. Wang, J. Zhu, L. Chen, Y. Zuo, X. Hu, Y. Yang, Autonomous and in situ ocean environmental monitoring on optofluidic platform, Micromachines 11 (1) (2020) 69.

[51] M. Tweedie, A. Macquart, J. Almeida, B. Ward, P. Maguire, Metered reagent injection into microfluidic continuous flow sampling for conductimetric ocean dissolved inorganic carbon sensing, Meas. Sci. Technol. 31 (6) (2020) 065104.

[52] G.X. Wang, D. Huang, J.H. Ji, C. Völker, F.R. Wurm, Seawater-degradable polymers: seawater-degradable polymers—fighting the marine plastic pollution, Adv. Sci. 8 (1) (2021) 2170004 (adv. Sci. 1/2021).

[53] G.X. Wang, D. Huang, J.H. Ji, C. Völker, F.R. Wurm, Seawater-degradable polymers—fighting the marine plastic pollution, Adv. Sci. 8 (1) (2021) 2001121.

[54] H. Hu, J. Li, Y. Tian, C. Chen, F. Li, W.B. Ying, R. Zhang, J. Zhu, Experimental and theoretical study on glycolic acid provided fast bio/seawater-degradable poly (butylene succinate-co-glycolate), ACS Sustain. Chem. Eng. 9 (10) (2021) 3850–3859.

[55] K. Formela, Ł. Zedler, A. Hejna, A. Tercjak, Reactive extrusion of bio-based polymer blends and composites-Current trends and future developments, Express Polym. Lett. 12 (1) (2018).

[56] Q. Hu, E.K. Wujcik, A. Kelarakis, J. Cyriac, X. Gong, Carbon-based nanomaterials as novel nanosensors, J. Nanomater. (2017) 1–2.

[57] F. Ahmed, N. Arshi, S. Kumar, S.S. Gill, R. Gill, N. Tuteja, B.H. Koo, Nanobiotechnology: scope and potential for crop improvement, in: Crop Improvement Under Adverse Conditions, Springer, New York, NY, 2013, pp. 245–269.

[58] M. Khodakovskaya, E. Dervishi, M. Mahmood, Y. Xu, Z. Li, F. Watanabe, A.S. Biris, Carbon nanotubes are able to penetrate plant seed coat and dramatically affect seed germination and plant growth, ACS Nano 3 (10) (2009) 3221–3227.

[59] R. Nair, S.H. Varghese, B.G. Nair, T. Maekawa, Y. Yoshida, D.S. Kumar, Nanoparticulate material delivery to plants, Plant Sci. 179 (3) (2010) 154–163.

[60] P. Kumar, P. Fennell, A. Robins, Comparison of the behaviour of manufactured and other airborne nanoparticles and the consequences for prioritising research and regulation activities, J. Nanoparticle Res. 12 (5) (2010) 1523–1530.

[61] Y. Wu, J.A. Phillips, H. Liu, R. Yang, W. Tan, Carbon nanotubes protect DNA strands during cellular delivery, ACS Nano 2 (10) (2008) 2023–2028.

[62] R. Wang, J. Zhang, H. Kang, L. Zhang, Design, preparation and properties of bio-based elastomer composites aiming at engineering applications, Compos. Sci. Technol. 133 (2016) 136–156.

[63] Y. Poirier, C. Nawrath, C. Somerville, Production of polyhydroxyalkanoates, a family of biodegradable plastics and elastomers, in bacteria and plants, Biotechnology 13 (2) (1995) 142–150.

[64] S. Muhammadi, M. Afzal, S. Hameed, Bacterial polyhydroxyalkanoates-eco-friendly next generation plastic: production, biocompatibility, biodegradation, physical properties and applications, Green Chem. Lett. Rev. 8 (3–4) (2015) 56–77.

[65] D.K. Chattopadhyay, K.V.S.N. Raju, Structural engineering of polyurethane coatings for high performance applications, Prog. Polym. Sci. 32 (3) (2007) 352–418.

[66] I. Banik, M.M. Sain, Role of refined paper fiber on structure of water blown soy polyol based polyurethane foams, J. Reinforc. Plast. Compos. 27 (14) (2008) 1515–1524.

[67] I. Banik, M.M. Sain, Nanoclay modified water-blown polyurethane foams derived from bifunctional soybean oil-based polyol, Polym. Plast. Technol. Eng. 49 (7) (2010) 701–706.

[68] A. Younas, Z. Yousaf, M. Rashid, N. Riaz, S. Fiaz, A. Aftab, S. Haung, Nanotechnology and plant disease diagnosis and management, in: Nanoagronomy, Springer, Cham, 2020, pp. 101–123.

[69] A.S. Nezhad, Future of portable devices for plant pathogen diagnosis, Lab Chip 14 (16) (2014) 2887–2904.

[70] D. Pimentel, Invasive plants: their role in species extinctions and economic losses to agriculture in the USA, in: Management of Invasive Weeds, Springer, Dordrecht, 2009, pp. 1–7.

[71] S. Karayılan, Ö. Yılmaz, Ç. Uysal, S. Naneci, Prospective evaluation of circular economy practices within plastic packaging value chain through optimization of life cycle impacts and circularity, Resour. Conserv. Recycl. 173 (2021) 105691.

[72] J. Korol, A. Hejna, K. Wypiór, K. Mijalski, E. Chmielnicka, Wastes from agricultural silage film recycling line as a potential polymer materials, Polymers 13 (9) (2021) 1383.

[73] D. Orjuela, D.A. Munar, J.K. Solano, A.P. Becerra, Assessment of the thermal properties of a rice husk mixture with recovered polypropylene and high-density polyethylene, using sulfur-silane as a coupling agent, Chem. Eng. Transact. 87 (2021) 565–570.

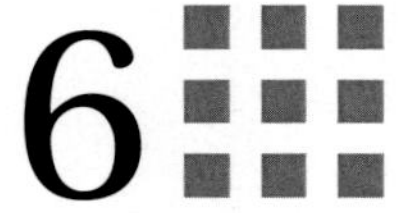

Rheology of elastomer blends and composites

Mariacristina Gagliardi

NEST, ISTITUTO NANOSCIENZE-CNR AND SCUOLA NORMALE SUPERIORE, PISA, ITALY

6.1 Introduction

The elastomer industry is among the most important in the petrochemical industries. The continuous discovery of new polymerization methods, additives, and processing techniques has led to a variety of applications for all polymeric materials and also for elastomers. Despite this, applied research on the rheology of elastomers is still struggling to find its way. After the theories proposed by Mooney [1] and Scott Blair [2], there have been no other significant advances in this science, and a consistent rheological theory is still lacking today.

Rheology is the science of deformation and flow that materials show under the effect of applied external forces. Materials studied by rheology are from elastic solids (Hookean) to viscous liquids (Newtonian). Rheological analysis finds applications not only in material development but also in their transport, handling, and storage.

The study of rheological properties comprises ad hoc analyses able to identify the flow law and the mechanical constitutive models of tested materials. Such rheological analyses are also useful in the optimization of material properties, being able to predict the long-term behavior of materials under specific working conditions.

In elastomer research, rheological analysis is relevant to understand properties of raw materials and final products. In particular, rheology can give so much information related to production processes, storage, ranges of optimal working conditions, and durability.

The first part of this chapter illustrates some basic aspects of rheology and the theory behind rheological tests. The fundamental terminology and the most common flow rules related to polymeric materials will be here explained. In the second part, the most important industrial applications of rheology of elastomer will be illustrated to underline the importance of this science in the industrial world.

Elastomer Blends and Composites. https://doi.org/10.1016/B978-0-323-85832-8.00018-3

6.2 Basic aspects of rheology

Rheology studies the flow of those materials that, under the application of external forces, show a plastic flow rather than elastic deformation. Considering the large variety of materials that can have plastic characteristics, rheological studies interest liquid, soft, and solid matter materials.

Properties investigated by rheology are those deriving from the combination of nonlinear viscous and plastic behaviors. Thus rheology aims at extend the domain of continuum mechanics (Fig. 6.1).

A complete rheological characterization can provide the prediction of the mechanical behavior of a system at the continuum-scale level.

Results can be also interpreted on the basis of material properties at the microscale and at the nanoscale, involving the study of such effects related to the molecular weight of polymers and the macromolecular architecture.

In elastomers, rheology helps in the development of production processes, in the optimization of cross-linking and vulcanization, in the analysis of the finished products.

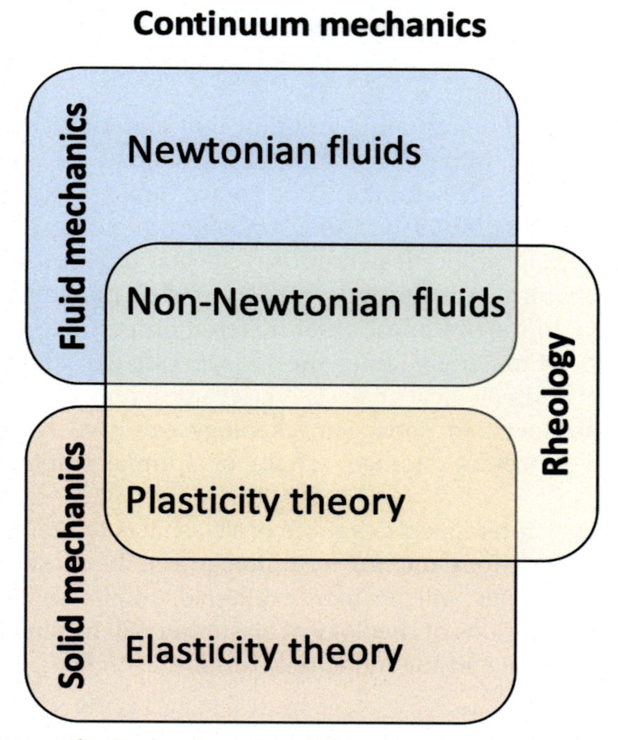

FIGURE 6.1 The continuum mechanics domain contains fluid mechanics and solid mechanics, and they share the domain of rheology.

In blends, rheology can give important information on the change of mechanical properties as a function of blend components, their compatibility after mixing, and the nonadditive laws.

In composite systems, rheological measurements are useful to understand the interactions between fillers and matrices in solutions and blends, the effect of filler geometry (e.g., particles, fibers, …), and the arrangement of fillers in the matrix.

In general, rheological studies are well established in the following fields:

Materials science, in particular for polymers, polymer melts, swellable polymer networks, gels and sol–gel systems, and colloids;
Food science and *cosmetics*, to study their stability with time, to develop suitable packaging and making them easy to use;
Geology, to understand and predict the flow of geological materials;
Physiology, for a better knowledge of body fluid motion, in both physiological and pathological conditions.

Systems with "fluid" characteristics behave as a liquid or flow under the effect of external loads. An external load generates a stress state in the fluid, defined as the applied force per area. Similarly to the solid mechanics approach, we can identify different stresses, such as torsional or shear, which can be applied to fluids. Thus, materials respond in different ways under the effect of different stresses.

Material response is strictly related to fundamental mechanical properties of the system, mainly depending on stress and strain tensors. It means that each stress type induces different deformations. The main aim of rheology is the identification of a relationship between external loads, internal stresses, and deformation.

In addition to the experimental rheometry, several studies can be completed with theoretical rheology principles. The theoretical rheology aims at mathematically explaining the relationship between the external forces acting on the system and the generated internal stresses and strain gradients.

Following is a basic list of some key terms and concepts related to rheology.

6.3 Basic key terms

We now list in this paragraph some introductory key terms. Their physical significance is following better clarified.

The term *rheology*, the study of matter flow and deformation, derives from the Greek words "rheo" (to flow) and "logia" (the study of). This name fully circumscribes the domain of this science, focused on the flow rules of solid-like liquids and liquid-like solids.

Rheological tests performed to identify the rheological law of a material are generally indicated as *rheometry*, comprising all experimental tests performed to discover the flow rule of a material.

In the analysis of rheological systems, the most important material parameter to characterize is called *viscosity*. The viscosity, generally indicated with the symbol h, is the resistance of the material to flow, or the ratio between the shear stress and the shear rate (following defined).

Rheometry tests are designed to simulate both a standard analysis, for the calculation of key parameters, and the actual processing and working conditions that the system undergoes.

To cover all the aspects of how the system behaves under specific conditions, rheometric tests use several kinds of mechanical loadings. A deep knowledge of the stress states in the material is strictly important to maximize the knowledge of the system gained by rheological characterization.

The most important amounts that can be calculated in a rheometric test are as follows:

The *shear stress*, that is the force applied to a body in tangential direction;
The shear strain, or the deformation arising after the application of a shear stress;
The shear rate, or the rate of (temporal) change in strain.

Results of rheometry tests are generally represented as plots of the measured physical quantity as a function of the applied external load. The most important plots obtained in rheometry are as follows:

The *flow curve*, that is the plot of the shear stress versus the shear rate.
The *viscosity curve*, or the plot of the viscosity versus the shear rate.

6.4 Rheological models

Rheological models are mathematical laws that generally describe the rheological behavior of materials and systems.

Such mathematical laws were developed to simplify the interpretation of experimental data from rheometry tests but also to calculate, in most cases, the rheological parameters characteristic of the system.

Empirical and semiempirical models were developed, based on the coupling of elastic and viscous elements, to catch both material characteristics.

Before going in detail, it is fundamental to understand a basic classification in Newtonian and non-Newtonian fluids, widely used in rheology.

6.5 Newtonian fluids (viscous liquids)

Isaac Newton formulated the first hypothesis on the resistance of a fluid to the motion promoted by its deformation [3].

In his hypothesis, Newton stated that "the resistance which arises from the lack of slipperiness of the parts of the fluid, other things being equal, is proportional to the velocity with which the parts of the liquid are separated from one another." This hypothesized "lack of slipperiness" was attributed to an internal friction generated during the fluid motion. Newton was the first who named the internal friction as *viscous* friction.

About 150 years later, George Stokes perfected the Newton's theory. He studied the forces involved in a system composed of two sliding plates separated by a fluid thin layer. Stokes discovered that the force required to slide the plates over each other is linearly proportional to the differential plate velocity and inversely proportional to the thickness of the fluid layer between them, the viscosity, and the overall surface of the plates involved in the contact with the fluid. Moreover, the stress profile across the thickness of fluid layer linearly changed, without any dependency on the spatial coordinate.

Those results indicated that the forces applied to the thin fluid layer were proportional to the rate of velocity change in the fluid across all the thickness. Fluids with this linear behavior are known as Newtonian fluids.

Generally speaking, a Newtonian fluid is a fluid characterized by a linear dependency of the shear rate upon the shear stress. The mathematical law used to describe the behavior of a Newtonian fluid contains a constant parameter, the so-called shear viscosity, that is used to relate shear stress and shear strain.

According to the given definition, in Newtonian fluids all the applied forces are proportional to the rates of velocity vector in the fluid. Thus, viscous stress tensors and the strain rate are related to each other by the constant viscosity tensor.

Now we consider a single element of fluid surrounded by other equal elements of fluid. Our single element of fluid undergoes a stress state given by the forces exerted by the motion of surrounding elements. Such external forces are, mainly, viscous forces. So, our single element of fluid gets gradually deformed over time.

Forces acting on this single element can be mathematically approximated with a first-order law describing the viscous stress tensor $\boldsymbol{\tau}$. (Note that the conventional bold symbol is used to indicate a tensor).

At the same time, the deformation of the single element of fluid over time can be mathematically approximated with a first-order time-dependent strain tensor $\boldsymbol{\gamma}$.

The strain tensor has a first-order time derivative called strain rate tensor $\dot{\boldsymbol{\gamma}}$. The $\dot{\boldsymbol{\gamma}}$ tensor describes the change with time of the deformation in the fluid element.

Elements of the tensor $\dot{\boldsymbol{\gamma}}$ are also expressed as velocity gradient ∇v.

Elements composing cited tensors depend on the spatial coordinates. Thus, tensors are expressed by 3×3 matrices in the selected coordinate system.

In Newtonian fluids, the stress tensor and the velocity gradient tensor are related by Eq. (6.1):

$$\tau = \mu(\nabla v) \tag{6.1}$$

$\boldsymbol{\mu}$ is a fourth-order tensor, and components are not affected by fluid velocity or its stress state.

A particular case of Newtonian fluids is represented by *isotropic* Newtonian fluids. "Isotropic" means that one or more physical properties of a material do not change when measured in different directions. In the case of isotropic Newtonian fluids, the properties that do not vary are mechanical properties, e.g., the elastic modulus, or the viscosity. Thus, the viscosity tensor is only composed of two real coefficients: the resistance to continuous shear deformation and the resistance to continuous expansion/compression.

On the other hand, also anisotropic Newtonian fluids can exist.

On the other hand, "anisotropic" means that physical properties of the material significantly change with the direction of measurements. In anisotropic Newtonian fluids, the tensor $\boldsymbol{\mu}$ relates the stress tensor with the deformation rate tensor. The last tensor contains the spatial derivatives of deformation rate as elements.

6.6 Non-Newtonian fluids

Unlike Newtonian fluids, the rheological behavior of a non-Newtonian fluid is not described by the linear Newton's law of viscosity. Moreover, viscosity depends on stress or, most commonly, on shear rate or shear rate history.

In other terms, the relation between the shear stress and the shear rate in non-Newtonian materials is not linear, and the value of the internal shear stress in the absence of applied shear rate could be non-zero.

Non-Newtonian fluids can also show a time-dependent viscosity, in some cases. For this reason, the viscosity is often not considered as a key parameter to adequately describe the behavior of these materials. To overcome this limitation, the behavior of such fluids is generally described by using the *apparent viscosity.*

Non-Newtonian fluids have a really complex behavior, characterized by several nonlinearities in the constitutive law. Thus, particular conditions are requested to perform a good rheological characterization of these materials. One widely used technique is the extensional rheometry, in which the applied external loads are purely extensional, with no shear loads. Another common technique is to perform the characterization in an oscillatory shear regime, in which the applied external loads follow a time-dependent law.

In summary, the non-Newtonian fluids are classified as follows:

Systems with a nonlinear viscosity;
Systems with a time-dependent viscosity.

In the inventory of systems with a nonlinear viscosity there are: viscoplastic (or Bingham) fluids, shear thinning (or pseudoplastic) fluids, shear thickening (or dilatant) fluids, and generalized non-Newtonian fluids.

In *viscoplastic fluids*, the relationship between shear stress–shear rate is linear but the intercept of the curve, given by the value of the shear stress for a null shear rate, is not null. This characteristic indicates that a small but finite yield stress is required to activate the fluid flow.

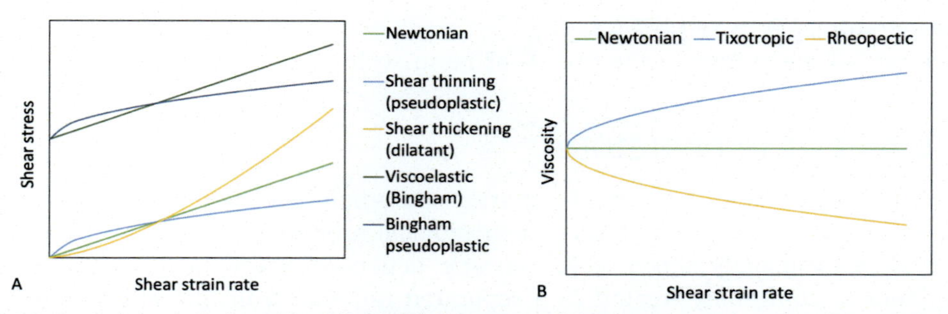

FIGURE 6.2 Classical rheological models of fluids: (A) flow curves of the most common classes of rheological fluids; (B) viscosity curves of Newtonian fluids versus time-dependent viscosity fluids.

The main characteristic of *shear thinning fluids* is that the apparent viscosity decreases with an increased stress. This nonlinear behavior indicates that the more is the applied external force, the simpler is the flow of the material.

The opposite behavior of shear thinning is the shear thickening. In *shear thickening fluids,* we can find an increasing apparent viscosity with increased applied stress. In other words, shear thickening materials increase their resistance to flow under high stresses, assuming a solid-like behavior.

In *generalized Newtonian fluid,* the viscosity is constant, the shear stress is a discrete or punctual function of the shear strain rate, and the internal stress state does not depend on the shear strain history.

Systems with a time-dependent viscosity are defined as: thixotropic fluids and rheopectic fluids.

In *thixotropic fluids,* the viscosity decreases with the external load application duration. It means that thixotropic materials have a tendency to become more fluids after a prolonged applied shear stress or a prolonged shear strain.

In *rheopectic fluids,* the apparent viscosity increases with the duration of the external load, thus they show a "more solid behavior," becoming less fluid, after prolonged applied loads.

The shear stress versus shear strain rate curves of non-Newtonian fluids with a nonlinear viscosity are illustrated in Fig. 6.2A, while the viscosity versus shear strain rate curves of non-Newtonian fluids with a time-dependent viscosity are reported in Fig. 6.2B.

6.7 Conditions affecting the rheological properties of materials

Rheological properties can be significantly affected by some specific conditions. In addition to the kind of external applied load and the loading frequency, the most relevant properties affecting rheology are as follows:

Temperature;
Structure of the system at the micro- or nano-scale.

6.8 Effect of temperature

The temperature is one of the most important parameters affecting all materials properties, and it has really strong effects on rheological properties.

The viscosity usually shows a very strong dependency on the temperature. This dependence is commonly treated as a separated function from the effect of all other parameters (e.g., the shear rate or the shear frequency). As a general rule, the viscosity depends on the temperature as:

$$\eta(\dot{\gamma}, T) = \eta(\dot{\gamma}) \cdot f(T) \tag{6.2}$$

and the function *f(T)* is defined for each material or system studied.

One of the simpler function form *f(T)* widely used is:

$$f(T) = \exp(-\alpha \cdot \Delta T) \tag{6.3}$$

Here, the parameter *a* is called temperature sensitivity of the viscosity, while the temperature difference is calculated between the working temperature and the temperature at which viscosity is known.

6.9 Effect of the system structure at the micro-/nano-scale

The structure of the system has a particular importance in macromolecular formulations, e.g., in polymer solutions or melts.

Basically, mechanical properties of a polymer at the macromolecular scale are determined by the intramolecular and intermolecular interactions of the polymer.

Polymeric chains are obtained by covalently attaching a large number of repeating units. Intramolecular forces acting in polymers are prevalently related to the C—C bond, while bonds between heteroatoms (e.g., C—N, C=N, C—O, C=O, C—S, and so on) are less frequent.

The energy of the C—C bond is ~350 kJ/mol, and this value leads to a theoretical stiffness of 300—400 GPa. Anyway, polymers show a considerably lower stiffness in experimental tests than the theoretical value. The difference between theoretical and actual stiffness is generally attributed to the macromolecular misalignment. This observation leads to consider also the intermolecular interactions as important factor in polymer strength. Intermolecular interactions are mainly accounted to be Van Der Waals forces, and their entity is inversely proportional to the six power of the distance between atoms. Thus, in the presence of solvents, plasticizers or when heated, the distance between two macromolecules increases, the free volume increases, thus intermolecular interactions and consequently stiffness decrease.

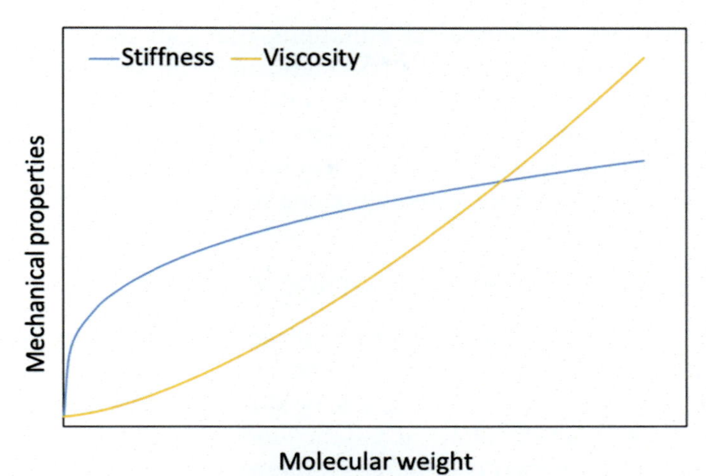

FIGURE 6.3 Trends of stiffness and viscosity of materials by increasing the polymer molecular weight.

The degree of polymerization, or the number of repeating units per macromolecule, is another factor affecting macromolecular interactions.

Polymers with a low degree of polymerization are liquid or very soft at room temperature, while increasing the degree of polymerization, they gradually become stiffer at the same temperature and with the same additives.

However, while the stiffness rapidly tends to a plateau, the viscosity steadily increases with molecular weight, generally with a power law (Fig. 6.3).

In systems containing macromolecular coils, soft particles, or vesicles, rheological properties are also affected by deformation of these components.

As a general rule, deformable structures dispersed in a liquid phase tend to get deformed in the direction of load application, while nondeformable structures tend to get aligned in the load direction.

In both cases, the most general rule is that systems become shear thinning during load application, while thixotropy dominates when loading is stopped, and systems tend to return thicker with time.

6.10 Applied rheology in elastomers, blends, and composites thereof

Rheological properties of elastomeric materials, their blends and composites can be assessed by means of several different tests. The selection of the proper characterization is given upon the sample properties and the information needed.

Experimental data gathered from rheometry are able to identify an appropriate mathematical model for the studied material, or in other words the *constitutive law* of the system. The constitutive law correlates the shear stress and the shear strain, or the shear strain rate, with each other.

Constitutive laws are mathematical equations that can analytically describe the material response, in terms of internal stress state, after the application of the external load. Thus, the main result of this calculation is the stress tensor.

The calculated stress tensor generally depends on the direction and orientation of the applied force in respect to the surface of application of the external load. To be defined, the velocity gradient needs that the direction along which the velocity varies is specified.

6.11 Static versus dynamic rheological tests

Rheological tests can be static, based on the application of a constant load, or dynamic, in which the applied load follows a cyclic function (e.g., sinusoidal).

In the case of static tests, a constant stress, or strain, is applied to study the creep of materials and to understand the effect of a static load for long-term applications. In the same kind of static test, the load can be removed and the materials studied in order to evaluate the molecular relaxation.

In steady shear flow analysis (Fig. 6.4A), the material lies between two flat plates. In this setup, one plate (generally, the top plate) can move and the other one (generally, the bottom plate) is fixed. The fluid is sheared by the motion of the moving plate in the direction of the motion.

If D_x is the displacement of the top plate in the x direction and is the distance between the two plates, the shear strain in the xy plane, indicated by xy, is given by:

$$\gamma_{xy} = \frac{D_x}{\delta} \tag{6.4}$$

Indicating with u_x the velocity of the top plate in the x direction, the velocity gradient $\dot{\gamma}$, also called shear rate, is given by:

$$\dot{\gamma} = \frac{dv_x}{dy} = \frac{u_x}{\delta} \tag{6.5}$$

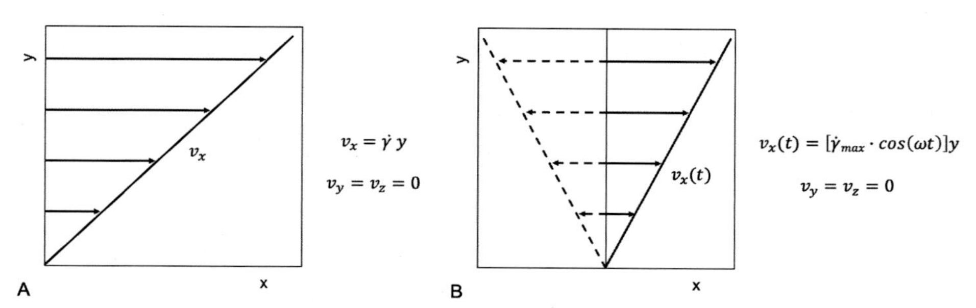

FIGURE 6.4 Schematics of: (A) steady shear flow analysis, the thin fluid layer is comprised between two flat plates and deformed, and the deriving velocity profile along the x direction is a function of the shear strain rate; (B) oscillatory measurements, in which the velocity profile along the x axis depends on the time and the frequency of the applied load.

The velocity profile across the fluid layer thickness v_x arises from the stretching of the fluid given by the relative motion of plates. Considering a linear velocity profile (see Fig. 6.4A), the velocity profile function is $v_x = v_0$ at $x = 0$ at the interface with the resting plate, and $v_x = D_{Rv0}$ at $x =$ at the interface with the moving plate, where D_R is the final displacement of the top plate.

In this test setup, the velocity profiles in directions y and z are zero.

The deformation between $x = 0$ and $x =$ is indicated with xx, and it is calculated as:

$$\gamma_{xx} = \frac{(D_R v_0)\Delta t}{\delta} \tag{6.6}$$

while the deformation in the y direction is given by $yy = -xx$.

The deformation is not a scalar. It is because this amount has multiple components in cartesian coordinates. Thus, it is described by the tensor:

$$\boldsymbol{\gamma} = \gamma_{ij} = \begin{bmatrix} \gamma_{xx} & \gamma_{xy} & \gamma_{xz} \\ \gamma_{yx} & \gamma_{yy} & \gamma_{yz} \\ \gamma_{zx} & \gamma_{zy} & \gamma_{zz} \end{bmatrix} \tag{6.7}$$

The shear strain rate tensor $\dot{\boldsymbol{\gamma}}$ is the time derivative of:

$$\dot{\boldsymbol{\gamma}} = \frac{\partial \gamma_{ij}}{\partial t} = \begin{bmatrix} \dot{\gamma}_{xx} & \dot{\gamma}_{xy} & \dot{\gamma}_{xz} \\ \dot{\gamma}_{yx} & \dot{\gamma}_{yy} & \dot{\gamma}_{yz} \\ \dot{\gamma}_{zx} & \dot{\gamma}_{zy} & \dot{\gamma}_{zz} \end{bmatrix} \tag{6.8}$$

Under the effect of the shear strain, a shear stress state with the same direction of the relative plate motion is generated within the material. Such shear stress state is another tensor, and its components are indicated with ij. The shear stress state generated by the applied deformation is also called deviatoric stress. The calculation of components of the deviatoric stress tensor involves the viscosity function. The mathematical equation used in this calculation is:

$$\tau_{ij} = -\eta \cdot \dot{\gamma}_{ij} \tag{6.9}$$

Also the overall stress state tensor comprises some additional components orthogonal to the surface over which the external load is applied.

The normal components of the stress represent the hydrostatic pressure in the fluid. There are two normal components, generally indicated with N_1 and N_2, that are calculated as the difference between tangential stresses in different directions:

$$N_1 = \tau_{xx} - \tau_{yy} \tag{6.10}$$

$$N_2 = \tau_{yy} - \tau_{zz} \tag{6.11}$$

These components N_1 and N_2 are also called primary and secondary normal stress differences, respectively.

Normal components of the stress N_1 and N_2 are related to the shear strain rate by means of the following equations:

$$N_1 = -\psi_1 \cdot \dot{\gamma}^2 \tag{6.12}$$

$$N_2 = -\psi_2 \cdot \dot{\gamma}^2 \tag{6.13}$$

where 1 and 2 are the primary and the secondary normal stress coefficient, respectively.

Parameters 1 and 2 are function of the shear rate and are experimentally evaluated.

In Eqs. (6.12) and (6.13), normal stress components are proportional to the square of the shear strain rate in order to prevent changes in the shear direction, in the case of negative *y*, or by changing the shear direction through the test.

Dynamic tests study the material response, in terms of stress or strain, under the effect of a cyclic load that can be applied as a steady shear or as an oscillatory regime. In addition to variable stresses or strain, also temperature and frequency can be cyclically varied.

In dynamic rheological measures in oscillatory regime (Fig. 6.4B), the externally imposed shear varies with time, following a predefined cyclic law. The oscillatory regime is useful in the calculation of the phase angle between the imposed shear strain rate and the measured shear stress.

The linear viscoelastic region (LVR) shows a linear correlation between shear stress and shear strain. It involves that the stress response function perfectly fits the imposed shear strain function. Also in this test, velocity profiles in directions *y* and *z* are zero.

In dynamic tests, the oscillatory velocity field is given by:

$$v_x(t) = [\dot{\gamma}_{max} \cdot \cos(\omega t)] \cdot y \tag{6.14}$$

where $\dot{\gamma}_{\max}$ is the maximum velocity gradient, and ω is the frequency of the cyclic applied shear strain.

The ratio between $\dot{\gamma}_{\max}$ and ω is the maximum value of the shear strain max. According to the previous definition of $v_x(t)$, the shear stress results:

$$\tau_{xy} = \tau_{\max} \cdot \cos(\omega t - \varphi) \tag{6.15}$$

where max is the maximum shear stress value registered in the test, and φ is the phase shift between the imposed strain and the measured stress.

The oscillatory stress measured in the test is a complex function in which one term is in-phase with the velocity and the other term is out of phase (orthogonal).

Now we define the complex viscosity * as follows:

$$\eta^* = \eta' - i\eta'' \tag{6.16}$$

where ′ and ″ are the two components used to define the complex viscosity. The complex viscosity is experimentally calculated as the ratio between values over time of the shear stress and the shear strain:

$$\eta^* = \frac{\tau_{xy}(t)}{\dot{\gamma}_{xy}(t)} \tag{6.17}$$

Combining Eq. (6.9) and Eq. (6.15), the shear stress becomes:

$$\tau_{xy} = -\eta' \cdot \dot{\gamma}_{max} \cos(\omega t) - \eta'' \cdot \dot{\gamma}_{max} \sin(\omega t) \tag{6.18}$$

If the shear strain rate approaches zero and the applied shear frequency is significantly low,′ is approximatively the values of viscosity calculated in steady shear flow regime. This amount is generally indicated as dynamic viscosity.

Now we define the complex shear modulus G^* as:

$$G^* = G' - iG'' \tag{6.19}$$

Here, G' and G'' are, respectively, the storage and the loss modulus. Such two components are related to the dynamic viscosity coefficients through the frequency of the applied shear strain:

$$G' = \omega\eta' \tag{6.20}$$

$$G'' = \omega\eta'' \tag{6.21}$$

Reformulating Eq. (6.16) as a function of the complex shear modulus, we obtain:

$$\tau_{xy} = -G' \cdot \gamma_{\mathrm{max}} \cos(\omega t) - G'' \cdot \gamma_{\mathrm{max}} \sin(\omega t) \tag{6.22}$$

We need to analyze the relationship between these physical amounts and the stress response upon the application of external loading to understand the importance such parameters.

G' linearly correlates with the component of the stress in-phase with the applied strain, and then it is strictly related to the recoverable elasticity of the material. On the contrary, G'' correlates to the orthogonal stress response to the applied strain, resulting in the dissipative energy losses in the material. Accordingly, the G' function is pretty similar to the shear modulus calculated for an ideal rubbery material, in which the stress response is perfectly in phase with the applied shear strain. Conversely, the G'' function has a physical meaning similar to that of the shear modulus in a purely viscous material, in which the elastic recovery is null and the energy resulting from applied external forces is fully dissipated by the viscous flow of molecules.

Components G' and G'' of the complex shear modulus generally depend on the frequency of the applied strain function, and this occurs more in macromolecular materials. Some materials tested under a frequency scan show a point called *crossover.* At the crossover, G' and G'' have the same value. In materials crossover where the G'/G'' crossover occurs, one of two modules dominates the other at low frequencies and vice versa at high frequencies.

Steady shear loads are generally used to study the rheology of the elastomer matrix before cross-linking. On the other hand, oscillatory tests are useful to understand the effects of cross-linking kinetics. Dynamic tests are particularly interesting because they can give several information on viscoelastic properties of the matrix. The only restriction to the application of dynamic tests is the identification, prior to the characterisation, of the linear viscoelastic range, in order to be sure that the molecular structure of the material does not change under the effect of the load and the material response does not depend on the magnitude of the load.

Rubbery materials can show a variety of rheomechanical properties, on the basis of their physical state but also at different production steps. So that, rheological studies can have their main applications in the analysis of the melting point of the pure rubber, the scorch time, and the cross-linking kinetics, the flow behavior during the molding of the manufact, the interface interactions in the presence of fillers, the Mooney viscosity, and several other properties.

6.12 Laboratory tests and instrumentations

The selection of the most useful experimental setup used to test rheological properties of materials depends on the macroscopic properties of the specimen.

The instrumentation used to evaluate rheological parameters of soft materials are known as rheometers.

A rheometer applies several kinds of mechanical loads to trigger the flow and, at the same time, it registers the response of the material.

Concerning applied loads, a rheometer that can control the applied shear stress or the applied shear strain (shear rheometer) or the extensional stress or extensional strain (extensional rheometers). The loading application is generally performed by applying constant or time-dependent shear rate functions.

Rheometry enables the calculation of all those functions and parameters, such as the apparent viscosity, the shear stress, the complex shear moduli, as a function of the shear rate, the time, or the frequency of the entity of the applied load.

Experimental techniques to measure the rheological properties of viscoelastic fluids are continuously under development. It is because reliable experimental data are vital to evaluate the accuracy of a constitutive equation to predict the actual properties of a fluid.

Moreover, as previously stated, a comprehensive understanding of rheological properties is useful to determine the optimum processing conditions (e.g., to control polymerisation reactions) and in the setup of desired physical/mechanical properties of the final product.

For all such reasons, several experimental methods have been developed. Main techniques currently used are based on steady-state or oscillatory cone-and-plate rheometers, steady-state shear flow in capillary rheometers, and elongational flow properties.

While the number of experimental apparatus for rheological analysis is really high, only a few of these are used for the analysis of elastomers. The most common rheometers used for elastomers are the cone-and-plate rheometer, the capillary viscometer, and the Mooney viscometer.

6.13 Cone-and-plate rheometer

This instrumentation studies the flow of an incompressible fluid placed between a cone and a plate, in which the cone has a small vertical angle, and it is centered over the horizontal flat plate. The space between the cone and the plate contains the fluid to be measured. Generally, the flat-bottom plate is fixed while the cone rotates around its axis. The cone rotation can be constant with a fixed rotation velocity, modulated on material response in the case of shear-thinning or shear-thickening materials to maintain a fixed deformation, ramped, or oscillatory. This aspect is crucial for material characterization, and it depends on what the operator needs to measure. A steady-state measure can be used to evaluate the rheological model of the material, the viscosity, and the elastic properties of the sample, while an oscillatory analysis enables the calculation of the viscoelastic parameters (e.g., G^*, G' and G'', $\tan(\delta)$, *, η' and η'').

6.14 Capillary viscometer

This apparatus is composed of a capillary in which the polymer, melt, or solution flows. The sample is loaded in a reservoir at the top of the capillary channel. The flow of the sample through the channel can occur under the effect of the gravity or by a plunger that pushes the sample with a known force F.

In gravimetric capillary rheometers, the operator measures the time necessary to the material to flow across the channel. On the basis of capillary section and length and time, the relative viscosity is calculated.

In plunger-type capillary rheometers, the force F applied to the fluid generates a longitudinal shear stress T_z on the fluid, given by $T_z = F/A$. This value is related to the pressure inside the capillary, and it can be used to calculate the flow curve by using the Bagley equation [4].

Both techniques use a small amount of sample, but measurements are time-consuming and allow calculating only shear viscosity.

6.15 Mooney viscometer

This apparatus is used to perform the most common rheological measure for uncured compounds. The Moony viscometer is composed of a knurled knob that rotates in a cavity filled with the liquid rubber. The sample is generally heated to be more fluid. The rotor starts rotating when the sample reaches a minimum temperature, depending on the compound. At the beginning of the test, the viscosity registered is generally very high and tends to decrease by temperature increasing, until values reach a plateau representing the characteristic of the sample. The apparatus gives an indirect measure of the torque in an arbitrary unit called Mooney unit. More viscous compounds have higher Mooney units.

6.16 Constitutive rheological models

Each class of materials has its own constitutive rheological model. The knowledge of the constitutive rheological model is really important for several practical industrial applications, and rheological analysis is fundamental to understand such aspect.

In the analysis or rubbers, their blends, and their composites, we can use rheology to study the uncured rubber melt, the cured rubber, blends, and the effect of fillers in the rubbery matrix. Each component or system has a typical constitutive rheological model.

6.17 Uncured rubber melts

The uncured rubber melt has a high viscous behavior and can be studied by means of the generalized Newtonian fluid in the low shear rate region.

The constitutive rheological law of Newtonian fluids has only one parameter, the viscosity. For this class of materials, the constitutive equation calculated by means of rheological tests is able to describe and predict the full flow behavior in every laminar flows. For Newtonian fluids, the constitutive law is:

$$\boldsymbol{\tau} = -\eta \cdot \dot{\boldsymbol{\gamma}} \tag{6.23}$$

Here $\boldsymbol{\tau}$ and $\dot{\boldsymbol{\gamma}}$ are both tensors.

As already reported in Eq. (6.9), the tensor, also called deviatoric stress tensor, is given by:

$$\boldsymbol{\tau} = \tau_{ij} = \begin{bmatrix} \tau_{xx} & \tau_{xy} & \tau_{xz} \\ \tau_{yx} & \tau_{yy} & \tau_{yz} \\ \tau_{zx} & \tau_{zy} & \tau_{zz} \end{bmatrix} \tag{6.24}$$

While, according to Eq. (6.1), we can define the strain rate tensor $\dot{\boldsymbol{\gamma}}$ as:

$$\dot{\boldsymbol{\gamma}} = \nabla\mathbf{v} + (\nabla\mathbf{v})^T \tag{6.25}$$

Here, ∇ is the gradient operator, **v** is the velocity gradient tensor, and the superscript T is the transpose operator. The velocity gradient is given by:

$$\nabla \mathrm{v} = \begin{bmatrix} \frac{\partial v_x}{\partial x} & \frac{\partial v_x}{\partial y} & \frac{\partial v_x}{\partial z} \\ \frac{\partial v_y}{\partial x} & \frac{\partial v_y}{\partial y} & \frac{\partial v_y}{\partial z} \\ \frac{\partial v_z}{\partial x} & \frac{\partial v_z}{\partial y} & \frac{\partial v_z}{\partial z} \end{bmatrix} \tag{6.26}$$

Components of the strain rate tensor (Eq. 6.25) are defined as:

$$\dot{\gamma}_{ij} = \frac{\partial v_i}{\partial x_j} + \frac{\partial v_j}{\partial x_i} \tag{6.27}$$

where v_i and v_j are the cartesian components of the velocity vector $\boldsymbol{v}$, and x_i and x_j are the components of the position vector $\boldsymbol{x}$. Combining Eqs. (6.27) and (6.9) for a steady shear flow, we obtain:

$$\tau_{xy} = -\eta\dot{\gamma}_{xy} = -\eta\frac{\partial v_x}{\partial y} \tag{6.28}$$

in which the component v_y is zero for this test (see the previous Section).

Once calculated, Eq. (6.28) can be applied to experimental data in the prediction of the Newtonian material behavior in each laminar flow.

The Newtonian constitutive rheological law can be generalized to be applied to non-Newtonian fluids. The generalization introduces a function of the shear rate as the viscosity parameter. This generalization needs to calculate the three invariants of the deformation rate tensor, and the second is the most important invariant. While the first invariant is empty for incompressible fluids, the second invariant of the deformation rate tensor contains both hydrostatic pressure and shear stresses. Finally, the third invariant is empty in shear-governed flows.

The second invariant of the shear rate tensor can be calculated as follows:

$$J_2 = \sum_i \sum_j \dot{\gamma}_{ij} \cdot \dot{\gamma}_{ji} \tag{6.29}$$

The viscosity is calculated as the square root of the halved second invariant, resulting:

$$\eta = \sqrt{\frac{1}{2} J_2} = \sqrt{\frac{1}{2}\sum_i \sum_j \dot{\gamma}_{ij} \cdot \dot{\gamma}_{ji}} \tag{6.30}$$

Combining Eqs. (6.28) and (6.30), we can calculate the constitutive equation in laminar flow for the majority of generalized Newton fluids.

There are some generalized Newton fluids for which the use of the proposed model is significantly hard to apply, mainly because of the difficult mathematical implementation. It is the case of transient flows or in fluids with not negligible normal stresses. To overcome this limitation, some semiempirical generalized flow laws, with a reduced number of parameters, were developed. Parameters of such semiempirical models are generally calculated by fitting experimental rheometric data.

One of the most used generalized flow models for the prediction of the viscosity is:

$$\eta = K\dot{\gamma}_{xy}^{(n-1)} \tag{6.31}$$

where K and n are fitting parameters, evaluated from experiments.

A simple two-parameter power law is easy to manage, and it is also reliable enough to be applied in the analysis of a large number of systems, in particular for polymer solutions.

However, this model fails for low flow rate regimes. In such cases, the power law model can be replaced by the Carreau model:

$$\frac{\eta - \eta_\infty}{\eta_0 - \eta_\infty} = \left[1 + \left(\lambda\,\dot{\gamma}_{xy}\right)^2\right]^{\frac{n-1}{2}} \tag{6.32}$$

In this equation, η_∞ and η_0 are viscosities at high and low shear rate respectively, is a fitting parameter calculated at the shear rate of interest, n is a fitting parameter of the overall set of data. The empirical viscous Carreau model has been widely used to analyze uncured rubber melts, mainly for its simplicity.

6.18 Elastomer blends

Elastomer blends can be broadly classified as miscible blends, partially miscible blends, and nonmiscible blends. Such differences are mainly based on the chemical similarities between macromolecules. When dissimilarities occur, the complexity of the system derives from some differences of the thermodynamics states of the blend. On the basis of this classification, the rheological behavior can be strongly different and needs to be studied with a variety of experimental approaches.

Completely miscible blends of two or more elastomers are relatively poorly interesting for technological and industrial applications. The rheological behavior of completely miscible elastomer blends can be tackled with a molecular approach.

Conversely, two- or multiphase elastomer blends are more interesting for industry. The rheological analysis of such systems should be based on more complex phenomenological studies, such as fluid mechanics. As a general rule, rheological properties of a nonmiscible elastomer blend are nonadditive function of the melt flow properties measured for the homopolymers composing the blend.

It is often quite difficult to deeply understand the rheological behavior of polymer blends, due to their complexity. The rheological properties of nonmiscible blends completely depend on the chemical composition and properties of the single components, their morphology, and the morphology of the blend at a microscale level, interactions between phases, interface forces between domains, and the strain history during the blend production.

Measurements obtained in the linear viscoelastic domain, e.g., in oscillatory regime and with small amplitude of deformations, are highly useful for the comprehension of some rheological characteristics of the blend, and the multiphase structure of the mixture is expected to be maintained for small strains. For this reason, rheological measurements with very small strains are used for the evaluation of the complex modulus, enabling the description of the equilibrium properties of the blend.

On the contrary, the evaluation of the nonlinear viscoelastic domain, and related measurements in this range, is very complex while being fundamental for a basic rheological analysis, due to the loss of the initial microstructure of the material.

At the same time, rheological measures at high temperature are critical. It is because the high temperature can lead to the coalescence of the phases and to the rearrangement of macromolecules.

6.19 Elastomer composites

Nanoparticles used as rubber fillers can improve the mechanical performances of elastomers and broaden their applications. Carbon black is widely used as filler thanks to its low price and its capability to form hierarchical structures in soft elastomeric matrices.

Vulcanized elastomers filled with carbon black find applications in automotive and transport or as damping materials. The interest in such elastomers for cited applications is related to their nonlinear viscoelasticity. When vulcanized and filled elastomers are subjected to cyclic loadings, their storage modulus G' rapidly decreases with the increase of the strain amplitude 0. The decrease of G' is followed by an increase of the loss modulus G''.

Usually, the G'' function has a maximum. The G'' maximum depends on the filler fraction. This particular behavior, with a strain amplitude-dependent G' in filled elastomers, was described, for the first time, as Payne effect [5,6].

The Payne effect can be interpreted according to several theories. Major theories to explain the Payne effect are based on the agglomeration/deagglomeration of nanoparticle aggregates [7], disruption/reconstruction of the hierarchical filler structure [8,9], matrix debonding from fillers [10], formation of a thin glassy layer around fillers [11], strain-softening of elastomer matrix around fillers [11]. Concerning the last phenomenon, damages occurring at the interfacial region between matrix and filler and the entanglements present in the cross-linked matrix can explain the strain softening.

6.20 Conclusions

Rheology is the science of fluid flow, and it mainly studies the behavior of plastic and non-Newtonian fluids. Rubbers and elastomers are considered fluid of interest for rheology.

The rheological analysis of a fluid is often quite complex and needs the knowledge of several experimental parameters. In the case of rubbers, this analysis can be performed on rubber melts, to support the industrial production processes, and on the final manufacts, to predict the performances of the product.

It is clear that rheology is a helpful tool for the development of elastomer materials. Although the accurate measure of rheological properties seems to be hard to be done, beneficial aspects of this characterization are important and should be advisable to be performed in view of the production process optimization.

References

[1] M. Mooney, Theory of the Non-Newtonian rheology of raw rubbers consisting of supermolecular rheological units, J. Appl. Phys. 27 (7) (1956) 691–696.

[2] G.S. Blair, On the use of power equations to relate shear-rate to stress in non-Newtonian liquids, Rheol. Acta 4 (1) (1965) 53–55.

[3] I. Newton, Philosophiæ Naturalis Principia Mathematica, 1687.

[4] E.B. Bagley, The separation of elastic and viscous effects in polymer flow, Trans. Soc. Rheol. 5 (1) (1961) 355.

[5] A. Zhao, X.Y. Shi, S.H. Sun, Insights into the Payne effect of carbon black filled styrene-butadiene rubber compounds, Chin. J. Polym. Sci. 39 (2021) 81–90.

[6] J. Zou, X. Wang, Rheological responses of particle-filled polymer solutions: the transition to linear-nonlinear dichotomy, J. Rheol. 65 (1) (2021).

[7] A.P. Meera, S. Said, Y. Grohens, S. Thomas, Nonlinear viscoelastic behavior of silica-filled natural rubber nanocomposites, J. Phys. Chem. C 113 (2009) 17997–18002.

[8] G.J. Kraus, Mechanical losses in carbon black filled rubbers, J. Appl. Polym. Sci. 39 (1984) 75–92.

[9] J. Yang, C. Han, Dynamics of silica-nanoparticle-filled hybrid hydrogels: nonlinear viscoelastic behavior and chain entanglement network, J. Phys. Chem. C 117 (2013) 20236–20243.

[10] S.S. Sternstein, A.-J. Zhu, Reinforcement mechanism of nanofilled polymer melts as elucidated by nonlinear viscoelastic behavior, Macromolecules 35 (2002) 7262–7273.

[11] H. Montes, F. Lequeux, J. Berriot, Influence of the glass transition temperature gradient on the nonlinear viscoelastic behavior in reinforced elastomers, Macromolecules 36 (2003) 8107–8118.

7

Morphological characteristics of elastomer blends and composites

A.V. Kiruthika

DEPARTMENT OF PHYSICS, SEETHALAKSHMI ACHI COLLEGE FOR WOMEN, KARAIKUDI, TAMIL NADU, INDIA

List of abbreviations

CNT carbon nanotube
NR Natural rubber
PALF Pineapple leaf fiber
PLA Poly lactic acid
rCR recycled chloroprene rubber
SBR Styrene-butadiene rubber
SR Silicone rubber
TM Tensile modulus
TS Tensile strength
vCR virgin chloroprene rubber

7.1 Introduction

Elastomers are elastic materials belonging to a member of polymeric family. These are typically amorphous with lightly cross-linked and flexible in nature. Elastomers are ideal polymers with weak intermolecular forces between the polymeric chains that allow them to have high degrees of freedom. Moreover the glass transition temperature was below the room temperature [1]. Elastomers exhibited low Young's modulus and very high elongation at break, often termed as rubber. The molecules in elastomers stretched, some crystalline regions occur within the structure. The cross-links provide the ability to regain the molecules into original configuration even after the removal of applied force. Elastomers have large toughness, abrasion resistance, permeability and also have better adhesion behavior to various fillers. These properties layout the elastomers in day-to-day applications. Even, they could modified by blending with other polymers to improve their hardness and strength and make them into useful materials. The industrial applications of elastomers as electrical insulators, tubes, belts, and tires, whereas in consumer type as sporting goods and casual wears [2]. It is used in nuclear industry as for

Elastomer Blends and Composites. https://doi.org/10.1016/B978-0-323-85832-8.00013-4

electrical wire insulation and sealing applications. Among the various commercially available materials, elastomers are widely used due to its renewability, better specific properties, compatibility, recyclability, and cost-effectiveness [3]. Based on the applications, the elastomers are classified in different grades. More than 10 types of elastomers are there(Rubber Type 61 to Type 70), depending upon the oil, heat, and weather resistance. Type 61 rubber is inexpensive, easiest processing methods, good mechanical properties, and elastomers that belong to this type are natural rubber (NR), polyisoprene rubber (IR), styrene butadiene rubber (SBR), etc. Type 62 has good ozone and weather resistance; they are IIR, CIIR, BIIR. Nitrile rubber (NBR) is of rubber type 63. Type 70(EPDM, EPM) is similar to type 62, in addition to that it has heat and poor oil resistance [4].

7.2 Morphology

Phase morphology played an important role in the evaluation of polymer blend characteristics. Optical microscopy, TEM, SEM, AFM, FESEM are the various characterization techniques that are employed to evaluate the morphological studies of elastomeric blends and its composites. The surface roughness, interfacial adhesion between fiber and matrix, fiber dislocation, microstructure, fracture surface of blends were analyzed by using the above techniques.

7.2.1 Optical microscopy(OM)

Among the several microscopic methods, OM is the 2D imaging technology and the oldest one, which provides information on a length scale of 10 μm. It is a technique used in polymer engineering research to analyze the samples, because it is a very simple method that does not require much sample preparation. OM uses converging lenses and visible light to get magnified images, thus it is also called as light microscopy. It helps to perceive various properties including granule size, voids, shape, microstructure, failure analysis of polymer samples, rubber blends and its composites. Further, polarized light and phase contrast microscopy are used for elastomer blend analysis. Also it is difficult to characterize the difference, as the light microscopy showed very little contrast among the different phases in elastomer blends. Because of the contrast in common light microscopy, a special form of OM was called phase contrast OM(PCOM), which has typically taken advantages of variation in refractive indices between one or more phases. Two types of phases could be analyzed in elastomers via OM. In general, the lower concentration of elastomers between dispersion phase and higher concentration as continuous phase, where for equal concentration, the blend has in continuous phases. The surface analogy of chitosan(CS) dispersion on PLA and PLA/ENR composites was illustrated by OM. The occurrence of white particles was noticed on the composite surface for PLA/CS and PLA/CS/ENR composites. These composites had white spots on the surface(Fig. 7.1). Compared to PLA/CS, the other composites have surface roughness with many number of small holes throughout the composites. This result was well

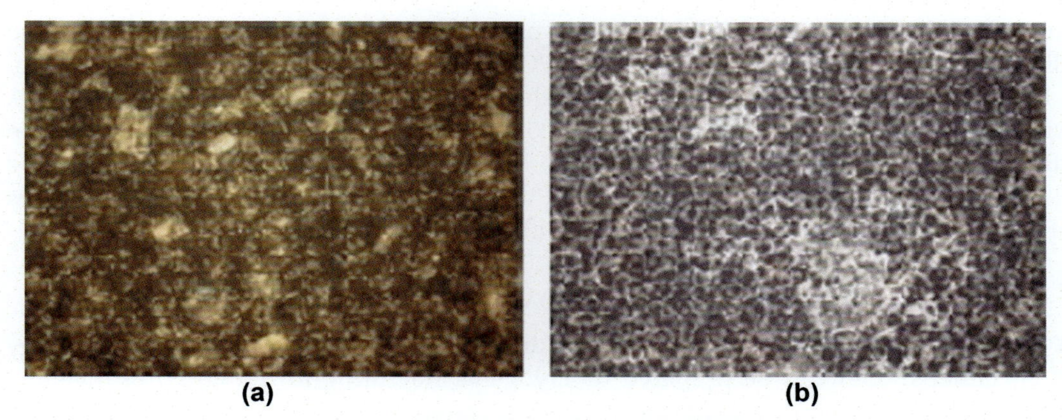

FIGURE 7.1 Optical microstructures (A) PLA/CS, and (B) PLA/CS/ENR [5].

coincided with the SEM analysis in which the agglomeration of CS particles was clearly obtained [5]. The OM studies of NR/BIIR blends filled CNTs with and without the self-healable material based on butyl imidazole at 35°C and 80°C were investigated. Blends with imidazole(at 35°C) exhibited the largest online conductance, which was found by better CNT dispersion, whereas reverse the case at 80°C without imidazole, i.e., smallest online conductance. Many undispersed CNT agglomerations, size of 40 μm, and a macrodispersion (15%) were noticed for the blends without imidazole. In the case of blends with imidazole (at 35°C), a reduction in macrodispersion (3%) with greater online conductance (4.6×10^{-3} S/cm) was noted. A significant decrement in agglomeration with macrodispersion (8%) and conductance as 4.6×10^{-11} S/cm was observed for blends at 80°C [6]. Ghosh et al. [7] prepared the blends from SR/fluoro rubber(FR) and studied its morphological characteristics with the effect of ground rubber vulcanizate powder(VP). From the OM analysis, the SR vulcanizate powder has the average particle size of about 33 μm, whereas for FR it was 1.0 μm. Also the obtained result was in better coincidence with AFM studies(tapping mode). A significant changes in surface roughness were observed in the SR/FR blends due to the vulcanizate powder. Studies have shown that the hysteresis loss and mechanical properties results also follow the surface roughness order. Blending of rubber (EPDM, NBR and NR) with the polystyrene(PS) polyaniline(PANI)[PS@PANI] core–shell approach was determined with individual mixing of PS, PANI, and rubber. Conductivity, electromagnetic interference (EMI), shielding analysis (30 dB: 188 GHz), thermal and mechanical properties were higher for core–shell blends than the individuals. PS@PANI has uniform dispersion in the individual rubber, and the region of white and dark spots was observed for PS/rubber and PANI, respectively. In PS/PANI-loaded rubber blends, it was observed that an irregular dispersion with agglomeration of PANI was noticed [8]. This blends provided an effective alternative as EMI shielding material for commercial application purposes. Jianhua et al. [9] proposed a new mechanical blending techniques to prepare the methyl vinyl silicone rubber(MVQ) composites with low-melting-point alloy(SnBi). From the OM analysis, the alloy particles were well dispersed in the MVQ matrix without any physical contact. Also, a large number of alloy particles were transferred from spherical to wrinkled structures with the increment in tensile strain from 100% to 250%. Thus, the average length of the alloy particles has increased, this is due to larger strain accompanied by large tensile stress. The surface and volume resistivity of the prepared composites had greater values of $9 \times 10^7\ \Omega$ and $3 \times 10^9\ \Omega$m respectively with a larger strain of 250%.

7.2.2 Scanning electron microscopy(SEM)

It is a versatile technique used to study the direct evaluation on the surface morphology of specimens. SEM can be prepared by cryogenic fracturing in the liquid nitrogen or selective solvent etching or kept under the T_g of the sample, and then it was coated with conductive layer. It can be utilized in various areas of material science including research and industrial fields. The tensile fractured and fatigue fractures samples with different blends of NR/vCR(95/5, 85/15, 75/25, 65/35, 50/50 phr) and NR/rCR were reported. Highest tensile strength with less torn surface was observed in 75/25 phr of NR/vCR blends. The incorporation of rCR in NR induced lower torn surface roughness and poor tensile strength. Compared to NR/vCR, NR/rCR have smooth surfaces, which is a clear indication of faster fatigue life failure, and this was attributed to brittle crack propagation [10]. Using the same blend ratios, the influence of SBR/vCR and SBR/rCR, on the morphological analysis, was studied. A good adhesion, more roughness on the surface and uniform dispersion of SBR/rCR was noticed. The inclusion of rCR in the above blends showed the existence of number of holes on the failure surface. The overall performance indicates that SBR/vCR and SBR/rCR were stiffer and more brittleness for the increasing ratios of blends [11]. Epoxidized natural rubber(ENR50), a modified form of NR, was blended with miscible acrylonitrile-butadiene rubber(NBR) at different proportions. SEM was recorded for the fractured surface of NR/NBRr and ENR 50/NBRr blends at composition of 95/5 and 65/35, respectively. Due to the separation of agglomerated NBRr from the matrix, a few holes and high matrix tearing were noticed in ENR50/NBR blends. The introduction of ENR50 in the blends has better interfacial bonding and effective wetting behavior, and hence it enhances the tensile strength of the same blends [12]. In another work, [13] prepared PLA/NR as well as PLA/ENR blends with different proportions at 90/10, 80/20, 70/30 by weight and studied its morphological studies. The particle size of rubber is 5–10 μm for PLA/NR(90/10). With increasing NR content, the dispersed particles has same size and shape together with phase inversion. In the case of PLA/ENR, good compatibility was observed as compared to PLA/NR. The partially compatible between PLA and ENR was due to coarse surface of ENR. Fracture topography of ethylene-propylene-diene monomer(EPDM) and silicone rubber(SR) blends was observed. The second phase with layer dimension was noted in EPDM/SR(80/20 wt%), whereas the phase inversion was noted for 60/40 wt% blends. A considerable reduction in dispersed phase size, because of lowering the concentration of

EPDM, was observed for 20/80 wt%. A greater immiscibility between EPDM and SR(40/ 60 wt%) was reached with low thermodynamic affinity and was noticed from the morphological analysis [14]. Xiaobo et al. [15] prepared and studied the characteristics of NR/ultrafine full-vulcanized powdered styrene-butadiene rubber(UFPSBR) blends. The outcome of this research work was increase in cross-link density(n), volume fraction of polymer in the swollen mass, shore A hardness, Akron abrasion loss, and decrease in heat buildup temperature(ΔT) was noted with increase in UFPSBR(0%–10%). The SEM studies showed diameter of UFPSBR aggregation was 10 μm. In general, the results indicated that the above blends can be utilized to optimize the elastomer properties. Blending of two elastomers NR/SBR was done by two methods: roll-mill and solution mixing. From the result, small holes were observed in roll-mill processing samples, whereas in the other process, holes were not noticed, i.e., a clear indication of a better enhancement in the miscibility between the two blends was noted [16]. The fractured surface after the tensile test and compatibility of chlorinated polyethylene rubber(CPR) and nitrile rubber(NR) blends were studied. The boundaries of the cavities were unclear and showed the good interfacial bonding between the two blends. With increasing the NBR, the domain size of NBR was increased. Similarly, after the aging period, the tensile strength and hardness of blends were raised with NBR content [17]. The compatibility agent chlorinated polyethylene (CM) was used for the blends of bromo butyl rubber(BIIR)/acrylonitrile-co-butadiene rubber(NBR). BIIR was highly dispersed in NBR phase, and a sharp reduction in dispersed size of 40–50 μm was noted for the blends with compatibilizer. The presence of holes and phase separation were available in the blends without a compatibilizer [18]. Elastomeric blends consisting of chlorosulfonated polyethylene(CSM)/isobutylene-co-isoprene(IIR) and CSM/CIIR(chlorinated IIR) were prepared by two roll mill methods. The cross-link density ($v = 847 \times 10^2$/mol dm^{-3}), molecular weight ($M_c = 831$/g mol^{-1}), and volume fraction ($\varphi_p = .3400$) were greater for CSR/CIIR blends(at 20/80 wt%). This blends also exhibited the surface roughness with tear lines and a poor compatibility between CSM and CIIR rubbers [19]. Zainoha et al. [5] prepared PLA/chitosan(CS)/ENR blend composites via solution casting techniques with the influence of CS content of 5, 10, 15wt%. The SEM analysis was attempted for the samples after the tensile test. The introduction of CS into PLA could be decreased the micro-holes and fibers in the PLA matrix, resulting in the good compatibility between PLA and CS. Contrarily, poor adhesion also noted due to increasing the CS content of 10%. The fractured surface contained micro-holes with clear gaps, which were observed in PLA/ENR composites (Fig. 7.2). An increase in miscibility was noticed when the CS content increased up to 10%; hence improved its compatibility too. The influence of graphite and rubber plasticizers on ceramizable SBR-based composites was evaluated [20]. Flat and large graphite parts were seen in SBR matrix, and this is attributed to the dispersion between a filler and the matrix. The particle size pull-puts and results in reduction in tensile behavior by initiating micro-crack generation during the elongation of the composites. A simple mixing process was used to design graphene nanoplates (GNP at 0.3, 0.5, 0.7 and 1 phr)/NR latex composites. The surface roughness with the

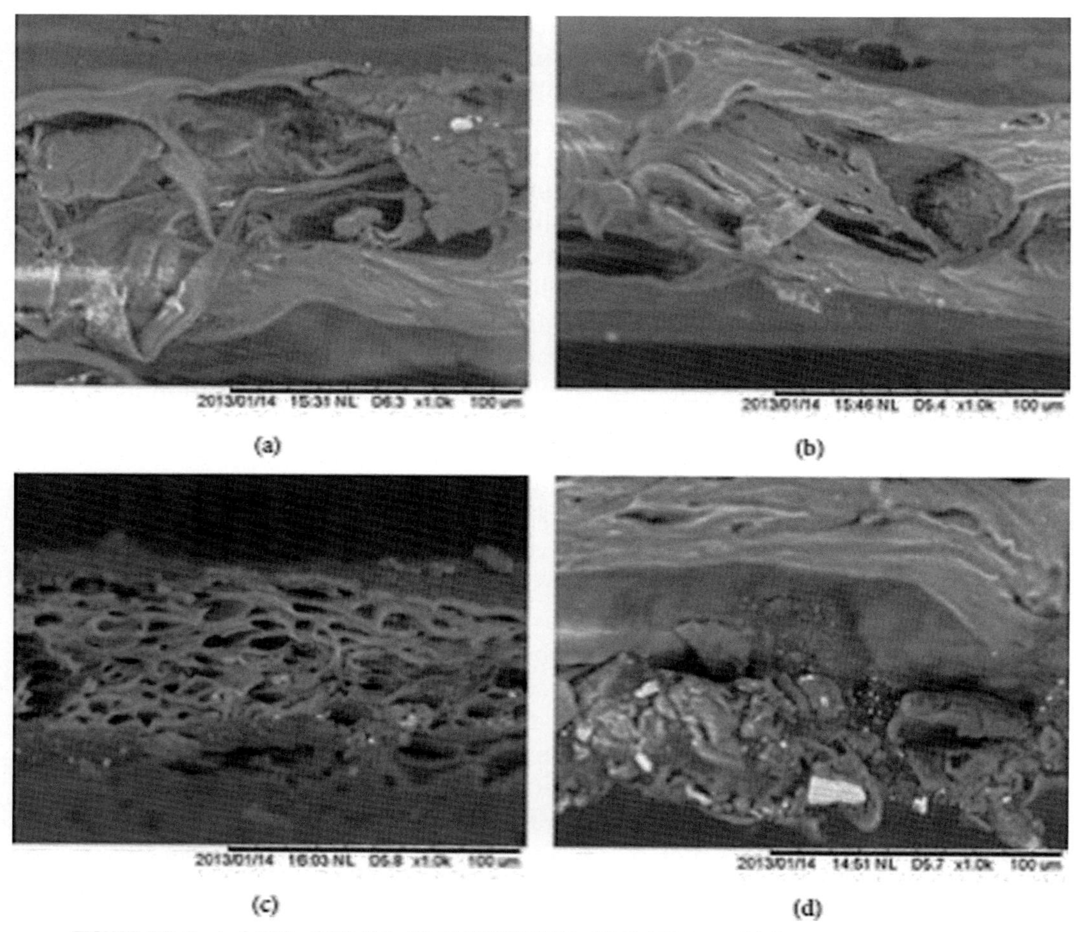

FIGURE 7.2 Typical SEM of (A) PLA, (B) PLA/CS(10%CS), (C) PLA/ENR, and (D) PLA/CS/ENR(10%CS) [5].

existence of GNP in NR was noticed in the fractured surface of composites [21]. The uniform distribution of GNP in NR suggests the good interfacial adhesion between the composites and NR, as evidence by cross-sectional view of SEM images (Fig. 7.3) The introduction of ammonium polyphosphate(APP) in the SR/PC composites was illustrated by Ref. [22]. The composites have the highest TS of 3.8 MPa, with the inclusion of 15%APP and 5%PC. The mechanical property results were evidenced by the SEM fractured surface of the composites with no holes and no deformed portions. Also a scattered filler was depicted in the surface of the composites. The effects of aluminum powder(0 to 40wt%) on the morphological characteristics of recycled rubber reinforced with polymer composites were studied. Microscopic observations revealed that the best dispersion and adhesion behavior of aluminum particle present in the rubber caused the enhancement in the composite materials [23]. In order to reduce the pollution in the

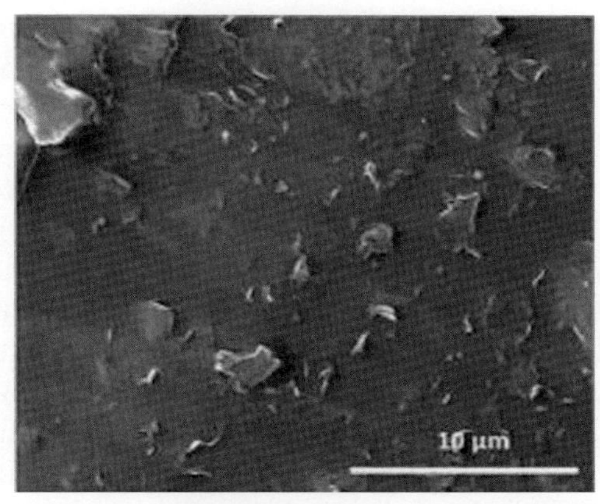

FIGURE 7.3 Cross-sectional SEM images of the graphene nanoplates@NR latex composite [21].

environment, [24] has suggested a novel approach for the development of composites. The authors used recycled rubber(RR) (obtained from waste tires) and wood flour(WF) for the composite preparation and studied their characteristics. In the composites, the TS was decreased from 3.3 to 1.35 MPa and T_g shifted from 15°C to 10°C because of incompatibility of two phases. In the SEM analysis, the microcells size was first increased from 10 to 20phr and then decreased with higher filler contents. The scattering of wood flour is good at 10 phr, but at 20–40phr some agglomeration was seen, resulting in the poor interfacial bonding between RR and WF particles. The mechanical behavior and life cycle of composites used in outdoor applications depend on their surface morphology [25]. Fresh samples have smoothness and uniformity in surfaces, contrarily surface roughness and cracks were seen in aged (720, 1440, 2160, 2880, 3600, 4320, 5000 h) silicone rubber(SR) composites. The surface crack was significantly large, which reduced their surface hydrophobicity for the unfilled SR and their composites. Highest TS(3.43 MPa) and highest percentage increase in hardness(8.66%) were noted in composites with 2%nano SiO_2+20%micro SiO_2. The findings from this work will help the electric companies in being able to find a superior alternative of porcelain insulators that are more efficient for high-voltage insulation in power lines.

7.2.3 Atomic force microscopy(AFM)

An invaluable technique is used to analyze the surface characterization of polymers, nanofibrillar structure, compositional maps of elastomers, and morphology of rubber blend samples with exceptional clarity. This techniques can also be used to determine the filler distribution and dispersion in blends. AFM offered a flat surface for the characteristics of the samples. Generally AFM provides high-resolution images(than SEM) at angstrom and nanoscale level with some physicochemical properties such as adhesion, friction, and elasticity. Compared to other electron microscopes, AFM is a powerful tool for the study of morphology, because it needs lower costs of equipment maintenance and the preparation of samples is too easy. AFM is also called as scanning probe microscopy(SPM), which can be operated in contact, noncontact, and tapping modes(TM). Among the three modes of AFM, tapping and contact modes are the most commonly used methods for the rubber blend samples. Experimental investigation on the NR/SBR/organo clay nanocomposites(NCs) with reference to compatibilizer(maleated EPDM or EPDM-g-MAH and ENR) through melt mixing techniques was studied. Tapping mode AFM was used to evaluate the microstructure analogy of these composites. In this mode, difference in phase angle of the excitation signal and cantilever response evaluated the difference in composition, phase stiffness, and viscoelastic response of the blends. The compatibilized NC had a surface roughness with improved interaction between NR and organoclay. Compared to uncompatibilized NC, the higher surface roughness(R_a-126nm) and root mean square(R_q-161nm) values were observed for the compatibilized NCs [26]. According to Ref. [27], the phase morphology of NR/SBR blends, with the influence of vinyl content(34.2wt% and 67.2wt%), curing temperature(140°C and 160°C), and curing time were investigated by AFM. From Fig. 7.4, it was noted that the morphology of the blends with various vinyl content was not same. The images were clearly revealed that NR/SBR with vinyl content at 67.2 wt% exhibited smallest domain size compared to NR/SBR with vinyl content at 34.2 wt%. The miscibility was also enhanced between the two blends. The obtained results were reflected in the TEM (Fig. 7.5), DSC and BDS measurements. Blends of SBR/NBR(100/0, 80/20, 20/80, 0/100) nanocomposites with respect to nanoparticles(cloisite 15A-C15A and Cloisite 30B—C30B) and ENR were developed through melt mixing process. TM-AFM of SBR/NBR blends indicated that a good dispersion and scattering of nanoparticles(C15A) in SBR and near the interface, with the droplet matrix surface of the blends. The addition of ENR with C15A results in a finer droplet size of the nanoparticles, and this was the reason to the improvement of SBR/NBR blends and interaction of polymers into nanoparticles. Hence the existence of ENR provided the better compatibility between the two polymers [28]. Composites based on multiwalled carbon nanotubes(MWCNTs)/NR were developed by two roll mill mixer techniques. Nanomechanical mapping AFM was

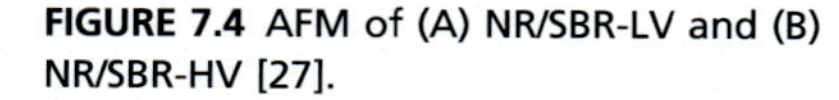
FIGURE 7.4 AFM of (A) NR/SBR-LV and (B) NR/SBR-HV [27].

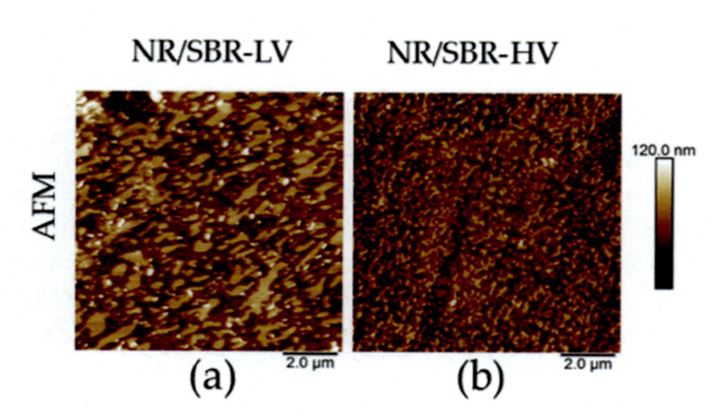

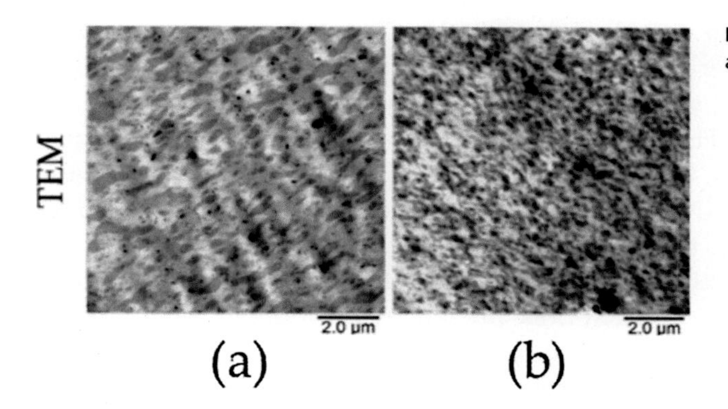

FIGURE 7.5 TEM images of (A) NR/SBR-LV and (B) NR/SBR-HV [27].

used to analyze the structure and nanoscale properties of prepared composites with reference to CNT loadings(1, 3, 5, 10, 20, 60 phr). CNTs were dispersed in NR, for low CNT loading and homogeneous dispersion for higher loading. A considerable reduction in adhesive energy and increase in mechanical properties were noted for the same loadings [29]. Tapping mode AFM was taken for TiO_2 filled chlorobutyl rubber composites, in this work, the TS and TM of the composites were increased with increase in TiO_2 loading. Similarly, the tear strength(21.54 kN/m), hardness(47), thermal conductivity(0.125 W/m K), and diffusivity(4.94 $m^2/s \times 10^{-8}$) respectively were higher with higher filler contents. The values of diffusion coefficient($8.836 \times 10^{-5} cm^2/s$), sorption coefficient(4.112 g/g), and permeability coefficient($3.634 \times 10^{-4} cm^2/s$) were decreased with lower TiO_2 contents. AFM with tapping mode showed a better dispersion of TiO_2 particles in the rubber matrix [30]. Using the same tapping mode, AFM was recorded for calcium sulfate($CaSO_4$) filled ENR composites. Single crystal formation was determined with the help of AFM image. A considerable enhancement in mechanical, thermal, dynamic-mechanical properties of these composites associated with some modification of fillers was found. This sis attributed to the aspect ratio of fillers [31]. Satyanarayana et al. [32] examined the phases of incompatible carboxylated nitrile rubber(XNOR) NR blends, rubber NC films, and rubber blend NC films using TM-AFM. Using solvent casting, the blends were prepared and with the effectiveness of nanoclays(Cloisite 15A and Cloisite 30B). AFM showed the spherical droplet size of NR as 4-5microns. The thickness of nanoclays lied in 50–60 nm for the blends(XNBR/NR) containing 8 phr of cloisite 15A. Further, AFM revealed the average thickness of cloisite 30B particles existed in NR as in the range of 500–550 nm, whereas in XNBR it was 50–60 nm. The obtained results were good in agreement with the TEM results as suggested by the authors. Various analytic techniques such as MDSC, DMA, ultrasonic velocity measurements, and AFM(contact mode) were used to characterize the NBR/EPDM(50/50) blends with the inclusion of composites (chlorinated polyethylene, CM). The phase image of NR/EPDM blends without compatibilizer had surface roughness with wriggles greater than 10 µm and inhomogeneous dispersed phase with agglomeration.

In case of blends with compatibilizer indicated that EPDM was evenly dispersed in NBR matrix with less number of agglomeration and decrease in phase size of 0.5–1 μm [33]. Zhao et al. [34] and his coworkers studied the morphology of NR/NBR/ENR ternary composites using SEM, AFM, and TEM. They have used the ENR as a compatibilizer. From the AFM results, the ternary composites had interfacial overlapping of phase boundary between the two blends(NR/NBR) due to the presence of ENR. The mechanical properties results showed that TS(22.8 MPa), tear strength(22.8 kN/m), and hardness(43) were higher for NR/NBR/ENR composites. Thus ENR acted as a good compatibilizer in ternary composites, which was proved in morphological and mechanical studies. It was proposed that these composites have potential applications in higher-performance tire tread products. The evaluation of graphene oxide(GO) particle size(40 μm and 40 nm) on the TS and stability of NR graphene composites was demonstrated [35]. AFM was used to measure thickness of the GO sheet and its composites. The thickness of the sheets decreased as the content of nanosize graphite increased. This results were good coincidence with the particle size analyzed by dynamic light scattering method. Also, the composite reinforced with nanosize GO and NR had a highest TS(22.03 MPa), better aging stability, and has the excellent barrier property, which maintains the quality of rubber films.

7.2.4 Transmission electron microscopy(TEM)

The most powerful tool and well-established technology to analyze the sample of thickness <100 nm. TEM is similar to light microscopy, the only difference is TEM uses electrons to pass through the sample to form a TEM image. Compared to PCOM, TEM has higher resolution because of smaller wavelength of electrons. It provides information on crystallinity, orientation, size, and shape of the dispersed phase of the samples. The complex aspect of TEM is the sample preparation, because a very thin sample is needed to pass sufficient electrons through with minimal energy loss. Thus, special attention is required for the size of the sample preparation to acquire better TEM images. The major advantages of TEM over SEM are: it provides high resolution equal to the dimension of atomic level and excellent image quality. TEM has some demerits, i.e., expensive, takes high vacuum environment, and it require experts to operate it. Kotani et al. [36] evaluated the distribution ratio of carbon black(CB, 0–70 phr) in poly isobuytrene rubber(IIR) and polyisoprene rubber(IR) blends using TEM. The prepared blends were cut by cryo-microtome with the thickness of <100 nm. Through TEM, it was found that IIR is in light gray color, IR as dark gray, and CB as black spots for the IIR/IR blends(CB- 10 phr). The calculated distribution ratio of CB as 17.83%, which was higher compared to other proportions. The presence of CB content was higher in IR than IIR phase. In the blend morphology, the area of IR was increased with increasing content of CB, and it was also difficult to find the distribution ratio of CB. However, CB was more easily dispersed into the IR than IIR phase. Nanocomposites were prepared from chlorinated(Cl)SBR/ZnS blends by using a two roll mill mixing process with the effect of various loadings of ZnS nanoparticles(0, 3, 5, 7, 10, 15 phr). The amorphous nature of Cl-SBR was reduced with the introduction of ZnS, and the attachment of ZnS in Cl-SBR was evidenced by XRD and EDX studies. Highest dielectric constant, less solvent as well as oil absorption, was observed for samples of Cl-SBR/ZnS(7 phr). As per the TEM observations, ZnS had several spherical-shaped nanocrystals with particle size of 2—15 nm. This was due to the good compatibility between Cl-SBR and ZnS, the NC showed the uniform distribution of particles in the matrix [37]. Amit et al. [38] studied the nano-scale morphological analysis of graphene-based solution styrene butadiene rubber(SSBR) composites with 3D-TEM. 3D image analysis is a powerful techniques to investigate structure of nano objects. As expected, the exfoliation of multiple graphene sheets into single was facilitated with the existence of carbon black(CB), as evidenced from TEM and other analytical techniques. The presence of oligo-layer graphene sheets was investigated by 3D-TEM. The phase morphology of VMQ(silicone rubber)-silica master-batch/BR composites was characterized by TEM. The composites for the analysis were prepared by ultra-microtome [39]. Two-phase morphology of the above said composites has the darker areas, which mentioned the VMQ domains, whereas brightest indicated as the presence of silica in VMQ master-batch. This inferred that silica particles were present at the interface between BR and VMQ phase. The same trend was also observed in SEM images. Finally, the authors proposed that VMQ/BR(at 20/80) had the good antizone aging properties and better thermal oxidative aging resistance. Ao et al. [40] used a new technique(latex blending) to prepare the Kaolinite(a clay mineral)/SBR composites with the influence of kaolinite(kaol) particles size, content, flocculant types. Kaol particles showed lamellar like structure arrangement with the thickness of dozen meters and the diameter ranging from 50 to 500 nm. The lattice distribution fringe of kaol has the distance ranged from 0.33 to 0.75 nm. Kaol particles were dispersed into SBR and existed in parallel orientation. Experimental investigation on the thermal and mechanical properties of NR composites reinforced with CNC(southern pine) was studied [41]. From the TEM analysis, the CNC showed slender rods and wide distribution in length(30—100 nm). The diameter lied from 6 to 10 nm. The average length and width found to be 77 ± 21 and 9 ± 2 nm respectively with the aspect ratio of CNC at 8.5wt%. A comparative study of novel slurry and latex blending techniques, new methods were adopted to prepare the carbon nanotubes(CNTs)/NR composites. TEM analysis inferred that, the CNTs were not dispersed in NR matrix by latex blending but it was well dispersed in slurry blending. Then the TS(raised by 15.2%), permittivity, and thermal conductivity were higher for CNTs/NR composites by slurry method. Payne effect($\Delta G'$) defined as storage modulus reduced with the increment of strain for the composites, and it was higher for composites done by latex blending process. That is, $\Delta G'$ was 287.18 and 255.53 respectively for latex and slurry blending method [42]. The effect of different filler loading(2, 4, 6, 8, 10wt%) of magnetic(Ni Zn ferrite) nanoparticles(used as a nucleating agent) with particle size of 10—30 nm was used to prepare PP/NR composites [43]. Ball milling cum melt blending techniques were used for the preparation of composites. Magnetization measurements(by vibrating sample magnetometer) such as remanence, susceptibility, and thermal properties were increased with increasing filler contents. TEM showed that the magnetic nanoparticles were evenly distributed in both PP and NR blends, at the same time few aggregations of nanoparticles existed in the NR. This is attributed to high viscousness of NR, which limiting the dispersion of nanoparticles during the manufacturing process. The strain sensing performance of self-assembled and solution blending effect of CNTs/graphene(GR) in silicone rubber(VMQ) composites were studied by Ref. [44]. A reduction of 53% in percolation threshold, superior strain sensing behavior, and repeatability with monotonic response for the self-assembled were observed. From the TEM studies, it was noticed that, CNTs were linked in the VMQ matrix, which are ready to synergize, and also the introduction of GR was enhanced the dispersion of SNTs. The self-assembly improved the blending of GR and CNTs, which forms a bridging structure. This was the reason for low percolation and enhanced stable resistance response. Two roll mixing techniques were used for the manufacture of composites from silicone rubber(QM), exfoliated graphite(EG-filler), and DCP(dicumyl peroxide as a curing agent). The composites with SR/EG/DCP(100/7/1.5) have exhibited the highest TS(6.811 MPa) and tear strength(30.22 N/mm). High-resolution TEM showed the better dispersed phase of EG with layered structure in SR matrix. This was due to the good adhesion between EG and SR, here EG act as reinforcing agent in the matrix [45].

7.2.5 Field emission scanning electron microscope(FESEM)

The surface morphology using FESEM for untreated and surface treated(alkali, silane, stearic acid) jute fiber (JF)-filled NR composites was studied. The untreated JF consists of smooth surface, because of the presence of waxy substances such as lignin and hemicellulose. Several cavities were produced for the treated fibers owing to the removal of waxy substance. In the case of composites, many holes were created for the untreated fibers/NR. The alkali and stearic acid treated JF/NR composites had good wetting and dispersion behavior, whereas in silane-treated, the composites had smooth and continuous surface (Fig. 7.6). The cure characteristics such as maximum torque(5.23Nm) and torque difference(4.98Nm), mechanical properties such as hardness(62), TS(17.04 MPa), and cross-link density(10.88×10^5mol/cm^3) were higher for alkali cum silane-treated JF/NR composites. Hence the authors suggested that this treated fiber is a suitable filler for the improvement of commercially viable and eco-friendly green rubber technology [47]. Melt blending and hot pressing techniques were used to develop biocomposites from PLA/NR/kenaf core powder(KF, 0, 5, 10, 15, 20 phr). The existence of surface roughness with line cavitation was observed for the fractures surface of PLA/NR(without KF). At the same time, KF was pulled out from PLA/NR matrix and was evidenced by FESEM analysis. Due to poor compatibility, between KF and matrix adhesion in the biocomposites, resulted the decrease in mechanical properties [48]. Kumarjyoti et al. [47] investigated the reinforcement of waste egg-shell derived calcium carbonate nanoparticles(WESNCC) in NR composites with and without the presence of maleated NR(MNR as a compatibilizing agent). $CaCO_3$ nanoparticles have weak compatibility in NR matrix for the composites without MNR. The agglomeration was

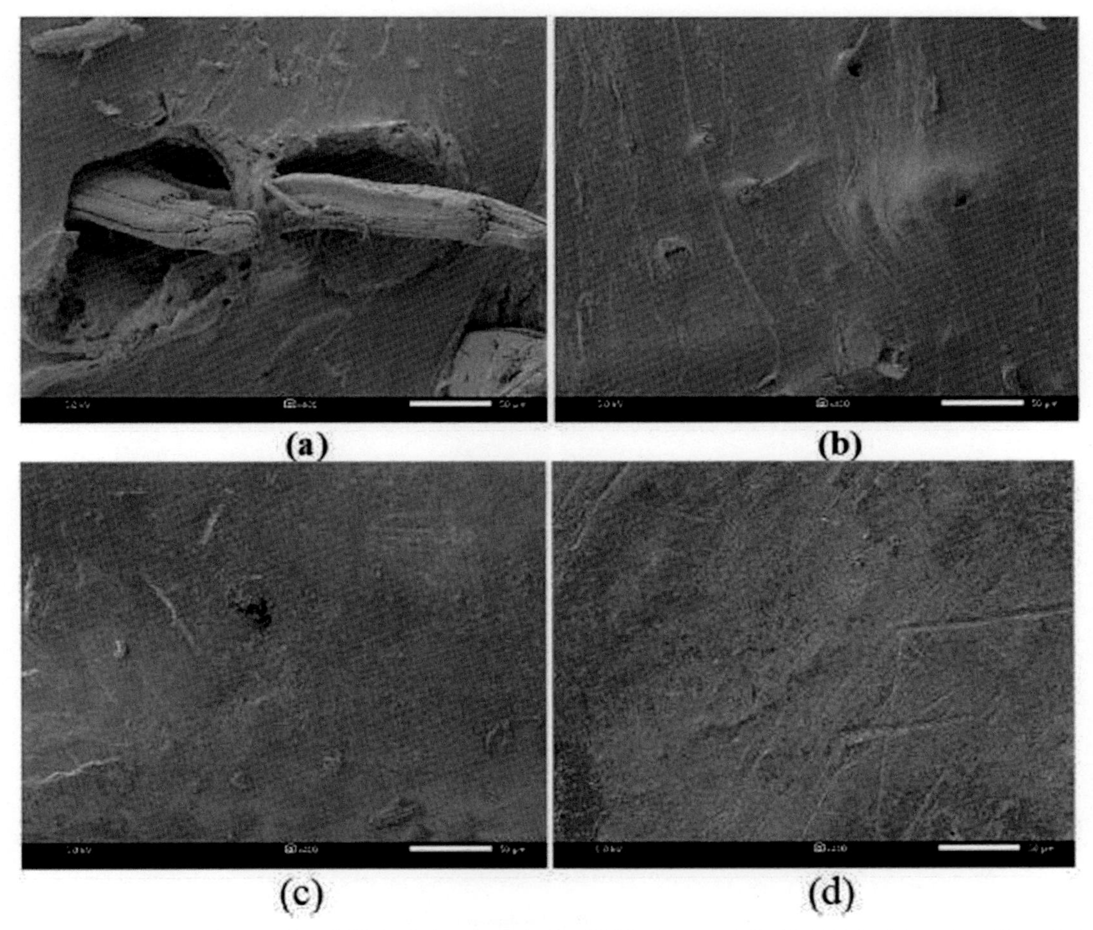

FIGURE 7.6 FESEM micrographs of (A) NR/JF(untreated), (B) NR/Alkali treated JF, (C) NR/Alkali Stearic acid treated JF and (D) NR/Alkali Silane treated JF [46].

larger for the composites without the MNR and also with the addition of nanoparticle content. Despite, at 5 phr of fillers, the nanoparticles were homogeneous in the rubber matrix, with the presence of MNR. Hence, the mechanical properties of NR/5%nanoparticles/MNR had superior than NR/5%nanoparticles. The authors of this work provided a better processing method for the utilization of egg-shell waste in industrial and engineering fields. Biocomposites from chicken feathers(CFs as 40, 50, 60, 75wt%) and poly(urea-urethanes-TPUU-SS) were prepared by using hot compression techniques. The density of the composites lied between the density of TPUU and CFs, and it was decreased as the CF content was increased. Similarly, the residual mass and Young's modulus(MPa) were higher for increasing CFs content. The fractured surface of biocomposites with 40wt% and 60wt% of CFs was analyzed using FESEM (Fig. 7.7A and B).

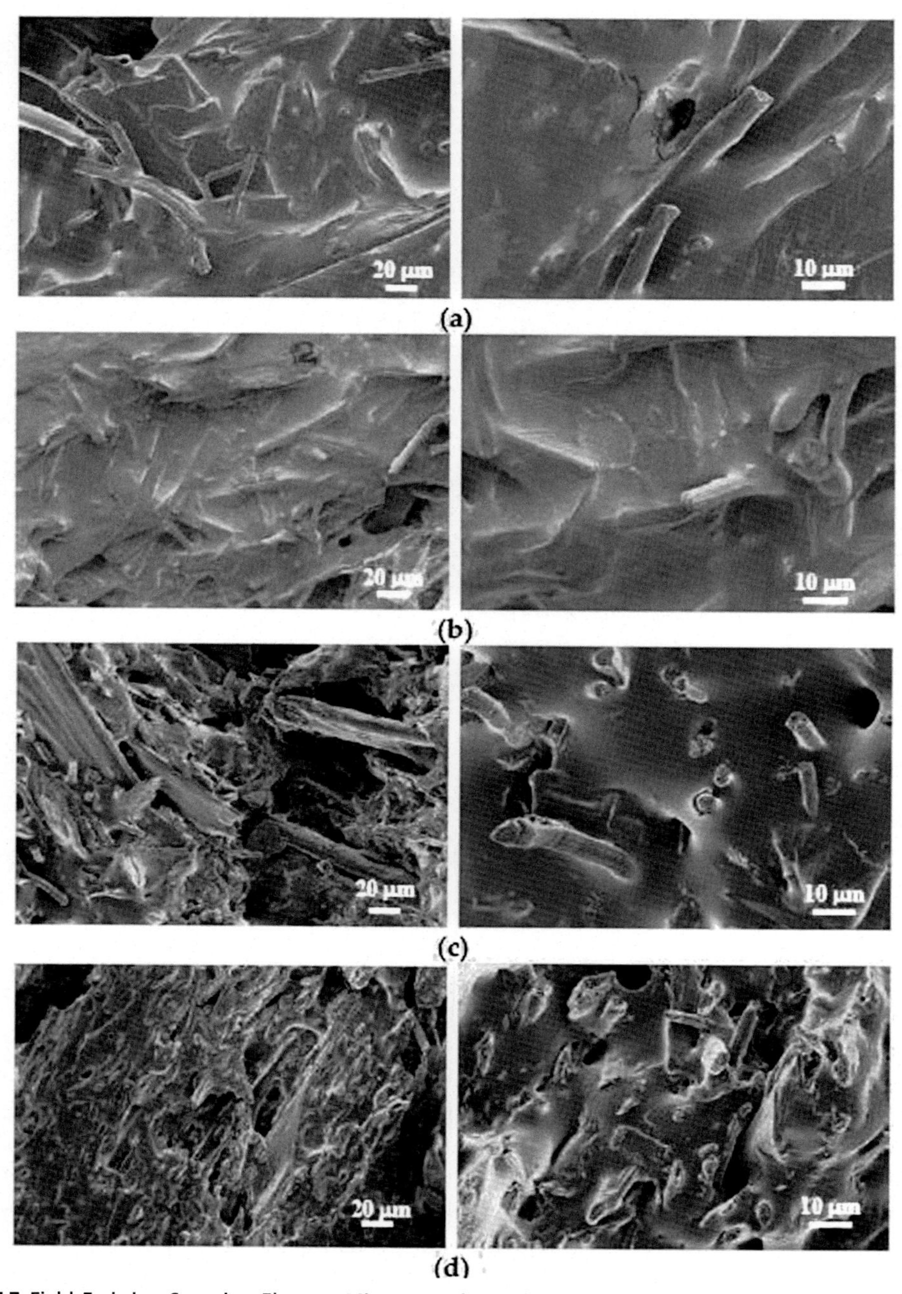

FIGURE 7.7 Field Emission Scanning Electron Microscope (FESEM) images of (A) TPUU-SS + 40 wt% CF; (B) TPUU-SS + 60 wt% CF; (C) TPUU + 40 wt% CF and (D) TPUU + 60 wt% CF [49].

The raw fibers were dispersed well with the matrix, which provided good compatibility as it was reflected in tensile results. A considerable interaction between the CF and TPUUs where the CF finely dispersed in the matrix was observed in TPUU/CF(60/40) composites. When the matrix decreased[TPUU/CF(40/60)], there was a poor interaction with uncovered fibers and voids from fiber pull-outs were noted(Fig. 7.7C and D). Studies have shown that the matrix(TPUU) contained disulfide linkages, which interact well with chicken feathers, a type of keratin fibers [49]. Nanosilica(from silica sand) and micro carbon black(MCB) were used as a reinforcing agents to enhance the characteristics of SBR/NR(70/30wt%) blends with and without δ–irradiation. Using the above combination, NCs were prepared by using compression molding process with vinyl trimethoxysilane(VTMS) a coupling agent. From FESEM studies, the nanoparticles silica/MCB and MCB were highly dispersed for irradiated samples, compared to irradiated one. Further, the curing efficiency was related to filler nanoparticles adhesiveness in SBR/NR matrix. The introduction of VTMS in the filler particles enhanced the curing of blends by heating the mixing and radiation process. Active radicals were formed with the coupling agents, which induced the cross-linking networks of blends. Thus, the δ-irradiation has potential on the interfacial adhesion between nanoparticles/rubber blend composites [50]. Using environmental scanning microscope(ESEM) and FESEM, the morphology of epoxidized NR(ENR) latex and high ammonia NR latex(HA-NRL) were described. The average size of the latex particles lied between 0.5 and 1.2 μm. From ESEM(at latex state), the agglomeration of latex particles was very large and spherical in shape, which was clustered by smaller latex particles. According to FESEM(for dry state), the average size of the particles was as 0.5–0.8 μm, a better agreement with the ESEM results [51].

7.3 Effect of plant fiber-reinforced elastomer composites

Different formulations of NR reinforced with untreated short jute fibers(1, 2.5, 5, 7.5 and 10 phr) composites were prepared using laboratory two roll mill method. The superior mechanical properties were attained with the effect of jute fiber content(at 10 phr) for modulus at 100%, 200%, and 300% elongation in the NR composites. Similarly, the hardness(55.2° Sh A), abrasion loss(0.075 cm^3), cross-link density(1.7152×10^{-4} mol/cm^3), and density(1.0210 g/cm^3) were higher for these composites. From SEM, the dust particles in the untreated jute fiber result in reduction in TS of the composites. The existence of fiber agglomeration and holes in the structure was also related to the reduction in tensile properties [52]. The morphology of composite samples from NR reinforced cereals straw fiber, where the fibers were modified with acetic acid and maleic anhydride. SEM images showed that the untreated straw fibers has a particle size of several tens of μm. In the case of treated composites(at 10 phr), the dispersion was enhanced in the matrix, which gave a better entrainment of straw particles, resulting in the good interfacial bonding between the two phases. The SEM results were coincided with mechanical properties and cross-link density of the vulcanizates [53]. Anuar and Zuraida [54] focused on the development of thermoplastic elastomer composites reinforced with kenaf bast fiber(20vol%). Using double melt blending and compression molding, the composites were prepared with the influence of modifiers in PP, i.e., thermoplastic

NR(TNPR) and PP/EPDM. In this study, the SEM results indicated that the presence of MAPP(coupling agent) and KF were improved the compatibility between the fiber and matrix. Thus the composites prepared by this method potentially increased the performance of mechanical characteristics. Santoprene, a thermoplastic elastomer, was used as the matrix for the PALF(0%–18%) composites. PALF could be impregnated with santoprene aligned successfully. The composites had flat and smooth fractured surfaces, although rather nonuniform surface. Also the influence of low PALF content and matrix orientation enhanced the mechanical properties of composites [55]. New technology(powdering method) was utilized for cellulose nanofibers(CNF), which allowed ZnO at the nanoscale [56]. CNF and ZnO were compounded with NR. The mechanical properties of composites with the effectiveness of CNF were increased than that of without CNF. From SEM, the ultrashort CNFs have thin and a few micron fibers, whereas middle and ultralong CNFs had high aspect ratio with 10 μm fibers. Cellulosic fibers with various diameters were noted in composites prepared by solvent exchange method and CNF powder particles with several tens to 100 μm were observed for composites without solvent exchange method. Office waste paper(OWP as 1%, 2%, 3%, 4%, 5%) from natural cellulosic fibers was considered as reinforcement for the polyurethane elastomer composites. The different loadings of OWP with physical blending technology were utilized for the manufacture of composites. The pretreated OWP fibers exhibited long and flat fibers with micron size ranging from 1 to 40 μm. The fractured surface of OWP/PUE composites had roughness because of the incorporation of cellulosic fibers. The good compatibility with diameter ranges below 5 μm was observed for OWP/PUE composites with 1 to 3wt% of fiber. In OWP/PUE at 4% and 5% fiber loadings, the diameter was increased to 10 and 20 μm, respectively with worst interfacial adhesion noted [57]. In another research work, the same matrix polyurethane(PU) was used to reinforce silkworm cocoon(6%) composites by dip molding and hot pressing techniques. The apparent density of cocoon/PU composites increased considerably from 1.17 to 1.30 g/cm^3. In the same way, the modulus and TS raised by 150% and toughness by 300% than that of cocoon. From SEM, the compact fibrous network was seen in native cocoon before pressing and after pressing composites, more compact structure were seen. The matrix does not fulfill the structure of cocoon due to the presence of holes in the structure, which was caused by the loss of liquid. Overall results of this work proposed that both cocoon and PU contributed the light weight composites and had superior mechanical properties with potential for multiscale energy-absorbing applications [58]. Compounds from bamboo charcoal powder(BCP, 10–50 phr) and NR were prepared with the effect of hexamethylenetetramine(bonding agent). The introduction of BCP in NR with the bonding agents enhanced Mooney viscosity, tensile modulus, hardness, and abrasion resistance. The SEM morphology of BCP contains different shapes, sizes, and lengths. The BCP particle shape was more regular, while the bonding between BCP and NR was required, in order for the reinforcement to be achieved [59]. The effects of various treatments(NaOH, $NaClO_2$, and KoH) were used to prepare sugancane bagasse cellulose(CB). SB was treated chemically with H_2SO_4 to produce acid-treated cellulose(CA). Composites were prepared using NR with SB, CA, and CB. The carbon content present in the surface of composites evaluated by EDS has followed the order of NR-CB(96.1)>NR-SB(80.7)>NR-CA(76.1)>NR(67.8). The presence of small aggregation in the H_2SO_4 treated and CB extracted NR composites. Moreover these composites absorbed less solvent(toluene, xylene) compared to neat NR and SB composites [60]. NR composites fabricated with jute fibers(JF), bleached JF(BJF), and CNCs at different filler concentrations of 5 and 10 phr were studied [61]. From the SEM images, the diameter of fibers was measured, and the values were higher for untreated JF(22 μm), CNC(16 μm), and BJF(7 μm) respectively. Studied showed that modulus and hardness values were higher for BJF/NR composites. A single nanocrystal with needle-like shape was observed for CNCs via TEM (Fig. 7.8A). The CNCs exhibited an agglomeration due to the drying

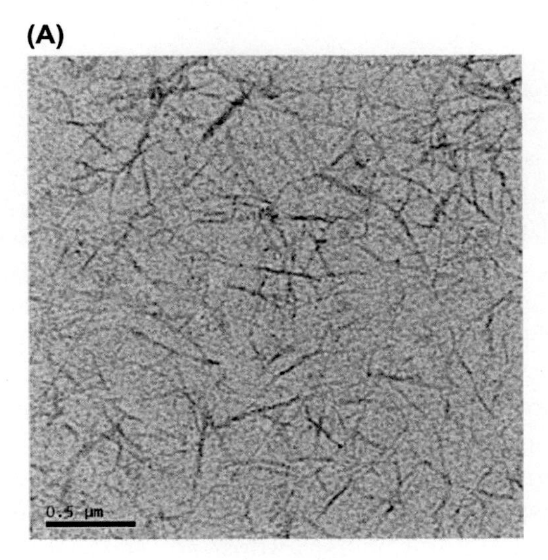

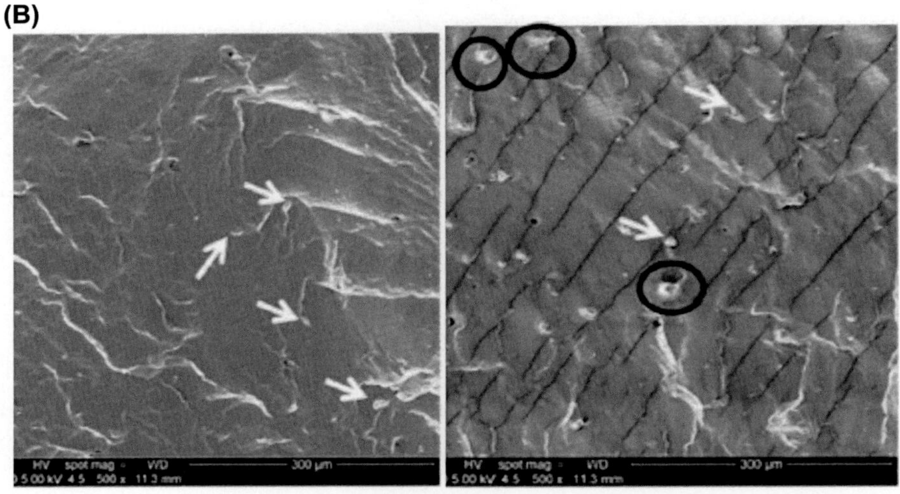

FIGURE 7.8 (A) TEM images of a single nanocrystals with for CNCs [61]. (B) SEM images of NR composites with (i) 5 phr of CNCs, (ii) 10 phr of CNCs [61].

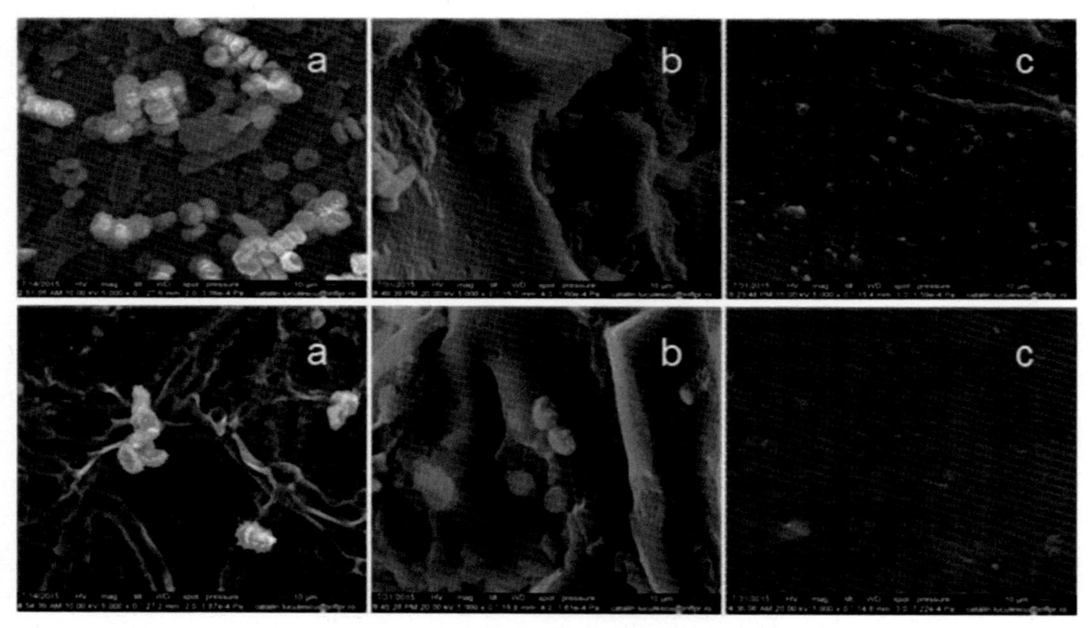

FIGURE 7.9 SEM of the NR/flax and NR/saw dust composite. (A) after 60 days of incubation with *Aspergillus niger*; (B) after 60 days of incubation with *Aspergillus niger* and after 72 h immersed in toluene; (C) the blanks (before incubation with the *Aspergillus niger*) and after 72 h immersed in toluene [62].

process at the time of preparation. Some grooves were noticed due to the pull-out of the JF from the NR, during cryogenic fractures. The composite consists of NR/CNC(5 phr), which had no agglomeration. At the same time, at 10 phr of CNCs, some black circles (Fig. 7.8B) were noted especially on higher concentration of 10 phr(CNC) [62] fabricated NR composites filled with natural fibers(sawdust and flax) with benzoyl peroxide as a cross-linking agent. The SEM characteristics of the prepared samples, before and after incubated with a fungus (*Aspergillus niger*), were studied. The production of bacterial colonies and elongating hyphae surrounded by a fibrillar extracellular matrix was found in all incubated NR composites (Fig. 7.9A). The existence of voids and cavities was observed in the incubation NR composites after immersion(72 h) in toluene(Fig. 7.9B). The smoothness in surface for the composites before incubation with fungus was noticed (Fig. 7.9C).

7.4 Effect of synthetic fiber-reinforced elastomer composites

Ozone treatment(0, 0.5, 1.5, 2.5, 3.5, 4.5 h) was given to short carbon fiber(SCF) and the treated SCF, acted as a reinforcement for NR composites. The existence of oxygen, silicon, and O/C ratio was higher for ozone-treated(4.5hr) fibers. The TS(26.9 MPa), permanent deformation(25%), elongation at break(4.99%) were higher for treated SCF(at 2.5 phr)/NR composites. The ozone-treated SCF enlarged the surface roughness and area of the fibers. In general, the outcomes of this work suggested that ozone treatment is an effective way to enhance the compatibility between fiber and NR, and hence increased the mechanical behavior of SCF/NR composites [63]. Rasaq et al. [64] utilized multi-walled carbon nanotubes(MWCNT, 0, 3, 6, 9, 12, 15 phr) as a filler for NR nano-composites. The morphological analysis showed that for MWCNT(3, 6, 12 phr) loading, more coarseness was observed for the NCs. The CNTs were adequately embedded in the NR phase. Also, there are some voids supported by the elimination of CNT, which indicated that there is a degree of weak interaction or inadequate wetting between CNT and matrix phase. In the same manner, the mechanical properties were higher for the same MWCNT loadings(3, 6, 12 phr). The results obtained from this studies suggested that CNT reinforced NR was suitable application for the development of anthropomorphic prosthetic foot. The influence of δ-irradiation(50–200 kGy) treatment on the surface morphological studies(SEM) of short glass fiber(GF, 10%–30%) reinforced with NBR(70%) and devulcanized waste rubber(DWR, 70%) composites were illustrated. NBR/DWR blends showed irregular sizes and porous texture. In the case of composites, the compatibility was better for the matrix and GF, and it has also no voids, with the presence of GFs. Further, the different δ-irradiation dosed composites had smoother in surface because of the formation of cross-links between the molecules in the elastomeric chains [65]. Four elastomers methacrylate-butadiene styrene(MBS), styrene maleic anhydride(SMA), ethylene methyacrylate(EMA), and SiR were used as toughening modifiers of GF reinforced polycarbonate composites. According to mechanical properties, the TS(68.55 MPa), flexural modulus(4639 MPa), and flexural strength (108.81 MPa) were higher for composites with a modifier SMA, whereas impact strength(218.85 J/m) was for composites with an SiR modifier. Brittle crack propagation in the flat surface and crack was impeded and branched in a scale surface was observed(SEM) with and without the introduction of SiR respectively. Hence the composite results indicated that SiR, the prominent agent to enhance the toughness of the polycarbonate composites, which was reflected in mechanical analysis [66].

7.5 Conclusions

Microscopy is of prime importance for the characterization of phase morphology, topography, chemical composition, surface properties, interfacial bonding between the matrix and fillers distribution of fillers/compatibilizers between the elastomeric chains. In this chapter, the morphological characteristics of elastomer blends and composites were presented. Various microscopic techniques(OM, SEM, AFM, TEM) with a brief introduction for the analysis of some scientific studies were discussed. OM served information at the scale of 10 μm. It is a simple method, does not require sample preparation and used to know about the granule size, voids, dispersed phase of blends and composites. SEM is a better option to analyze the surface morphology of the samples. AFM is a powerful tool to characterize the filer distribution and dispersion of blends at the nanoscale level. It provides high-resolution image to analyze the sample of thickness <100 nm. Special care needs for the size of the sample preparation to get quality TEM images. This microscopic techniques also give further information about aggregation, exfoliation, and agglomeration of smallest(at the nanoscale) particles in the composites. The various techniques have assisted research scientists and material engineers in the industrialization of elastomeric materials. The influence of different plant and synthetic fibers reinforced elastomer composites revealed the presence of aggregations, grooves, voids, surface roughness, and filler distribution in the matrix phase, etc., were discussed at the end of this chapter.

References

[1] O. Tonguç, Elastomeric Micro- and Nanocomposites for Neutron Shielding, Micro and Nanostructured Composite Materials for Neutron Shielding Applications, Woodhead Publishing Series in Composites Science and Engineering, 2020, pp. 125–137.

[2] A.A.A. Mariam, P. Deepa, A.E. Ali, Polymers to Improve the World and Lifestyle: Physical, Mechanical, and Chemical Needs, Polymer Science and Innovative Applications, Materials, Techniques, and Future Developments, 2020, pp. 1–19.

[3] P. Tasneem, S.A. Farooq, Mechanical properties, sealability, and recyclability of elastomeric materials in petroleum industry, reference module in materials science and materials engineering, Encycl. Renew. Sustain. Mater. 5 (2020) 131–147.

[4] H. Kalle, P. Minna, H. Tirila, Elastomeric Materials, Tampere University of Technology the Laboratory of Plastics and Elastomer Technology, 2007. https://laroverket.com/wp-content/uploads/2015/03/Elastomeric_materials.pdf.

[5] Z. Zainoha, I. Md Saiful, H. Azman, M.K. Mohamad, A. Reza, I.M. Inuwa, M. Hasan, Mechanical properties and morphological characterization of PLA/Chitosan/Epoxidized natural rubber composites, Adv. Mater. Sci. Eng. 2013 (2013), https://doi.org/10.1155/2013/629092. Article ID 629092.

[6] H.H. Le, S. Hait, A. Das, S. Wießner, K.W. Stöckelhuber, F. Böhme, R. Uta, K. Naskar, G. Heinrich, H. J. Radusch, Self-healing properties of carbon nanotube filled natural rubber/bromobutyl rubber blends, Express Polym. Lett. 11 (2017) 230–242.

[7] A. Ghosh, R.S. Rajeev, S.K. De, W. Sharp, S. Bandyopadhyaya, Atomic force microscopic studies on the silicone rubber–fluororubber blend containing ground rubber vulcanizate powder, J. Elastomers Plastics 38 (2006) 119–132.

[8] R. Panigrahi, K.S. Suneel, J. Pionteck, Fabrication of elastomer blends involving core (Polystyrene) @Shell (polyaniline) approach, their characterization and applications in electromagnetic shielding, Rubber Chem. Technol. 91 (2018) 97–119.

[9] G. Jianhua, Y. Xiwang, C. Xuming, Z. Xingrong, Improved antistatic properties and mechanism of silicone rubber/low-melting-point-alloy composites induced by high-temperature cyclic stretching, Compounds 739 (2018) 9–18.

[10] S.Z. Salleh, I. Hanafi, A. Zulkifli, Study on the effect of virgin and recycled chloroprene rubber (vCR and rCR) on the properties of natural rubber/chloroprene rubber (NR/CR) blends, J. Polym. Eng. 33 (2013) 803–811.

[11] A.A. Azrem, N.Z. Noriman, M.N. Razif, S.T. Sam, M.S. Saiful, Physical and morphological properties of styrene butadiene rubber/recycled chloroprene rubber (SBR/CRr) blends, Adv. Mater. Res. 795 (2013) 119–123.

[12] S.A. Hazwani, I. Hanafi, A.R. Azura, Tensile properties and morphology of epoxidized natural rubber/recycled acrylonitrile-butadiene rubber (ENR 50/NBRr) blends, Procedia Chem. 19 (2015) 359–365.

[13] K. Pongtanayut, C. Thongpin, O. Santawitee, The effect of rubber on morphology, thermal properties and mechanical properties of PLA/NR and PLA/ENR blends, Energy Procedia 34 (2013) 888–897.

[14] B. Rafael, E.O.B. Guilherme, H.K. Alexander, A.B.B. Diego, A.P. Rafael, D.A. José, Rheological study of EPDM/silicone rubber blends phase inversion and characterization of resultant mechanical and thermal properties, J. Appl. Polym. Sci. 138 (2021) app50140.

[15] L. Xiaobo, G. Ying, B. Lina, W. Zhong, Preparation and characterization of natural rubber/ultrafine full-vulcanized powdered styrene–butadiene rubber blends, Polym. Bull. 71 (2014) 2023–2037.

[16] M.A. Mansilla, A. Ghilarducchi, H. Salva, A.J. Marzocca H, Alpha (vitrea) transition in vulcanized natural rubber/styrene butadiene rubber blends prepared by mechanical and solution mixing, Solid State Phenom. 184 (2012) 405–410.

[17] X.Z. Zhen, H.C. Chun, W.G. Xin, K.K. Jin, X.X. Zhen, A study on the compatibility and physical properties of chlorinated polyethylene rubber/nitrile rubber blends, J. Appl. Polym. Sci. 120 (2010) 1180–1185.

[18] Y. Mohd, S. Pratibha, K.N. Pandey, V. Vishal, K. Vijay, Ultrasonic and viscometery compatibility studies of nitrile-butyl rubber blends, Indian J. Pure Appl. Phys. 51 (2013) 621–626.

[19] G. Marković, V. Jovanović, S. Samaržija-Jovanović, M. Marinović-Cincović, J. Budinski-Simendić, Curing and mechanical properties of chlorosulphonated polyethylene rubber blends, Chem. Ind. Chem. Eng. Q. 17 (2011) 315–321.

[20] I. Mateusz, A. Rafał, M.B. Dariusz, M. Marcin, P. Zbigniew, Z. Magdalena, R. Przemysław, S. Bartłomiej, Effect of graphite and common rubber plasticizers on properties and performance of ceramizable styrene–butadiene rubber-based composites, J. Therm. Anal. Calorim. 138 (2019) 2409–2417.

[21] D.L. Duong, A.N. Tuan, D.Q. Viet, T.N. Tham, A.N. Duy, N.P.D. Linh, P.T. Nghia, V.B. Sheshanath, A new approach of fabricating graphene nanoplates@natural rubber latex composite and its characteristics and mechanical properties, J. Carbon Res. 4 (3) (2018) 50. https://doi.org/10.3390/c4030050.

[22] G. Zhang, J. Wang, Study on application behavior of pyrolysis char from waste tires in silicone rubber composites, E-Polymers 16 (3) (2016) 255–264.

[23] M. Alloucha, K. Moez, M. Jamel, W. Mondher, D. Fakhreddine, Experimental investigation on the mechanical behavior of recycled rubber reinforced polymer composites filled with aluminium powder, Construct. Build. Mater. 259 (2020) 119845.

[24] M.J. Phiri, M.P. Mapoloko, M. Khotso, P.H. Shanganyane, Curing, thermal and mechanical properties of waste tyre derived reclaimed rubber–wood flour composites, Mater. Today Commun. 25 (2020) 101204.

[25] B. Mehmood, A. Mohammad, U. Rahmat, Accelerated aging effect on high temperature vulcanized silicone rubber composites under DC voltage with controlled environmental conditions, Eng. Fail. Anal. 118 (2020) 104870.

[26] M. Tavakoli, A.K. Ali, N. Hossein, NR/SBR/Organoclay nanocomposites: effects of molecular interactions upon the clay microstructure and mechano-dynamic properties, J. Appl. Polym. Sci. 123 (2012) 1853–1864.

[27] K. Darja, A.K. Hossein, L. Jorge, Phase morphology of NR/SBR blends: effect of curing temperature and curing time, Polymers 10 (2018) 510.

[28] A. Monfa, A. Jalali-Arania, Morphology and rheology of (styrene-butadiene rubber/acrylonitrile-butadiene rubber) blends filled with organoclay: the effect of nanoparticle localization, Appl. Clay Sci. 108 (2015) 1–11.

[29] W. Dong, F. So, N. Ken, N. Ken-ichi, I. Shigeki, U. Hiroyuki, M. Akira, N. Toru, E. Morinobu, N. Toshio, Production of a cellular structure in carbon nanotube/natural rubber composites revealed by nanomechanical mapping, Carbon 48 (2010) 3708–3714.

[30] A. Saritha, K. Joseph, A. Boudenne, S. Thomas, Mechanical, thermophysical, and diffusion properties of TiO_2-filled chlorobutyl rubber composites, Polym. Compos. 32 (2011) 1681–1687.

[31] S.B. Shib, H. Sakrit, S.N. Tamil, W. Sven, W.S. Klaus, J. Dieter, J. Andreas, F. Dieter, H. Gert, J.C.B. James, D. Amit, Water-responsive and mechanically adaptive natural rubber composites by in situ modification of mineral filler structures, J. Phys. Chem. B 123 (2019) 5168–5175.

[32] M.S. Satyanarayana, K.B. Anil, K. Dinesh, Preferentially fixing nanoclays in the phases of incompatible carboxylated nitrile rubber (XNBR)-natural rubber (NR) blend using thermodynamic approach and its effect on physico-mechanical properties, Polymer 99 (2016) 21–43.

[33] K.N. Pandey, D.K. Setu, G.N. Mathur, Determination of the compatibility of NBR-EPDM blends by an ultrasonic technique, modulated DSC, dynamic mechanical analysis, and atomic force microscopy, Polym. Eng. Sci. (2005) 1265–1276.

[34] X. Zhao, N. Kaijing, X. Yong, P. Zheng, J. Li, H. David, Z. Liqun, Morphology and performance of NR/NBR/ENR ternary rubber composites, Compos. B Eng. 107 (2016) 106–112.

[35] P.L. Lai, C.J. Joon, M.H. Nay, K.G. Leng, P.L. Fook, Y.L. Yi, Effect of graphene oxide particle size on the tensile strength and stability of natural rubber graphene composite, Mater. Sci. Eng. B 262 (2020) 114762.

[36] M. Kotani, D. Hidehiko, K. Hideaki, M. Kiyoshige, K. Hironori, Distribution ratio of carbon black in polyisobutyrene/polyisoprene rubber blends using high-resolution solid-state ^{13}C NMR, Polym. J. (2015) 1–6.

[37] V.C. Jasna, T. Anilkumar, A.N. Adarsh, M.T. Ramesan, Chlorinated styrene butadiene rubber/zinc sulfide: novel nanocomposites with unique properties- structural, flame retardant, transport and dielectric properties, J. Polym. Res. 25 (2018) 144.

[38] D.R.B. Amit, J. Rene, J. Dieter, F. Dieter, W.S. Klaus, H. Gert, Nano-scale morphological analysis of graphene–rubber composites using 3D transmission electron microscopy, RSC Adv. 4 (2014) 9300.

[39] L. Lin, H. Qing-Yuan, W. You-Ping, The aging properties and phase morphology of silica filled silicone rubber/butadiene rubber composites, RSC Adv. 10 (2020) 20272.

[40] Z. Ao, Z. Yinmin, Z. Yongfeng, Characterization of kaolinite/emulsion-polymerization styrene butadiene rubber (ESBR) nanocomposite prepared by latex blending method: dynamic mechanic properties and mechanism, Polym. Test. 89 (2020) 106600.

[41] Z. Chunmei, D. Yi, P. Jun, T. Lih-Sheng, S. Ronald, C. Craig, Thermal and mechanical properties of natural rubber composites reinforced with cellulose nanocrystals from southern pine, Adv. Polym. Technol. (2014) 21448.

[42] G. Jiang-Shan, L. Zhiming, Y. Zhengqi, H. Yan, A novel slurry blending method for a uniform dispersion of carbonnanotubes in natural rubber composites, Results Phys. 15 (2019) 102720.

[43] Y. Lih-Jiun, H.A. Sahrim, K. Ing, A.T. Mouad, B.B.A.R. Shamsul, N. Elango, K.A. Chun, Magnetic, thermal stability and dynamic mechanical properties of beta isotactic polypropylene/natural rubber blends reinforced by NiZn ferrite nanoparticles, Defence Technol. 15 (2019) 958–963.

[44] Y. Heng, Y. Li, Y. Xue, Z. Zhong, N.F. Dai, Monotonic strain sensing behavior of self-assembled carbon nanotubes/graphene silicone rubber composites under cyclic loading, Compos. Sci. Technol. 200 (2020) 108474.

[45] P.S. Sarath, V.S. Sohil, R. Rakesh, K.P. Mrituanjay, T.H. Józef, T. Sabu, C.G. Soney, Fabrication of exfoliated graphite reinforced silicone rubber composites - mechanical, tribological and dielectric properties, Polym. Test. 89 (2020) 106601.

[46] K. Roy, C.D. Subhas, T. Lazaros, P. Aphiwat, P. Pranut, Effect of various surface treatments on the performance of jute fibers filled natural rubber (NR) composites, Polymers 12 (2) (2020) 369.

[47] R. Kumarjyoti, C.D. Subhas, R. Natthaphon, P. Pranut, Understanding the reinforcing efficiency of waste eggshell-derived nano calcium carbonate in natural rubber composites with maleated natural rubber as compatibilizer, Polym. Eng. Sci. (2019) 1428–1436.

[48] N.F. Alias, H. Ismail, I.K.M. Ku, The effect of kenaf loading on water absorption and impact properties of polylactic acid/natural rubber/kenaf core powder biocomposite, Mater. Today Proc. 17 (2019) 584–589.

[49] A. Ibon, M. Sarah, A. Itxaso, R. Alaitz, G. Hans-Jürgen, Flexible biocomposites with enhanced interfacial compatibility based on keratin fibers and sulfur-containing poly(urea-urethane)s, Polymers 10 (2018) 1056.

[50] M.E. Hanan, S.M. Wael, M.E. Mai, Irradiated Rubber Composite with Nano and Micro Fillers for Mining Rock Application, De Gruyter, 2019, pp. 737–753.

[51] M.Y. Norhanifah, R.M.R. Fatimah, M. Asrul, Latex particles morphology of epoxidised and high ammonia natural rubber latex from electron microscopy techniques, Malays. J. Anal. Sci. 20 (2016) 1123–1128.

[52] J. Datta, M. Włoch, Preparation, morphology and properties of natural rubber composites filled with untreated short jute fibres, Polym. Bull. 74 (2017) 763–782.

[53] M. Marcin, M. Justyna, S. Krzysztof, Natural rubber composites filled with cereals straw modified with acetic and maleic anhydride: preparation and properties, J. Polym. Environ. 26 (2018) 4141–4157.

[54] H. Anuar, A. Zuraida, Improvement in mechanical properties of reinforced thermoplastic elastomer composite with kenaf bast fibre, Compos. B 42 (2011) 462–465.

[55] K. Asama, A. Taweechai, Mechanical properties of preferentially aligned short pineapple leaf fiber reinforced thermoplastic elastomer: effects of fiber content and matrix orientation, Polym. Test. 37 (2014) 36–44.

[56] H. Ryuji, N. Asahiro, Y. Yoshiaki, Development of powdering method for cellulose nanofibers assisted by zinc oxide for compounding reinforced natural rubber composite, Curr. Res. Green Sustain. Chem. 3 (2020) 100005.

[57] L. Wanqing, F. Changqing, Z. Xing, L. Yaguang, P. Mengyuan, Polyurethane elastomer composites reinforced with waste natural cellulosic fibers from office paper in thermal properties, Carbohydr. Polym. 197 (2018) 385–394.

[58] L. Fei, T. Yi, C. Lei, J. Lingxiao, W. Dayang, W. Tao, High fibre-volume silkworm cocoon composites with strong structure bonded by polyurethane elastomer for high toughness, Compos. A 125 (2019) 105553.

[59] I. Norazura, D. Norfahira, A.F.R. Nur, F.B. Sarul, S.M.M. Muhamad, Bamboo charcoal filled natural rubber vulcanizates: the effect of filler loading and bonding agent, AIP Conf. Proc. 1985 (2018) 040015. https://doi.org/10.1063/1.5047192.

[60] N.N. Sibiya, M.J. Mochane, T.E. Motaung, L.Z. Linganiso, S.P. Hlangothi, Morphology and properties of sugarcane bagasse cellulose- natural rubber composites, Wood Res. 63 (2018) 821–832.

[61] A.C. Carla, S.V. Ticiane, Cellulose nanocrystals and jute fiber-reinforced natural rubber composites: cure characteristics and mechanical properties, Mater. Res. 22 (Suppl. 1) (2019) e20190192.

[62] M. Stelescu, M. Elena, C. Gabriela, C. Corina, Development and characterization of polymer eco-composites based on natural rubber reinforced with natural fibers, Materials 10 (2017) 787.

[63] C. Junmei, Z. Shugao, Influence of ozone treatment on microstructure and mechanical properties of pitch-based short carbon fiber-reinforced natural rubber, J. Elastomers Plast. 49 (2016) 1–17.

[64] O.M. Rasaq, K.A. Oladiran, S.A. Ambali, A.M. Rasheed, S.A. Asipita, Carbon nanotube reinforced natural rubber nanocomposite for anthropomorphic prosthetic foot purpose, Sci. Rep. 9 (2019) 20146.

[65] A.R. Heba, G.A.G. Hend, M.H. Medhat, A. Abdel, H.M. Wagiha, Effect of gamma irradiation on the properties of short glass fibre-reinforced elastomers based on nitrile butadiene rubber and devulcanised waste rubber, Int. J. Environ. Anal. Chem. (2019). https://doi.org/10.1080/03067319.2019.1657853.

[66] L. Jun, L. Jiao, L. Xiuhong, G. Yanjin, W. Guilong, C. Liang, Flame retardancy and toughening modification of glass fiber-reinforced polycarbonate composites, Polym. J. (2019). https://doi.org/10.1038/s41428-019-0181-8.

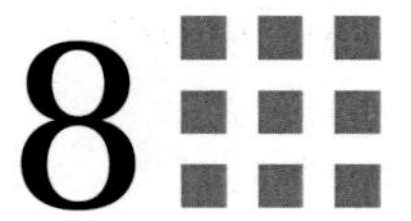

8 Mechanical behavior of elastomer blends and composites

G. Rajeshkumar, S. Arvindh Seshadri
DEPARTMENT OF MECHANICAL ENGINEERING, PSG INSTITUTE OF TECHNOLOGY AND APPLIED RESEARCH, COIMBATORE, TN, INDIA

8.1 Introduction

The mixing of polymeric materials is usually done to obtain better properties compared to the individual constituents. Elastomers are generally blended together for different reasons such as enhancing the physical and mechanical properties, reducing the cost of manufacturing, and achieving superior processing characteristics [40]. Elastomer blends can be either miscible or immiscible based on interactions present between the individual materials. Additionally, elastomers can be blended with themselves and also, with thermoplastic or thermoset polymer [39].

Rubber and thermoplastic polymers are blended together to form rubber-thermoplastic blend in which either the rubber phase is predominant or thermoplastic material is. These blends exhibit high flexibility combined with good impact properties, and they can be used as substitutes to conventional plastics. They are employed in a wide variety of applications such as automotive, biomedical, appliances, etc. [38a]. Some of the methods that can be utilized to produce rubber-thermoplastic blends include injection molding, compression molding, extrusion, blow molding, etc. [8,39].

Another type of elastomer blend is the rubber-thermoset blends. Here, thermoset polymers are blended with either core—shell rubber particles or miscible reactive rubbers. Epoxy is a widely used material when it comes to elastomer blends. It is used in various applications such as coating, adhesives, and other engineering applications. It possesses good mechanical properties along with corrosion resistance and chemical resistance. However, they have poor toughness. These brittle thermoset materials can be blended with liquid rubbers or by adding preformed rubbers in order to enhance their toughness [39].

An elastomeric composite is a class of material, which consists of two or more constituents having a specific interface between them. Generally, they are composed of a rubber matrix, reinforcements in the form of fillers, and other materials, which play a significant role in vulcanization, storage, etc. [11]. Elastomer composites can be

Elastomer Blends and Composites. https://doi.org/10.1016/B978-0-323-85832-8.00012-2

classified into macro, micro, and nano composites based on size of the filler utilized [10]. Some of the superior properties of elastomer composites include high strength and durability, resistant to corrosion, lightweight, and flexible to design and process changes [11].

Elastomer composite, like any other composite, has two major components: matrix and reinforcement. The elastomer matrix can be of two different types, namely general-purpose and specialty elastomers. Elastomers containing hydrocarbon chains such as natural, isoprene, butadiene, and styrene butadiene rubbers fall under the general-purpose category. On the other hand, specialty elastomers include acrylonitrile butadiene rubber, hydrogenated acrylonitrile butadiene rubber, silicone rubber, ethylene propylene diene terpolymer, etc. These materials have peculiar characteristics such as oil resistance, good thermal stability, and resistance to acids and other chemicals. In the case of reinforcement, they are categorized based on chemical structure, particle size, and reinforcing effect. Properties of fillers such as shape and size, presence of functional groups, specific structure, and surface area are some of the factors that influence the performance of elastomer composites. A wide variety of reinforcements are used to reinforce elastomers, and these include silica, carbon black, carbon nanotube, graphene, carbides, talc, etc. [11].

Elastomer composites find applications in a large number of industries. Some of these are mechanical, automobile, railways, electrical and electronics engineering-based applications, robotics, and smart textiles [12,16,17]. They are used in components such as couplings, shock absorbers, seals and membranes, pipes, and bushings. Additionally, they are also used in biomedical applications such as drug delivery, implants such as knee and joint replacements, dental implants, contact lenses, packaging, heart valves, and other medical devices [10].

8.2 Mechanical behavior of elastomer blends

In general, the mechanical properties are classified as static mechanical properties (SMPs) and dynamic mechanical properties (DMPs). In the case of SMP of elastomer blends, majority of studies focuses on properties such as tensile strength, tensile modulus, impact strength, and elongation at break as function of the blend ratio. As far as the DMP of elastomer blends are concerned, the main concentration is on storage modulus, loss modulus, and loss tangent [35,37].

8.3 SMP of elastomer blends

In order to enhance the toughness of poly-lactic acid (PLA), the thermoplastic polyurethane (TPU) elastomer at different weight ratios was blended together with PLA and their mechanical properties were explored. The specimens were manufactured by compounding followed by hot-pressing process. Properties such as tensile strength,

Table 8.1 SMP of PLA/TPU blends.

Specimen	Weight composition (wt.%)	Tensile strength (MPa)	Elongation at break (%)	Impact strength (kJ/m^2)
PLA/TPU	(100/0)	59	5	3.5
	(90/10)	50	185	5
	(80/20)	42	350	25
	(70/30)	35	350	58

elongation at break, and impact strength were determined (Table 8.1). The results indicated that the tensile strength gradually decreased with the increase in TPU quantity, whereas the elongation at break increased drastically with TPU addition. Furthermore, there was also a significant rise in the impact strength of elastomer blends [13].

Similar to the above study, another work focused on increasing the toughness of polyoxymethylene (POM) by blending it with TPU using a double screw extruder [22]. The resultant blends had a TPU content from 0 wt.% to 20 wt.%. Mechanical properties such as tensile strength and maximum elongation were computed. It was witnessed that at low concentrations of the elastomer (till 10 wt.%), the addition of TPU had minimal influence on the above properties. As the TPU content increased, an increase in elongation was observed along with a drop in the tensile strength (Table 8.2).

It can be interpreted that the elastomer blend exhibits brittle characteristics at low weight ratios of TPU, while it becomes ductile above 10 wt.% of TPU. The authors also mentioned that the inclusion of TPU leads to the improvement in the crack propagation resistance of POM as TPU hinders the crack growth; the microcracks are retained in the TPU elastomer particles [22].

The SMP of acrylonitrile butadiene rubber (NBR) and polyvinyl chloride (PVC) elastomer blends was analyzed (Table 8.3). The concentration of PVC addition is in terms of parts per 100 parts of rubber by weight (phr). The NBR was initially loaded with 40 phr fast extrusion furnace (FEF) carbon black nano powder, and then different concentrations of PVC were added [1].

Table 8.2 SMP of POM/TPU blends.

Specimen	Weight composition (wt.%)	Tensile strength (MPa)	Elongation at break (%)
POM/TPU	(100/0)	63	11
	(95/5)	62	10.5
	(90/10)	59	12
	(85/15)	50	15.2
	(80/20)	52	15

Table 8.3 SMP of 40FEF/(PVC-NBR) blends.

PVC content (phr)	Tensile strength (MPa)	Elongation at break (%)	Elastic modulus (MPa)	Change in degree of microhardness
0	12.6	1400	1.7	0
20	7.5	1200	1.8	3
40	6.5	1020	2.1	8.5
60	6	600	2.6	17
80	6.4	500	4	23
90	7	400	5.2	24.5

The drop in tensile strength is because of the poor interfacial bonding between PVC and NBR. However, above PVC concentrations of 50 phr, the tensile strength increases slightly due to minor improvement in bonding between PVC and NBR formed (covulcanization). This enhanced bonding will result in better stress transfer between the materials [36]. On the contrary, there is a sharp decrease in the elongation as PVC is added, and this is attributed to the rigid interphase between PVC and NBR.

Two different kinds of elastomers were added to polypropylene (PP), and the effect of elastomer type on the elongation of blends was investigated. The elastomers that were incorporated were polyolefin (POE) and propylene-based elastomer (PBE), and their contents were from 0 to 30 wt.%. For both the elastomers that were included (0–30 wt.%), the elongation at break was found to increase from 40% to 1200%. However, the inclusion of PBE exhibited better results. In particular, the addition of 10 wt.% PBE led to a 988.5% rise in the elongation, while the corresponding incorporation of POE resulted in only about 180% [14].

Ethylene-octane random copolymer POE elastomers with high crystallinity (EG8450G) and low crystallinity (EG8842) were blended together at different mass ratios, and the SMP of the resultant blends was investigated [28]. The specimens were subjected to tensile test and single "load–unload" cycle and parameters such as elastic modulus, elastic recovery, and residual strain were determined, at applied strains of 30%–1000% and also at three different thermal processes, namely: slow cooling, quenching, and annealing. Table 8.4 lists the SMP of POE elastomer blends as a function of mass fraction of EG8450 at 300% strain.

It can be interpreted that elastic modulus is highest for annealing followed by slow cool and quenching, while elastic recovery is maximum for quenching and minimum for annealed specimens. However, at lower strains, the elastic recovery is sensitive to thermal processing, but it is stable at higher applied strains. Additionally, the authors also noticed that at lower strains, elastic recovery is larger and residual strain is lower and vice versa [28].

The consequence of acrylonitrile (ACN) addition on the tensile and impact properties of PLA and NBR blend was explored [26]. The NBR with different ACN content was used,

Table 8.4 SMP of polyolefin elastomer blends [28].

Mass fraction of EG8450G	Thermal treatment	Elastic modulus (MPa)	Elastic recovery (%)	Residual strain (%)
0	Slow cool	3	59.5	50
	Anneal	3	60	45
	Quench	3	59	40
0.25	Slow cool	5	41.6	70
	Anneal	4.9	42	70
	Quench	5.1	40.2	80
0.5	Slow cool	15	29.8	125
	Anneal	16	25	140
	Quench	12	29.8	140
0.75	Slow cool	30	20.1	155
	Anneal	33	18.5	175
	Quench	20	20.1	175
1	Slow cool	46	18	195
	Anneal	72.5	16	205
	Quench	37	18.5	200

and these were named as NBR19, NBR33, and NBR51, where the numeral following NBR signifies the ACN content. The weight percent of rubber added to PLA to form elastomer blend was 10 wt.%. The tensile and impact properties of PLA/NBR blend with 10 wt.% of rubber were given in Table 8.5.

The impact strength of blends was found to be greater at low ACN content. This is because of the smaller size the rubber particles, the better interfacial bonding between PLA and NBR. The tensile modulus and tensile strength were found to reduce by nearly 15%–17% and 22%–27%, respectively, than the pure PLA. However, the authors pointed out that PLA/NBR blends showed better tensile strength and elongation as compared to other blends of PLA and rubbers [26].

Table 8.5 Tensile and impact properties of PLA/NBR blend with 10 wt.% of rubber [26].

Specimen	Izod impact strength (J/m)	Young's modulus (GPa)	Tensile strength (MPa)	Elongation at break (%)
PLA	22	3.5	65.36	4.5
PLA/NBR19	100	2.9	50.6	110.4
PLA/NBR33	90	2.74	49.03	82.2
PLA/NBR51	82	2.97	47.33	68.6

Table 8.6 Tensile properties of TPEE/PBT elastomer blends [20].

Specimen	Yield strength (MPa)	Tensile strength (MPa)	Young's modulus (MPa)	Elongation at break (%)
TPEE	8.27	16.81	56.09	899.1
TPEE90/PBT10	10.75	18.25	101.05	676.8
TPEE80/PBT20	12.75	18.77	187.61	617.04
TPEE70/PBT30	16.95	20.28	278.63	288.36
TPEE60/PBT40	20.28	–	464.57	53.28
TPEE50/PBT50	–	23.38	530.51	–
PBT	–	45.84	1550.67	5.3

Different wt.% of poly (butylene terephthalate) (PBT) was blended with thermoplastic polyester elastomer (TPEE), and their mechanical performance was investigated [20]. The elastomer blends were prepared using melt compounding, and 10–50 wt.% of PBT was added. The blends were subjected to tensile test, and the outcomes are listed in Table 8.6.

The authors concluded that the PBT content must be restricted to 30 wt.% as the phase inversion point for the TPEE/PBT system is nearly at 62.8 wt.% TPEE, and this means that as the PBT content approaches to 40 wt.%, it becomes the continuous phase rather than TPEE [20].

The mechanical properties (Table 8.7) of epoxidized natural rubber (ENR) containing 10 mol.% (ENR10), 25 mol.% (ENR25), and 50 mol.% (ENR50) epoxide groups blended with dodecanedioic acid (DA) were studied [32].

TPU elastomer was blended with poly(acrylonitrile-co-butadiene) rubber (NBR) comprising different ACN contents, and their mechanical performance was evaluated. For mechanical tests, NBR with 34 wt.% of AN was considered (Table 8.8). The authors suggested these elastomer blends for underwater sonar devices [21].

Polyamide-612 (PA-612) and maleic anhydride grafted poly (ethylene–octene) elastomer (POE-g-MA) elastomer blends were produced, and their tensile strength, tensile modulus, elongation at break, and impact strength were calculated [24]. The addition of POE-g-MA was in the range 0–35 wt.%. Tensile test and unnotched Izod impact tests were conducted to estimate the properties (Table 8.9).

Table 8.7 Mechanical properties of ENR/DA elastomer blends [32].

Sample	Elongation at break (%)	Resilience (%)	Hardness shore A
ENR10+DA	660	75	29
ENR25+DA	690	60	35
ENR50+DA	400	10	38

Table 8.8 Mechanical properties of TPU/NBR34 blends with various NBR34 content [21].

NBR34 content (wt.%)	Tensile strength at 100% elongation (MPa)	Tensile strength at break (MPa)	Elongation at break (%)
0	4.9	32.5	29
5	4.9	33.5	29.5
10	5	35	27
15	4.9	33.5	25.5
20	5.1	33	25
25	4.6	30	12
30	4	26.5	7

Table 8.9 Mechanical properties of PA-612/POE-g-MA blends [24].

POE-g-MA content (wt.%)	Tensile strength (MPa)	Tensile modulus (MPa)	Impact strength (J/m)	Elongation at break (%)
0	53	315	315	40
5	50	303	400	30
10	44	278	555	140
20	35	260	875	375
35	28	195	650	365

The stress–strain curve indicated that between the range of 5 and 10 wt.% of POE-g-MA, there was a brittle to ductile transition present. Further, the specimens with 10–20 wt.% POE-g-MA exhibited necking phenomena, and also the toughness was found to increase till 20 wt.% addition of POE-g-MA [24].

8.4 DMP of elastomer blends

Linear low-density polyethylene (LLDPE) and ethylene-co-methyl acrylate (EMA) were blended together, and dynamic mechanical properties were estimated for different blend composition. The blends consisting of various proportion of EMA (0 wt.%, 20 wt.%, 30 wt.%, 50 wt.%, 70 wt.%, and 100 wt.%) were prepared. DMA tests were conducted within the temperatures from −150 °C to 125 °C and at a frequency range of 0.4–40 Hz. According to the results, the storage modulus was found to increase as EMA content increased. Additionally, as the EMA content increased, the temperature corresponding to loss tangent peak shifted toward lower value. These observations were attributed to the reduction in crystallinity and improvement of free volume with the increase in EMA content [9].

Elastomer blends comprising natural rubber (NR) and PP were produced, and their DMP was found out. In addition, DMA tests were also carried out for maleic anhydride-grafted natural rubber (NR-g-MAH)/maleic anhydride-grafted polypropylene (PP-g-MAH) blends (Table 8.10) [6].

Table 8.10 DMP of NR/PP and NR-g-MAH/PP-g-MAH blends [6].

Blends	Storage modulus (MPa)	Loss modulus (MPa)	Loss tangent (tan δ)
NR/PP	5850	1248	0.702
NR-g-MAH/PP-g-MAH	3000	556.37	0.84

Table 8.11 Dynamic mechanical properties of PP/EPDM blends [31].

Type of specimen	Content of PP (wt.%)	Content of EPDM (wt.%)	Max. storage modulus (MPa)	Max. loss tangent (tan δ)
Untreated	70	30	5000	1.41
Silane cross-linked	90	10	4900	1.22
	80	20	4550	1.29
	70	30	4520	1.25
	60	40	4500	1.25

The dynamic mechanical analysis of elastomer blends containing silane cross-linked (SC) PP and ethylene propylene diene monomer (EPDM) was performed for different composition ranges [31]. The three-point bending mode was employed, and the heating was done between −80 °C and 80°C at 5 °C/min. The results are presented in Table 8.11.

The storage modulus was found to decrease with the addition of EPDM as EPDM has low storage modulus. Also, the silane cross-linked specimen with 70–30 ratio had lower storage modulus than untreated sample, and this is can be because of the degradation of PP. On looking at the loss tangent curves, it was concluded that the existence of silane cross-linking resulted in the improvement in the interfacial bonding of PP and EPDM [31].

The viscoelastic characteristics of cyclic olefin copolymer/polyolefin elastomer (COC/POE) blends were investigated for the entire composition range [23]. The blends were manufactured using a twin-screw compounder. The DMA was performed at a frequency of 1 Hz and between −100 °C and 120 °C, at 3 °C/min heating rate. The elastic and loss modulus of the blends are listed out in Table 8.12.

Table 8.12 Elastic and loss modulus of COC/POE blends [23].

COC content (wt.%)	POE content (wt.%)	Max. elastic modulus (MPa)	Max. loss modulus (MPa)
90	10	820	80
80	20	815	80
70	30	1000	36
60	40	690	42
50	50	340	12

Table 8.13 DMP of PP/SEBS blends [5].

PP content (wt.%)	SEBS content (wt.%)	Storage modulus (MPa)	Loss modulus (MPa)	Loss tangent
80	20	5200	–	–
60	40	5800	–	–
40	60	8800	3000	0.163

It was noticed that when the POE content was less than 30 wt.%, the storage modulus was higher than that for pure COC. This was due to the fact that, at low POE contents, the elastic POE chains diffuse into the amorphous rigid COC chains, and this results in better properties. The temperature corresponding to maximum loss modulus was more or less the same for all the blends, and this shows that the two polymers are immiscible in nature [23].

Elastomer blends of PP and styrene-(ethylene-butylene)-styrene (SEBS) were fabricated using material extrusion 3D printing and their DMP was investigated, and the outcomes were presented in Table 8.13 [5].

Polyamide-6 (PA6) and Ethylene-Butylene (EB) elastomers were blended together, and the blends were produced using two different methods, namely polymerization and extrusion. The effect of blending and the manufacturing method on the DMP was studied, and the results are given in Table 8.14 [30].

PP was blended with ethylene/octene copolymer elastomers (POE) and styrene-ethylene/butylene styrene triblock copolymer (SEBS) at different weight ratios (20 wt.%, 25 wt.%, 30 wt.%), and their dynamic mechanical properties were estimated (Table 8.15) [44].

Table 8.14 DMP of PA6/EB blends [30].

Blend preparation method	PA6 content (wt.%)	EB content (wt.%)	Storage modulus (MPa)	Loss tangent
Polymerization	100	0	7500	0.135
	85	15	7500	0.123
Extrusion	100	0	7500	0.16
	85	15	8100	0.175

Table 8.15 DMP of PP/POE and PP/SEBS blends [44].

PP content (wt.%)	POE content (wt.%)	SEBS content (wt.%)	Storage modulus (MPa)	Loss modulus (MPa)	Loss tangent
80	20	0	3200	114	0.1
75	20	0	3125	95	0.102
70	30	0	2800	85	0.11
0.80	0	20	3375	114	0.11
75	0	25	3225	95	0.105
70	0	30	3075	80	0.108

Table 8.16 DMP of PA-612/POE-g-MA blends [24].

PA-612 content (wt.%)	POE-g-MA content (wt.%)	Maximum storage modulus (MPa)	Loss tangent
95	5	1880	0.083
90	10	1860	0.085
80	20	1750	0.087
65	35	1700	0.094

With the addition of POE and SEBS, there was a decrease in the storage modulus, and this can be either due to the decline in crystallinity of PP or the low storage modulus of the elastomer (POE or SEBS) [44].

Polyamide-612 (PA-612) and maleic anhydride grafted poly (ethylene–octene) elastomer (POE-g-MA) were blended together, and their DMP was determined (Table 8.16). POE-g-MA was added from 0 to 35 wt.%. The DMA test was performed between −80 and 120 °C at a heating rate of 5 °C/min and frequency of 1 rad/s [24].

8.5 Mechanical behavior of elastomer composites

Similar to elastomer blends, research works investigate both SMP and DMP of elastomer composites. In the case of composites, the influence of fiber or reinforcement loading on the properties is analyzed.

8.6 SMP of elastomer composites

The PLA was blended together with TPU, and further, elastomer composites were produced by reinforcing these blends with glass fibers (GF). Various mechanical properties such as tensile strength ($\sigma_{tensile}$), flexural strength ($\sigma_{flexural}$), Young's modulus (E), Flexural modulus ($E_{flexural}$), elongation at break (ε), critical stress intensity factor (K_{IC}), and critical strain energy release rate (G_{IC}) were measured. The elastomer blend was prepared by adding 10 wt.% of TPU and the composites were produced by adding 15 wt.% of GF [41]. The SMP of PLA/TPU blends and composites are presented in Table 8.17.

Table 8.17 SMP of PLA/TPU blends and composites [41].

Samples	$\sigma_{Tensile}$ (MPa)	E (GPa)	$\sigma_{Flexural}$ (MPa)	$E_{flexural}$ (GPa)	ε (%)	G_{IC} (kJ/m^2)	K_{IC} (MPam$^{1/2}$)
PLA	58.6	2.85	97.9	3.6	4.88	5.44	3.43
PLA/10 TPU	53.4	2.49	88.9	3.03	18.07	9.46	6.96
PLA/10 TPU/15 GF	64.4	3.8	99.7	4.69	4.21	6.84	4.85

Table 8.18 Mechanical properties of TPNR-KF and PP/EPDM-KF [2].

Sample	Tensile strength (MPa)	Tensile modulus (GPa)	Flexural strength (MPa)	Flexural modulus (GPa)	Impact strength (MPa)
TPNR	19	0.25	21	1.25	3
TPNR-KF	21	0.52	25	1.95	4
PP/EPDM	17	0.5	28	1	5
PP/EPDM-KF	23	1.2	39	2.55	10

TPU is added to PLA to enhance its ductile properties as PLA is a brittle polymer. However, addition of TPU has led to a decline in strength and modulus. Therefore, glass fibers are added to the blend to provide superior mechanical properties. The inclusion of GF results in better load transfer between matrix and reinforcement and also, reduces the movement of polymer macromolecular chains. These factors contribute to the higher strength of the composite [41].

The PP was blended along with NR and EPDM separately, and the resultant blends were reinforced with 20 wt.% kenaf fiber (KF) to form thermoplastic natural rubber (TPNR)-KF and PP/EPDM-KF composites. Tensile, flexural, and impact properties of these composites were studied (Table 8.18) [2].

It was witnessed that the tensile strength of TPNR and PP/EPDM has improved upon kenaf fiber addition, and also, the tensile modulus has almost doubled for both the composites. This is because of the high mechanical properties that the reinforcement possesses and the efficient stress transfer between the constituents. Additionally, the better flexural properties of PP/EPDM-KF compared to TPNR-KF illustrate that PP/EPDM has superior interfacial bonding with kenaf fiber. The better impact strength of PP/EPDM-KF is also due to the better fiber-matrix adhesion, which decreases fiber pullouts, thus requiring greater amount of energy to cause failure [2].

The influence of montmorillonite (MMT) clay on the tensile and impact properties of amorphous polyamide (a-PA) and ethylene-1-octene copolymer (EOR) blends was analyzed [43]. The nanocomposites were manufactured by melt compounding followed by injection molding. Different weight percentage of MMT was added as reinforcement. The resultant tensile and impact properties of a-PA/EOR/MMT composites are provided in Table 8.19.

Table 8.19 Tensile and impact properties of a-PA/EOR/MMT composites [43].

a-PA content (wt.%)	EOR content (%)	MMT content (%)	Tensile modulus (GPa)	Tensile strength (MPa)	Izod impact strength (J/m)
80	20	0	1.85	48	80
		1.67	2.47	58	58
		3.4	2.92	62	40
		5.48	3.7	47	21
		7.5	4.06	38	18.5

Table 8.20 Young's modulus and strain at break of TPU/CNS and TPU/CNT composites [38].

Matrix	CNS content (wt.%)	Young's modulus (MPa)	Strain at break (mm/mm)	CNT content (wt.%)	Young's modulus (MPa)	Strain at break (mm/mm)
TPU	0	9	13.8	0	9	13.8
	1	20	11	1	8	13.8
	2	37.5	4.2	2	11.5	10.4
	3	43	2.4	3	17	8.7

Nanocomposites comprising TPU as matrix and different proportion of carbon nanostructures (CNS), carbon nanotubes (CNT) as reinforcements were fabricated. The samples were subjected to tensile tests to assess the Young's modulus and elongation at break [38]. The Young's modulus and strain at break of TPU/CNS and TPU/CNT composites are given in Table 8.20.

The Young's modulus of TPU/CNS is significantly higher when compared to TPU/CNT composites. However, when CNS is added, the elasticity reduces considerably. The reduction in elongation at break of TPU/CNS is ascribed to the interconnected mesh-like morphology of CNS, which limits the TPU penetration in the meshes, and this creates a local stress concentration upon loading. In the case of CNT, these meshes are not connected, and hence, they do not affect the elastic properties of TPU much. The authors suggested these composites to be used in strain sensors [38].

TPEE was reinforced with graphene nanoplatelets (GNP) and slag particles. Here, 0.1 wt.% of GNP and 5 wt.% of slag particles were bonded together using a suitable chemical technique and then were mixed with TPEE. This was termed as TPEEGS. In addition, TPEE composites were also prepared by physically mixing 0.1 wt.% GNP and 5 wt.% slag particles and adding them to TPEE (TPEEGS′), individually adding 0.1 wt.% GNP to TPEE (TPEEG) and 5 wt.% slag particles to TPEE (TPEES) [33]. The obtained SMP of TPEE and its composites are given in Table 8.21.

When bonded particles are used, there is a reduction in the Young's modulus compared to the specimens with physically mixed particles, and this is due to the low

Table 8.21 Mechanical properties of TPEE and its composites.

Specimen	Yield strength (MPa)	Elongation at break (%)	Young's modulus (MPa)
TPEE	6.9	9.06	37
TPEEG	7.92	9.21	45.74
TPEES	7.19	7.39	46.1
TPEEGS	9.73	8.59	49.82
TPEEGS′	7.72	6.38	61.13

Table 8.22 Tensile strength and elongation at break of TPU/WF composites [7].

Wood flour content (wt.%)	Tensile strength (MPa)	Elongation at break (%)
0	27.47	521.5
10	18.66	428.6
20	15.61	378.5
30	18.16	50.2
40	20.62	49.8

effective volume ratio. On the contrary, the TPEEGS composites have the maximum yield strength, and this is associated with the presence of synergistic effect between the GNP and slag particles bonds. When a large deformation is present, the particles tend to be pulled apart from each other, and due to this, more amount of energy is consumed by the composite. This results in a higher yield strength [33].

Elastomer composites comprising TPU and different contents of wood flour (WF) were fabricated using fused deposition molding, and their mechanical properties were determined (Table 8.22) [7].

One major parameter that determined the tensile strength was the efficient and uniform load transfer. When more WF was added, it resulted in issues such as poor wetting of resin and agglomeration, and this caused poor stress transfer [7].

Polyurea urethane (PUUR) was reinforced with different weight proportions of hollow glass micro spheres (HGS) to form elastomer composites, and their mechanical properties were found out [29]. These micro spheres were added instead of glass fibers as they have minimum surface energy, thus mitigating the issue of agglomeration. The mechanical properties of PUUR/HGS composites for different HGS content are presented in Table 8.23.

Table 8.23 SMP of PUUR/HGS composites for different HGS content [29].

HGS content (wt.%)	Tensile strength (MPa)	Young's modulus (MPa)	Elongation at break (%)	Hardness (shore A)
0	27.3	26	289	86.5
4.8	23.8	44	241	74.2
9.1	24.6	47	267	75.3
13	25.1	53	305	76.6
16.7	21.9	57	336	78.8
23.1	15.8	82	245	81.1
28.6	15.6	141	236	82.7
33.3	–	–	–	83
37.5	–	–	–	86.4

Table 8.24 Mechanical properties of XNBR composites [25].

Specimen	Reaction time with PCPA (hrs.)	Young's modulus (MPa)	Tensile strength (MPa)	Elongation at break (%)
XNBR	–	2.06	2.59	233.9
XNBR/BN	–	15.24	4.32	125.8
XNBR/BN-	3	14.65	8.2	122.9
PCPA	6	14.1	7.25	131.8
	9	13.2	6.65	130.1
	12	10.98	6.54	164.3

Even though the bonding between the PUUR matrix and HGS was good, there was a decline in tensile strength as HSG was added. The matrix here can undergo large deformations when compared to glass spheres. Therefore, when tensile loads were applied, the matrix deformed elastically, whereas the glass spheres did not. After a certain point, the HGS tends to get peeled off from the matrix, and this created micro voids around the reinforcement. This led to a reduction in the tensile strength and elongation at break. The authors also concluded that the optimum reinforcement content to be added is between 10 and 20 wt.% to achieve maximum strength ratios [29].

Boron nitride (BN) platelets were chemically modified using poly(catechol-polyamine) (PCPA) at different reaction times (3, 6, 9, or 12 h), and these modified BN platelets were reinforced to carboxylated acrylonitrile-butadiene rubber (XNBR) to form elastomer composites [25]. The amount of reinforcement that was added was 30 vol.%. The SMP of XNBR composites is given in Table 8.24.

As XNBR/BN-PCPA composites have better interfacial bonding between matrix and reinforcement, they exhibit better tensile strength than XNBR/BN composites. However, the tensile strength decreases with the increase in reaction time. This is because the reinforcing effect of BN-PCPA is lower when compared to pure BN platelets due to the PCPA coating. Additionally, the whole content of the polymer phase in the composites is increased due to the soft PCPA coating on boron nitride reinforcement [25].

A novel polyester bio-based engineering elastomer (BEE), poly(1,4-butanediol/1,3-propanediol/sebacate/itaconate/succinate), was produced using five different monomers that are obtained from biomass [42]. These elastomers were further reinforced with different contents (in parts per 100 rubber or phr) of nano-silica (SiO_2) to form bio-elastomer nanocomposites. The SMP of these BEE/SiO_2 nanocomposites is given in Table 8.25.

Table 8.25 SMP of these BEE/SiO_2 nanocomposites [42].

SiO_2 content (phr)	Tensile strength (MPa)	Stress at 100% strain (MPa)	Elongation at break (%)	Hardness (shore A)
30	14.8	5.7	223	72
40	18.9	7.4	216	75
50	20.5	10.1	189	81

The excellent mechanical properties of these BEE nanocomposites were because of the strong bonding present between BEE and nano-silica. The BEE synthesized have carboxyl, ester, and terminal hydroxyl groups. These groups tend to form hydrogen bonds with the silanol groups present on the nano-silica surface, and this contributes to the high performance of the BEE nanocomposites [42].

8.7 DMP of elastomer composites

Polycarbonate/ethylene methyl acrylate (PC/EMA) elastomeric composites were prepared by reinforcing it with different contents (in part per 100 rubber (phr)) of multiwalled carbon nanotubes (MWCNT) [3]. The DMP of these composites (PC/EMA-MWCNT) was further determined (Table 8.26).

The storage modulus increased with the addition of MWCNT, and this was because of the enhanced interfacial bonding between the constituents. On the contrary, storage modulus reduced as temperature was raised, and this is due to the rise in the movement of polymer chains. Similar to storage modulus, the loss modulus value also increased with MWCNT addition, and this was associated with the improvement in the dissipation of energy due to higher internal friction [3].

Thermoplastic copolyester elastomers (TCEs) were added with different contents of short glass fiber (SGF), short carbon fiber (SCF), and fillers such as silicon carbide (SiC), Alumina (Al_2O_3), and polytetrafluoroethylene (PTFE) [34]. The resultant composites were subjected to dynamic mechanical tests and the outcomes are given in Table 8.27.

The storage modulus was seen to be significantly greater for composites containing 40 wt.% of fillers and fibers, and this was ascribed to the stiff interface that was formed between the constituents. As the material becomes stiffer, its capability to store energy is improved. The inclusion of 32 wt.% reinforcement provided the least value of damping factor, and this was because the fillers and fibers restricted the mobility of the polymer chains, and hence, damping performance of the composite was reduced [34].

Diglycidyl ether of bisphenol-A (DGEBA) epoxy resin was cured with diamino diphenyl sulfone (DDS), and altered epoxy resin was blended with 10 parts per 100

Table 8.26 DMP of PE/EMA- MWCNT composites [3].

Content of MWCNT (phr)	Storage modulus at 130 °C (MPa)	Loss modulus (MPa)	Damping factor (tan δ)	Glass transition temperature (°C)
0	2801.73	485	1.62	143.6
0.5	3449.57	600	1.42	145.4
1	3505.44	625	1.3	148.1
3	3617.18	700	1.06	148.9
5	3717.74	760	1	150.3
10	3784.78	855	0.91	153

Table 8.27 DMP of TCE composites [34].

TCE content (wt.%)	Fiber content (wt.%)	Filler content (wt.%)	Storage modulus (MPa)	Loss modulus (MPa)	Damping factor
85	–	PTFE (15)	2505	93	0.075
68	SGF (20)	PTFE (12)	3151	109	0.06
60	SGF (17.5) + SCF (2.5)	PTFE (10) + SiC (5) + Al_2O_3 (5)	4381	171	0.0768

rubber (10 phr) of Poly(styrene-co-acrylonitrile) (SAN) [19]. The blends were further reinforced with different amounts of glass fiber and their DMP was investigated (Table 8.28).

PVC and ethylene vinyl acetate (EVA) were blended together and the elastomer blends were reinforced with ungrafted and grafted kenaf fibers [4]. Further, dynamical mechanical tests were conducted, and the results are presented in Table 8.29.

The storage modulus increases once kenaf fibers are added, and this is due to the better stress transfer that prevails between matrix and reinforcement, thus resulting in greater stiffness. Two loss modulus peaks were formed at −32 °C (β) and 77 °C (α), respectively. When fiber was added, the α peak was found to shift toward higher

Table 8.28 DMP of epoxy-SAN/glass composites [19].

Matrix	Content of glass fiber (vol.%)	Storage modulus at room temperature (GPa)	Loss modulus (MPa)	Damping factor
Neat epoxy	–	2.77	200	0.853
Epoxy/SAN	10	7.4	920	0.168
	20	9.6	970	0.162
	30	12.6	1030	0.148
	40	13.8	1120	0.136
	50	15.7	800	0.111
	60	14.2	890	0.085

Table 8.29 DMP of PVC/EVA/kenaf composites [4].

Matrix	Type of fiber	Fiber content (wt.%)	Storage modulus (MPa)	Loss modulus (MPa)	Damping factor
PVC/	–	0	5500	310	0.78
EVA	Ungrafted kenaf	30	6950	420	0.75
	Grafted kenaf	30	7100	450	0.71

temperature, and this is attributed to the reduction in molecular movement between matrix and reinforcement. The β peak, which refers to the movement of amorphous region side chains from the polymer backbone, was higher for composites with grafted kenaf fibers, and this means that it has greater amount of amorphous content [4].

Composites were produced by reinforcing sisal fibers with powdered tire rubber, and dynamic mechanical analysis of the composites was conducted. Sisal fibers were reinforced between the range of 5 and 30 wt.%. The curves indicated that the storage modulus was higher for greater fiber loading, and this was because of the enhanced transfer of stress between the matrix and fiber. The storage modulus increased up to 20 wt.% of fiber loading, and further, it was constant at 30 wt.%. The authors suggested that this trend may be because of the fiber–fiber interaction present as fiber loading was increased. Similar to storage modulus, loss modulus was also seen to increase with the addition of fibers. In addition, the storage and loss moduli of the composites were also estimated for different fiber lengths (2, 5, 10, 15 mm). The results indicated that the dynamic performance of the composites was maximum when fiber length was 10 mm, and they continued to decrease at longer fiber lengths. This reduction was associated with fiber entanglements [27].

PVC and NBR were blended together, and the blends were reinforced with different quantities (in parts per 100 rubber (phr)) of single-wall carbon nanotube (SWCNT) and nanoclay (NC) [18]. The composites were then subjected to dynamic mechanical analysis and outcomes are given in Table 8.30.

The reduction in the value of damping factor represents the occurrence of elastic interaction between the phases present. Therefore, the addition of SWCNT has resulted in the decrease of damping factor, and this shows the presence of elastic interaction between the polymer and SWCNT. Additionally, the inclusion of nanoclay further reduces the damping factor, and this indicates that the interaction is greater when nanoclay is added in comparison to SWCNT [18].

The DMP of NR-short coir fiber composites was studied. The quantity of coir fiber added was in terms of part per 100 rubber (phr). The loss modulus was found to be greater for the composites in comparison to NR, and this means that the composites

Table 8.30 DMP of PVC/NBR composites [18].

Content of SWCNT (phr)	Content of NC (phr)	Storage modulus (GPa)	Damping factor
0	0	1	0.7
0.5	0	1.2	0.63
1	0	1.75	0.6
1.5	0	2.2	0.58
0	3	2.2	0.42
0	5	2.3	0.4
0	7	2.25	0.39

Table 8.31 DMP of the natural rubber/HCNF composites [45].

Amount of HCNF (phr)	Storage modulus (MPa)	Damping factor
0	22.95	2.49
1	22.9	2.26
3	20.49	2.11
5	18.79	1.69

have better heat dissipation capacity. This is because as fiber content increases in the composite, the surface area at the interface is greater, and hence, the energy that is dissipated is more. Additionally, the storage modulus also increased with the addition of coir fibers. Unlike loss modulus and storage modulus, damping factor value declined as fiber content was higher. However, from the storage modulus curves, the authors concluded that the composites have better thermal stability than NR [15].

NR was reinforced with different amounts (1–5 phr) of helical carbon nanofibers (HCNF), and the DMP of the composites was estimated and presented in Table 8.31 [44].

The value of damping factor tends to reduce as fibers are added to the matrix, and this is because of the fiber-matrix adhesion, which prevents the movement of macromolecules. Hence, when more reinforcement is added, there is a tendency for greater number of immobile rubber layers to be formed around the reinforcement. This also indicated the efficient bonding between HCNF and rubber matrix [45].

8.8 Conclusions

The elastomer blends and composites with improved mechanical properties found its applications in automobile, railways, electrical and electronics engineering, and bio medical applications. It is possible to alter the different properties of elastomer composites by varying the type, content, or size of reinforcements. This flexibility supports the industries to use the elastomer composites for wide range of applications. Huge variety of reinforcements are available in the market worldwide, but only limited numbers were utilized to fabricate elastomer composites. Therefore, future research in the development of new elastomeric composites serves the industrial and market needs.

References

[1] M. Abu-Abdeen, I. Elamer, Mechanical and swelling properties of thermoplastic elastomer blends, Mater. Des. 31 (2) (2010) 808–815, https://doi.org/10.1016/j.matdes.2009.07.059.

[2] H. Anuar, A. Zuraida, Improvement in mechanical properties of reinforced thermoplastic elastomer composite with kenaf bast fibre, Compos. B Eng. 42 (3) (2011) 462–465, https://doi.org/10.1016/j.compositesb.2010.12.013.

[3] N. Bagotia, D.K. Sharma, Systematic study of dynamic mechanical and thermal properties of multiwalled carbon nanotube reinforced polycarbonate/ethylene methyl acrylate nanocomposites, Polym. Test. 73 (2019) 425–432, https://doi.org/10.1016/j.polymertesting.2018.12.006.

[4] N.A. Bakar, C.Y. Chee, L.C. Abdullah, C.T. Ratnam, N.A. Ibrahim, Thermal and dynamic mechanical properties of grafted kenaf filled poly (vinyl chloride)/ethylene vinyl acetate composites, Mater. Des. 65 (2015) 204–211, https://doi.org/10.1016/j.matdes.2014.09.027.

[5] S.S. Banerjee, S. Burbine, N.K. Shivaprakash, J. Mead, 3D-printable PP/SEBS thermoplastic elastomeric blends: preparation and properties, Polymers 11 (2) (2019), https://doi.org/10.3390/polym11020347.

[6] S. Benmesli, F. Riahi, Dynamic mechanical and thermal properties of a chemically modified polypropylene/natural rubber thermoplastic elastomer blend, Polym. Test. 36 (2014) 54–61, https://doi.org/10.1016/j.polymertesting.2014.03.016.

[7] H. Bi, Z. Ren, R. Guo, M. Xu, Y. Song, Fabrication of flexible wood flour/thermoplastic polyurethane elastomer composites using fused deposition molding, Ind. Crop. Prod. 122 (2018) 76–84, https://doi.org/10.1016/j.indcrop.2018.05.059.

[8] A. Boonmahitthisud, A. Mongkolvai, S. Chuayjuljit, Toughness improvement in bio-based poly(-lactic acid)/epoxidized natural rubber blend reinforced with nanosized silica, J. Polym. Environ. (2021), https://doi.org/10.1007/s10924-021-02063-z.

[9] J.S. Borah, T.K. Chaki, Dynamic mechanical, thermal, physico-mechanical and morphological properties of LLDPE/EMA blends, J. Polym. Res. 18 (4) (2011) 569–578, https://doi.org/10.1007/s10965-010-9450-0.

[10] P. Deepalekshmi, P.M. Visakh, A.P. Mathew, A.K. Chandra, S. Thomas, Advances in elastomers: their composites and nanocomposites: state of art, new challenges and opportunities, in: P. Visakh, S. Thomas, A. Chandra, A. Mathew (Eds.), Advanced Structured Materials, Springer, 2013. https://doi.org/10.1007/978-3-642-20928-4_1.

[11] N. Dishovsky, M. Mihaylov, Elastomer-Based Composite Materials: Mechanical, Dynamic and Microwave Properties, and Engineering Applications, CRC Press, 2018.

[12] P.A. Eutionnat-Diffo, A. Cayla, Y. Chen, J. Guan, V. Nierstrasz, C. Campagne, Development of flexible and conductive immiscible thermoplastic/elastomer monofilament for smart textiles applications using 3d printing, Polymers 12 (10) (2020) 1–31. https://doi.org/10.3390/polym12102300.

[13] F. Feng, L. Ye, Morphologies and mechanical properties of polylactide/thermoplastic polyurethane elastomer blends, J. Appl. Polym. Sci. 119 (5) (2011) 2778–2783. https://doi.org/10.1002/app.32863.

[14] Y. Gao, J. Li, Y. Li, Y.Q. Yuan, S.H. Huang, B.X. Du, Effect of elastomer type on electrical and mechanical properties of polypropylene/elastomer blends, in: Proceedings of the International Symposium on Electrical Insulating Materials, vol. 2, Institute of Electrical Engineers of Japan, 2017, pp. 574–577. https://doi.org/10.23919/ISEIM.2017.8166554.

[15] V.G. Geethamma, G. Kalaprasad, G. Groeninckx, S. Thomas, Dynamic mechanical behavior of short coir fiber reinforced natural rubber composites, Compos. Appl. Sci. Manuf. 36 (11) (2005) 1499–1506. https://doi.org/10.1016/j.compositesa.2005.03.004.

[16] A. Georgopoulou, F. Clemens, Piezoresistive elastomer-based composite strain sensors and their applications, ACS Appl. Electr. Mater. 2 (7) (2020) 1826–1842. https://doi.org/10.1021/acsaelm.0c00278.

[17] A. Georgopoulou, S. Michel, B. Vanderborght, F. Clemens, Piezoresistive sensor fiber composites based on silicone elastomers for the monitoring of the position of a robot arm, Sensor Actuator Phys. 318 (2021) 112433. https://doi.org/10.1016/j.sna.2020.112433.

[18] A. Hajibaba, G. Naderi, E. Esmizadeh, M.H.R. Ghoreishy, Morphology and dynamic-mechanical properties of PVC/NBR blends reinforced with two types of nanoparticles, J. Compos. Mater. 48 (2) (2014) 131–141. https://doi.org/10.1177/0021998312469242.

[19] N. Hameed, P.A. Sreekumar, B. Francis, W. Yang, S. Thomas, Morphology, dynamic mechanical and thermal studies on poly(styrene-co-acrylonitrile) modified epoxy resin/glass fibre composites, Compos. Appl. Sci. Manuf. 38 (12) (2007) 2422–2432. https://doi.org/10.1016/j.compositesa.2007.08.009.

[20] J. Huang, J. Wang, Y. Qiu, D. Wu, Mechanical properties of thermoplastic polyester elastomer controlled by blending with poly(butylene terephthalate), Polym. Test. 55 (2016) 152–159. https://doi.org/10.1016/j.polymertesting.2016.08.020.

[21] H.G. Im, K.R. Ka, C.K. Kim, Characteristics of polyurethane elastomer blends with poly(-acrylonitrile- co-butadiene) rubber as an encapsulant for underwater sonar devices, Ind. Eng. Chem. Res. 49 (16) (2010) 7336–7342. https://doi.org/10.1021/ie100975n.

[22] K. Pielichowski, A. Leszczynska, Structure-Property relationships in polyoxymethylene/thermoplastic polyurethane elastomer blends, J. Polym. Eng. (2005). https://doi.org/10.1515/POLYENG.2005.25.4.359.

[23] H.A. Khonakdar, S.H. Jafari, M.N. Hesabi, Miscibility analysis, viscoelastic properties and morphology of cyclic olefin copolymer/polyolefin elastomer (COC/POE) blends, Compos. B Eng. 69 (2015) 111–119. https://doi.org/10.1016/j.compositesb.2014.09.034.

[24] S. Kumar, B.K. Satapathy, S.N. Maiti, Correlation of morphological parameters and mechanical performance of polyamide-612/poly (ethylene-octene) elastomer blends, Polym. Adv. Technol. 24 (5) (2013) 511–519. https://doi.org/10.1002/pat.3113.

[25] X. Liu, Q. Han, D. Yang, D. Yang, Y. Ni, Y. Ni, L. Yu, L. Yu, Q. Wei, Q. Wei, L. Zhang, Thermally conductive elastomer composites with poly(catechol-polyamine)-modified boron nitride, ACS Omega 5 (23) (2020) 14006–14012. https://doi.org/10.1021/acsomega.0c01404.

[26] M. Maroufkhani, A.A. Katbab, W. Liu, J. Zhang, Polylactide (PLA) and acrylonitrile butadiene rubber (NBR) blends: the effect of ACN content on morphology, compatibility and mechanical properties, Polymer 115 (2017) 37–44. https://doi.org/10.1016/j.polymer.2017.03.025.

[27] M.A. Martins, L.H.C. Mattoso, Short sisal fiber-reinforced tire rubber composites: dynamical and mechanical properties, J. Appl. Polym. Sci. 91 (1) (2004) 670–677. https://doi.org/10.1002/app.13210.

[28] L.A. Maynard, B.L. DeButts, J.R. Barone, Mechanical and thermal properties of polyolefin thermoplastic elastomer blends, Plast. Rubb. Compos. 48 (8) (2019) 338–346. https://doi.org/10.1080/14658011.2019.1625633.

[29] K. Mizera, M. Chrząszcz, J. Ryszkowska, Thermal and mechanical properties of ureaurethane elastomer composites with hollow glass spheres, Polym. Compos. 39 (6) (2018) 2019–2028. https://doi.org/10.1002/pc.24162.

[30] T.S. Omonov, C. Harrats, N. Moussaif, G. Groeninckx, S.G. Sadykov, N.R. Ashurov, Polyamide 6/ethylene-butylene elastomer blends generated via anionic polymerization of ε-caprolactam: phase morphology and dynamic mechanical behavior, J. Appl. Polym. Sci. 94 (6) (2004) 2538–2544. https://doi.org/10.1002/app.21249.

[31] H. Peng, M. Lu, H. Wang, Z. Zhang, F. Lv, M. Niu, W. Wang, Comprehensively improved mechanical properties of silane crosslinked polypropylene/ethylene propylene diene monomer elastomer blends, Polym. Eng. Sci. 60 (5) (2020) 1054–1065. https://doi.org/10.1002/pen.25361.

[32] M. Pire, S. Norvez, I. Iliopoulos, B. Le Rossignol, L. Leibler, Epoxidized natural rubber/dicarboxylic acid self-vulcanized blends, Polymer 51 (25) (2010) 5903–5909. https://doi.org/10.1016/j.polymer.2010.10.023.

[33] Y. Qiu, D. Wu, W. Xie, Z. Wang, S. Peng, Thermoplastic polyester elastomer composites containing two types of filler particles with different dimensions: structure design and mechanical property control, Compos. Struct. 197 (2018) 21–27. https://doi.org/10.1016/j.compstruct.2018.05.035.

[34] H. Rajashekaraiah, S. Mohan, P.K. Pallathadka, S. Bhimappa, Dynamic mechanical analysis and three-body abrasive wear behaviour of thermoplastic copolyester elastomer composites, Adv. Tribol. (2014). https://doi.org/10.1155/2014/210187.

[35] G. Rajeshkumar, V. Hariharan, S. Indran, M.R. Sanjay, S. Siengchin, J.P. Maran, N.A. Al-Dhabi, P. Karuppiah, Influence of sodium hydroxide (NaOH) treatment on mechanical properties and morphological behaviour of phoenix sp. fiber/epoxy composites, J. Polym. Environ. 29 (3) (2021) 765–774. https://doi.org/10.1007/s10924-020-01921-6.

[36] G. Raju, C.T. Ratnam, N.A. Ibrahim, M.Z.A. Rahman, W.M.Z.W. Yunus, Enhancement of PVC/ENR blend properties by poly(methyl acrylate) grafted oil palm empty fruit bunch fiber, J. Appl. Polym. Sci. 110 (1) (2008) 368–375. https://doi.org/10.1002/app.28662.

[37] S. Ramakrishnan, K. Krishnamurthy, G. Rajeshkumar, M. Asim, Dynamic mechanical properties and free vibration characteristics of surface modified jute fiber/nano-clay reinforced epoxy composites, J. Polym. Environ. 29 (4) (2021) 1076–1088. https://doi.org/10.1007/s10924-020-01945-y.

[38] Z. Sang, K. Ke, I. Manas-Zloczower, Effect of carbon nanotube morphology on properties in thermoplastic elastomer composites for strain sensors, Compos. Appl. Sci. Manuf. 121 (2019) 207–212. https://doi.org/10.1016/j.compositesa.2019.03.007.

[38a] K.R. Sumesh, V. Kavimani, G. Rajeshkumar, S. Indran, G. Saikrishnan, Effect of banana, pineapple and coir fly ash filled with hybrid fiber epoxy based composites for mechanical and morphological study, J. Mater. Cycles Waste Manag. (2021). https://doi.org/10.1007/s10163-021-01196-6.

[39] S. Thomas, Advances in elastomers: their blends and interpenetrating networks-state of art, new challenges and opportunities, in: Advances in Elastomers I: Blends and Interpenetrating Networks, vol. 1, 2013.

[40] D. V Rosato, D. V Rosato, Plastics Processing Data Handbook, Springer Science & Business Media, 2012. https://doi.org/10.1007/978-94-010-9658-4.

[41] S.D. Varsavas, C. Kaynak, Effects of glass fiber reinforcement and thermoplastic elastomer blending on the mechanical performance of polylactide, Compos. Commun. 8 (2018) 24–30. https://doi.org/10.1016/j.coco.2018.03.003.

[42] T. Wei, L. Lei, H. Kang, B. Qiao, Z. Wang, L. Zhang, P. Coates, K.C. Hua, J. Kulig, Tough bio-based elastomer nanocomposites with high performance for engineering applications, Adv. Eng. Mater. 14 (1–2) (2012) 112–118. https://doi.org/10.1002/adem.201100162.

[43] Y. Yoo, L. Cui, P.J. Yoon, D.R. Paul, Morphology and mechanical properties of rubber toughened amorphous polyamide/MMT nanocomposites, Macromolecules 43 (2) (2010) 615–624. https://doi.org/10.1021/ma902232g.

[44] C.J. Zheng, J.M. Yang, H. Zhao, Q.C. Chen, AC performance, physical and mechanical properties of polypropylene/polyolefin elastomers blends, in: Proceedings of the IEEE International Conference on Properties and Applications of Dielectric Materials, vols. 2018, Institute of Electrical and Electronics Engineers Inc, 2018, pp. 910–913. https://doi.org/10.1109/ICPADM.2018.8401203.

[45] X. Zheng, Y. Jin, J. Chen, B. Li, Q. Fu, G. He, Mechanical properties and microstructure characterization of natural rubber reinforced by helical carbon nanofibers, J. Mater. Sci. 54 (19) (2019) 12962–12971. https://doi.org/10.1007/s10853-019-03771-7.

9 Thermal behavior of elastomer blends and composites

Atul Kumar Maurya, Rupam Gogoi, Gaurav Manik

DEPARTMENT OF POLYMER AND PROCESS ENGINEERING, INDIAN INSTITUTE OF TECHNOLOGY ROORKEE, SAHARANPUR, UP, INDIA

9.1 Introduction

The thermoplastic elastomer industry is growing day by day due to their rigorous need for the industry and is gaining attention even beyond when compared to the plastic industry. As per data reported by Maynard et al. [1], the elastomeric sector is going to increase by 25% from the period 2016 to 2022. This hot segment of the plastic industry can manipulate and change the various polymers and properties of different plastics by blending them with elastomer. The history of blends is utterly long and rigorously correlated to the elastomers and polymers themselves. Sometimes, for specific purpose and properties, there is a need of new materials that can be fulfilled by blending of two polymers, which may deliver new system with exceptional and desired additional properties. The blended polymers may have different thermal and mechanical properties compared to their original components [2]. However, two different elastomers might be immiscible and incompatible with each other, which makes blending more challenging. Mixing or compounding relatively immiscible and incompatible elastomers together may lead to bad dispersion and distribution of particles of different phases in the matrix. Consequently, a coupling agent or compatibilizer is needed between di- or multiphase elastomer blends, and thus, a decrease in interfacial tension coefficient and formation and stabilization of the aimed morphology occur [2]. Nanoparticles have also shown to provide coupling effect between two immiscible elastomers and polymers [2].

Utilization of any substance, including elastomeric blends and composites, depends a lot on its thermal stability and thermal degradation profile when exposed to high temperatures. The thermal behavior of these blends and composites may be significantly affected by their compositions, but synergistically it may vary and show anomalous behavior compared to their components' individual behavior. Change in the thermal behavior of blends and composites is attributable to the extent of interaction between different species and groups of their constituents. These interactions may alter the thermal behavior of elastomer blends and composites when compared to their original

Elastomer Blends and Composites. https://doi.org/10.1016/B978-0-323-85832-8.00008-0

constituents. Applicability (outdoor/indoor) of these elastomers and composites is dependent on their behavior against thermal exposures. The thermal behavior (stability and degradation) of these elastomeric blends has been studied and reviewed by many researchers. For instance, Ahmed et al. [3] compatibilized marble waste (MW)-filled polypropylene (PP) and acrylonitrile butadiene (NBR) rubber blend with chlorinated polyethylene (CPE) to achieve improved thermal aging in comparison to uncompatibilzed MW-PP/NBR blend. Likewise, Mantia et al. [2] reviewed different polymeric combinations and their thermal degradation behavior. The key finding from this study was the impact of reactions between the various functional groups of the blends component; this interaction may lead to acceleration, degradation, or stabilization of the blend. Other thermal studies based on elastomer–elastomer and elastomer–plastic blends have been reported in past, for instance, millable polyurethane (MPU)/ethylene-co-vinyl acetate (EVA) [4], natural rubber/butadiene rubber [5], isotactic polypropylene (iPP) with poly (styrene-*b*-ethylene-butylene-*b*-styrene) [6], ethylene–octene copolymer [6], ethylene–propylene copolymers [7], and polyisobutylene [7].

Like elastomeric blends, composites have been developed when specific applications from two totally different phases were needed. In particular, this requires the use of synergistic properties from two different materials, where continuous phase called matrix and the discontinuous phase termed as reinforcement are mixed together. The newly developed composite possesses noticeable and different properties than their original components. For instance, short natural fiber-reinforced polymer composites [8,9], where matrix phase provides protection to the natural fiber from moisture and degradation and short natural fiber, strengthens the composite. Based on the requirement of the application, various composites have been developed by researchers and scientists previously.

Likewise, improved thermal stability and thermal degradation of the various elastomeric composites have been reviewed and reported. For instance, MWCNT-filled fluoroelastomer nanocomposite showed improved thermal decomposition temperature [10]. Perez et al. [11] studied that the addition of MWCNT to the NBR and SBR showed improved thermal degradation and stability of the composites. Similarly, enhanced thermal properties for the elastomeric composites have been studied for the crumb rubber/natural rubber/organo montmorillonite [12], organo-modified layered silicates in the ethylene propylene diene monomer (EPDM)/chlorobutyl rubber (CIIR) blends [13], and carbon black filled natural [14] and nitrile [15] rubber. The current study briefly discusses the effect of temperature on stability, degradation, and thermal profile of the elastomeric blends and composites.

9.2 Thermodynamics of the rubber–rubber and rubber–polymer blends

Before going to an in-depth study of the blends, it is necessary to learn the thermodynamics aspect of the binary rubber blends. The thermal behavior of the blends relies on the thermodynamic miscibility, temperature, and composition of the rubber. Fig. 9.1A–C shows the rubber blends at different composition and temperature, where X and Y axes represent the composition of the rubber and temperature, respectively.

In Fig. 9.1, upper critical solution temperature (UCST) (A), lower critical solution temperature (LCST) (B), and simultaneous effect of UCST and LCST (C) have been illustrated.

Miscibility of a polymer is dependent on temperature and composition. The temperature above which a polymer blend becomes miscible is referred as UCST, whereas temperature below which miscibility occurs is called LCST as can be seen in Fig. 9.1A and B, respectively. It is to be noted here that UCST is a rarely occurring phenomenon in case of polymer blends. Besides temperature, the composition of the blend system is also critical for the miscibility and phase separation behavior. Some blends show simultaneous effect of these two behaviors in polymer pair and can be observed in Fig. 9.1C [16].

Miscibility of blends of two different types (rubber–rubber and rubber–polymer) depends on their intend miscible and immiscible segments in them. A miscible blend will show homogeneous morphology, whereas in another case there would be a heterogeneous phase and forms two separate components of individual elastomer upon blending. Due to the high molecular weight of rubbers, a miscible rubber rarely exists. Thermodynamically, the miscibility of the blends can be explained by the following equations:

$$\Delta G = \Delta H - T\Delta S \tag{9.1}$$

Where,

ΔG = Gibbs free energy, ΔH = enthalpy change, ΔS = change in entropy of mixing and $T\Delta S$ = combinatorial change in entropy of mixing.

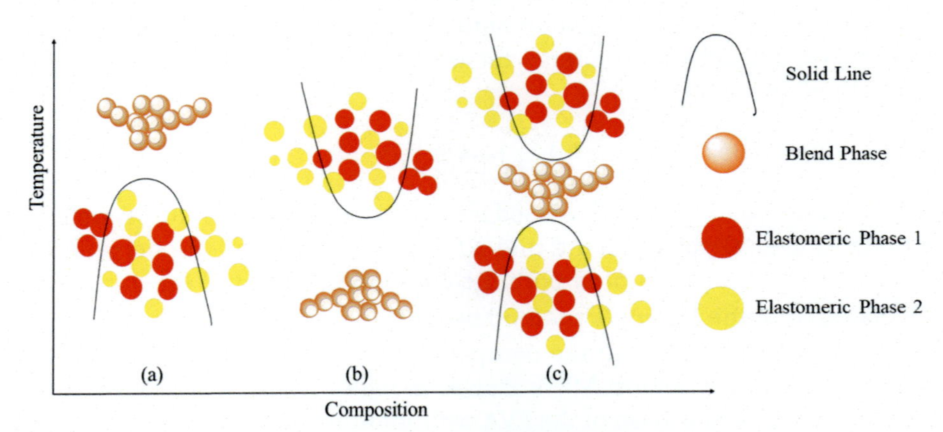

FIGURE 9.1 Illustration of (A) upper critical solution temperature (UCST), (B) lower critical solution temperature (LCST), and (C) simultaneous effect occurrence of UCST and LCST [15].

For a miscible rubber—rubber and rubber—polymer blend, ΔG must be negative. Again, immiscible rubbers can be categorized into compatible and incompatible blends. During blending, interactions between different species might be possible, which decide the thermal properties, stability, and degradability of the blends. Such interactions may lead to the reactions between:

Macroradicals and molecules
Two different macroradicals
Macro- and micromolecules
Macromolecule and microradicals
Macroradicals and micromolecules
Two different micromolecules

9.3 Thermal behavior of blends

Basic concepts relating to the thermal behavior of the blends have been studied and explained by many researchers through use of the differential scanning calorimetry (DSC), dynamic mechanical thermal analysis (DMA/DMTA), and thermogravimetric analysis (TGA). Blending of di- or multiphase elastomer to the polymer may alter the thermal degradation temperature, glass transition temperature, melting point, and viscoelastic behavior. Interaction of the functional groups from different phases synergistically affects the thermal properties of the blends. Additionally, developed blends may possess increase or decrease in the aforementioned thermal properties, depending upon their behavior against thermal exposure. Likewise, a strong compatibility at the interface induces different thermal properties not similar to their constituents materials. For instance, Shafeeq et al. [4] prepared a blend of millable polyurethane (MPU) and ethylene-co-vinyl acetate (EVA) in open mill mixing and cured with dicumyl peroxide (DCP). DSC analysis showed that both the components' glass transition temperatures could be distinguishable, confirming that both the elastomers retain their identity. In the same experiment, the activation energy for the crystallization has been calculated by Kissinger method, given as:

$$\ln\left(\frac{\varnothing}{T_P^2}\right) = k - \frac{\Delta E}{RT_P} \tag{9.2}$$

Where $\varnothing$ represents cooling rate, T_p is used for peak temperature of the crystallization, R represents the universal gas constant, and ΔE represents activation energy. A blend of MPU and EVA with equal amount of both the components showed the maximum value of ΔE calculated from the Kissinger method using slope of $\ln(\varnothing/T_P^2)$ *vs* $1/T_P$ curve. This increase in ΔE was attributable to the competeness of the EVA phase toward crystallization in an order manner against MPU. However, higher amount of EVA or MPU in the blend leads to easy crystallization of the EVA without much hindrance in first case and in a tiny region in later case. Similarly in the degradation curve, all the blends went through a two-step degradation curve unlike of MPU. Degradation behavior of the blends has been dependent on the composition of the blends, in particular, a blend of 50:50 of each component showed slowest degradation rate.

DMA studies demonstrated that beyond 0°C, all the blends were temperature-independent of the storage modulus. Above −53°C, EVA-rich blends, specially 40 wt.% and 60 wt.%, showed superior value of storage modulus. Some other elastomeric and elastomer–polymeric blend's thermal behavior has been explained in later section briefly based on their thermal characterization such as DSC, TGA, and DMA.

9.3.1 Thermal behavior analysis of elastomeric blends by DSC technique

DSC technique reveals glass transition temperature (T_g), melting point (T_m), crystallization temperature, and percentage crystallinity of the blend.

Blends prepared by Dutta et al. [17] via melt mixing of EVA and TPU show almost a single T_g from DSC analysis. However, a unique T_g does not demonstrate that blends are miscible, but it can be concluded that blends are technologically compatible. Theoretically, Fox equation given below can be used for calculating T_g of a blend.

$$\frac{1}{T_g} = \frac{W_1}{T_{g_1}} + \frac{W_2}{T_{g_2}} \tag{9.3}$$

Where, W = weight fractions of the components

T_g = glass transition temperature of the polymer

Subscripts 1 and 2 represent components 1 and 2.

A minimal difference in experimental and theoretically calculated T_g of the blends was observed through use of this equation earlier [17].

In another study, Maynard et al. [1] compounded ethylene–octene copolymer polyolefin thermoplastic elastomers (POEs) with different degrees of crystallinity. DSC studies confirm that blends are immiscible and quenching of the different blends suppresses the crystallinity percentage of the blends while annealing thickens the crystals. More crystals of the same thickness improve modulus and decrease the elastic recovery.

Zaharescu et al. [18] prepared a blend of EPDM and NR and reported that DSC characterization shows the presence of a lot of double bonds in NR making the blend unstable.

DSC analysis of isotactic polypropylene (iPP) and nitrile rubber (NBR) (iPP/NBR) blend reported by George et al. [19] suggests a decrease in the crystallinity of the blend on increasing NBR (elastomer) content. This decrease in crystallinity was exhibited due to the presence of rubber particles in the inter- and intraspherulitic region of the crystalline plastic phase reported by Martuscelli [20].

Consequently, a decrease in T_m of blend was also observed due to nitrile rubber. In the case of incompatible blends, crystallinity decreases as the effect of noncrystalline rubber retards the crystal growth. Fig. 9.2 represents the effect on crystallinity percentage on increasing NBR content in the iPP/NBR Blend.

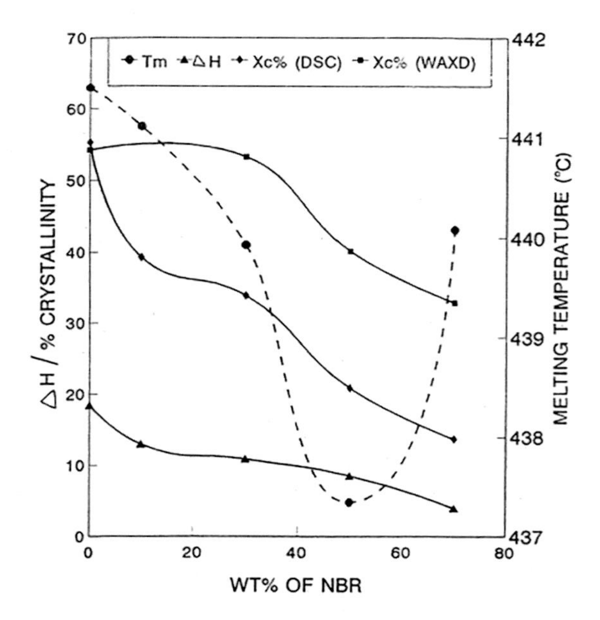

FIGURE 9.2 Illustration of variation of crystallinity, ΔH, and T_m with wt% of NBR. *Adapted from the reference S.* George, K.T. Varughese, S. Thomas, Polymer 41 (14) (2000b), 5485–5503. *https://doi.org/10.1016/S0032-3861(99)00719-3. Reprinted with the permission from Elsevier.*

Asaletha et al. [21] prepared natural rubber/polystyrene blends and reported even with the addition of suitable compatibilizer thermodynamically blend was incompatible. This was confirmed from the presence of two T_g corresponding to NR and PS with the help of DSC analysis. This result was also in agreement with Paul et al. [22] that if two polymers are **far from miscible**, no compatibilizer can make one system (like in composites).

A nano fiber made of crystallizable poly (ε-caprolactone) (PCL) and liquid NBR homogeneous blend was prepared by Maccaferri et al. [23] with the help of electrospinning.

The authors reported a single T_g for the blend, which was in between the T_g of the reference polymers. The fabrication process has been illustrated in Fig. 9.3. In some particular condition of electrospinning technique, two partially miscible polymers can be blended by using homogeneous solution of both the components. A sudden evaporation of the solvent freezes blend constituents that would not be separated afterward. Technologically, kinetics of this method is prominent over the thermodynamically driven outcomes of two immiscible blends [23].

Lai et al. [24] melt-blended polycaprolactone (PCL) and natural rubber (NR) and fabricated self-healing polymer. DSC report from this research suggests that a blend of 60:40 ratio of PCL/NR shows a slight increase in the crystallization temperature. An addition of 0.26 wt.% of acrylic acid (AA) improves crystallization temperature, indicating the nucleating role of AA moiety. However, a minimal variation has been found in the T_m from the pristine PCL confirming the similar packing structure and size of all the blends and polymer system. The crystallinity of the PCL/NR with 0.26 wt.% of acrylic acid (AA) showed an increment of ~2% that was exhibited due to the nucleation effect of PCL grafted AA.

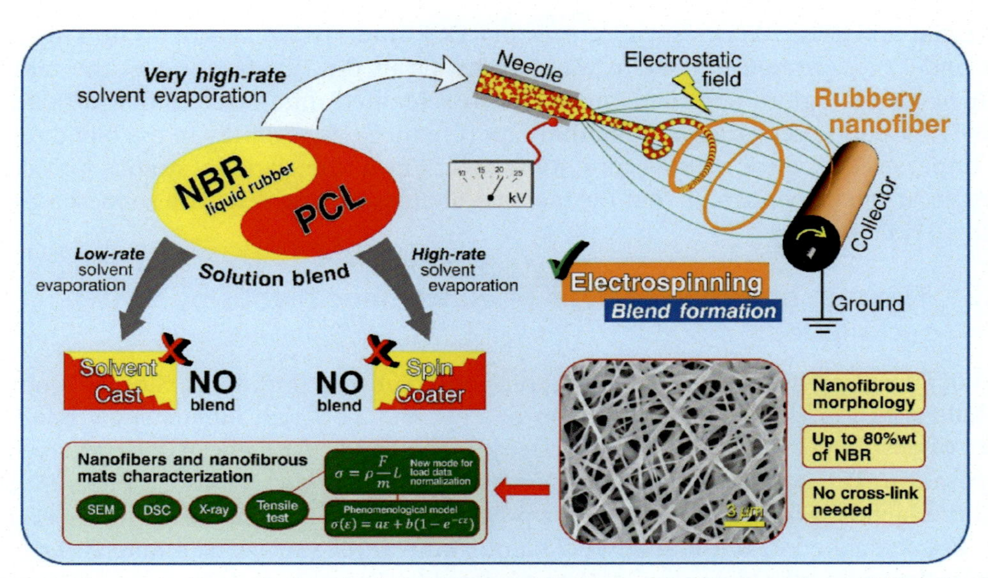

FIGURE 9.3 Electrospinning method to obtain homogeneous blend of a crystallizable polymer (PCL) and a liquid rubber (NBR). *Reprinted with the permission from Elsevier. E. Maccaferri, L. Mazzocchetti, T. Benelli, Rubbery nanofibers by co-electrospinning of almost immiscible nbr and pcl blends, Mater. Design Elsevier. (2020). https://www.sciencedirect.com/science/article/pii/S0264127519306483 (Accessed 28 January 2021).*

The application of waste tire rubber (WTR) in polymeric blends has been reviewed by Ramarad et al. [25]. Ground tire rubber (GTR)/reclaimed tire rubber (RTR) did not influence these properties much upon mixing or blending with the polymer. Articles reported for the LDPE/GTR [26], EVA/GTR [27], and PP/RTR [28] blend claimed that there was no convincing change in the melting temperature and crystallinity index of the blends. The crystallinity of blends with 500 μm particle size has further strengthened the claim. However, a reduced T_m has been reported when SBS was used as a compatibilizer for the LLDPE/GTR blend. A narrow melting region for this blend system might be attributable to the imperfect crystal or thinner lamella formation confirming the better compatibilization between LLDPE and GTR [29].

A blend of PP/RTR showed a decrease in melting enthalpy and crystallinity exhibited due to the restrictions created by sulfur cross-links in PP chains' packing [28]. In another study, a blend of PP/GTR coupled by dynamic vulcanization in the presence of *trans*-polyoctylene rubber (TOR) reported slight change in the T_m of the blend. This minor change in the melting point might be attributable to the decrease in lamella thickness at the interface due to the presence of covulcanized TOR [30]. The crystallinity of the GTR/HDPE blend did not change much as reported by Sonnier et al. [31], in the presence of DCP (up to 0.01 DCP to HDPE ratio) subjected to free radical cross-linking to improve interfacial adhesion. Work executed utilizing DCP and HVA-2 separately to cross-link GTR and PP confirmed that there was no change in the PP structure as the melting point of blends before and after modification remained unaffected. Other articles on GTR/EVA blend upon gamma irradiation report reduced crystallinity in the blend due to the imperfect/defective crystal formation [32]. From the various studies presented herewith, it can be concluded that the increase of cross-linking density also increases the blend's crystallinity.

9.3.2 Thermal behavior analysis of elastomeric blends by DMA technique

DMA of a material helps explore the stiffening and damping properties. Storage Modulus (E') values in the glass transition region of TPU and TPU-rich blends show relatively lower values than EVA and EVA-rich system for an EVA/TPU blend. However, at room temperature, TPU offers a much higher E' value than EVA and EVA-rich blends. E' value of all the blends was in between that for pristine EVA and TPU at ambient temperature [17]. The calculated T_g of the EVA/TPU blends from DMA showed a unique and shifted peak in the same way like using DSC as reported by Dutta et al. [17].

A DMA study on PCL/NR self-healing blend polymer suggests that the storage moduli of all the blends will lie in between the E' of their reference components. The use of PCL grafted AA improves the E' value of the blend, but still, its value is smaller than PCL [24]. Interestingly, tan δ curve of all the blends shifts toward a lower temperature than PCL and NR (shifts toward each other in immiscible case and makes single T_g in case of a miscible blend). The reason speculated behind this shift was due to the dilatant stress from crystallized PCL on the NR molecular chains, which had increased its mobility. A similar effect has also been reported in the EVA (ethylene-vinyl acetate)/mPE (metallocene polyethylene) blends [33].

The storage modulus of an elastomer—elastomer blend increases due to the increment of the cross-linking density and carbon black content as observed earlier with the addition of the GTR/RTR content in the rubber blend [34,35] by restricting the chain mobility. The same observation was reported for RTR/SBR and WTR/SBR, where storage modulus in later case was more significant [36]. Loss modulus follows the same trend as of storage modulus, which increases with the increment of RTR/WTR content and suggests an increase in the blends' viscoelasticity [34,35]. The tan δ peaks show an increment in the blends' glass transition temperature, and a decrease in peak height suggests that lower heat is required for molecular chain deformation. RTR compounded SBR and NR blend showed lower peak height of tan δ than blend from WTR [36,37].

An improved storage modulus of the blend of GTR/HDPE has been reported by Kim et al. due to the addition of acrylamide modification compared to unmodified one throughout the temperature range [38]. The peak height of the modified GTR/HDPE also decreased confirming good interfacial adhesion and reinforcing efficiency of GTR in HDPE. Mészáros et al. used polyethylene/ethylene-vinyl acetate copolymer as a matrix for reinforcing GTR and to improve compatibility, further electron beam (EB) irradiation was applied. T_g of both phases shifted closer with EB application, confirming improved interfacial adhesion [39]. 40 wt.% GTR blended with HDPE has been studied by Li et al. and reported reduced storage modulus compared to HDPE. This change in storage modulus and peak height of tan δ was further reduced when GTR was added in the presence of additives EPDM, dimethyl silicon oil, and dicumyl peroxide. A shift in the tan δ peak from 44°C to -53°C was observed in the same experiment confirming enhanced interfacial interaction between different phases [40].

9.3.3 Thermal behavior analysis of elastomeric blends by TGA

TGA gives the thermal degradation and thermal stability properties of any materials. For instance, a blend of TPU and EVA prepared earlier using melt blending technique was analyzed by TGA [17]. The analysis suggests that TPU is thermally less stable than EVA and that blends showed stability in between those of individual components. Both TPU and EVA exhibited a two-step degradation. A similar study has been already discussed earlier in Section 9.3, where MPU has been blended with EVA and degradation behavior reported [4].

A blend of isotactic polypropylene (iPP) and nitrile rubber (NBR) has been prepared by George et al. [19], and a thermal study from their research suggests that a 30 wt% loading of NBR to iPP shows maximum thermal stability in comparison to 50 wt.% loading. The key finding from this research was that the degradation stability builds upon morphology and degree of interaction between two phases. So, a decrease in one component's interfacial area may affect the overall morphology and interaction of phases forming a binary blend.

Interfacial area per unit volume for a blend can be calculated from the equation [41].

$$A = n \times 4\pi R^2 \tag{9.4}$$

Where A is the occupied area of dispersed phase, and R is the radius of the phase particles, n is the number of particles of the lesser phase per unit volume and can be calculated from the equation

$$n = \varnothing_d 4/3\pi R^3 \tag{9.5}$$

Where $\varnothing_d$ is the volume fraction of the dispersed phase.

An improvement in the thermal degradation of the phenolic and MA compatibilized blend of the NBR/iPP was also reported in the same study. A schematic has been given in Figs. 9.4 and 9.5 for both the compatibilized methods.

Likewise, the blend between NR and PS was found to be incompatible as reported from previously conducted DSC studies, but still the addition of a suitable compatibilizer enhanced the thermal degradation temperature of the blend confirmed from the TGA curve [21].

Other blends prepared from Nylon 6 with acrylate rubber [42] and ethylene–propylene rubber [43] have been reported earlier. In both cases, the compatibilizer in such blends was found to play an essential role in increasing the thermal stability compared to the uncompatibilized system.

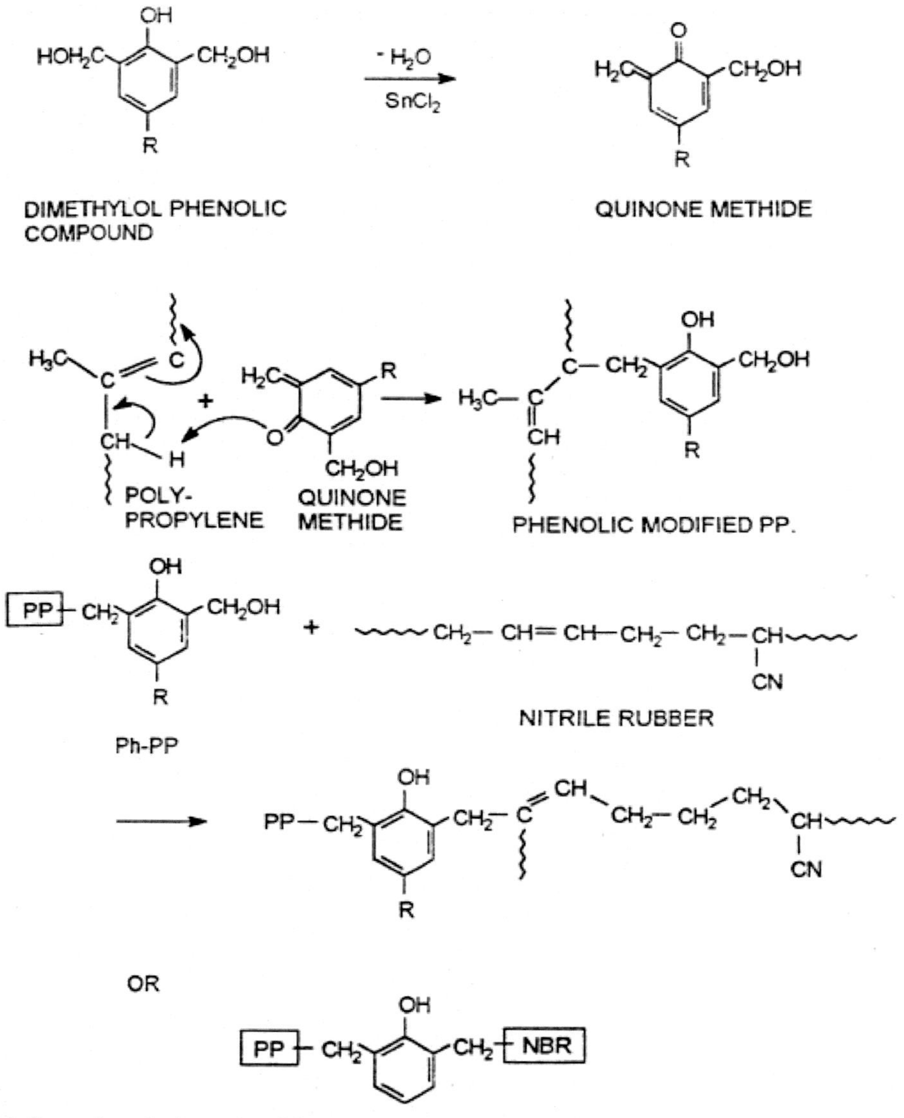

FIGURE 9.4 Illustration of schematics of formation of phenol modified PP and a graft copolymer between PP and NBR. *Reprinted with the permission from Elsevier. S. George, K.T. Varughese, S. Thomas, Thermal and crystallisation behavior of isotactic polypropylene/nitrile rubber blends, Polymer 41 (2000) 5485–5503. https://doi.org/10.1016/S0032-3861(99)00719-3.*

Another study of waste tire rubber, ground tire rubber (GTR), and reclaimed tire rubber (RTR) indicated that these being degradable materials were supposed to influence the blends' thermal stability significantly. Although there was a decrease in the onset of the GTR/RTR blended elastomer's thermal degradation due to the presence of volatile materials, but still at 50 and 70 wt.% loss, thermal stability was improved substantially on increasing GTR/RTR content [25,44]. Thermal stability of the GTR/RTR blended gum rubber was found to improve compared to gum rubber. A similar increase in thermal stability has been observed earlier in other cases for the NR/RTR [34] and NR/PBR/RTR [35] blends as well. This increase in the blends' thermal stability was exhibited due to the increase in cross-linking density of the blends with the increase of RTR content in the blends. Also, the inclusion of a coupling agent essentially maleic anhydride results in enhanced thermal degradation temperature of the blends by virtue of the improved compatibility and interfacial adhesion between virgin polymer and RTR [45].

A blend of latex modified PP with GTR did not show any significant improvement in the thermal degradation temperature of the blend although there was an apparent increase in mechanical properties, but that might be attributable to the physical wetting of the intermolecular phases [46].

The addition of MA grafted elastomer/polymer to the blend system increases compatibility between them further. For instance, when PP-g-MA was used instead of PP for blending 65 wt.% of the GTR, the thermal stability increased significantly in the former case. Likewise, when SEBS-g-MA and SEBS were used as compatibilizer for PP/amine modified GTR blend, the former compatibilizer with MA group increased thermal stability substantially. Here, a noticeable catch is that the EB block of SEBS or SEBS-g-MA provides compatibility with PP, whereas MA reacts with GTR [47]. A schematic reaction between MA and amine modified GTR has been given Fig. 9.5.

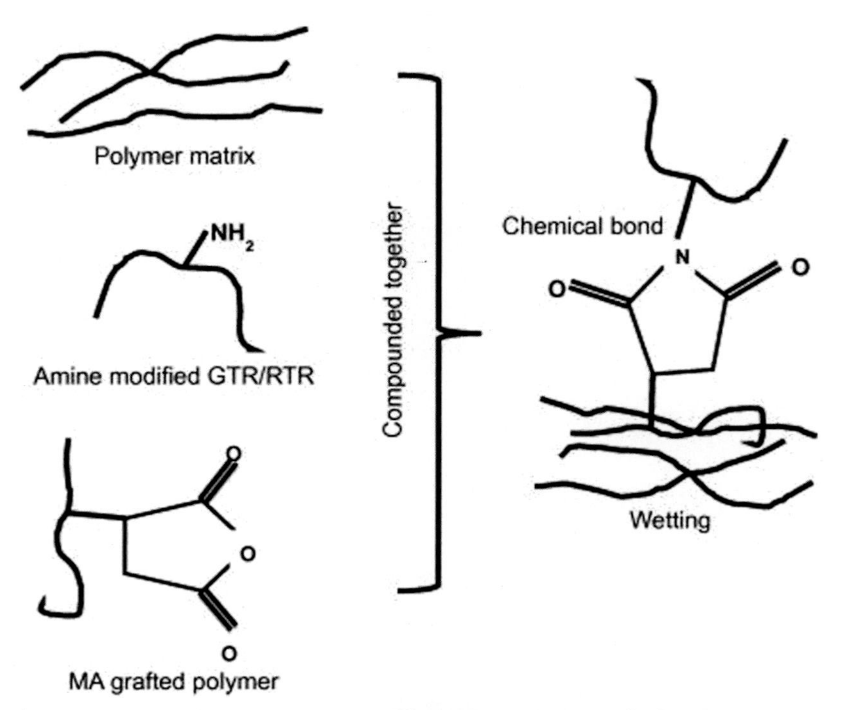

FIGURE 9.5 Schematic reaction between amine modified GTR and MA grafted polymer. *Adapted from the S. Ramarad, M. Khalid, C.T. Ratnam, A.L. Chuah, W. Rashmi, Waste tire rubber in polymer blends: a review on the evolution, properties and future, Prog. Mater. Sci. 72 (2015) 100–140. https://doi.org/10.1016/j.pmatsci.2015.02.004. Reprinted with the permission from Elsevier.*

Schematic reaction between amine modified GTR and MA grafted polymer. Adapted from the reference number [25]. [Reprinted withthe permission from Elsevier].

Other compatibilizers such as trans-polyoctylene rubber [48], dicumyl peroxide, and *N,N′-m*-phenylenebismaleimide [49] have also been used as a compatibilizer for PP/ WRT blends and have achieved improved thermal stability. A proposed mechanism has been given in Fig. 9.6.

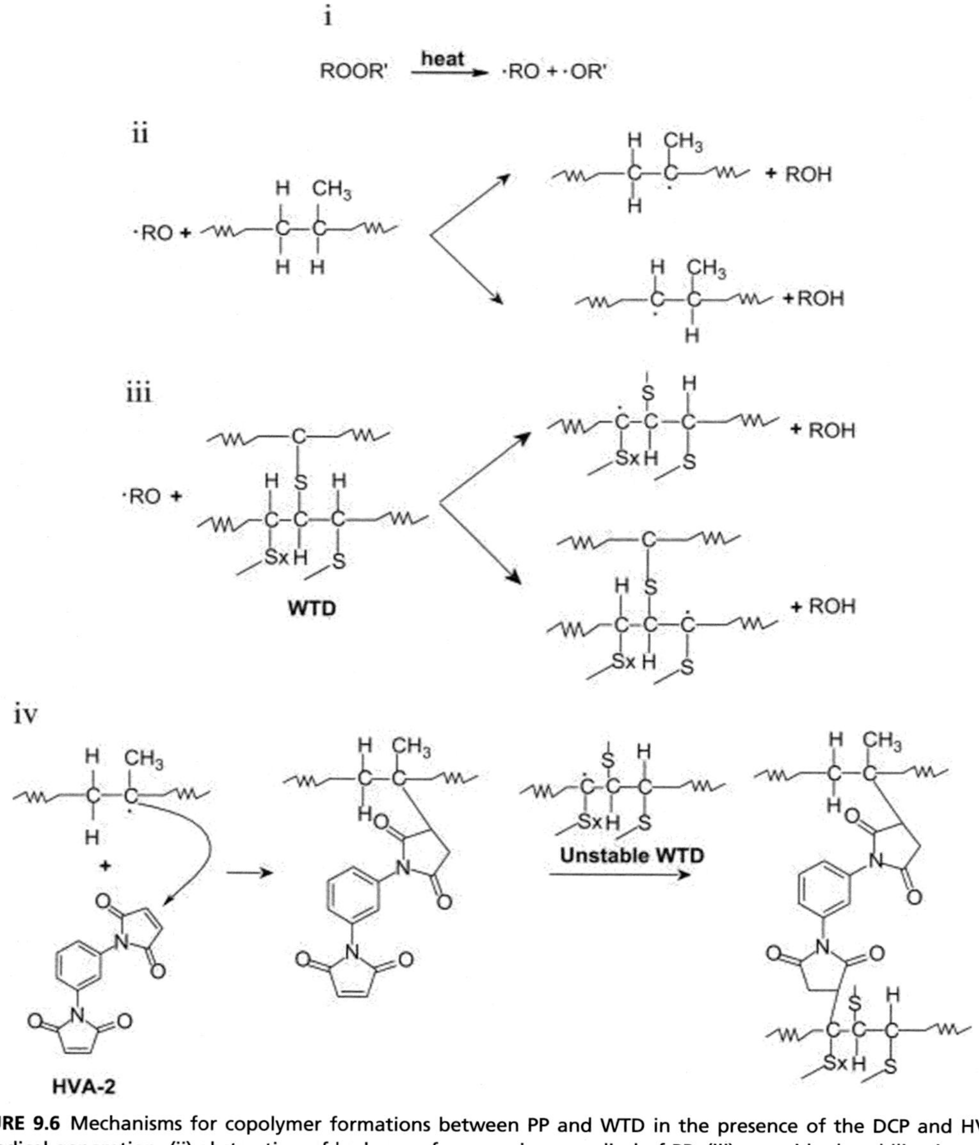

FIGURE 9.6 Mechanisms for copolymer formations between PP and WTD in the presence of the DCP and HVA-2. (i) Radical generation, (ii) abstraction of hydrogen from a primary radical of PP, (iii) peroxide destabilization of WTD, (iv) copolymer formation in the presence of HVA-2. *Adapted from the reference M. Awang, H. Ismail, Preparation and characterization of polypropylene/waste tyre dust blends with addition of DCP and HVA-2 (PP/ WTDP-HVA2), Polym. Test. 27 (2008) 321–329. https://doi.org/10.1016/j.polymertesting.2007.12.001.*

Some other common examples of elastomeric blends with improved thermal stability characterized through thermogravimetric analysis and reported in the literature include iPP/NBR (PP-g-MA and Ph-PP coupling agent) [50], PVC/EVA [51], Nylon 6/ACM [52], Nylon 66/EPR-g-MA [53], EPDM/SAN [54], HIPS/SBS [55], NR/EVA [56], and PP/EOC [57]. The analysis clearly suggested an increase in thermal stability when a suitable coupling agent was inserted in the blend system.

9.4 Thermal behavior of elastomeric composites

Thermal behavior of particulate and fibrous reinforced elastomeric composite has been studied by many researchers and reviewed and reported in this section. Like elastomeric blends, for a specific purpose, sometimes a particulate or fiber needs to be added to complete the particular task. Reinforcement of a particular type of filler or fiber to the elastomer may alter the thermal degradation temperature, glass transition temperature, melting point, and viscoelastic behavior. Interaction of the functional groups from different phases synergistically affects the thermal properties in composite. Additionally, fillers/fibers may increase or decrease the aforementioned thermal properties, depending upon their behavior against thermal exposure. Likewise, a strong compatibility at the interface induces different thermal properties not similar to their constituents materials. For instance, nano clay reinforced EPDM rubber composite for outdoor application fabricated by Rana et al. [58] was supposed to provide more thermal stability in comparison to pristine EPDM due to the inclusion of a more thermally stable nano clay. In the coming section, thermal and thermomechanical behavior of such composites has been studied with the help of DSC, DMA, and TGA.

9.4.1 Thermal behavior of elastomeric composites analyzed by DSC technique

Differential scanning calorimetry (DSC) has been explored effectively to study percentage crystallinity, melting behavior, and heat of the elastomeric composites' enthalpy. Carbon nano tube (CNT) reinforced elastomeric composites have been found to display higher thermal degradation temperature, due to the inclusion of more thermally stable CNTs. The increase in the thermal stability was also exhibited by them due to the higher dissipation of the heat inherently due to the higher thermal conductivity of the CNTs [10]. Pham et al. [59] recorded enhancement in thermal decomposition property of the fluoroelastomer-filled MWCNT composite. This improvement attributed to the development of weak sites of fluorocarbon backbone inactivated due to physical and chemical interactions between MWCNTs and fluoroelastomer. The antioxidant nature of MWCNT also helped in improving decomposition temperature.

T_g and T_m of the elastomeric composites increase on the addition of MWCNT, and this effect is more prominent when functionalization of the MWCNT is inserted. Zhan et al. found an increment of ~20°C when polytetramethylene ether glycol (PTMEG) and OH grafted MWCNT has been used as reinforcement with the TPU. The T_g of MWCNT-g-PTMEG filled TPU composite enhances to a lower degree in comparison to MWCNT-g-OH, which might be exhibited due to the microphase separation on inclusion of reinforcement [60]. Likewise, T_g of all the composites increased with increment of the nano clay content into the EPDM/nano clay composite [58].

9.4.2 Thermal behavior of elastomeric composites analyzed by DMA technique

DMA studies showed that E' of the MWCNT-g-PTMEG filled TPU composite was enhanced in comparison to TPU at room temperature [60]. The tan δ curve also decreases with the increasing grafted MWCNT loading content. This might be attributable to the diminished chain mobility owing to physical and chemical adsorption of the PU molecules on the MWCNT surface during dynamic mechanical deformation. The reduction in tan δ peak will lower the heat development and damping capability for PU/MWCNT composites, which may be applicable in enhancing the dynamic fatigue properties of materials [60]. Same effect of increased E' and decreased tan δ has been reported by Raja et al. for the functionalized MWCNT filled TPU composites. A decrease in tan δ peak, which is related to the T_g of soft segment, exhibited due to the microphase separation with the incorporation of MWCNT; as a result there are lesser hard segments present in the delicate phase, which can block the motion of the soft segment [61]. Also, better compatibility and interaction between filler and matrix lead to an increased E' value, which is analogous to tensile modulus. Fig. 9.7 shows a schematic of interaction between MWCNT-g-PTMEG and TPU.

In an another study, Ashok et al. used EPDM/chlorobutyl rubber blend as matrix for organomodified layered silicate (OMLS)-filled composites. At 5 phr content of OMLS mobility of chain restriction and E' was highest for the composite system. This increase was attributable to high aspect ratio and formation of the intercalated and exfoliated structure of OMLS, resulting in better interfacial interaction with the base matrix [13].

EPDM rubber reinforced with nano clay shows the same effect of improvement in E′ and decreased tan δ of the composite [58]. Enhancement in the interfacial bonding between nano clay and EPDM was found to be the reason behind improved E'. Estimation of the activation energy of the EPDM and composites has been calculated using Arrhenius equation given as

$$f = z\exp\left(-\frac{E_a}{RT}\right) \tag{9.6}$$

where f is the test frequency; z is the preexponential factor; R is the gas constant taken as 8.314×10^{-3} kJ/K mol.

A convincing improvement in the activation energies captured from the loss tangent T_g of the nano clay composites has been reported. The addition of nano clay to composites protects the structure to higher expansion against the thermal degradation compared to pristine EPDM [58].

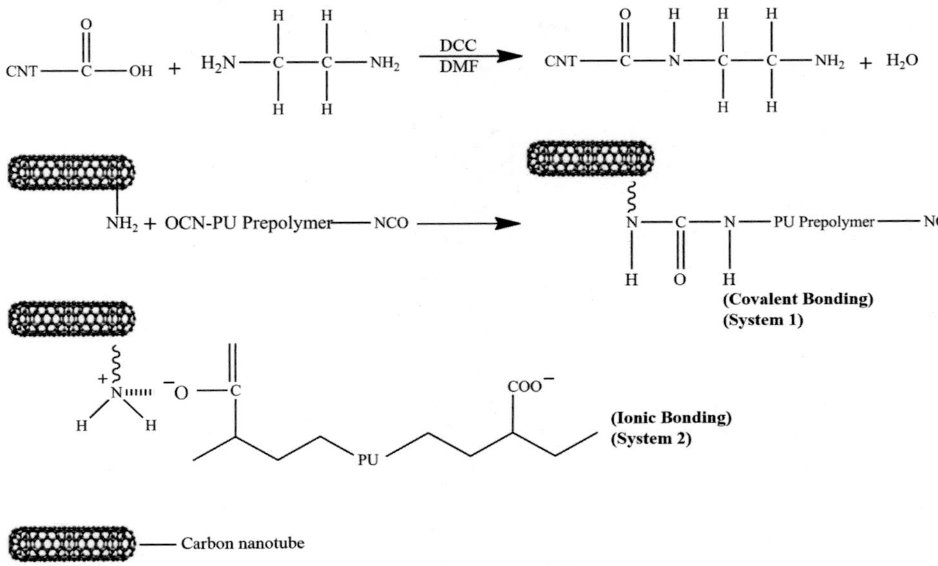

FIGURE 9.7 Covalent bonding and ionic bonding between the carbon nanotube and waterborne polyurethane. *Adapted from the reference H.C. Kuan, C.C.M. Ma, W.P. Chang, S.M. Yuen, H.H. Wu, T.M. Lee, Synthesis, thermal, mechanical and rheological properties of multiwall carbon nanotube/waterborne polyurethane nanocomposite, Compos. Sci. Technol. 65 (2005) 1703–1710. https://doi.org/10.1016/j.compscitech.2005.02.017 with the permission from Elsevier.*

In another study, a blend of 85 wt.% of PP, 10 wt.% of SEBS, and 5 wt.% of SEBS-g-MA was used as matrix. A reinforcement of fly ash and sisal fiber to this base matrix improved E' and decreased tan δ value like in other cases [9]. A similar study has been reported by Panaitescu et al. [63] where they used 15 wt.% of SEBS with 85 wt.% of PP and used this blend as a matrix for reinforcing hemp fiber. They also reported improvement in the E' of the composite in comparison to pristine PP.

9.4.3 Thermal behavior of elastomeric composites based on TGA technique

Thermal degradation of various composites and blends can be studied using TGA technique.

Thermal degradation temperature of the MWCNTs-filled SBR and NBR composites improved, on increasing content and modification of the filler, which significantly affected degradation behavior of the composites [11]. In this regard, acid-treated CNT reinforced to NR, and silicon rubber increased the composite's thermal degradation temperature by ~10°C compared to unmodified one [64,65]. Fig. 9.8 gives the degradation behavior against the temperature of the pristine silicon rubber and different CNT-filled silicon rubber composite.

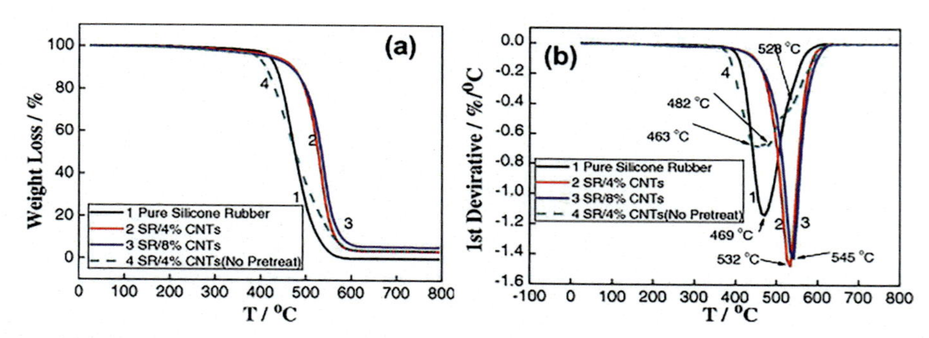

FIGURE 9.8 (A) TGA curves and (B) corresponding DTG curve of pure silicone rubber and SR/MWCNTs nanocomposites. *Adapted from the reference S. Shang, L. Gan, M.C. wah Yuen, Improvement of carbon nanotubes dispersion by chitosan salt and its application in silicone rubber, Compos. Sci. Technol. 86 (2013) 129–134. https://doi.org/10.1016/j.compscitech.2013.07.010 with the permission from Elsevier.*

Zhan et al. [60] fabricated MWCNT and PTMEG-functionalized MWCNT-filled TPU composites through extrusion process and reported improved thermal degradation stability in comparison to unfunctionalized one. This improvement was related to the capture and stabilization of the macroradicals formation amid the thermal degradation process by CNT and nanotube barrier effect. Same effect for the MWCNT-reinforced TPU has been reported by Kuan et al. [62] and Raja et al. [61].

In another study, a blend of natural rubber and crumb rubber in 100:10 ratio was used as a matrix and filled with montmorillonite (organo modified nanoclay) with enhanced thermal stability on increasing OMMT loading content [66]. This improvement was attributed to the increase in cross-linking density as a result, rigidity of the rubber mix/blend increases.

The TGA of nano clay, EPDM, and EPDM/nanoclay composite shows that wt.% loss against the temperature was the lowest by nano clay and highest by EPDM, where all the composite's wt.% lies in between the reference materials. An increment in the thermal stability of all the composites was reported [58]. Likewise, an addition fly ash to the sisal fiber reinforced fly ash hybrid impact modified polypropylene composite increases the composites' thermal degradation temperature [9].

In another study, nano silica reinforced NR/RTR [67] and talc reinforced recycled-PE/GTR [68] blend showed improved thermal stability on increasing silica content proportionally. However, addition of talc and wood flour to the PE-g-MA/GTR blend reported a decreased thermal stability [69], which is attributable to the low thermal degradation of wood flour particles.

9.5 Conclusion

Exact prediction of the thermal properties such as glass transition temperature, melting point, activation energy, crystallinity, storage modulus, damping factor, and degradation properties of the elastomeric blends and composite cannot be calculated using simple additive rules. Indeed, the blend and composites composition affects these behaviors depending upon the interaction between various functional groups. Besides, a coupling agent's presence may also decide interactions at the interface between them. Glass transition temperature and the melting point of a blend may decrease or increase depending upon the chain restrictions created by the dispersed phase both in blends and composites. Reinforcement may try to protect contraction against heat deformation and increase the melting point of the composites. Also, the blends or composite's degradation rate may evolve to a greater, intermediary, or lower rate than that of the pure components. In most cases, both blends and composites demonstrated a storage modulus that is intermediate to that of their original components.

References

[1] L.A. Maynard, B.L. DeButts, J.R. Barone, Mechanical and thermal properties of polyolefin thermoplastic elastomer blends, Plast. Rubber Compos. 48 (2019) 338–346, https://doi.org/10.1080/14658011.2019.1625633.

[2] F.P. La Mantia, M. Morreale, L. Botta, M.C. Mistretta, M. Ceraulo, R. Scaffaro, Degradation of polymer blends: a brief review, Polym. Degrad. Stabil. 145 (2017) 79–92, https://doi.org/10.1016/j.polymdegradstab.2017.07.011.

[3] K. Ahmed, Eco-thermoplastic elastomer blends developed by compatibilizing chlorinated polyethylene into industrial-waste-filled polypropylene/acrylonitrile butadiene rubber system, Arabian J. Sci. Eng. 40 (2015) 2929–2936, https://doi.org/10.1007/s13369-014-1561-1.

[4] V.H. Shafeeq, G. Unnikrishnan, Experimental and theoretical evaluation of mechanical, thermal and morphological features of EVA-millable polyurethane blends, J. Polym. Res. 27 (2020), https://doi.org/10.1007/s10965-020-2027-7.

[5] A.F.M.S. Amin, A. Lion, P. Höfer, Effect of temperature history on the mechanical behaviour of a filler-reinforced NR/BR blend: literature review and critical experiments, ZAMM Zeitschrift Fur Angew. Math. Und Mech. 90 (2010) 347–369, https://doi.org/10.1002/zamm.200900365.

[6] N. Fanegas, M.A. Gómez, C. Marco, I. Jiménez, G. Ellis, Influence of a nucleating agent on the crystallization behaviour of isotactic polypropylene and elastomer blends, Polymer 48 (2007) 5324–5331, https://doi.org/10.1016/j.polymer.2007.07.004.

[7] E. Martuscelli, C. Silvestre, G. Abate, Morphology, crystallization and melting behaviour of films of isotactic polypropylene blended with ethylene-propylene copolymers and polyisobutylene, Polymer 23 (1982) 229–237, https://doi.org/10.1016/0032-3861(82)90306-8.

[8] R. Gogoi, N. Kumar, S. Mireja, S.S. Ravindranath, G. Manik, S. Sinha, Effect of hollow glass microspheres on the morphology, rheology and crystallinity of short bamboo fiber-reinforced hybrid polypropylene composite, JOM 71 (2019) 548–558, https://doi.org/10.1007/s11837-018-3268-3.

[9] A.K. Maurya, R. Gogoi, G. Manik, Study of the moisture mitigation and toughening effect of fly-ash particles on sisal fiber-reinforced hybrid polypropylene composites, J. Polym. Environ. (2021) 1–16, https://doi.org/10.1007/s10924-021-02043-3.

[10] B. Mensah, H.G. Kim, J.H. Lee, S. Arepalli, C. Nah, Carbon nanotube-reinforced elastomeric nanocomposites: a review, Int. J. Smart Nano Mater. 6 (2015) 211–238. https://doi.org/10.1080/19475411.2015.1121632.

[11] L.D. Perez, M.A. Zuluaga, T. Kyu, J.E. Mark, B.L. Lopez, Preparation, characterization, and physical properties of multiwall carbon nanotube/elastomer composites, Polym. Eng. Sci. 49 (2009) 866–874. https://doi.org/10.1002/pen.21247.

[12] B.K. Saleh, S.F. Halim, M.H. Khalil, Evaluation of thermal and mechanical properties of crumb/natural rubber nanocomposites, Egypt, J. Chem. 63 (2020) 2523–2532. https://doi.org/10.21608/ejchem.2019.5522.1493.

[13] N. Ashok, M. Balachandran, N.C. Das, S. Remanan, Nanoreinforcement mechanism of organo-modified layered silicates in EPDM/CIIR blends: experimental analysis and theoretical perspectives of static mechanical and viscoelastic behavior, Compos. Interfac. 28 (2021) 35–62. https://doi.org/10.1080/09276440.2020.1736879.

[14] N.J. Nkengafac, A. Alegria, S. Arrese-igor, A. Edgengele, E. Eugene, A comparative study on the thermal behaviour of natural rubber filled with carbon black and plant residues, J. Mat. Sci. Res. Rev. 6 (2020) 21–30.

[15] U. Shankar, S. Bhandari, D. Khastgir, Carbon black-filled nitrile rubber composite as a flexible electrode for electrochemical synthesis of supercapacitive polyaniline, Polym. Compos. 40 (2019) E1537–E1547. https://doi.org/10.1002/pc.25069.

[16] A. Vayyaprontavida Kaliyathan, K. Varghese, A.S. Nair, S. Thomas, Rubber–rubber blends: a critical review, Prog. Rubber Plast. Recycl. Technol. 36 (2020) 196–242. https://doi.org/10.1177/1477760619895002.

[17] J. Dutta, K. Naskar, Investigation of morphology, mechanical, dynamic mechanical and thermal behaviour of blends based on ethylene vinyl acetate (EVA) and thermoplastic polyurethane (TPU), RSC Adv. 4 (2014) 60831–60841. https://doi.org/10.1039/c4ra07823c.

[18] T. Zaharescu, V. Meltzer, Thermal properties of EPDM/NR blends, Polym. Degrad. Stabil. (2000). Elsevier, https://www.sciencedirect.com/science/article/pii/S0141391000001154. (Accessed 28 January 2021).

[19] S. George, K.T. Varughese, S. Thomas, Thermal and crystallisation behavior of isotactic polypropylene/nitrile rubber blends, Polymer 41 (2000) 5485–5503. https://doi.org/10.1016/S0032-3861(99)00719-3.

[20] Z. Bartczak, A. Gał??ski, E. Martuscelli, Spherulite growth in isotactic polypropylene-based blends: energy and morphological considerations, Polym. Eng. Sci. 24 (1984) 1155–1165. https://doi.org/10.1002/pen.760241502.

[21] R. Asaletha, M.G. Kumaran, S. Thomas, Thermal behaviour of natural rubber/polystyrene blends: thermogravimetric and differential scanning calorimetric analysis, Polym. Degrad. Stabil. 61 (1998) 431–439. https://doi.org/10.1016/S0141-3910(97)00229-2.

[22] D. Paul, Polymer Blends, vol. 1, 2012. https://books.google.com/books?hl=en&lr=&id=TUrNGS3rKWgC&oi=fnd&pg=PP1&ots=38DFvsCSEz&sig=D-P30gS8A2XCJuekw7Bi5jtA6NI. (Accessed 28 January 2021).

[23] E. Maccaferri, L. Mazzocchetti, T. Benelli, Rubbery nanofibers by co-electrospinning of almost immiscible nbr and pcl blends, Mater. Design (2020). Elsevier, https://www.sciencedirect.com/science/article/pii/S0264127519306483. (Accessed 28 January 2021).

[24] S.-M. Lai, J.-L. Liu, Y.-H. Huang, Preparation of self-healing natural rubber/polycaprolactone (NR/PCL) blends, J. Macromol. Sci. Part B 59 (2020) 587–607. https://doi.org/10.1080/00222348.2020.1757218.

[25] S. Ramarad, M. Khalid, C.T. Ratnam, A.L. Chuah, W. Rashmi, Waste tire rubber in polymer blends: a review on the evolution, properties and future, Prog. Mater. Sci. 72 (2015) 100–140. https://doi.org/10.1016/j.pmatsci.2015.02.004.

[26] N. Sunthonpagasit, Scrap tires to crumb rubber: feasibility analysis for processing facilities, Resour. Conserv. Recycl. (2004). Elsevier, https://www.sciencedirect.com/science/article/pii/S0921344903000739. (Accessed 27 January 2021).

[27] R. Mujal-Rosas, J. Orrit-Prat, X. Ramis-Juan, M. Marin-Genesca, A. Rahhali, Study on dielectric, thermal, and mechanical properties of the ethylene vinyl acetate reinforced with ground tire rubber, J. Reinfor. Plast. Comp. 30 (2011) 581–592. https://doi.org/10.1177/0731684411399135.

[28] S. Tantayanon, S. Juikham, Enhanced toughening of poly(propylene) with reclaimed-tire rubber, J. Appl. Polym. Sci. 91 (2004) 510–515. https://doi.org/10.1002/app.13182.

[29] J. Qin, H. Ding, X. Wang, M. Xie, Z. Yu, Blending LLDPE and ground rubber tires, Polym. Plast. Technol. Eng. 47 (2008) 199–202. https://doi.org/10.1080/03602550701816217.

[30] M. Awang, H. Ismail, Polypropylene-based blends containing waste tire dust: Effects of trans-polyoctylene rubber (TOR) and dynamic vulcanization, Polym. Testing (2007). Elsevier, https://www.sciencedirect.com/science/article/pii/S0142941807000633. (Accessed 27 January 2021).

[31] R. Sonnier, E. Leroy, L. Clerc, A. Bergeret, J.M. Lopez-Cuesta, A.S. Bretelle, P. Ienny, Compatibilizing thermoplastic/ground tyre rubber powder blends: efficiency and limits, Polym. Test. 27 (2008) 901–907. https://doi.org/10.1016/j.polymertesting.2008.07.003.

[32] C.T. Ratnam, S. Ramarad, M.K. Siddiqui, A. Sakinah, Z. Abidin, L.T. Chuah, Irradiation cross-linking of ethylene vinyl acetate/waste tire dust: effect of multifunctional acrylates, J. Thermoplast. Compos. Mater. 29 (2016) 464–478. https://doi.org/10.1177/0892705713518814.

[33] S.-M. Lai, P.-H. Huang, H.-C. Kao, L.-C. Liu, Shape memory properties of melt-blended ethylene vinyl acetate (Eva)/Metallocene polyethylene eco-blends, J. Macromol. Sci. Part B 56 (2017) 97–113. https://doi.org/10.1080/00222348.2016.1273175.

[34] D. De, D. De, G.M. Singharoy, Reclaiming of ground rubber tire by a novel reclaiming agent. I. Virgin natural rubber/reclaimed GRT vulcanizates, Polym. Eng. Sci. 47 (2007) 1091–1100. https://doi.org/10.1002/pen.20790.

[35] D. De, P. Panda, M. Roy, Reinforcing effect of reclaim rubber on natural rubber/polybutadiene rubber blends, Mater. Design (2013). Elsevier, https://www.sciencedirect.com/science/article/pii/S0261306912007042. (Accessed 27 January 2021).

[36] Y. Li, S. Zhao, Y. Wang, Microbial desulfurization of ground tire rubber by Sphingomonas sp.: a novel technology for crumb rubber composites, J. Polym. Environ. (2012). https://doi.org/10.1007/s10924-011-0386-1. Springer.

[37] Y. Li, S. Zhao, Y. Wang, Improvement of the properties of natural rubber/ground tire rubber composites through biological desulfurization of GTR, J. Polym. Res. 19 (2012). https://doi.org/10.1007/s10965-012-9864-y.

[38] J.I. Kim, S.H. Ryu, Y.W. Chang, Mechanical and dynamic mechanical properties of waste rubber powder/HDPE composite, J. Appl. Polym. Sci. 77 (2000) 2595–2602. https://doi.org/10.1002/1097-4628(20000919)77:12<2595::AID-APP60>3.0.CO;2-C.

[39] L. Mészáros, T. Bárány, T. Czvikovszky, EB-promoted recycling of waste tire rubber with polyolefins, Radiat. Phys. Chem. 81 (2012) 1357–1360. https://doi.org/10.1016/j.radphyschem.2011.11.058.

[40] Y. Li, Y. Zhang, Y. Zahang, Mechanical properties of high-density polyethylene/scrap rubber powder composites modified with ethylene-propylene-diene terpolymer, dicumyl peroxide, and silicone oil, J. Appl. Polym. Sci. 88 (2003) 2020–2027. https://doi.org/10.1002/app.11907.

[41] M. Matos, B.D. Favis, P. Lomellini, Interfacial modification of polymer blends-the emulsification curve: 1. Influence of molecular weight and chemical composition of the interfacial modifier, Polymer 36 (1995) 3899–3907. https://doi.org/10.1016/0032-3861(95)99784-R.

[42] A. Jha, A.K. Bhowmick, Thermal degradation and ageing behaviour of novel thermoplastic elastomeric nylon-6/acrylate rubber reactive blends, Polym. Degrad. Stabil. 62 (1998) 575–586. https://doi.org/10.1016/S0141-3910(98)00044-5.

[43] E. Martuscelli, F. Riva, C. Sellitti, Crystallization, morphology, structure and thermal behaviour of nylon-6/rubber blends, Polymer (1985). Elsevier, https://www.sciencedirect.com/science/article/pii/0032386185900400. (Accessed 28 January 2021).

[44] M. Hassan, R. Aly, S. Aal, Mechanochemical devulcanization and gamma irradiation of devulcanized waste rubber/high density polyethylene thermoplastic elastomer, J. Industr. Eng. Chem. (2013). Elsevier, https://www.sciencedirect.com/science/article/pii/S1226086X13000750. (Accessed 27 January 2021).

[45] M.M. Hassan, G.A. Mahmoud, H.H. El-Nahas, E.S.A. Hegazy, Reinforced material from reclaimed rubber/natural rubber, using electron beam and thermal treatment, J. Appl. Polym. Sci. 104 (2007) 2569–2578. https://doi.org/10.1002/app.25297.

[46] M. Awang, H. Ismail, Processing and properties of polypropylene-latex modified waste tyre dust blends (PP/WTDML), Polym. Testing (2008). Elsevier, https://www.sciencedirect.com/science/article/pii/S0142941807001420. (Accessed 27 January 2021).

[47] S.H. Lee, M. Balasubramanian, J.K. Kim, Dynamic reaction inside Co-rotating twin screw extruder. II. Waste ground rubber tire powder/polypropylene blends, J. Appl. Polym. Sci. 106 (2007) 3209–3219. https://doi.org/10.1002/app.26490.

[48] M. Awang, H. Ismail, M.A. Hazizan, Polypropylene-based blends containing waste tire dust: effects of trans-polyoctylene rubber (TOR) and dynamic vulcanization, Polym. Test. 26 (2007) 779–787. https://doi.org/10.1016/j.polymertesting.2007.04.007.

[49] M. Awang, H. Ismail, Preparation and characterization of polypropylene/waste tyre dust blends with addition of DCP and HVA-2 (PP/WTDP-HVA2), Polym. Test. 27 (2008) 321–329. https://doi.org/10.1016/j.polymertesting.2007.12.001.

[50] S. George, K. Varughese, Thermal and crystallisation behaviour of isotactic polypropylene/nitrile rubber blends, Polymer (2000). Elsevier, https://www.sciencedirect.com/science/article/pii/S0032386199007193. (Accessed 29 January 2021).

[51] C. Thaumaturgo, E.C. Monteiro, Thermal stability and miscibility in PVC/EVA blends, J. Therm. Anal. 49 (1997) 247–254. https://doi.org/10.1007/bf01987445.

[52] A. Jha, Thermal degradation and ageing behaviour of novel thermoplastic elastomeric nylon-6/acrylate rubber reactive blends, Elsevier, Polym. Degrad. Stabil., (1998). (Accessed 29 January 2021). https://www.sciencedirect.com/science/article/pii/S0141391098000445.

[53] A. Choudhury, A. Balmurulikrishnan, G. Sarkhel, Polyamide 66/EPR-g-MA blends: mechanical modeling and kinetic analysis of thermal degradation, Polym. Adv. Technol. 19 (2008) 1226–1235. https://doi.org/10.1002/pat.1116.

[54] O. Chiantore, M. Guaita, Thermal oxidative degradation of AES, Polym. Degrad. Stabil. (1995). Elsevier, https://www.sciencedirect.com/science/article/pii/014139109400097R. (Accessed 29 January 2021).

[55] T.H. Grgurić, V. Rek, Ž. Jelčić, D. Hace, Z. Gomzi, Determination of the kinetic parameters of the thermal oxidative degradation of styrene/butadiene copolymers, Polym. Eng. Sci. 39 (1999) 1394–1397. https://doi.org/10.1002/pen.11529.

[56] P. Jansen, Effect of compatibilizer and curing system on the thermal degradation of natural rubber/EVA copolymer blends, Polym. Degrad. Stabil. (1996). Elsevier, https://www.sciencedirect.com/science/article/pii/0141391095002383. (Accessed 29 January 2021).

[57] K. Wang, F. Addiego, N. Bahlouli, S. Ahzi, Y. Rémond, V. Toniazzo, R. Muller, Analysis of thermomechanical reprocessing effects on polypropylene/ethylene octene copolymer blends, Polym. Degrad. Stabil. 97 (2012) 1475–1484. https://doi.org/10.1016/j.polymdegradstab.2012.05.005.

[58] A.S. Rana, M.K. Vamshi, K. Naresh, R. Velmurugan, R. Sarathi, Mechanical, thermal, electrical and crystallographic behaviour of EPDM rubber/clay nanocomposites for out-door insulation applications, Adv. Mater. Process. Technol. 6 (2020) 54–74. https://doi.org/10.1080/2374068X.2019.1703339.

[59] T.T. Pham, V. Sridhar, J.K. Kim, Fluoroelastomer-MWNT nanocomposites-1: dispersion, morphology, physico-mechanical, and thermal properties, Polym. Compos. 30 (2009) 121–130. https://doi.org/10.1002/pc.20521.

[60] Y.H. Zhan, R. Patel, M. Lavorgna, F. Piscitelli, A. Khan, H.S. Xia, H. Benkreira, P. Coates, Processing of polyurethane/carbon nanotubes composites using novel minimixer, Plast. Rubber Compos. (2010) 400–410. https://doi.org/10.1179/174328910X12777566997496.

[61] M. Raja, A.M. Shanmugharaj, S.H. Ryu, Influence of surface functionalized carbon nanotubes on the properties of polyurethane nanocomposites, Soft Mater. 6 (2008) 65–74. https://doi.org/10.1080/15394450802046895.

[62] H.C. Kuan, C.C.M. Ma, W.P. Chang, S.M. Yuen, H.H. Wu, T.M. Lee, Synthesis, thermal, mechanical and rheological properties of multiwall carbon nanotube/waterborne polyurethane nanocomposite, Compos. Sci. Technol. 65 (2005) 1703–1710. https://doi.org/10.1016/j.compscitech.2005.02.017.

[63] D.M. Panaitescu, Z. Vuluga, C.G. Sanporean, C.A. Nicolae, A.R. Gabor, R. Trusca, High flow polypropylene/SEBS composites reinforced with differently treated hemp fibers for injection molded parts, Compos. B Eng. 174 (2019) 107062. https://doi.org/10.1016/j.compositesb.2019.107062.

[64] S. Shang, L. Gan, M.C. wah Yuen, Improvement of carbon nanotubes dispersion by chitosan salt and its application in silicone rubber, Compos. Sci. Technol. 86 (2013) 129–134. https://doi.org/10.1016/j.compscitech.2013.07.010.

[65] A.A. Abdullateef, S.P. Thomas, M.A. Al-Harthi, S.K. De, S. Bandyopadhyay, A.A. Basfar, M.A. Atieh, Natural rubber nanocomposites with functionalized carbon nanotubes: mechanical, dynamic mechanical, and morphology studies, J. Appl. Polym. Sci. 125 (2012) E76–E84. https://doi.org/10.1002/app.35021.

[66] B. Saleh, S. hanna, M.H. Khalil, Evaluation of thermal and mechanical properties of crumb/natural rubber nanocomposites, Egypt. J. Chem. (2020). https://doi.org/10.21608/ejchem.2019.5522.1493.

[67] D. De, P.K. Panda, M. Roy, S. Bhunia, A.I. Jaman, Reinforcing effect of nanosilica on the properties of natural rubber/reclaimed ground rubber tire vulcanizates, Polym. Eng. Sci. 53 (2013) 227–237. https://doi.org/10.1002/pen.23255.

[68] M.M. Hassan, R.O. Aly, J.A. Hasanen, E.S.F. El Sayed, Influence of talc content on some properties of gamma irradiated composites of polyethylene and recycled rubber wastes, J. Appl. Polym. Sci. 117 (2010) 2428–2435. https://doi.org/10.1002/app.32120.

[69] A.R. Kakroodi, D. Rodrigue, Reinforcement of maleated polyethylene/Ground tire rubber thermoplastic elastomers using talc and wood flour, J. Appl. Polym. Sci. 131 (2014). https://doi.org/10.1002/app.40195.

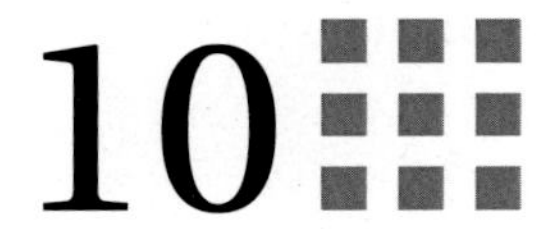

Viscoelastic behavior of elastomer blends and composites

Rupam Gogoi[1], Gaurav Manik[1], Sushanta K. Sahoo[2]

[1]DEPARTMENT OF POLYMER AND PROCESS ENGINEERING, INDIAN INSTITUTE OF TECHNOLOGY ROORKEE, SAHARANPUR, UP, INDIA; [2]MATERIALS SCIENCE AND TECHNOLOGY DIVISION, CSIR – NATIONAL INSTITUTE FOR INTERDISCIPLINARY SCIENCE AND TECHNOLOGY, THIRUVANANTHAPURAM, KERALA, INDIA

10.1 Introduction

The word elastomer was derived from the combination of two words: elastic and polymer. Elastomers are loosely cross-linked polymers having rubber-like flexibility and elasticity. Elastomers have extremely small modulus of elasticity (in the order of ~1–10 MPa) and significantly high deformability [1]. Different types of elastomers such as natural rubbers, ethylene propylene diene rubber, styrene-butadiene copolymers, ethylene propylene rubber, polyisoprene, polybutadiene, silicone elastomers, fluoro elastomers, nitrile rubber, polyurethane elastomers, bio-based toughened epoxies, etc., are used for various industrial applications. These elastomers exhibit a typical material property called viscoelasticity. It is important to note here that not only elastomers but many polymers like soft matters, metals at high temperature also exhibit viscoelasticity behavior of varying degree. These properties are very-crucial for the automotive and structural components requiring damping ability under different strain and frequencies.

10.1.1 Viscoelasticity: a property of materials

Viscoelasticity is a property of material and a dynamical concept, which involves aspects of two types of common natural responses upon application of force; (a) classical elasticity and (b) classical fluid. Fig. 10.1 represents a typical difference in the flow behavior of viscoelastic fluid and a viscous fluid.

In a classical elastic solid, the linear elastic behavior is demonstrated using the Hooke's law in which the stress is proportional to the strain as shown in Eq. (10.1).

$$\sigma = E\varepsilon \tag{10.1}$$

Where σ is the stress, E is elastic modulus, and ε is the strain. On the other hand, for a classical fluid having linear viscous behavior, the stress is given by the Newton's law as

Elastomer Blends and Composites. https://doi.org/10.1016/B978-0-323-85832-8.00014-6

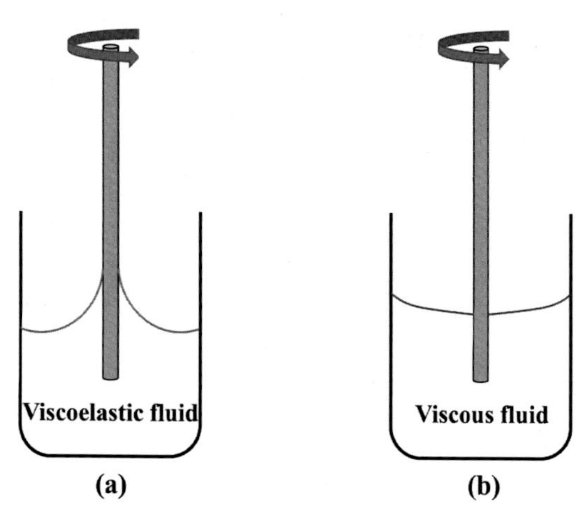

FIGURE 10.1 Illustration of comparison of the Weissenberg effect due to a rotating rod in a (A) viscoelastic fluid and a (B) viscous fluid (Newtonian fluid) [1].

shown in Eq. (10.2).

$$\sigma = \eta \frac{d\varepsilon}{dt} \tag{10.2}$$

Here, η is the viscosity of the fluid.

Unlike classical elastic solids and classical viscous fluids, the stress and strain relationship in a viscoelastic material is time-dependent and can be physically represented using a combination of spring and dashpot. The spring can be assumed similar to the capacitance of a circuit, which stores energy and the viscosity of a dashpot can be related to the resistance, which dissipates energy. Various constitutive models using different combinations of spring and dashpot as shown in Table 10.1 have been developed over the years to approximately represent and describe the viscoelastic behavior of polymers. Creep recovery and stress relaxation are the two general tests to define the viscoelastic behavior of polymers. In a creep test, a constant load is applied on the material over a period of time while in a stress relaxation test, a constant deformation or strain is subjected to the test material. Fig. 10.2 shows strain response of classical elastic solid, classical viscous fluid, and viscoelastic material to a constant stress input.

10.1.2 Constitutive models of linear viscoelasticity

Mathematical models provide a deeper understanding of the physics behind the nature of viscoelastic material. Maxwell and Kelvin—Voigt are the simplest constitutive models describing the linear viscoelasticity. However, the creep and stress relaxation behavior of viscoelastic material, which are inherent to them, cannot be explained fully by both these models. The Maxwell model can explain relaxation but not creep, whereas the Kelvin—Voigt does the opposite. Both the models correspond to materials that show only

Table 10.1 Physical and mathematical representation of viscoelastic behavior of material using combination of spring and dashpot through various models developed over the years.

Model		**Representation**
Maxwell model		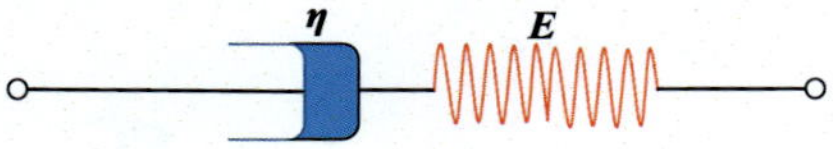 $\sigma + \frac{\eta}{E}\dot{\sigma} = \eta\dot{\varepsilon}$
Kelvin–Voigt model		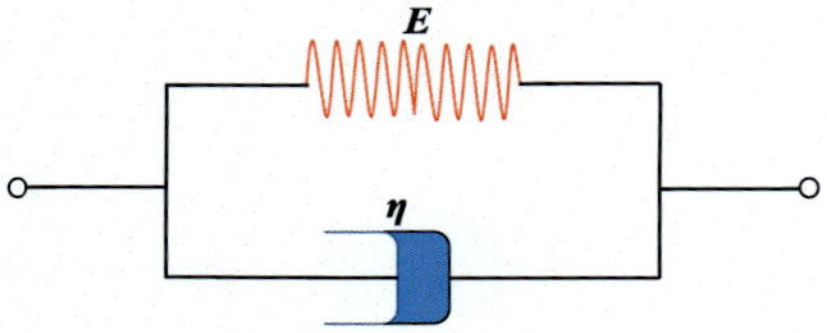 $\sigma = E\varepsilon + \eta\dot{\varepsilon}$
Standard linear solid model	Maxwell representation	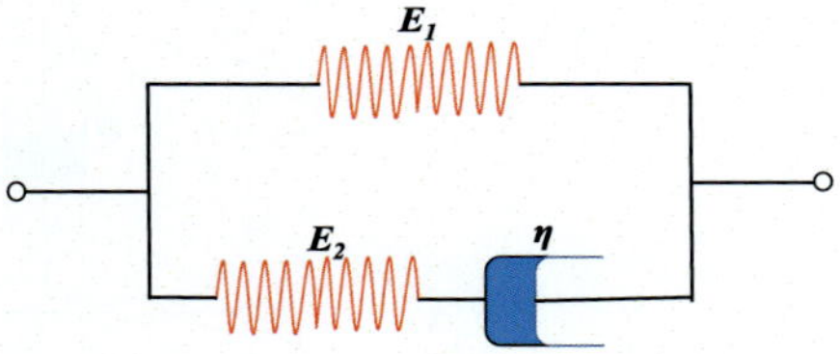 $\sigma\frac{\eta}{E_2}\dot{\sigma} = E_1\varepsilon + \frac{\eta(E_1 + E_2)}{E_2}\dot{\varepsilon}$
	Kelvin representation	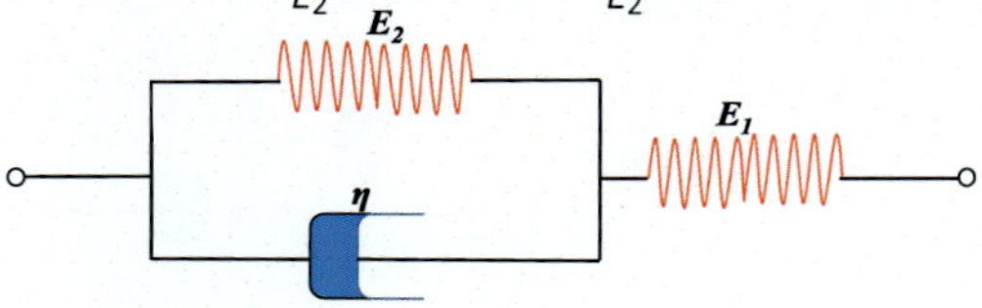 $\sigma + \frac{\eta}{E_1 + E_2}\dot{\sigma} = \frac{E_1 E_2}{E_1 + E_2}\varepsilon + \frac{E_1\eta}{E_1 + E_2}\dot{\varepsilon}$
Burgers model	Maxwell representation	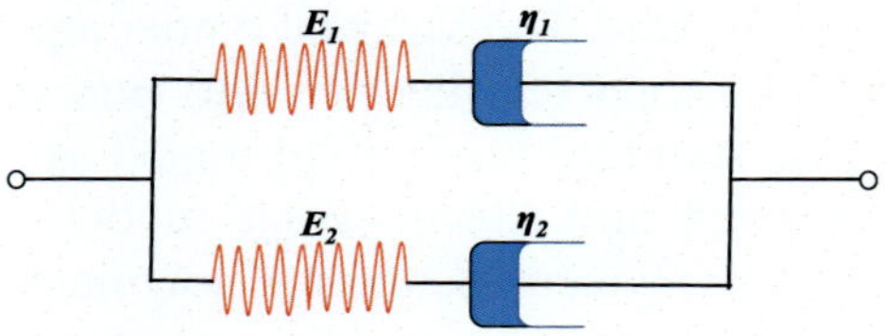 $\sigma + \left(\frac{\eta_1}{E_1} + \frac{\eta_2}{E_2}\right)\dot{\sigma} + \frac{\eta_1\eta_2}{E_1 E_2}\ddot{\sigma} = (\eta_1 + \eta_2)\dot{\varepsilon} + \frac{\eta_1\eta_2(E_1 + E_2)}{E_1 E_2}\ddot{\varepsilon}$

Continued

Table 10.1 Physical and mathematical representation of viscoelastic behavior of material using combination of spring and dashpot through various models developed over the years.—cont'd

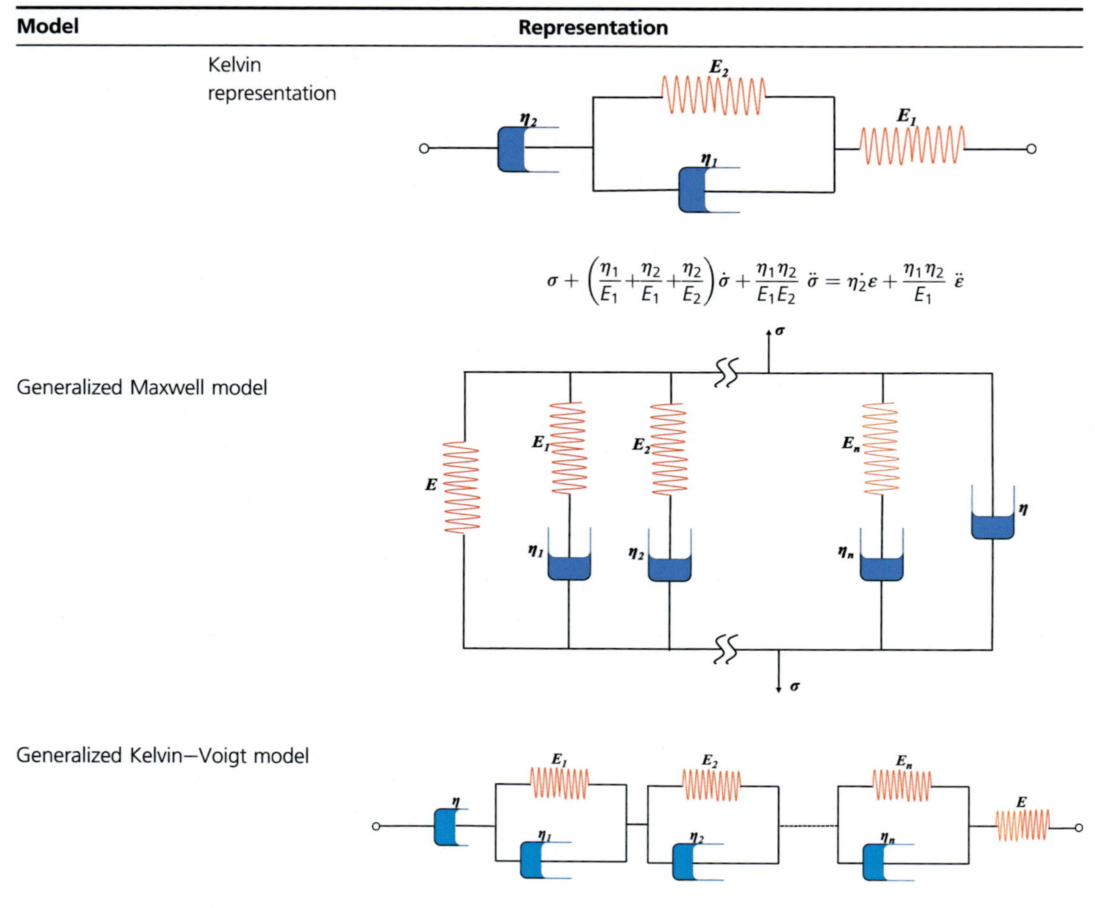

one characteristic time. The Maxwell model represents a viscoelastic material as a series combination of spring (elastic element) and dashpot (viscous element) as shown in Table 10.1. The standard linear solid model as shown in Table 10.1, which is a combination of Maxwell and Kelvin–Voigt model, offers assessment of creep and stress relaxation for most linear viscoelastic polymers. The creep and relaxation in model are represented by an exponential curve of rise and decay, respectively, involving time constant or relaxation time(s), as a function of spring stiffness and dashpot viscosity. The Burgers model and the series combination of Maxwell and Voigt models, generally such models are used fit for experimental behavior of the materials having several

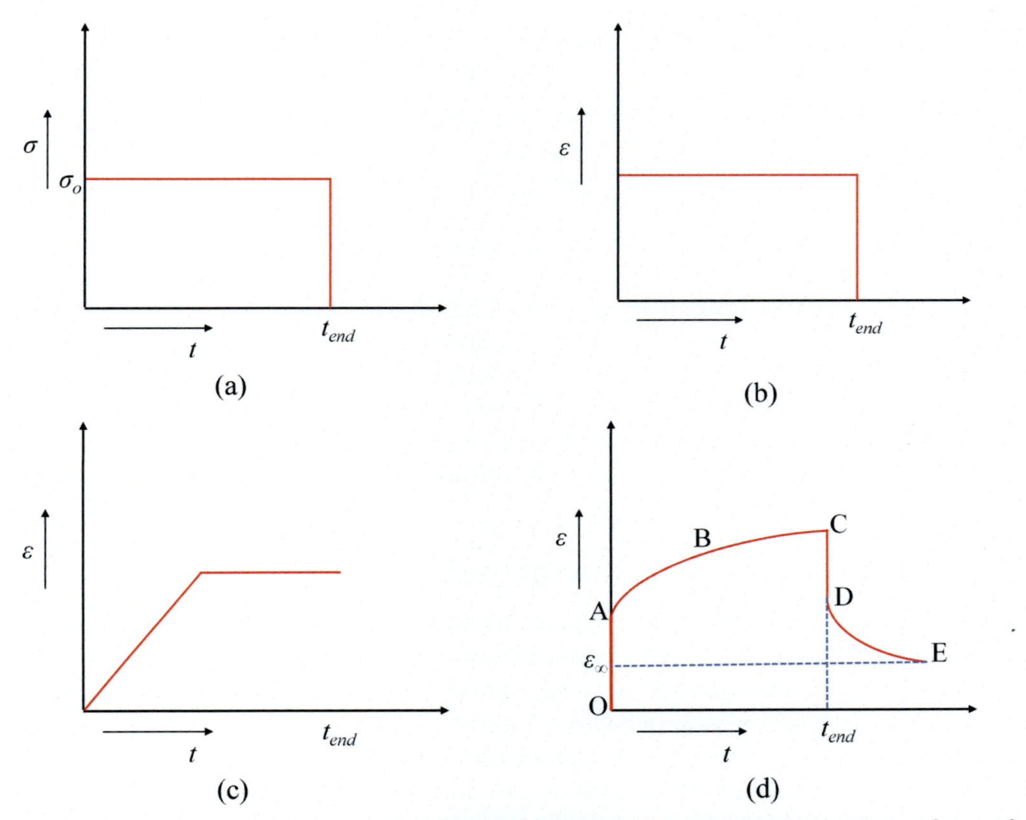

FIGURE 10.2 Illustration of strain, ε response to (A) constant stress (σ_0) input followed by release of stress for (B) classical elastic solid, (C) classical viscous fluid, and (D) viscoelastic material. For the viscoelastic material, *ABC*, continued straining under constant stress (consist of both elastic and viscous response); *CD*, instantaneous strain recovery; and *DE*, delayed recovery; *OA*, instantaneous response/elastic response.

characteristic times. A complex viscoelastic rheological model will usually be of the form of the generalized Maxwell model or the generalized Kelvin–Voigt model [2]. The generalized Maxwell model also known as Wiechert model, consists of n number of Maxwell elements as shown in Table 10.1, which are connected in parallel arrangement [3]. Similarly, the generalized Kelvin–Voigt model consists of n number of elements in series. In a realistic condition, a polymer does not possess a single relaxation rather has polymer chains of various lengths, which leads to a distribution of relaxation times spreading over a longer time period depending upon the type of polymer and complexity in the system. In general, a linear viscoelastic constitutive equation can be represented in the following general form using Eq. (10.3) [2].

$$P_o\sigma + P_1\dot{\sigma} + P_2\dddot{\sigma} + \cdots = q_o\varepsilon + q_1\dot{\varepsilon} + q_2\dddot{\varepsilon} + \cdots \tag{10.3}$$

In short Eq. (10.3) can be expressed as

$$\mathbf{P}\sigma = \mathbf{Q}\varepsilon \tag{10.4}$$

Where **P** and **Q** are the linear differential operators.

$$\mathbf{P} = \sum_{i=0}^{n} p_i \frac{\partial^i}{\partial t^i}, \ \mathbf{Q} = \sum_{i=0}^{n} q_i \frac{\partial^i}{\partial t^i} \tag{10.5}$$

The constitutive model of a polymer can be created using Eq. (10.3) by replacing the coefficients p_i, and q_i.

Elastomers such as rubber, on the other hand, are categorized under hyperelastic material, in which the stress–strain relationship is derived from the strain energy density function, and they follow a nonlinear viscoelastic behavior. The most widely used hyperelastic models are Neo-Hookean, Ogden, and Mooney–Rivlin model. While the Neo-Hookean model predicts stress most accurately up to a strain of 100% under uniaxial straining condition, the Mooney–Rivlin and Ogden models predict best up to a strain of 200% and 700%.

10.1.3 Dynamic loading and responses

Creep and relaxation cannot give comprehensive information on mechanical behavior of viscoelastic polymers. For a practical application of such materials often subjected to dynamic loading conditions, one need to evaluate responses related to short load cycles. Hence, predicting viscoelastic behavior of a material at such high rates of loading condition is a prerequisite for different applications. Unlike creep and relaxation test, an oscillatory load is subjected to the material to obtain such dynamic response. Within the linear viscoelastic region, if a material is subjected to an oscillating stress, then it will respond sinusoidally as shown in Fig. 10.3.

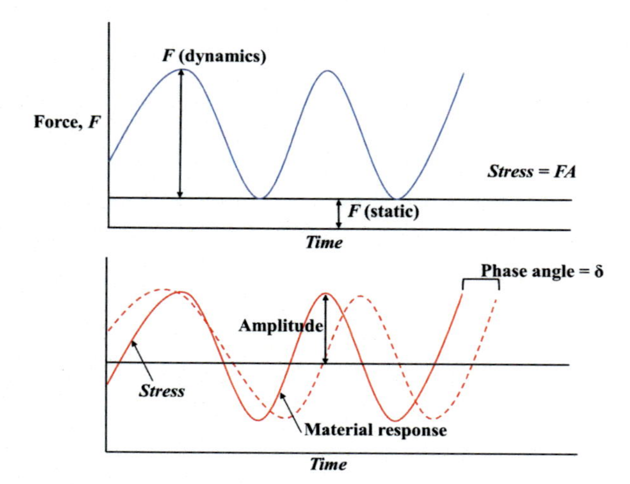

FIGURE 10.3 Illustration of input and output wave pattern when a sample is subjected to a sinusoidal oscillating stress.

Depending upon the type of material discussed here, the strain wave output may be different. For instance, in case of a classical elastic solid, there will be no phase lag, meaning $\delta = 0°$, whereas $\delta = 90$ degrees $= \pi/2$ for classical viscous material. Viscoelastic materials respond in between these two extreme limits, and thus, have a finite value of δ within 0 and 90 degrees.

For a viscoelastic material under the application of an oscillating strain, ε given by Eq. (10.6), the stress response at any time can be given by Eq. (10.7).

$$\varepsilon = \varepsilon_0 \sin\omega t \tag{10.6}$$

$$\sigma(t) = \sigma_0 \sin(\omega t + \delta) \tag{10.7}$$

Where ε_0 is the maximum strain and σ_0 is the maximum stress at the maximum strain. Eq. (10.7) can be rewritten as

$$\sigma(t) = \sigma_0(\sin\omega t \cos\delta + \cos\omega t \sin\delta) \tag{10.8}$$

For a small strain amplitudes and time-independent polymers (linear viscoelastic regime), the resulting stress in Eq. (10.8) can be written in terms of the dynamic storage modulus (E') and the dynamic loss modulus (E'') as shown in Eq. (10.9).

$$\sigma(t) = \varepsilon_o(E' \sin\omega t + E'' \cos\omega t) \tag{10.9}$$

Where,

$$E' = \frac{\sigma_o}{\varepsilon_o}\cos\delta \tag{10.10}$$

and

$$E'' = \frac{\sigma_o}{\varepsilon_o}\sin\delta \tag{10.11}$$

The tangent of the phase angle, tan δ, is one of the most basic properties measured. The loss tangent, tan δ is also called damping and is independent of geometry effects. It is the ratio of loss to storage modulus as shown in Eq. (10.12) and indicates how efficiently a material loses energy when subjected to stress through molecular rearrangements and internal friction.

$$\tan\delta = \frac{E''}{E'} \tag{10.12}$$

The above properties, E', E'', and tan δ, are measured for viscoelastic materials such as elastomers and polymers using an instrument called dynamic mechanical analyzer (DMA). Fig. 10.4 shows an illustrative example of relationship between loss and storage. Temperature scanning of polymer in DMA helps researchers to investigate their relaxation processes. The change in free volume [4] or relaxation times [5] are used in a polymer to describe its thermal transitions.

Fig. 10.5 shows the variation in free volume of polymer with respect to temperature and plot of a typical DMA for polymer.

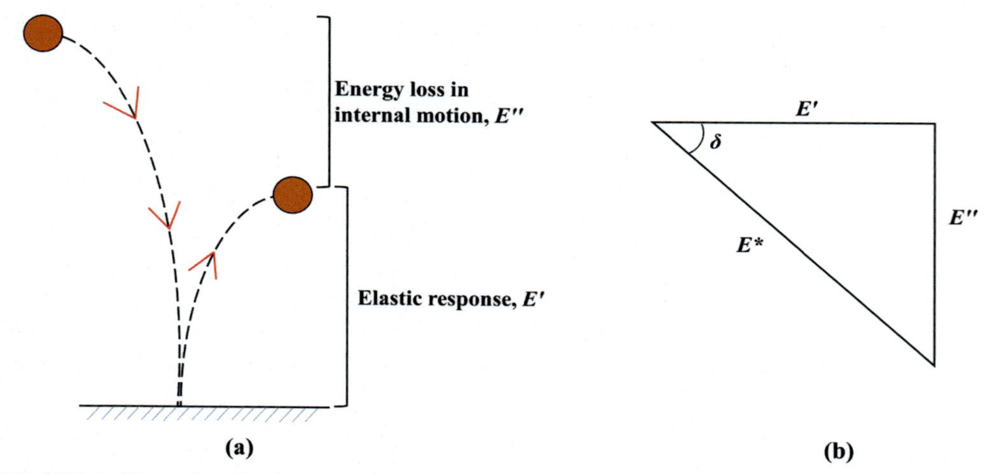

FIGURE 10.4 An illustration of relation between loss and storage. (A) Response of a ball made out of viscoelastic material when dropped from a height. The height recovered by the ball can be described as the storage, E′, whereas the difference between initial and the final height can be termed as the loss, E″ is energy due to friction and internal motions. (B) The relationship between the phase angle δ, E*, E′, and E″ is graphically shown.

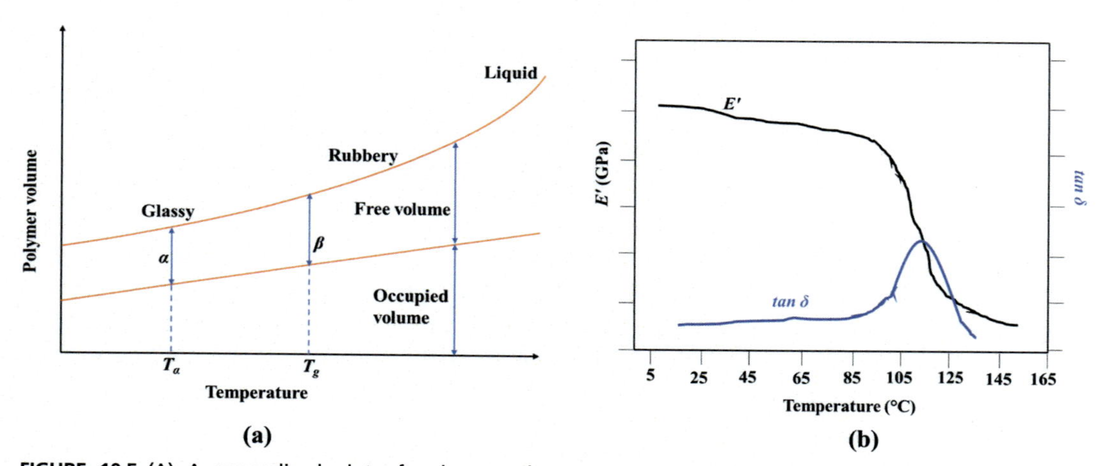

FIGURE 10.5 (A) A generalized plot of polymer volume versus temperature. Below the glassy and rubbery transitions, the free volume region is represented by α and β [6] (B) Typical representation of output after running a temperature scan of polymer in DMA.

As the name suggests, the free volume of a polymer is described as the space available for the molecules to have internal movement. In relation to the thermal transitions over a range of temperatures, the free volume can be explained by considering a crankshaft mechanism for the movement of the polymer side chain, small groups, etc. In this mechanism it is considered that the polymer chain consists of various mobile segments, which can have free movement to some degree. It is believed that with increasing free volume, this free movement of the side chains or small groups of the polymer also increases, which results in decreasing modulus. Heijboer [7] studied this motion of polymer chains and classified the type of motions into β and γ transitions. The study of the softening of material at specific temperature and frequency helps to find the end application of the material. Fig. 10.6 shows an ideal transition of polymer, showing changes in modulus over a range of temperature from very low to high.

Starting at the low temperature, as the polymer is exposed to an increasing temperature, it expands and the free volume increases, which leads to localized movements of chain including bending and stretching of the chains. At this stage, the γ transition occurs, which is characterized by the transition temperature T_γ. With increasing temperature, the whole of the side chain and localized atom groups start to have free motion, which develop some toughness in the material. This transition is called β transition and is characterized by the transition temperature T_β. With a further increase in temperature, the glassy polymer shows a transition to rubbery phase having a large-scale motion, and the transition temperature is called glass transition temperature, T_g. During this transition, the chains in the amorphous regions begin to coordinate large-scale motions, and hence, the amorphous regions begin to melt. A 100% crystalline material does not show T_g while for a semicrystalline and some amorphous polymer, T_α^* and T_{ll} can be observed as shown in Fig. 10.6. The former occurs in crystalline or semicrystalline polymer due to crystal slippage, whereas the latter is a movement of coordinated segments in the amorphous phase that relates to reduced viscosity. As the temperature increased, a flat rubbery plateau is observed. Beyond the rubbery plateau, melting of polymer occurs, which is characterized by large-scale chain slippage. For a thermoset material, no T_m is observed, whereas, for a cured elastomer beyond T_g chain,

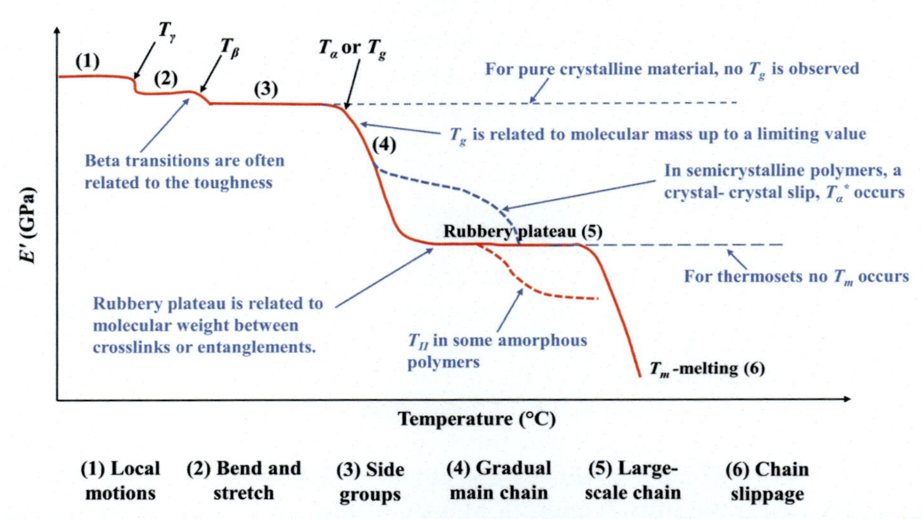

FIGURE 10.6 A generalized behavior of polymer over a range of temperature from very low to high [8]. The E′ decreases with increasing temperature and shows different transitions related to the chain movement and free volume. The different chain movement or motions at each stage are shown below the figure and numbered from (1) to (6).

slippage does not occur due to the presence of cross-link structure, and hence, it begins to burn or degrade with increasing temperature. Such temperature scan of a polymer using DMA is an effective tool to determine various viscoelastic properties of polymer and elastomers. From the output data of DMA, the T_g can be defined as the temperature at which (i) maximum of the tan δ occurs or (ii) maximum change of E'' occurs or (iii) the middle point of E' versus temperature curve, or (iv) the region where E' increases with increasing frequency at constant temperature [9].

In the following sections, viscoelastic properties exclusively of elastomer blends [10–13] and composites analyzed through DMA are discussed. Factors affecting these properties such as phase compatibility and morphology, type and ratio of individual elastomer in blends, type, size, and wt.% of filler, additives, or reinforcement in elastomeric composites are taken into considerations.

10.2 Viscoelasticity of elastomer blends

Polymer blends are a mix of two or more different polymers or elastomers, which are not covalently bonded together. The unique feature of each polymer can be obtained in a synergistic and economical form by preparing blends of polymers or elastomers. The unique attributes may be from the point of view of cost, processing, thermal and mechanical properties, etc. A critical issue in the formation of blend is the issue of compatibility among the blending polymers [14]. Thermodynamically an immiscible blend will have free energy, $\Delta G_m \approx \Delta H_m > 0$, (where H_m is enthalpy), but it can be made compatible if an efficient way of stress transfer among the different phases can be achieved. Most common procedure is to use compatibilizers [15], which help to increase the interfacial interaction between two different polymer phases, by reducing the interfacial tension [14,16–21]. DMA is an important technique to obtain insights of compatibility and viscoelasticity of such blends and hence, has been followed in various earlier studies [22–25].

Singh et al. [26] prepared two different (20:80) blends of ethylene propylene diene terpolymer (EPDM) and poly(vinyl chloride) (EPDMPVC) along with methyl methacrylate-grafted EPDM and PVC (MMAgEPDM–PVC). The grafting contents considered were 4%, 13%, 21%, and 32%. The viscoelastic properties of the different blends were evaluated at 1.0 Hz using a Du-Pont 983 DMA in the temperature range of −100 to 150 °C and at a temperature ramp of 5 °C/min. A sample dimension of 13 × 11 × 3 mm was used. In case of the EPDMPVC blend, multiple tan δ peaks were observed, which clearly indicated the incompatibility of the blend. Three relaxation peaks at −18, 34, and 114 °C exhibited the γ, β, and α transitions. The blend showed two T_g, one at −43°C, which was due to the EPDM phase, and another at 94 °C, which was due to PVC. With the introduction of MMA grafting, the MMAgEPDM–PVC blends displayed relaxation curve different than the neat EPDMPVC blend. At lower grafting contents (4% and 13%) of MMA, the T_g reduced while at higher contents (21% and 32%), it increased. The grafting of EPDM by MMA at lower content produced polarity, and hence, an improved interaction within the blend was observed, which resulted in shifting of T_g. At higher grafting wt.%, the high-molecular-weight MMA brought in rapid chain entanglements and prominent plateau region was observed for the blends. The delayed viscous flow due to the chain entanglements might have increased the T_g, as was observed in the DMA plots. The T_g due to the PVC portion of the MMAgEPDM–PVC blend at 4% and 13% was 100 and 103 °C, respectively, with a broad tan δ peak, and it closely approached the T_g of PVC. While for 21% and 32% grafting, it increased to higher temperatures of 108 and 114 °C, respectively. The shifting of the secondary relaxation peaks of the blends with higher grafting content to higher temperature along with reduction in the peak intensity can be explained as the increase in chain entanglement brought in by PMMA chains.

Mousa et al. [27] studied the viscoelastic behavior of dynamically vulcanized PVC-epoxidized natural-rubber (ENR) thermoplastic elastomers blend. The effect of the concentration of sulfur (curatives) on the viscoelastic behavior was investigated using a Monsanto MDR2000. A gradual increase in the E' with increase in sulfur dosage was observed. This could be due to the formation of intermolecular cross-links. The damping behavior is a very sensitive indicator of cross-linking. The E'' of all the tested specimens decreases with an increased degree of cross-linking, and thus, the tan δ decreases to a minimum value at maximum sulfur dosage. This again could be attributed to the microstructural changes that have occurred through the cross-link formation via dynamic vulcanization. Khanra et al. [14] studied the compatibilization of silicone rubber (MVQ) and fluoroelastomer (FKM)-based high-performance blend. Fluorosilicone rubber (FSR) was used as a compatibilizer alone and also with silica filler to observe their synergistic effect on the blend compatibility. For a 50/50 blend of FKM and MVQ having conventional curing system, two peaks of tan δ curve were obtained. One at −44.5°C for MVQ and another at 1.6°C for FKM. With the addition of only 2.5 phr FSR, the peaks shifted to −42.1°C and 3.5°C for MVQ and FKM, respectively, and hence, showed a 4 °C shift of both the peaks toward each other. Furthermore, adding 15 phr silica filler along with 2.5 phr FSR resulted in a significant 8 °C shift of the peaks, proving the increased compatibilization between both the phases through enhancement of the interfacial interaction as well as better adhesion. On the other hand, increasing FSR content to 5 phr in the 50/50 blend of FKM and MVQ resulted in no shift of T_g, rather it was negligible in comparison to the neat blend. This may be attributed to the fact of the formation of another phase of FSR, which may hamper the compatibilization. Shafeeq and Unnikrishnan [28] prepared blends of ethylene-co-vinyl acetate (EVA) and millable polyurethane (MPU) in different EVA/MPU ratios of 20/80 (E20), 40/60 (E40), 50/50 (E50), 60/40 (E60), and 80/20 (E80). Dicumyl peroxide (DCP) of 2 phr was used in each blend as the curing agent. The blends were prepared in an open mill mixing. The analysis indicated that neat EVA exhibits transition from glassy to rubbery phase at −21°C while neat MPU shows a sharp drop in E' at −66°C. In case of blends, the transitions or drop in E' was observed in between the above two temperatures, and it increased from −64°C to

−4 °C with increasing EVA content in the blends. The neat MPU shows tan δ peak at −33°C, corresponding to the glassy to rubbery transition. The MPU-rich blends exhibit a broad peak between −60 and 20 °C with shoulder peaks at 13 °C for E40 and at −8°C for E20. This can be due to the merging of T_g values of the individual elastomers in the blend. Since merging of peaks corresponding to glass transition is a consequence of miscibility, it was observed that the blends with higher MPU content may have behaved as miscible systems. EVA as a major phase exhibits two distinct tan δ peaks as shown by E50, E60, and E80, and hence, they form an immiscible blend system. Apart from this, the shift of T_g to a more positive region was a consequence of increasing crystallinity in the blends.

Lai et al. [28] prepared self-healing natural rubber (NR)/polycaprolactone (NR/PCL) blends in the ratio of 40/60 and 60/40. Different blends were prepared both with neat PCL and 2 and 4 phr grafted-acrylic acid PCL (PCL-g-2AA and PCL-g-4AA). From the DMA analysis it was observed that neat PCL exhibited a higher E' to that of NR, which was due to the high crystallinity of the former. All the blends showed E' in between the neat PCL and NR, within which the blend having grafted PCL had higher value compared to the blend with neat PCL. The 40/60 blend of NR/PCL-g-4AA showed the highest E' among all the blend systems. From the tan δ plot, the T_g of NR and PCL was obtained at about −66°C and −55°C, respectively. However, in the case of NR/PCL blends, the shifting of the tan δ peaks was not toward each other (which should had been the case for an indication of miscibility), instead it shifted to a lower temperature. In a miscible blend, the T_g values of individual components shift and merge with each other. But it was observed that the T_g of the blends reduced to even lower values than those of NR. Possibly, a dilatant stress on the NR due to the crystallized PCL increased the molecular mobility of the NR and caused the depression in T_g of the NR component in all blends [28].

Tomova et al. [29] studied ternary blends based on polyamide 6 (PA 6), PA 66, and elastomer. Ethene–propene copolymer (EPM), ethene-octene copolymer grafted-maleic anhydride (EOgMA), and ethene-propene-diene copolymer grafted-MA (EPDM-g-MA) were used as the three different elastomers in the blends. In the ternary blend, the PA 6/PA 66 ratio was fixed at 50:50 while the elastomer content was varied from 10 to 40 wt.%. The T_g (α transition) of neat PA 6 and PA 66 was around 63 and 74 °C, while EPDM-g-MA, EO-g-MA, and EPM exhibited at −53, −43, and −33°C, respectively. As can been seen in Fig. 10.7, tan δ peaks of the ternary blends do not form a single peak indicating absence of complete miscibility. From Fig. 10.7. A and E it can be observed that the T_g of the blends with increasing EO-g-MA and EPM decreases and reaches β transition of the binary blend of PA 6/PA 66. This can be explained by the difference in the thermal expansion coefficient of the constituents within the blend, which increased the thermal stress in the elastomer phase creating a negative pressure. In case of the blends having EPDM-g-MA, the interfacial tension is smaller compared to that of EO-g-MA with polyamide. Thus, there does not exist any shift of T_g as can be observed in Fig. 10.7C. The E' of the all the ternary blends decreases with increasing rubber content as expected [29].

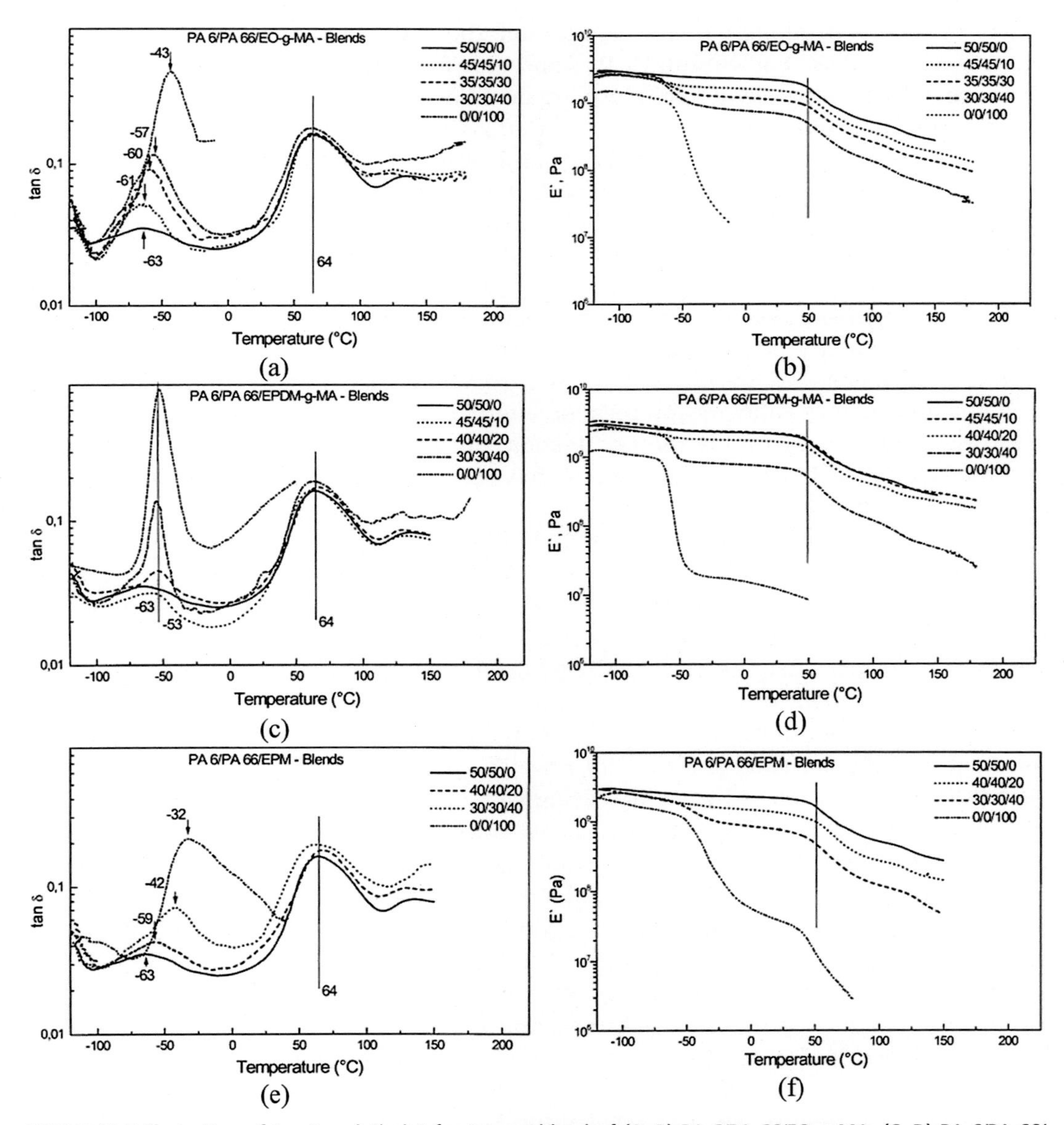

FIGURE 10.7 Illustration of tan δ and E′ plot for ternary blend of (A, B) PA 6/PA 66/EO g MA, (C, D) PA 6/PA 66/EPDM-g-MA, and (E, F) PA 6/PA 66/EPM respectively [29]. *From D. Tomova, J. Kressler, H.J. Radusch, Phase behavior in ternary polyamide 6/polyamide 66/elastomer blends, Polymer 41 (21) (2000) 7773–7783. https://doi.org/10.1016/S0032-3861(00)00127-0. Reprinted with permission from Elsevier.*

Peng et al. [30] prepared silane cross-linked polypropylene (PP)/EPDM elastomer blends and investigated the structure, properties, and viscoelastic behaviors using DMA. For the PP/EPDM blends E' decreased gradually with increasing EPDM concentration due to the low E' of rubbery phase. In addition, the E' of blend having DCP as a curing agent was lower than that without DCP as observed for the 70:30 blend of PP/EPDM. The presence of DCP in the blend might have triggered degradation of the PP chains. The tan δ curve indicated the T_g of PP at 20.5°C, whereas the PP/EPDM blends exhibited two distinct T_g, one around 20 °C (PP phase) and other at −30°C (EPDM phase). It was observed that compared to the neat blend, the T_g shifted in case of the silane cross-linked blend. Since EPDM is cross-linked easily and exhibits high cross-linking degree, T_g of EPDM shifts from −30.5°C to −27.4°C, while T_g of PP has no obvious change with the content of EPDM increasing [30].

The growing demand in the automotive sector is not only creating air pollution but in addition a secondary issue of elastomeric waste from tires is being added. Landfills created from waste tires are not only fire hazardous but also are increasingly becoming breeding ground for deadly mosquitoes. An alternative method of reusing such valuable waste is to prepare useful elastomeric blends. But compatibility of such blends often becomes an issue. Hence, a lot of research has been carried out in this direction to find an optimum solution [31–33].

Hejna et al. [34] investigated the compatibility in reclaimed rubber (RR) and poly(-εcaprolactone) (PCL) blends. Initially, RR was prepared by mixing ground tire rubber with bitumen in a 100:10 ratio in a two roll-mill for 15 min. Next RR/PCL blend was prepared in a Brabender mixer at 120 °C, with screw rotation of 80 rpm and mixing time of 8 min. Two types of PCL were used Capa™ 6800 and Capa™ FB100. The PCL content in the blends was varied from 10 to 50 wt.%. In the tan δ plot of the blends, only a single peak was observed indicating a good compatibility and homogeneity among the blending ingredients. The T_g of the 90:10 RR: PCL blend was almost 15 °C higher than pure PCL. The T_g increased with increasing wt.% of PCL. Comparing the two types of PCL, the blend with Capa™ FB100 possessed relatively higher T_g values for the same ratio of blends. In addition, the E' values improved with PCL incorporation due to their semicrystalline nature, which offered stiffness to the blends.

Mészáros et al. [35] carried out an electron beam (EB) promoted recycling of ground tire rubber (GTR) with polyethylene (PE). A 20 wt.% EVA was used as a compatibilizing agent. Blends of PE/GTR and PE/GTR/EVA were prepared in 70/30 and 50/30/20 wt.%, respectively. Blends were premixed mechanically followed by internal mixing in a Barbender PlastiCorder at 10 rpm and temperature in between 165 and 175°C. The extrudate was injection molded to form samples. For the EB treatment, the prepared samples were exposed to electron accelerator, in air, at absorbed doses of 50, 100, 150, and 200 kGy (or kJ/kg).

It can be observed from Fig. 10.8A that pure EVA exhibits a significant relaxation peak corresponding to its T_g of −13.2°C. However, flat trend in the tan δ curve was observed for both PE and its blends with GTR. A relaxation peak at −50°C was observed for the blend, which slightly shifted upon EVA incorporation, as can be observed for blend PEGTR30 in Fig. 10.8A. This phenomenon refers to the increased compatibility of the

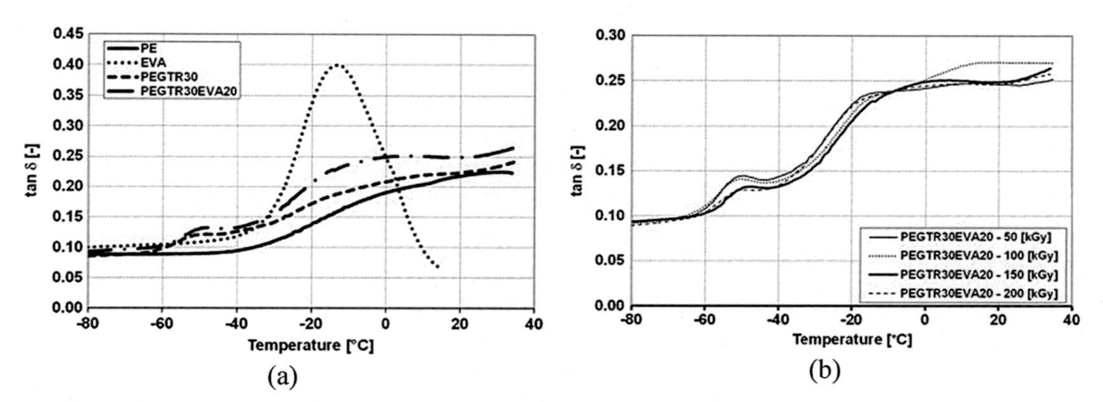

FIGURE 10.8 Illustration of tan δ plot for (A) binary and ternary blends of PE, GTR, and EVA and (B) EB-treated ternary blends of PE, GTR, and EVA [35]. *From L. Mészáros, T. Bárány, T. Czvikovszky, EB-promoted recycling of waste tire rubber with polyolefins, Radiation Physics and Chemistry 81 (9) (2012) 1357–1360. https://doi.org/10.1016/j.radphyschem.2011.11.058. Reprinted with permissions from Elsevier.*

components. Furthermore, it can be observed from Fig. 10.8B that exposing the samples of ternary blend to EB dosage shifted the tan δ peaks or the T_g to a higher temperature. The moderate decrease in the tan δ peak height with increasing dosage of EB explains the limitation of chain flexibility to such exposure.

10.3 Viscoelasticity of elastomer composites

Composites in general are type of material in which reinforcing material or fillers are incorporated to a continuous phase such as polymer, elastomer, ceramics, or metal. The synergism among the components offers different and improved physical and thermal properties of the composite, which are useful in many applications. In case of elastomer composites, based on the type of base elastomer, reinforcing filler, its geometry, size, fraction, etc., play a major role in deciding the properties of the composites [16,17,36–40]. Viscoelasticity of such composites, which is dependent majorly on the chain mobility, is affected by the nature of filler incorporation and requires a critical analysis. For instance, Praveen et al. [41] investigated the effect of filler geometry on viscoelastic damping of graphite, short aramid, and carbon-fiber-filled styrene butadiene rubber (SBR) composites. DMA was conducted using a rectangular sample with dimensions $25 \times 10 \times 2$ mm in a DMA machine of GABO, Germany, in tension mode. For a comparative study and use as a base matrix, a control sample of 20 parts per hundred-gram rubber (phr) N330 carbon-black-filled SBR with conventional sulfur-based curing systems was used. The three different sets of composites were prepared by mixing 15, 30, and 55 phr of graphite and 10, 20, and 30 phr of both aramid and carbon fibers separately to the controlled base matrix. Irrespective of the loading content, the tan δ peak height and the width of the graphite filled SBR were similar while it reduced compared to the control SBR sample. Absence of a good interfacial adhesion restricted the polymer chains to take part in the relaxation process. The basic concept of damping involves the absorption of external energy through molecular friction. The large internal surface area of graphite due to its layered structure helps in damping by producing friction due to shear, which adds to the energy dissipation. This extra dissipation due to the structure of graphite is in addition to that of the base polymer matrix. The E' value of the graphite-filled SBR was higher than the controlled SBR sample for the entire temperature region. With increasing graphite filler content, E' increased; however, the improvement was more pronounced in the rubbery region. For a 55 phr loading of graphite, E' was the highest. A similar observation of improved E' was observed in case of short carbon fiber and short aramid fiber-filled SBR composites. With increasing loading, E' increased similar to the graphite-filled SBR composites. For a 30 phr loading of short carbon and aramid fiber separately in SBR composites showed the highest E'. It is worth noticing here that the variation of E' for the same sample at glassy and rubbery region was higher for the control sample compared to the carbon and aramid fiber-filled samples. This can be attributed to the combination of the hydrodynamic effects of the fibers embedded in a viscoelastic medium and to the mechanical restraint introduced by the filler at the high concentrations, which reduce the mobility and deformability of the matrix. Out of all the three incorporation, aramid fiber-filled composites showed highest E' values. Carbon fiber and graphite-filled composites showed similar values of E', since the former had a breakage of fiber during the roll-mill process to prepare the mix. While the aramid fibers were intact during processing and hence their composites were able to bear more load. Furthermore, the T_g of the composites shifted toward a positive temperature, with composites of aramid (−30°C) showing the highest among the other two with (carbon fiber composite: −32°C and graphite-filled composite: −36°C). Effective reinforcing effect of aramid fiber was hence observed. The effective interphase created around the aramid fibers supported the increase in T_g.

In another study, Praveen et al. [42] studied the effect of incorporating carbon black and nanoclay fillers in SBR matrix. Two different sets of samples were prepared using two roll mills, one with nanoclay (Montmorillonite) and other with both the nanoclay and carbon black (high abrasion furnace black, HAF N330). In both the sets, the nanoclay concentration was varied as 5, 10, 15, and 20 phr, and all the samples had conventional curing formulations. In the hybrid composites, the carbon black content was fixed at 20 phr.

In the nanoclay filled composites without carbon black, as seen in Fig. 10.9A and B, the E' increased with increasing concentration of nanoclay, whereas the tan δ peak decreased due to improved filler–polymer interaction, which reduced the relaxation process. A good dispersion of the nanoclay filler within the SBR matrix demonstrated a pseudo-cross-linking point and remarkably helped in improving composites modulus. A similar observation was made for the hybrid composites of carbon black and nanoclay as observed in Fig. 10.9C and D.

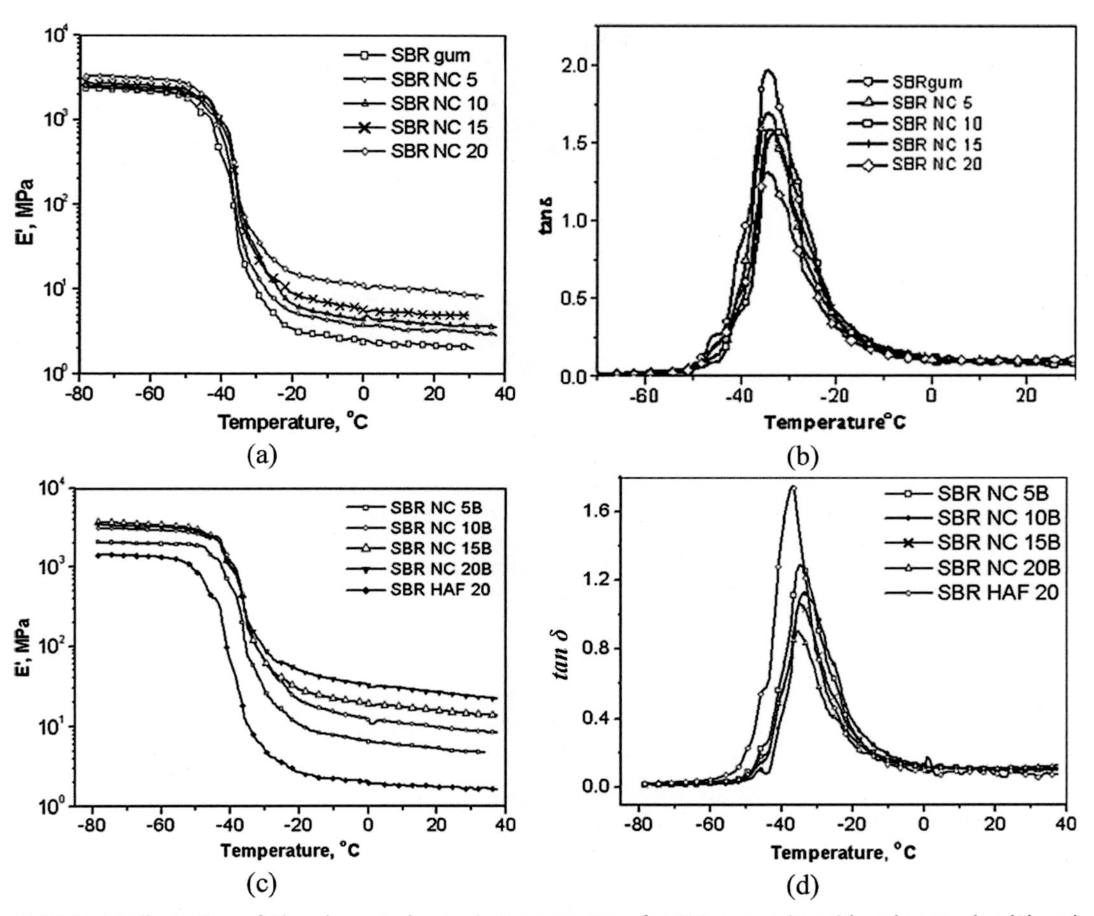

FIGURE 10.9 Illustration of E′ and tan δ plot against temperature for SBR composite with only nanoclay (A) and (B) and hybrid SBR composite of carbon black and nanoclay (C) and (D) respectively [42]. *From S. Praveen, P.K. Chattopadhyay, P. Albert, V.G. Dalvi, B.C. Chakraborty, S. Chattopadhyay, Synergistic effect of carbon black and nanoclay fillers in styrene butadiene rubber matrix: Development of dual structure, Composites Part A: Applied Science and Manufacturing, 40 (3) (2009) 309–316. https://doi.org/10.1016/j.compositesa.2008.12.008. Reprinted with permission from Elsevier.*

Compared to the control sample of SBR filled with only carbon black, the E' of the hybrid composites improved significantly explaining the synergistic effect of incorporation. Moreover, the reduced segmental mobility of the polymer chains was caused due to their adsorption on the filler surface attributed to good filler–polymer interactions. Sankaran et al. [43] prepared a 90:10 blend of bromobutyl rubber (BIIR)/polyepichlorohydrin rubber as the base matrix for nanoclay filled rubber nanocomposite. The nanoclay content was varied as 3, 5, and 10 phr. The DMA curves obtained for the nanocomposites are illustrated in Fig. 10.10. The incorporation of nanoclay reduced both the tan δ peak and its width implying reduction in damping and ability of composites to

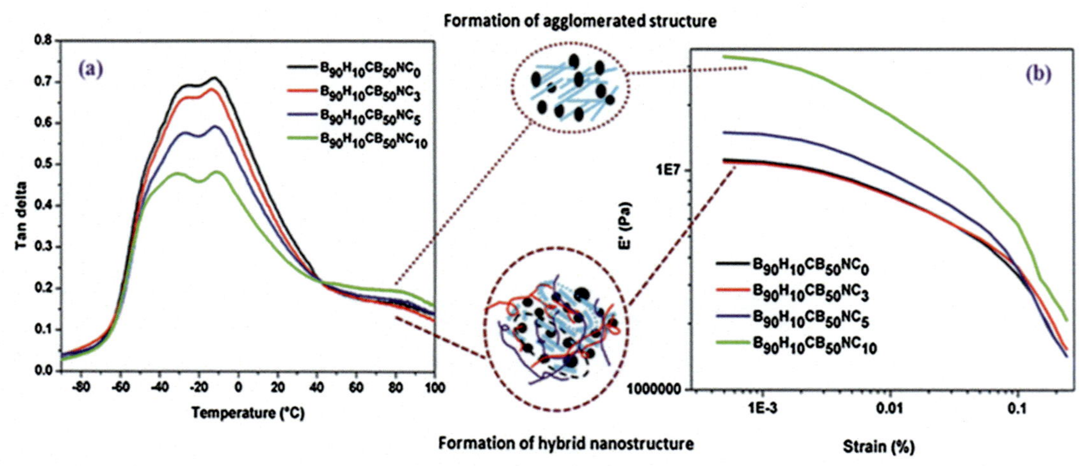

FIGURE 10.10 Variation of (A) tan δ and (B) E′ for different BIIR/CO blend-based nanocomposites [43]. *B*, BIIR; *CB*, carbon black; *H*, polyepichlorohydrin rubber; *NC*, nanoclay. *From K. Sankaran, G.B. Nando, P. Ramachandran, S. Nair, U. Govindan, S. Arayambath, S. Chattopadhyay, Influence of hybrid nanostructures and its tailoring mechanism on permeability, rheology, conductivity, and adhesion properties of a novel rubber blend nanocomposite, RSC Advances, 5 (107) (2015) 87,864–87,875. https://doi.org/10.1039/c5ra17178d. Reprinted with permission from RSC.*

dissipate energy as can be clearly observed from Fig. 10.10A. Additionally, the T_g values did not show any shift in temperature compared to the composite without nanoclay. It was observed that the E' value of composite with 3 phr nanoclay ($B_{90}H_{10}CB_{50}NC_3$) was relatively lower compared to the control sample with no nanoclay ($B_{90}H_{10}CB_{50}NC_0$). This can be attributed to the hybrid nanostructures that were formed without affecting chain mobility of the base polymer matrix. However, at higher content of nano filler, the E' improved, and it was also concluded that agglomeration at higher loading dominated over the formation of hybrid nanostructure. Conducting a strain sweep in DMA helped in determining the Payne effect of the composites, which was used to analyze the filler–filler interactions as depicted in Fig. 10.10B. It was observed that the Payne effect increased with increasing filler content, which was due to the nanoclay agglomeration, and at concentration of 3 phr, the composite $B_{90}H_{10}CB_{50}NC_3$ exhibited a lower Payne effect compared to $B_{90}H_{10}CB_{50}NC_0$, which again confirmed the structural breakdown of hybrid nanofillers. It was hence concluded that formation of hybrid nanostructure is very effective in $B_{90}H_{10}CB_{50}NC_3$ and reaches percolation threshold in $B_{90}H_{10}CB_{50}NC_5$. Importantly, this balances the adhesion properties up to the percolation threshold. Higher dosage of nanoclay deteriorates adhesion due to the formation of more agglomerated structures.

Goswami et al. [15] optimized the graphene content in a carbon-black-filled nitrile butadiene rubber (NBR). The graphene content was varied as 2,4, 6, and 8 phr. An increment of 9% in tensile strength of the carbon-black-filled hybrid system was observed with a graphene loading of 4 phr without hampering the ultimate strain.

Further addition of graphene resulted in degraded properties. At low temperatures, composite without graphene loading possessed a high E' compared to graphene-filled NBR composites. It is due to the ineffective reinforcement of graphene in the low-temperature range. However, with the incorporation of grapheme, the E' increased in the rubbery state. Unfilled composite exhibited an E' 10.09 MPa at 10 °C, while a 30% increment in respective value is achieved with an addition of 8 phr of graphene. The tan δ curve showed no transition in the T_g, but the peak height reduced remarkably indicating restricted molecular movement.

Sallat et al. [44] prepared composites of silica-particle-filled bromo butyl rubber (BIIR). Pristine and surface-treated silica particles with alkoxysilane were used for the composites. The base polymer BIIR was ionically modified (BIIR-i). The DMA analysis of the composites revealed an increase in E' compared to BIIR-i. The silane treatment did not show any significant effect in improving the E' of the composites, rather E' of the composites with pristine silica was higher than the modified one. Among the silane-treated, composite filled with silica modified with silane having alkyl functional groups exhibited a slightly better result. The relaxation peaks of base polymer and composites exhibit a typical T_g (around −22°C) of BIIR. For the composites, the *tan* δ peak height and width reduced compared to BIIR-i. The *tan* δ curves of composites with modified silica fillers reside between the neat silica-filled composite and BIIR-i. Both the E' and relaxation indicate prevailing network density in the composites, which increases with silica incorporation compared to the BIIR-i.

Ashok et al. [45] carried out experimental and theoretical analysis on the nano-reinforcement mechanism of organomodified layered silicates (OMLS) in EPDM/chlorobutyl rubber (CIIR) blends. To ensure dispersion of the OMLS, two-step compounding process was adopted by initially preparing 3:1 weight ratio of EPDM—OMLS and CIIR—OMLS master batches in an internal mixer and then compounding the same with calculated EPDM/CIIR blend in a two-roll laboratory mill. CIIR was compounded and precured before blending with EPDM to ensure cure compatibility. In the prepared composites, the EPDM to CIIR blend ratio was fixed at 80:20, and the OMLS content was varied at 2.5, 5, 7.5, and 10 phr. Compared to unfilled blend, the E' increased in case of composites filled with OMLS. The highest increase in E' was obtained for the composites having 5 phr OMLS in both the glassy and rubbery regions. However, further increase in the nanoreinforcement led to agglomeration, and hence, the E' decreased due to the rubber chain slippage past each other. The relaxation behavior and the energy dissipation of the blend and the nano composites were observed from the tan δ plot. It was observed that the tan δ peak reduced in case of the composites. A significant reduction and broadening of tan δ were observed for the composites with 5, 7.5, and 10 phr of OMLS along with a reduction in the T_g. This reduction can be attributed to the increase in extensibility provided by the organomodifier group in nanoclays. The exfoliated platelets of OMLS separate polymer chains and reduce chain to chain interaction and thereby reduce T_g.

Kanbur and Tayfun [46] developed different multifunctional thermoplastic polyurethane (TPU) elastomer composites containing fullerene (C_{60}) and studied their damping properties. The C_{60} content in TPU was varied from 0.5 to 2 wt.%. Surface modification of C_{60} (m-C_{60}) was done by refluxing it in 3:1 v/v solution of concentrated H_2SO_4 and HNO_3. DMA results indicated that incorporation of C_{60} improved the E' of all the composites, and the highest value was exhibited by TPU/0.5 m- C_{60}. The formation of tri-dimensional network among the filler particles and TPU restricted the chain mobility and hence the improvement in E' was observed for the composites compared to base TPU. From the tan δ curve it was observed that in case of the composites broadening of the peaks compared to neat TPU explained the induced vibration-damping and sound-deadening properties [46]. In addition, the T_g of the all the composites shifted toward lower temperature compared to TPU. It indicates that the composites gain characteristic structural features such as resistance for cyclic deformation, shape recoverability, grip and rolling resistance as the application parameters on lower temperature environment.

10.4 Conclusion

Viscoelasticity is an important material property particularly for elastomers, their blends and composites, since it correlates crucial parameters such as T_g, E', and E'' with temperature and frequency. The applicability of elastomeric composites can be determined by evaluating such parameters using DMA. In addition, compatibility or miscibility of new and modified blends can also be verified from the viscoelastic properties. In the present work, the theoretical understanding behind viscoelasticity evolved in literature over years is presented. Since viscoelastic material behaves very differently compared to classical fluid and classical elastic solid, hence, different mathematical equations in the form of constitutive models have been developed over the years using springs and dashpots as the representation of elastic and viscous components. Furthermore, from previous experimental studies it was observed that the damping characteristics of elastomer blends significantly rely on the blend ratio and the use of compatibilizers. The compatibility within the blends can be expressed as the merging of T_g of the individual components. In case of a completely compatible blend, the tan δ peaks of individual polymer or elastomer shift toward each other and form a single peak defining the T_g of the blend, whereas multiple T_g's are observed for immiscible blends. The peak intensity and broadening of the tan δ curve also point out the relaxation behavior. A decreasing tan δ generally describes an improvement in storage property and reduction of the loss component, whereas the area under the tan δ curve can be related to the energy dissipation ability of the material. Similarly, in case of elastomer composites based on the reinforcing efficiency, the pattern of the tan δ curve changes. Stiff reinforcing materials help in improving the E' value and also sometimes increase the T_g of the composites as a result of restricted chain mobility. This effect is more prominent where the interfacial filler–polymer adhesion is significant. Nonetheless, a conclusive remark on the trend of viscoelastic properties of elastomer blends and composites cannot be made based solely on theoretical understandings. An in-depth analysis using characterization tool such as DMA must be carried out to assist the same. We propose that this chapter would not only help in developing new materials but also provide insights in understanding compatibility of waste elastomers such as rubber with other polymer for finding better alternative of reuse and waste management.

References

[1] H.F. Brinson, L.C. Brinson, Polymer Engineering Science and Viscoelasticity: An Introduction, second ed., Springer US, 2015 https://doi.org/10.1007/978-1-4899-7485-3.

[2] P. Kelly, Viscoelasticity, in: Solid Mech. Part I An Introduction. To Solid Mech., University of Auckland, 2013, pp. 283–342.

[3] D. Roylance, Engineering Viscoelasticity, 2001, pp. 1–37. https://web.mit.edu/course/3/3.11/www/modules/visco.pdf.

[4] P.J. Flory, Principles of Polymer Chemistry, Cornell University Press, 1953.

[5] R. Bird, C. Curtis, R. Armstrong, O. Hassenger, Dynamics of Polymer Fluids, second ed., vols. 1 and 2, 1987.

[6] R.P. Padbury, Bulk Property Modification of Fiber Forming Polymers Using Vapor Phase Techniques, North Carolina State University, 2014.

[7] J. Heijboer, Secondary loss peaks in glassy amorphous polymers, Int. J. Polym. Mater. Polym. Biomater. 6 (1977) 11–37, https://doi.org/10.1080/00914037708075218.

[8] K.P. Menard, N.R. Menard, Dynamic mechanical analysis in the analysis of polymers and rubbers, in: Encyclopedia of Polymer Science and Engineering, Wiley, 2015, pp. 1–33, https://doi.org/10.1002/0471440264.pst102.pub2.

[9] V.G. Geethamma, G. Kalaprasad, G. Groeninckx, S. Thomas, Dynamic mechanical behavior of short coir fiber reinforced natural rubber composites, Compos. Part A Appl. Sci. Manuf. 36 (2005) 1499–1506, https://doi.org/10.1016/j.compositesa.2005.03.004.

[10] N. Jalalifar, B. Kaffashi, S. Ahmadi, The synergistic reinforcing effects of halloysite nanotube particles and polyolefin elastomer-grafted-maleic anhydride compatibilizer on melt and solid viscoelastic properties of polylactic acid/polyolefin elastomer blends, Polym. Test. 91 (2020) 106757. https://doi.org/10.1016/j.polymertesting.2020.106757.

[11] N. Ashok, K. Prakash, D. Selvakumar, M. Balachandran, Synergistic enhancement of mechanical, viscoelastic, transport, thermal, and radiation aging characteristics through chemically bonded interface in nanosilica reinforced EPDM-CIIR blends, J. Appl. Polym. Sci. 138 (2021) 50082. https://doi.org/10.1002/app.50082.

[12] B. Liu, T. Jiang, X. Zeng, R. Deng, J. Gu, W. Gong, L. He, Polypropylene/thermoplastic polyester elastomer blend: crystallization properties, rheological behavior, and foaming performance, Polym. Adv. Technol. 32 (2021) 2102–2117. https://doi.org/10.1002/pat.5240.

[13] A. Kasgoz, Mechanical, tensile creep and viscoelastic properties of thermoplastic polyurethane/polycarbonate blends, Fibers Polym. 22 (2021) 295–305. https://doi.org/10.1007/s12221-021-0113-z.

[14] S. Khanra, D. Ganguly, S.K. Ghorai, D. Goswami, S. Chattopadhyay, The synergistic effect of fluorosilicone and silica towards the compatibilization of silicone rubber and fluoroelastomer based high performance blend, J. Polym. Res. 27 (2020) 1–17. https://doi.org/10.1007/s10965-020-02062-z.

[15] M. Goswami, B.S. Mandloi, A. Kumar, S. Sharma, S.K. Ghorai, K. Sarkar, S. Chattopadhyay, Optimization of graphene in carbon black-filled nitrile butadiene rubber: constitutive modeling and verification using finite element analysis, Polym. Compos. 41 (2020) 1853–1866. https://doi.org/10.1002/pc.25503.

[16] R. Gogoi, N. Kumar, S. Mireja, S.S. Ravindranath, G. Manik, S. Sinha, Effect of hollow glass microspheres on the morphology, rheology and crystallinity of short bamboo fiber-reinforced hybrid polypropylene composite, JOM 71 (2019) 548–558. https://doi.org/10.1007/s11837-018-3268-3.

[17] R. Gogoi, G. Manik, B. Arun, High specific strength hybrid polypropylene composites using carbon fibre and hollow glass microspheres: development, characterization and comparison with empirical models, Compos. B Eng. 173 (2019) 106875. https://doi.org/10.1016/j.compositesb.2019.05.086.

[18] R. Gogoi, N. Kumar, S. Mireja, S.K. Sethi, G. Manik, Natural fibre based hybrid polypropylene composites: an insight into thermal properties, in: International Conference on Composite Materials, 2019. Melbourne, Australia, https://www.researchgate.net/publication/341579412. (Accessed 7 February 2021).

[19] R. Gogoi, S.K. Sethi, G. Manik, Surface functionalization and CNT coating induced improved interfacial interactions of carbon fiber with polypropylene matrix: a molecular dynamics study, Appl. Surf. Sci. 539 (2021) 148162. https://doi.org/10.1016/j.apsusc.2020.148162.

[20] R. Gogoi, G. Manik, Development of thermally conductive and high-specific strength polypropylene composites for thermal management applications in automotive, Polym. Compos. (2021). https://doi.org/10.1002/pc.25947 pc.25947.

[21] A.K. Maurya, R. Gogoi, G. Manik, Study of the moisture mitigation and toughening effect of fly-ash particles on sisal fiber-reinforced hybrid polypropylene composites, J. Polym. Environ. (2021) 1–16. https://doi.org/10.1007/s10924-021-02043-3.

[22] S.K. Sahoo, S. Mohanty, S.K. Nayak, Toughened bio-based epoxy blend network modified with transesterified epoxidized soybean oil: synthesis and characterization, RSC Adv. 5 (2015) 13674–13691. https://doi.org/10.1039/c4ra11965g.

[23] S.K. Sahoo, S. Mohanty, S.K. Nayak, Synthesis and characterization of bio-based epoxy blends from renewable resource based epoxidized soybean oil as reactive diluent, Chinese J. Polym. Sci. 33 (2015) 137–152. https://doi.org/10.1007/s10118-015-1568-4.

[24] S.K. Sahoo, S. Mohanty, S.K. Nayak, Study of thermal stability and thermo-mechanical behavior of functionalized soybean oil modified toughened epoxy/organo clay nanocomposite, Prog. Org. Coating 88 (2015) 263–271. https://doi.org/10.1016/j.porgcoat.2015.07.012.

[25] S.K. Sahoo, S. Mohanty, S.K. Nayak, A study on effect of organo modified clay on curing behavior and thermo-physical properties of epoxy methyl ester based epoxy nanocomposite, Thermochim. Acta 614 (2015) 163–170. https://doi.org/10.1016/j.tca.2015.06.021.

[26] D. Singh, V.P. Malhotra, J.L. Vats, The dynamic mechanical analysis, impact, and morphological studies of EPDM–PVC and MMA-g-EPDM–PVC blends, J. Appl. Polym. Sci. 71 (1999) 1959–1968.

[27] A. Mousa, U.S. Ishiaku, Z.A. Mohd Ishak, Rheological and viscoelastic behavior of dynamically vulcanized poly(vinyl chloride)–epoxidized natural-rubber thermoplastic elastomers, J. Appl. Polym. Sci. 74 (1999) 2886. https://doi.org/10.1002/(SICI)1097-4628(19991213)74:12<2886::AID-APP12>3.0.CO;2-Q.

[28] V.H. Shafeeq, G. Unnikrishnan, Experimental and theoretical evaluation of mechanical, thermal and morphological features of EVA-millable polyurethane blends, J. Polym. Res. 27 (2020) 1–11. https://doi.org/10.1007/s10965-020-2027-7.

[29] D. Tomova, J. Kressler, H.J. Radusch, Phase behaviour in ternary polyamide 6/polyamide 66/elastomer blends, Polymer 41 (2000) 7773–7783. https://doi.org/10.1016/S0032-3861(00)00127-0.

[30] H. Peng, M. Lu, H. Wang, Z. Zhang, F. Lv, M. Niu, W. Wang, Comprehensively improved mechanical properties of silane crosslinked polypropylene/ethylene propylene diene monomer elastomer blends, Polym. Eng. Sci. 60 (2020) 1054–1065. https://doi.org/10.1002/pen.25361.

[31] R. Gogoi, K.P. Biligiri, N.C. Das, Performance prediction analyses of styrene-butadiene rubber and crumb rubber materials in asphalt road applications, Mater. Struct. Constr. 49 (2016) 3479–3493. https://doi.org/10.1617/s11527-015-0733-0.

[32] S. Ramarad, M. Khalid, C.T. Ratnam, A.L. Chuah, W. Rashmi, Waste tire rubber in polymer blends: a review on the evolution, properties and future, Prog. Mater. Sci. 72 (2015) 100–140. https://doi.org/10.1016/j.pmatsci.2015.02.004.

[33] V. Venudharan, K.P. Biligiri, N.C. Das, Investigations on behavioral characteristics of asphalt binder with crumb rubber modification: rheological and thermo-chemical approach, Construct. Build. Mater. 181 (2018) 455–464. https://doi.org/10.1016/j.conbuildmat.2018.06.087.

[34] A. Hejna, Ł. Zedler, M. Przybysz-Romatowska, J. Cañavate, X. Colom, K. Formela, Reclaimed rubber/poly(ε-caprolactone) blends: structure, mechanical, and thermal properties, Polymers 12 (2020) 1204. https://doi.org/10.3390/polym12051204.

[35] L. Mészáros, T. Bárány, T. Czvikovszky, EB-promoted recycling of waste tire rubber with polyolefins, Radiat. Phys. Chem. 81 (2012) 1357–1360. https://doi.org/10.1016/j.radphyschem.2011.11.058.

[36] U. Shankar, D. Oberoi, S. Avasarala, S. Ali, A. Bandyopadhyay, Design and fabrication of a transparent, tough and UVC screening material as a substitute for glass substrate in display devices, J. Mater. Sci. 54 (2019) 6684–6698. https://doi.org/10.1007/s10853-018-03285-8.

[37] S.K. Sethi, S. Kadian, G. Anubhav, R.P. Chauhan, G. Manik, Fabrication and analysis of ZnO quantum dots based easy clean coating: a combined theoretical and experimental investigation, ChemistrySelect 5 (2020) 8942–8950. https://doi.org/10.1002/slct.202001092.

[38] S.K. Sethi, G. Manik, A combined theoretical and experimental investigation on the wettability of MWCNT filled PVAc-g-PDMS easy-clean coating, Prog. Org. Coating 151 (2021) 106092. https://doi.org/10.1016/j.porgcoat.2020.106092.

[39] U. Shankar, C.R. Gupta, D. Oberoi, B.P. Singh, A. Kumar, A. Bandyopadhyay, A facile way to synthesize an intrinsically ultraviolet-C resistant tough semiconducting polymeric glass for organic optoelectronic device application, Carbon N. Y. 168 (2020) 485–498. https://doi.org/10.1016/j.carbon.2020.07.015.

[40] U. Shankar, S. Bhandari, D. Khastgir, Carbon black-filled nitrile rubber composite as a flexible electrode for electrochemical synthesis of supercapacitive polyaniline, Polym. Compos. 40 (2019) E1537–E1547. https://doi.org/10.1002/pc.25069.

[41] S. Praveen, B.C. Chakraborty, S. Jayendran, R.D. Raut, S. Chattopadhyay, Effect of filler geometry on viscoelastic damping of graphite/aramid and carbon short fiber-filled SBR composites: a new insight, J. Appl. Polym. Sci. 111 (2009) 264–272. https://doi.org/10.1002/app.29064.

[42] S. Praveen, P.K. Chattopadhyay, P. Albert, V.G. Dalvi, B.C. Chakraborty, S. Chattopadhyay, Synergistic effect of carbon black and nanoclay fillers in styrene butadiene rubber matrix: Development of dual structure, Compos. Part A Appl. Sci. Manuf. 40 (2009) 309–316. https://doi.org/10.1016/j.compositesa.2008.12.008.

[43] K. Sankaran, G.B. Nando, P. Ramachandran, S. Nair, U. Govindan, S. Arayambath, S. Chattopadhyay, Influence of hybrid nanostructures and its tailoring mechanism on permeability, rheology, conductivity, and adhesion properties of a novel rubber blend nanocomposite, RSC Adv. 5 (2015) 87864–87875. https://doi.org/10.1039/c5ra17178d.

[44] A. Sallat, A. Das, J. Schaber, U. Scheler, E.S. Bhagavatheswaran, K.W. Stöckelhuber, G. Heinrich, B. Voit, F. Böhme, Viscoelastic and self-healing behavior of silica filled ionically modified poly(-isobutylene-co-isoprene) rubber, RSC Adv. 8 (2018) 26793–26803. https://doi.org/10.1039/c8ra04631j.

[45] N. Ashok, M. Balachandran, N.C. Das, S. Remanan, Nanoreinforcement mechanism of organo-modified layered silicates in EPDM/CIIR blends: experimental analysis and theoretical perspectives of static mechanical and viscoelastic behavior, Compos. Interfac. 28 (2021) 35–62. https://doi.org/10.1080/09276440.2020.1736879.

[46] Y. Kanbur, U. Tayfun, Development of multifunctional polyurethane elastomer composites containing fullerene: mechanical, damping, thermal, and flammability behaviors, J. Elastomers Plastics 51 (2019) 262–279. https://doi.org/10.1177/0095244318796616.

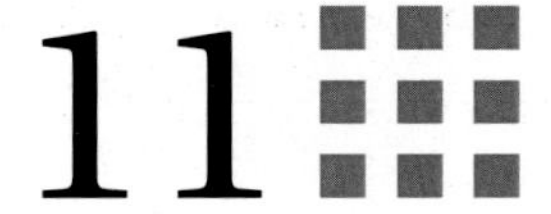

11 Spectroscopy of elastomer blends and composites

Sabarish Radoor[1], Jasila Karayil[2], Amritha Bemplassery[3], Aswathy Jayakumar[1], Jyotishkumar Parameswaranpillai[4], Suchart Siengchin[1]

[1]MATERIALS AND PRODUCTION ENGINEERING, THE SIRINDHORN INTERNATIONAL THAI-GERMAN GRADUATE SCHOOL OF ENGINEERING (TGGS), KING MONGKUT'S UNIVERSITY OF TECHNOLOGY NORTH BANGKOK, BANGKOK, THAILAND; [2]GOVERNMENT WOMEN'S POLYTECHNIC COLLEGE, CALICUT, KERALA, INDIA; [3]DEPARTMENT OF CHEMISTRY, NATIONAL INSTITUTE OF TECHNOLOGY, CALICUT, KERALA, INDIA; [4]DEPARTMENT OF SCIENCE, FACULTY OF SCIENCE & TECHNOLOGY, ALLIANCE UNIVERSITY, BENGALURU, KARNATAKA, INDIA

11.1 Introduction

Material science is one of the well-established disciples that design and develop new and advanced materials with superior performance. The evolution of material science dates back to centuries and, by no doubt, revolutionized our human civilization. The never-ending demand to harness new materials to meet the challenges of the modern world urges material scientists over the globe to innovate novel materials. Elastomer composite or blends are an important class of composite materials, which has been widely exploited to design industrial products. Recently researchers have developed conductive elastomer blend, which serves as flexible bioelectronic. Some interesting applications of elastomer blend, namely biomedical application and bionic locomotive application, are also reported [1].

Elastomers are amorphous polymers, which are capable of attaining their original form. The unique property of elastomer includes adhesive nature, abrasion resistance, and impermeability. Owing to its cost-effective nature, recyclability, and renewability, elastomers have been widely used to design wide variety of materials such as seals, adhesives, gaskets, fire-proofing materials, conveyor belts, footwear, etc. [1–3].

The properties of elastomers could be improved, and its defects could be minimized by blending it with different materials. Different methods such as simple mixing, solution blending, melt bending, mechanical blending, and mechanical chemical mixing have been employed to develop elastomer blends [4–7]. Elastomer blend and composites have superior properties than pure elastomer; therefore they have been

Elastomer Blends and Composites. https://doi.org/10.1016/B978-0-323-85832-8.00016-X

commercially used in various industrial products. The physical properties of elastomer blends and composites could be accessed by techniques such as FT-IR, Raman, fluorescence, NMR, etc. This chapter discusses the important spectroscopic techniques used to characterize elastomeric blend/composite.

11.2 FT-IR and Raman spectroscopy

TPU (thermoplastic polyurethane) is one of the promising elastomers that is widely for different applications. The properties of TPU could be enhanced by mixing it with different materials such as wood fiber, polyoxymethylene, natural rubbers, PLA, PVC, chitosan, etc. [8–13]. The flame retardancy of TPU is improved by blending it with functionalized polyphosphazene. The successful preparation of phosphazene was confirmed by FT-IR; with characteristic band appearing at 1087.9 cm^{-1} (P–O–C bond.), 1293.8 cm^{-1} (N=P bond), 1458.6 cm^{-1} (C=C bond), 2957.9 cm^{-1} (C–H bond) and 1623.3 cm^{-1} (C–F bond). Pure TPU elastomer has band at 3324.06, 1053.54, 1730.17, and 1528 cm^{-1} assigned to the N–H, C–O–C, and C=O stretching and due to the presence of benzene, respectively. However, upon blending with phosphazene, a broad –NH bond with low intensity was observed. This confirms the hydrogen bond between the –NH group of TPU and the fluorine of the phosphazene [8]. The miscibility of chitosan/gelatin/TPU blend nanofibers was studied using FT-IR analysis. The blend has characteristic bands of chitosan 3350 cm^{-1}, 2922 cm^{-1}, 1066 cm^{-1}, (1640 cm^{-1} (amide I), 1534 cm^{-1} (amide II), and 1234 cm^{-1} (amide III)), gelatin, and TPU. Thus, confirming the successful incorporation of TPU in chitosan/gelatin blend [14]. Aparna et al. [15] compared the mechanical and thermal properties of TPU/NR-CSP (thermoplastic polyurethane/natural rubber/composite incorporated with coconut shell powder). In addition to the characteristic adsorption bands, the silane-treated coconut shell powder displayed peak at 2966, 2941, 2885, 2843, 1159, 1086, 1603, 1093, and 1906 cm^{-1}. This is assigned to the epoxy and silyl groups. Thus, FT-IR analysis confirms the successful grafting of coconut fiber. León and coworkers [16] employed FT-IR and Raman imaging to understand the chemical compatibility between ABS (acrylonitrile-butadiene-styrene) and TPU (thermoplastic polyurethane). Hydrogen-bonded supramolecular interaction between ABS and TPU is quite evident from the FT-IR and Raman spectrum. The spectroscopic results show a homogeneously distributed ABS in the system (Fig. 11.1). In another work, Jeong et al. studied PVC/TBU using spectroscopic technique. FT-IR spectra of PVC/TBU blend revealed that the chemical structure of TBU is retained in the blend. The results further indicate that the blend has both amorphous and crystalline domains [17].

Previous study led by Tan and team [18] shows that POE (polyolefin elastomer) acts as compatibilizer for TPU/TPS blend and consequently improves the miscibility and interfacial interaction between TPU and TPS (thermoplastic starch). This could be due to better inter-hydrogen bond between TPU and TPS. Meanwhile, strong chemical interaction between modified natural rubber NRs and TPU phases was observed in the FT-IR spectra of NR/TPU blend [10]. Prochon et al. [19] reported that the presence of protein (keratin) improves the thermomechanical properties of styrene-butadiene rubber (SBR)

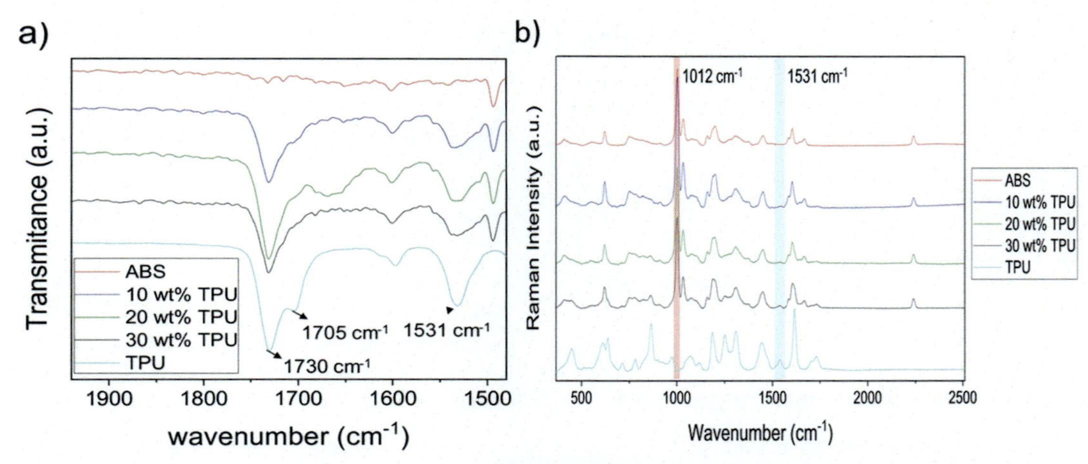

FIGURE 11.1 FT-IR spectra and Raman spectra of pure ABS (*red*), ABS:TPU blends containing 10 wt% TPU (*blue*), 20 wt% TPU (*green*), 30 wt% TPU (*black*), and pure TPU (*light blue*). *From A.S. de León, A. Domínguez-Calvo, and S.I. Molina, Materials with enhanced adhesive properties based on acrylonitrile-butadiene-styrene (ABS)/thermoplastic polyurethane (TPU) blends for fused filament fabrication (FFF). Materials & Design 182 (2019).*

elastomers. The FT-IR results shows a decrease in the intensity of characteristic band of elastomers ((1545 cm^{-1}, 970 cm^{-1} (C=C trans)), 700 cm^{-1} (vibration from styrene ring)), which could be due to the cross-linking between SBR and protein. A similar result was reported by Fang and coworkers [20] for PEU (polyester urethane)/PAN (polyacrylonitrile) system. The blend has high intensity for band at 1622, 1620, 1537 cm^{-1}, and 1001 cm^{-1} which could be due to the cross-linking between the isocyanate cross-linker and the polymer. Recently, Bolados and coworkers [21] grafted poly(styrene-*b*-ethylene-butylene-*b*-styrene) (SEBS) with itaconic acid and employed it for the reparation of nanocomposite containing $BaTiO_3$(barium titanate). In addition to the characteristic peak of SEBS (2200,1100, 1300, 1500, and 1601 cm^{-1}), a new peak emerged at 1713,1743 cm^{-1} and 1770 cm^{-1} for polymer blends, thus indicating successful grafting of SEBS with itaconic acid. Liu et al. [22] improved the properties of polymethylvinylsiloxane (PMVS) by simultaneously introducing a carboxy group by thiol-ene click reaction and introducing an epoxy group functionalized TiO_2 (E-TiO_2) particles. Disappearance of C=C peak and the appearance of new peak at 1720 cm^{-1} (C=O stretching) for modified elastomer could be taken as strong evidence for the successful grafting of carbonyl group on PMVS. Fillers and activators are generally introduced in elastomer composites to improve its physical properties [23]. Menes and coworkers [7] prepared thermoplastic polyurethane (TPU)/ultrathin graphite (UTG) composite by solution compounding strategy. The chemical interaction between TPU and UTG or UTGO (ultrathin graphite oxide) was confirmed by Raman and FT-IR analysis (Fig. 11.2). Although most of the characteristic bands of TPU are retained even after blending with UTG, a slight red shift was observed for N—H band. This hints the possibility of chemical interaction between the TPU and UTG. Furthermore, Raman spectra correlate well with

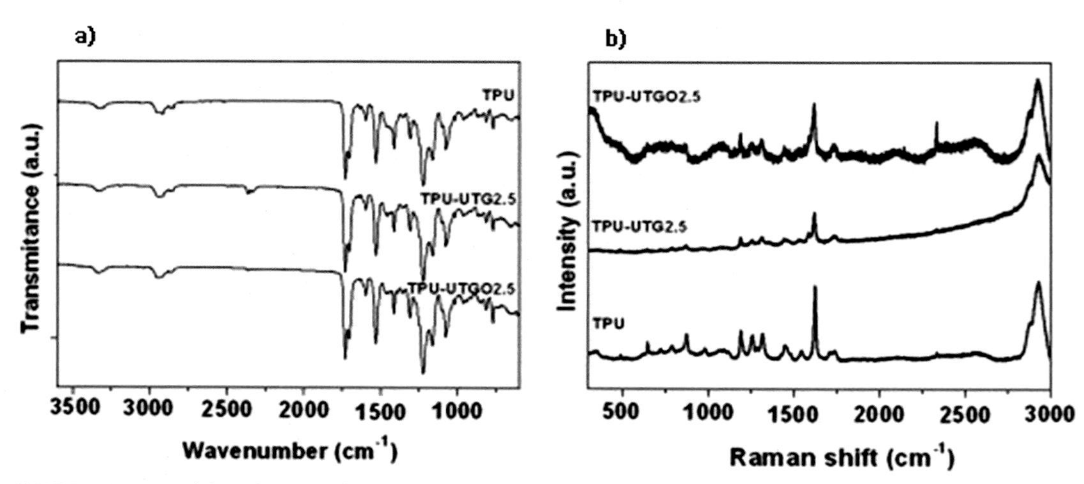

FIGURE 11.2 FT-IR (A) and Raman (B) spectra of neat TPU, TPU-UTG, and TPU-UTGO composite. *Reproduced with permission from Elsevier, License Number: 5016610543511. From O. Menes, M. Cano, A. Benedito, E. Giménez, P. Castell, W.K. Maser, and A.M. Benito, The effect of ultra-thin graphite on the morphology and physical properties of thermoplastic polyurethane elastomer composites. Composites Science and Technology 72(13) (2012) 1595–1601.*

the FT-IR results. A decrease in the intensity of the characteristic peak of TPU was noted with UTG content. The graphite (G) band was shifted to 12 cm^{-1}, probably due to the interaction of UTG with TPU. Owing to the presence of oxygen atom, there will be greater interaction between the UTGO and the elastomer, and hence a significant upfield of the characteristic peaks of TPU was observed in the composites. The interaction between CNT and natural rubber was studied using Raman spectroscopy. An upward shift in G mode and the absence of any change in the RB band of graphite in the composite indicate direct coupling of carbon nanotubes and the natural rubber. The Raman result also shows that carbon nanotube carries most of the load in the composite [24]. Shanmugharaj et al. [25] employed chemically modified CNT to generate a composite with natural rubber. The successful functionalization of CNT was confirmed by FT-IR and Raman spectroscopy. A peak that appeared at 1110 and 1043 cm^{-1} could be due to the formation of Si–O–Si and Si–O–C linkage between silane and CNT, respectively. The defect observed for CNT is due to the presence of carboxyl functional group. This will enhance the polymer–filler interaction, and the composites exhibit high tensile strength and modulus (Fig. 11.3). Strankowski et al. [26] employed reduced graphene oxide (rGO) as filler in polyurethane composite. Both Raman and the FT-IR analysis reveal that the characteristic band of polymer remained even after introduction of rGO. The decreased intensity of N–H, C–O bond and the shift in the frequency of C=C bond in modified composite indicate filler–matrix interaction. Mishra and coworkers [27] studied the chemical interaction between the epoxy resin and POSS filler by FT-IR analysis. The result indicates good dispersion of filler in the matrix. The broad hydroxyl peak (3500 cm^{-1}) and absence of characteristic epoxy peak at 1024 cm^{-1} suggest the polymerization of the epoxy resin spectra.

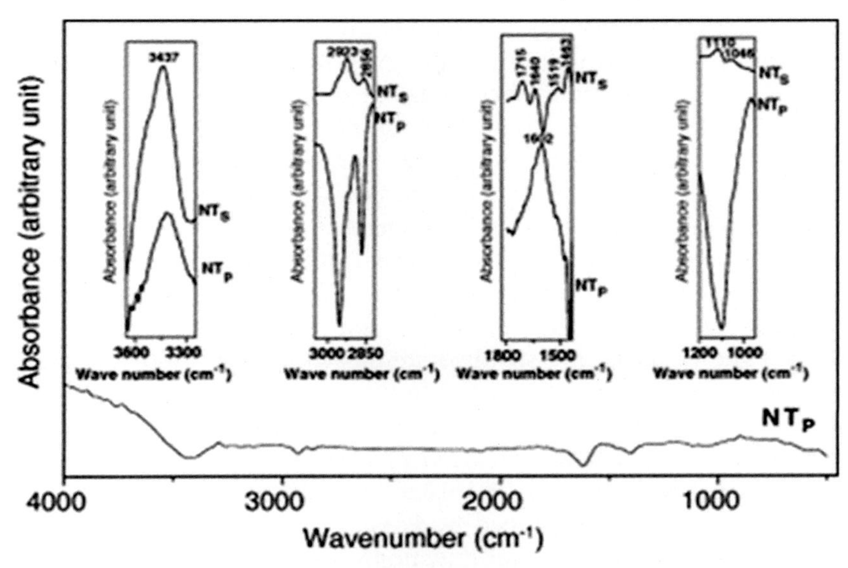

FIGURE 11.3 FT-IR spectra of carbon nanotubes. *Reproduced with permission from Elsevier, License Number: 5016611116807. From A. Shanmugharaj, J. Bae, K. Lee, W. Noh, S. Lee, and S. Ryu, Physical and chemical characteristics of multiwalled carbon nanotubes functionalized with aminosilane and its influence on the properties of natural rubber composites. Composites Science and Technology 67(9) (2007) 1813–1822.*

11.3 Fluorescence spectroscopy

Fluorescence Spectroscopy though does not present structural evidence in detail; it provides an efficient and simplistic way of characterization of polymeric blends. One of the most striking advantages of fluorescence is that it can avail high spatial resolution characterizations of compounds with ranges lower than even picosecond [28]. From the fluorescence data photophysical properties of a material can be studied, which would pave way for the structural and morphological characterizations of the compound. This technique is particularly useful in the study of fluorescent conjugated polymers, which find importance in electronic fields. Nonconjugated polymeric compounds though nonemissive can be labeled with fluorophores in such a way that the properties of polymers are not affected [28]. Luminescence studies provided detailed information about the electronic energy states and symmetry of the organic polymers. In order to obtain an absolute understanding of lifetimes, intensities, excitation, and emission properties, etc., a combined study of different fluorescence techniques may be needed. Jablonski diagram forms the basis of fluorescence spectroscopy. Fluorescence can be defined as the radiative transitions from singlet excited state S1 of a molecule to ground state. The prime aspect that differentiates fluorescence from phosphorescence is that the former process is accompanied by retention of spin multiplicity and that is a spin allowed process [28].

Steady-state fluorescence can be considered as the simplest type of fluorescence spectroscopy that records emission spectrum when sample is illuminated with a continuous light beam. Sophisticated methods are developing that could amplify the obtainable timescale of studying processes to even femtoseconds. These types of time-resolved measurements are highly beneficial in analyzing the dynamics of excitons produced in conjugated polymer films. This method is superior to other conventional ones in availing more information regarding polymer materials [29]. Time-resolved measurements help in the resolution of emission produced from two different fluorophores in the case of blends. These studies outshine the steady-state measurements as time-resolved data can deliver precise individual decay times. Pulse fluorometry and phase-modulation fluorometry are two extensive methods for measuring lifetimes. The former method is characterized by pulsed light source, which is used to excite sample molecules [28]. Time-resolved and steady-state measurements though share similarities in excitation and detection systems, they differ in the nature of these constituents.

Fluorescence spectroscopy is important in the characterization of polymer blends owing to the use of nonradiative transition [30,31] (NRET) and excimer fluorescence [32,33] especially in the detection of miscibility and small-scale phase separation. Excimer formation was first introduced by Förster and is now popular in the case of many aromatic molecules such as naphthalene [34], styrene [32], etc. An excimer could be defined as a short-lived molecule produced between an aromatic molecule in electronic excited state and its identical molecule. Extent of polymer aggregation influences changes in excimer and fluorescence intensity. When polymer chains diffuse each other, fluorophore concentration gets diluted, which causes diminished probability of formation of excimers. This phenomenon was observed in the case of miscible polymer blends. Thus, as excimer formation requires a distance between fluorophores of about 3–4 A^0, decrease in its formation could indicate the miscibility of polymers. First attempt to study the miscibility of polymer blends through nonradiative energy transfer (NRAT) technique was of Armani et al. [35]. When emission spectra of donor and acceptor fluorophores got overlapped, an energy transfer from donor to acceptor takes place. Thus, when energy transfer efficiency was experimentally calculated as the ratio of donor to acceptor emission intensity, this could be an indication of polymer chain miscibility up to a scale of about 3 nm [35].

Conjugating polymers are an interesting group of materials, which exhibit semiconductor properties and some displayed fluorescent behavior. These fluorescent conjugated polymeric compounds play a pivotal role in polymer-based LEDs. A wide variety of applications in diodes, photovoltaics, transistors, and also bioelectronics are associated with conjugated polymeric blends. Different processes such as excited state energy transfer, excitation energy transfer, luminescence quenching, etc., may occur in optoelectronic devices. The fact that the efficiency of polymer blend-based devices is better compared to that of single polymer-based ones is mostly favored in LED devices [36]. First studies of fluorescent polymers were about light-emitting diodes and pioneering studies focused on poly(p-phenylene vinylene)s [28]. The polymers that are most widely studied could be named as poly(9,9-dioctylfluorene) (PFO) and a copolymer poly(9,9-dioctylfluorene-alt-benzothiadiazole) (F8BT). Both are harmonious in terms of charge mobilities and blends based on both the polymers and are efficient and successful. They exhibit green emission accordingly by a fantastic energy transfer between F8BT and PFO. The energy transfer between PFO and F8BT was observed to decrease considerably as the amount of F8BT was reduced, which was experimentally analyzed by Bradley and coworkers in 2001 [37]. Similar studies were conducted by Voigt and coworkers [38] spin casting PFO/F8BT blends with chloroform and toluene solvents. It was observed that phase-separation length scale was remarkably greater for that blend, which was spin-coated from toluene compared to chloroform (more volatile). They analyzed residual emission of PFO to assess the energy transfer between PFO and F8BT. It was concluded that effective energy transfer from PFO to F8BT occurs in films forming chloroform solutions, which possessed better mixing of two components. Only 5 wt% of F8BT was sufficient to bring about complete energy transfer (Fig. 11.4).

Clough et al. [39] studied the factors affecting the stress-softening effect of elastomers using emission spectroscopy. They exhibited using mechanoluminescence that covalent bond scission plays a major role in stress softening effect. Mechanophore, Bis(adamantyl)-1,2-dioxetane after delivering excited ketones on cleaving, emits light on relaxing to ground state. This low-intensity light emitted was transferred to 9,10-diphenylanthracene (DPA), which plays the role of fluorescent acceptor via a FRET mechanism.

In the case of nonconjugated and nonfluorescent polymers, an extrinsic fluorescent probe was made use of in studying phase behavior [40,41]. Careful selection of size and shape of probe could minimize possible perturbation problem on the probed region. Labeling different parts of components of blend as donor and acceptor fluorophore will be effective, and time-resolved studies could assess their spatial distribution. Atvars et al. time-correlated single photon counting as well as steady-state fluorescence measurements on PVA, PVAc, and their blends [42]. Fluorescein and anthracene were used as fluorescent probes. They calculated interface thickness and concluded that in blends with larger PVAc content, there was a higher surface-to-volume ratio of PVAc observed.

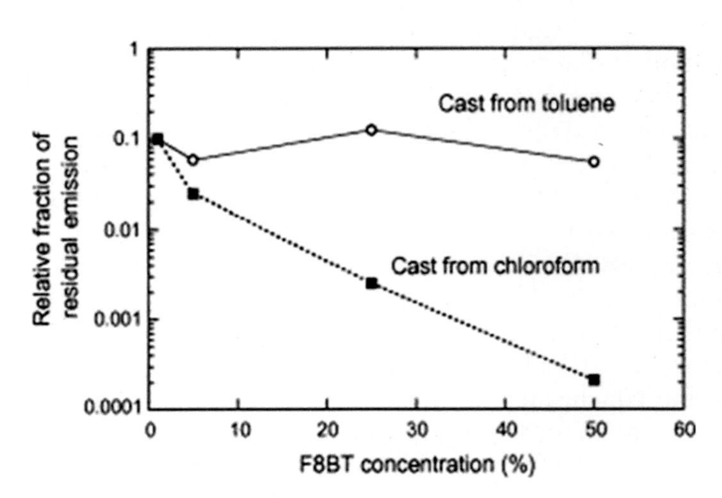

FIGURE 11.4 The fractional emission of residual F8 fluorescence compared to the total fluorescence emission as a function of the relative F8BT concentration. *Reproduced with permission from Elsevier, License Number: 5016620062300. From M. Voigt, J. Chappell, T. Rowson, A. Cadby, M. Geoghegan, R. Jones, and D. Lidzey, The interplay between the optical and electronic properties of light-emitting-diode applicable conjugated polymer blends and their phase-separated morphology. Organic Electronics 6(1) (2005) 35–45.*

11.4 NMR spectroscopy

Solid-state nuclear magnetic resonance (NMR) spectroscopy is a robust characterization technique in investigating the phase structures, heterogeneity, and miscibility of polymer blends [43,44]. Solid-state ^{13}C NMR spectra provide two very useful spin relaxation times, which are spin lattice relaxation times in the rotating frame ($T_{1\rho}{}^{H}$) and in the laboratory frame ($T_1{}^{H}$). Values of $T_{1\rho}{}^{H}$ and $T_1{}^{H}$ can avail approximate calculation of length scale of heterogeneity up to about nanometers [45]. NMR spectroscopy was first introduced by Kwei and coworkers to analyze the microstructure of polymer blend by evaluating the ^{1}H spin-lattice and spin–spin relaxation times, T_1 and T_2 [46]. They used polystyrene (PS)/polyvinyl methyl ether (PVME) blend. McBrierty et al. [47] and Nishi et al. [48] also exhibited that the spin relaxation can be made use of in examining the morphology of blends. Nuclear Overhauser effect spectroscopy was found to be beneficial in liquid-state NMR in interpretation of structures of polymer molecules. ^{13}C NOE spectroscopy was proved to be effective in probing the miscibility in PS/PVME blends [49]. These experiments evidenced an interaction between phenyl group of PS and methyl group of PVME. Miscibility was attributable to intermolecular interactions, and such interactions of polybutadiene/polyisoprene blends above Tg can be explored using NOESY technique under magic angle spinning (MAS) [50]. Weak interactions between vinyl side chain of polybutadiene and polyisoprene methyl group were confirmed, which proved the effectiveness of this technique.

^{1}H spin-lattice relaxation experiments were also used in analyzing the miscibility of polymer blends. Protons of a miscible blend undergo thorough spin–spin coupling and provide same relaxation rates. On the contrary, protons of different environments or those not in a close proximity may undergo independent relaxation. Accordingly, partial or no averaging of relaxation rates would indicate immiscible or partially miscible blends. Asano et al. [51] investigated the phase separation and miscibility of Bisphenol A polycarbonate/poly(methyl methacrylate) (PC/PMMA) blend. They examined $T_{1\rho}{}^{H}$ and $T_1{}^{H}$ values obtained from well-resolved ^{13}C signal to study miscibility and domain size. Thermally induced phase separation of the blend was also analyzed from the phase diagram. Moreover, kinetics of phase separation of PC/PMMA blend was also determined by examining the change of composition of blend. The values of $T_1{}^{H}$ observed for PC are in accordance with that of PMMA within the experimental error limits for all compositions. However, the $T_{1\rho}{}^{H}$ value is different for PC and PMMA. The relaxation times observed are reformed by spin diffusion. The diffusive path length $<L^2>^{1/2}$ was calculated for both relaxation times. It was found that $<L^2>^{1/2}$ was about 130–300 A^0 for a $T_1{}^{H}$ value of about 0.3 s and for a $T_{1\rho}{}^{H}$ value of 8 ms, it was about 22–50 A^0. This indicates that the blend can be considered homogeneous on a scale of 200–300 A^0 if single $T_1{}^{H}$ values are observed for the two polymers. The criteria are similar for $T_{1\rho}{}^{H}$ experiments at 20–30 A^0. Thus, the conclusions drawn from results are that the PC/PMMA blends are homogeneous on a scale of 200–300 A^0, however heterogeneous on a scale of 20–30 A^0 at all compositions. The $T_{1\rho}{}^{H}$ values were observed to be much closer for PC and PMMA for 5:5 and 7:3 blends rather than 3:7 blend. This signifies that homogeneity is more for PMMA-rich blend than for PC-rich blend.

Averaging of $T_1^{\ H}$ rates due to spin diffusion was confirmed as calculated and observed values are closely correlated. This also denotes that $T_1^{\ H}$ values of pure polymers are not changed on blending. The local motion of phenyl group of PC will be restricted on blending, whereas no change for the local motion of PMMA was observed. However, O'Gara et al. [52] pointed out that motion of phenyl group does not much contribute to the spin-lattice relaxation process of solid PC. The fact that the values of $T_{1\rho}^{\ H}$ of PC and PMMA are not fully averaged indicates that mixing caused the alteration of molecular motion responsible for $T_{1\rho}^{\ H}$ relaxations. Observed $T_{1\rho}^{\ H}$ values lie between those of pure polymers and ^1H spin diffusion averages values of $T_1^{\ H}$, which suggests partially miscibility of 20–30 A^0.

Phase separation of PC/PMMA blend occurs when heated above its bimodal curve, which leads to the formation of two coexistent phases: PMMA-rich and PC-rich phases. Phase separation temperature for 3:7, 5:5, and 7:3 blends were roughly estimated. The $T_1^{\ H}$ recovery curves after heating at 30 min at different temperatures were analyzed. At heating temperatures within 140°C, same $T_1^{\ H}$ values are obtained for PC and PMMA. As the temperature of heating increases, $T_1^{\ H}$ values of polymer reach that of pure polymers, which signals the occurrence of phase separation.

In order to have a clear understanding of influence of blending on motions on a polymer blend, $T_{1\rho}^{\ H}$ measurements are very important [53]. The ^{13}C $T_{1\rho}^{\ H}$ relaxation times are beneficial in providing information on each specific site motion. Slower spin diffusion and no averaging of relaxation times for ^{13}C nuclei help in retaining information on motion compared to H-1 nuclei. A new technique known as Two-dimensional Wide Line Separation NMR (WISE NMR) is a sophisticated technique that correlates carbon chemical shifts with proton line shape and avails details regarding polymer chain dynamics.

McGhee and coworkers used solid-state NMR spectroscopy to present a deep insight into the structures of poly2,5-bis(3-hexadecylthiophene-2-yl)thieno[3,2-b]thiophene (PBTTT): fullerene blends [54]. X-ray diffraction could not provide satisfactory data regarding this blend. Solid-state 2D ^{13}C heteronuclear correlation (HET-COR) NMR technique was made use of in exploring dipole–dipole coupled ^1H and ^{13}C nuclei of neighboring molecules. This could bring about concise information on the molecular interactions between PBTTT-C16 and $PC_{71}BM$ [55]. Strong intensity correlations between ^1H and ^{13}C signals from fullerene derivative and conjugated polymer as evidenced from HETCOR spectrum prove their close proximities and interactions. The ^{13}C signals from $PC_{71}BM$ side group (Ex: 172 ppm from C=O group) are in correlation with ^1H signal at 1.4 ppm of aliphatic PBTTT-C16 polymer side chain. There are also correlations observed between fullerene and conjugated polymer. 2D NMR spectroscopy also verified and confirmed strong interactions and molecular proximities. These provide clear evidences for intercalative interactions of fullerene derivative into polymer network.

11.5 Conclusion

This chapter explains the spectroscopic techniques commonly employed to study the elastomer blend/composites. Fluorescence spectroscopic analysis could be used to understand the miscibility of elastomer and its stress softening effect. It is also employed to understand the structural and morphological properties of the materials. NMR is one of the widely used spectroscopic techniques to characterize elastomers. As spin relaxation time is related to the morphology, NMR technique was used to access the morphology of the blend. Moreover, 2D NMR gives a clear picture about the chemical interaction in the elastomer. Two-dimensional Wide Line Separation NMR is used to understand the polymer chain dynamics. FT-IR and Raman spectroscopic were commonly employed to identify the chemical interaction in the blend/composites. The presence of hydrogen bond and the successful grafting in the composite/blend could be easily accessed by FTIR analysis. The effect of different fillers on the elastomers, the distribution of fillers, and the miscibility of the elastomer blends could be easily understood by FT-IR and Raman analysis.

Acknowledgments

Authors gratefully thank for financial support by the King Mongkut's University of Technology North Bangkok (KMUTNB), Thailand, through the Postdoctoral Program (Grant No. KMUTNB-63-Post-03 and KMUTNB-64-Post-03 to SR) and (Grant No. KMUTNB-BasicR-64-16).

References

[1] T. Özdemir, Elastomeric micro- and nanocomposites for neutron shielding, in: Micro and Nanostructured Composite Materials for Neutron Shielding Applications, 2020, pp. 125–137.

[2] Z. Wang, T.J. Pinnavaia, Nanolayer reinforcement of elastomeric polyurethane, Chem. Mater. 10 (12) (1998) 3769–3771.

[3] K. Wündrich, A review of radiation resistance for plastic and elastomeric materials, Radiat. Phys. Chem. 24 (5–6) (1984) 503–510.

[4] H. Anuar, A. Zuraida, Improvement in mechanical properties of reinforced thermoplastic elastomer composite with kenaf bast fibre, Compos. B Eng. 42 (3) (2011) 462–465.

[5] D. Yang, S. Huang, M. Ruan, S. Li, Y. Wu, W. Guo, L. Zhang, Improved electromechanical properties of silicone dielectric elastomer composites by tuning molecular flexibility, Compos. Sci. Technol. 155 (2018) 160–168.

[6] T. Liu, F. Yu, X. Yu, X. Zhao, A. Lu, J. Wang, Basalt fiber reinforced and elastomer toughened polylactide composites: mechanical properties, rheology, crystallization, and morphology, J. Appl. Polym. Sci. 125 (2) (2012) 1292–1301.

[7] O. Menes, M. Cano, A. Benedito, E. Giménez, P. Castell, W.K. Maser, A.M. Benito, The effect of ultra-thin graphite on the morphology and physical properties of thermoplastic polyurethane elastomer composites, Compos. Sci. Technol. 72 (13) (2012) 1595–1601.

[8] K.P. Singh, A. Mishra, N. Kumar, D.N. Tripathi, T.C. Shami, Evaluation of thermal, morphological and flame-retardant properties of thermoplastic polyurethane/polyphosphazene blends, Polym. Bull. 75 (6) (2017) 2415–2430.

[9] J. Yang, W. Yang, X. Wang, M. Dong, H. Liu, E.K. Wujcik, Q. Shao, S. Wu, T. Ding, Z. Guo, Synergistically toughening polyoxymethylene by methyl methacrylate–butadiene–styrene copolymer and thermoplastic polyurethane, Macromol. Chem. Phys. 220 (12) (2019).

[10] E. Kalkornsurapranee, N. Vennemann, C. Kummerlöwe, C. Nakason, Novel thermoplastic natural rubber based on thermoplastic polyurethane blends: influence of modified natural rubbers on properties of the blends, Iran. Polym. J. 21 (10) (2012) 689–700.

[11] J.-J. Han, H.-X. Huang, Preparation and characterization of biodegradable polylactide/thermoplastic polyurethane elastomer blends, J. Appl. Polym. Sci. 120 (6) (2011) 3217–3223.

[12] V.J.R.R. Pita, E.E.M. Sampaio, E.E.C. Monteiro, Mechanical properties evaluation of PVC/plasticizers and PVC/thermoplastic polyurethane blends from extrusion processing, Polym. Test. 21 (5) (2002) 545–550.

[13] X. Liu, X. Gu, J. Sun, S. Zhang, Preparation and characterization of chitosan derivatives and their application as flame retardants in thermoplastic polyurethane, Carbohydr. Polym. 167 (2017) 356–363.

[14] S. Samimi Gharaie, S. Habibi, H. Nazockdast, Fabrication and characterization of chitosan/gelatin/thermoplastic polyurethane blend nanofibers, J. Text. Fibr. Mater. 1 (2018).

[15] A.K. Balan, S. Mottakkunnu Parambil, S. Vakyath, J. Thulissery Velayudhan, S. Naduparambath, P. Etathil, Coconut shell powder reinforced thermoplastic polyurethane/natural rubber blend-composites: effect of silane coupling agents on the mechanical and thermal properties of the composites, J. Mater. Sci. 52 (11) (2017) 6712–6725.

[16] A.S. de León, A. Domínguez-Calvo, S.I. Molina, Materials with enhanced adhesive properties based on acrylonitrile-butadiene-styrene (ABS)/thermoplastic polyurethane (TPU) blends for fused filament fabrication (FFF), Mater. Des. 182 (2019).

[17] H.M. Jeong, J.H. Song, S.Y. Lee, B.K. Kim, Miscibility and shape memory property of poly(vinyl chloride)/thermoplastic polyurethane blends, J. Mater. Sci. 36 (22) (2001) 5457–5463. https://doi.org/10.1023/A:1012481631570.

[18] L. Tan, Q. Su, S. Zhang, H. Huang, Preparing thermoplastic polyurethane/thermoplastic starch with high mechanical and biodegradable properties, RSC Adv. 5 (98) (2015) 80884–80892.

[19] M. Prochon, G. Janowska, A. Przepiorkowska, A. Kucharska-Jastrzabek, Thermal properties and combustibility of elastomer–protein composites, J. Therm. Anal. Calorim. 109 (3) (2011) 1563–1570.

[20] J. Fang, T. Lin, W. Tian, A. Sharma, X. Wang, Toughened electrospun nanofibers from crosslinked elastomer-thermoplastic blends, J. Appl. Polym. Sci. 105 (4) (2007) 2321–2326.

[21] H. Aguilar-Bolados, R. Quijada, M. Yazdani-Pedram, S. Maldonado-Magnere, R. Verdejo, M.A. Lopez-Manchado, SEBS-grafted itaconic acid as compatibilizer for elastomer nanocomposites based on BaTiO3 particles, Polymers 12 (3) (2020).

[22] X. Liu, H. Sun, S. Liu, Y. Jiang, B. Yu, N. Ning, M. Tian, L. Zhang, Mechanical, dielectric and actuated properties of carboxyl grafted silicone elastomer composites containing epoxy-functionalized TiO2 filler, Chem. Eng. J. 393 (2020).

[23] D. Basu, A. Das, K.W. Stöckelhuber, U. Wagenknecht, G. Heinrich, Advances in layered double hydroxide (LDH)-based elastomer composites, Prog. Polym. Sci. 39 (3) (2014) 594–626.

[24] M.A. López-Manchado, J. Biagiotti, L. Valentini, J.M. Kenny, Dynamic mechanical and Raman spectroscopy studies on interaction between single-walled carbon nanotubes and natural rubber, J. Appl. Polym. Sci. 92 (5) (2004) 3394–3400.

[25] A. Shanmugharaj, J. Bae, K. Lee, W. Noh, S. Lee, S. Ryu, Physical and chemical characteristics of multiwalled carbon nanotubes functionalized with aminosilane and its influence on the properties of natural rubber composites, Compos. Sci. Technol. 67 (9) (2007) 1813–1822.

[26] M. Strankowski, D. Włodarczyk, Ł. Piszczyk, J. Strankowska, Polyurethane nanocomposites containing reduced graphene oxide, FTIR, Raman, and XRD studies, J. Spectr. 2016 (2016) 1–6.

[27] K. Mishra, G. Pandey, R.P. Singh, Enhancing the mechanical properties of an epoxy resin using polyhedral oligomeric silsesquioxane (POSS) as nano-reinforcement, Polym. Test. 62 (2017) 210–218.

[28] F.P. La Mantia, M. Morreale, L. Botta, M.C. Mistretta, M. Ceraulo, R. Scaffaro, Degradation of polymer blends: a brief review, Polym. Degrad. Stabil. 145 (2017) 79–92.

[29] J.E. Loefroth, Time-resolved emission spectra, decay-associated spectra, and species-associated spectra, J. Phys. Chem. 90 (6) (1986) 1160–1168.

[30] H. Morawetz, Fluorescence study of polymer chain interpenetration and of the rate of phase separation in incompatible polymer blends, Polym. Eng. Sci. 23 (12) (1983) 689–692.

[31] C.T. Chen, H. Morawetz, Characterization of polymer miscibility by fluorescence techniques. Blends of styrene copolymers carrying hydrogen bond donors with polymethacrylates, Macromolecules 22 (1) (1989) 159–164.

[32] R. Xie, B. Yang, B. Jiang, Excimer fluorescence study on the miscibility of poly(vinyl methyl ether) and styrene-butadiene-styrene triblock copolymer, Polymer 34 (24) (1993) 5016–5019.

[33] L. Dong, D.J.T. Hill, A.K. Whittaker, K.P. Ghiggino, Miscibility studies in blends of poly(methyl methacrylate) with poly(styrene-co-methacrylonitrile) using solid-state NMR and fluorescence spectroscopy, Macromolecules 27 (20) (1994) 5912–5918.

[34] C.W. Frank, M.A. Gashgari, Excimer fluorescence as a molecular probe of polymer blend compatibility. 1. Blends of poly(2-vinylnaphthalene) with poly(alkyl methacrylates), Macromolecules 12 (1) (1979) 163–165.

[35] F. Amrani, J.M. Hung, H. Morawetz, Studies of polymer compatibility by nonradiative energy transfer, Macromolecules 13 (3) (1980) 649–653.

[36] E. Moons, Conjugated polymer blends: linking film morphology to performance of light emitting diodes and photodiodes, J. Phys. Condens. Matter 14 (47) (2002) 12235–12260.

[37] A.R. Buckley, M.D. Rahn, J. Hill, J. Cabanillas-Gonzalez, A.M. Fox, D.D.C. Bradley, Energy transfer dynamics in polyfluorene-based polymer blends, Chem. Phys. Lett. 339 (5–6) (2001) 331–336.

[38] M. Voigt, J. Chappell, T. Rowson, A. Cadby, M. Geoghegan, R. Jones, D. Lidzey, The interplay between the optical and electronic properties of light-emitting-diode applicable conjugated polymer blends and their phase-separated morphology, Org. Electron. 6 (1) (2005) 35–45.

[39] J.M. Clough, C. Creton, S.L. Craig, R.P. Sijbesma, Covalent bond scission in the mullins effect of a filled elastomer: real-time visualization with mechanoluminescence, Adv. Funct. Mater. 26 (48) (2016) 9063–9074.

[40] M. Beija, M.-T. Charreyre, J.M.G. Martinho, Dye-labelled polymer chains at specific sites: synthesis by living/controlled polymerization, Prog. Polym. Sci. 36 (4) (2011) 568–602.

[41] J. Wu, J.K. Oh, J. Yang, M.A. Winnik, R. Farwaha, J. Rademacher, Synthesis and microstructure characterization of dye-labeled poly(vinyl acetate-co-dibutyl maleate) latex for energy transfer experiments, Macromolecules 36 (21) (2003) 8139–8147.

[42] T.D.Z. Atvars, I. Esteban, B. Illera, B. Serrano, M.R. Vigil, I.F. Piérola, Structural study of polymer blends by fluorescence spectroscopy, J. Lumin. 72–74 (1997) 467–469.

[43] D.J.T. Hill, A.K. Whittaker, K.W. Wong, Miscibility and specific interactions in blends of poly(4-vinylphenol) and poly(2-ethoxyethyl methacrylate), Macromolecules 32 (16) (1999) 5285–5291.

[44] J. Wang, M.K. Cheung, Y. Mi, Miscibility of poly(ethyl oxazoline)/poly(4-vinylphenol) blends as investigated by the high-resolution solid-state 13C NMR, Polymer 42 (5) (2001) 2077–2083.

[45] J.-M. Huang, S.-J. Yang, Studying the miscibility and thermal behavior of polybenzoxazine/poly(ε-caprolactone) blends using DSC, DMA, and solid state 13C NMR spectroscopy, Polymer 46 (19) (2005) 8068–8078.

[46] T.K. Kwei, T. Nishi, R.F. Roberts, A study of compatible polymer mixtures, Macromolecules 7 (5) (1974) 667–674.

[47] V.J. McBrierty, D.C. Douglass, T.K. Kwei, Compatibility in blends of poly(methyl methacrylate) and poly(styrene-co-acrylonitrile). 2. An NMR study, Macromolecules 11 (6) (2002) 1265–1267.

[48] T. Nishi, T.T. Wang, T.K. Kwei, Thermally induced phase separation behavior of compatible polymer mixtures, Macromolecules 8 (2) (1975) 227–234.

[49] J.L. White, P. Mirau, Probing miscibility and intermolecular interactions in solid polymer blends using the nuclear overhauser effect, Macromolecules 26 (12) (1993) 3049–3054.

[50] S.A. Heffner, P.A. Mirau, Identification of intermolecular interactions in 1,2-polybutadiene/polyisoprene blends, Macromolecules 27 (25) (1994) 7283–7286.

[51] A. Asano, K. Takegoshi, K. Hikichi, Solid-state NMR study of miscibility and phase-separation of polymer blend: polycarbonate/poly(methyl methacrylate), Polym. J. 24 (6) (1992) 555–562.

[52] J.F. O'Gara, A.A. Jones, C.C. Hung, P.T. Inglefield, Temperature dependence of local motions in glassy polycarbonate from carbon and proton nuclear magnetic resonance, Macromolecules 18 (6) (1985) 1117–1123.

[53] K. Schmidt-Rohr, J. Clauss, B. Blümich, H.W. Spiess, Miscibility of polymer blends investigated by1H spin diffusion and13C NMR detection, Magn. Reson. Chem. 28 (13) (1990) S3–S9.

[54] N.C. Miller, E. Cho, M.J.N. Junk, R. Gysel, C. Risko, D. Kim, S. Sweetnam, C.E. Miller, L.J. Richter, R. J. Kline, M. Heeney, I. McCulloch, A. Amassian, D. Acevedo-Feliz, C. Knox, M.R. Hansen, D. Dudenko, B.F. Chmelka, M.F. Toney, J.-L. Brédas, M.D. McGehee, Use of X-ray diffraction, molecular simulations, and spectroscopy to determine the molecular packing in a polymer-fullerene bimolecular crystal, Adv. Mater. 24 (45) (2012) 6071–6079.

[55] J. John, D. Klepac, M. Didović, C.J. Sandesh, Y. Liu, K.V.S.N. Raju, A. Pius, S. Valić, S. Thomas, Main chain and segmental dynamics of semi interpenetrating polymer networks based on polyisoprene and poly(methyl methacrylate), Polymer 51 (11) (2010) 2390–2402.

12

Wide-angle X-ray diffraction and small-angle X-ray scattering studies of elastomer blends and composites

Angel Romo-Uribe

RESEARCH & DEVELOPMENT, ADVANCED SCIENCE & TECHNOLOGY DIVISION, JOHNSON & JOHNSON VISION CARE INC., FL, JACKSONVILLE, UNITED STATES

12.1 Focus

The addition of particles of μm to nm-scale to elastomeric polymers induces changes to thermal and mechanical properties. The aim is to produce high-performance lightweight composites for high-demanding applications. However, the changes to bulk physical properties reflect changes of macromolecular dynamics induced by the presence of fillers into the polymeric matrix. Depending on the concentration and state of dispersion of the fillers, polymer chain confinement effects may play a predominant role in the chain dynamics and dynamics changes will eventually be reflected in the bulk properties. Furthermore, filler–polymer interactions may be coupled as the systems are usually not athermal.

Therefore, the design of polymeric composites with exceptional properties involves the understanding of the state of dispersion of the fillers and the role that the particles spatial arrangement plays under complex thermomechanical history. Thus, short- and long-range order need to be characterized and X-ray scattering with spatial resolution ranging from Å to nm-scale is an optimum technique for multiscale structural characterization. Here the focus is on the application of wide-angle and small-angle X-ray scattering (WAXS and SAXS, respectively) to the structural characterization of polymeric materials with emphasis on elastomeric materials filled with microparticles.

The combination of wide-angle and small-angle X-ray scattering (WAXS and SAXS) enables the access from Å- to nm-scale in structure analysis. High-intensity source based on synchrotron radiation enables spatial and time-resolved resolution. A great advantage is that X-ray scattering is a nondestructive technique and requires little if any sample preparation and therefore enables studies not only of research specimens but of as-finished parts. Furthermore, hyphenated techniques can be implemented, that is, the microstructure of specimens can be studied under shear flow, uniaxial deformation,

Elastomer Blends and Composites. https://doi.org/10.1016/B978-0-323-85832-8.00010-9

electric/magnetic fields, controlled atmosphere, and temperature. Therefore, X-ray scattering is a very versatile analytical tool to understand the microstructure of materials, including elastomeric composites.

The fundamentals and the sophisticated (and often fastidious) mathematical treatment of X-ray scattering and the methods of production of X-rays are amply covered by numerous textbooks, and several are listed as further reading at the end of this chapter. Hence, here the author rather focuses on practical aspects including instrumentation, recording techniques, and the kind of information extracted from this powerful analytical technique. The chapter contains throughout key aspects for the study of elastomeric composites using X-ray diffraction with a section of applications to nanostructured molecular networks. The choice of the material is largely based on the author's experience and his results collected over years of experience in National and Industrial R&D laboratories, most of these results are yet unpublished.

12.2 X-ray diffraction

12.2.1 The beginnings of WAXD

Historically, X-ray scattering was first demonstrated using wide-angle and transmission mode by the now famous experiment of Paul Knipping, Walter Friedrich, and Max von Laue in 1912. These results not only demonstrated conclusively the wave nature of the X radiation but also opened the door to the structure elucidation of matter by the pioneering work of W. H. Bragg and L. H. Bragg immediately after Laue's discovery. Since the discovery of the X-radiation by W. Röntgen in 1895, there was controversy regarding the corpuscular and wave nature of this extraneous and highly penetrating radiation. The duality particle wave, at the heart of quantum physics, was not recognized at the time. Even without the certainty on the nature of the X-rays, the pioneering work of Röntgen immediately opened medical applications when he demonstrated the penetration of this radiation through the human body where high atomic number elements (e.g., bones) provided contrast. That is, in the period from 1895 to 1897, Röntgen fundamentally determined the production and properties of X radiation and established that most materials are transparent to X-ray radiation.

The next years and up to 1911 investigations followed on the determination of the absolute energy of X-rays, and polarization and absorption effects among others. At this stage, a meeting between Peter Paul Ewald and Max von Laue in January 1912 discussing the PhD dissertation of Ewald (under Prof. A. Sommerfeld) on the optical properties of an anisotropic arrangement of isotropic resonators opened the door to the possible diffraction of X radiation by a spatially periodic atomic structure. After all, the theoretical calculations of Ewald were wavelength-independent. Laue envisioned the experiment, but he was a theoretician and lacked experimental skills and X-ray instrumentation. Therefore, he approached his Group leader, Prof. Sommerfeld, who had a well-equipped X-ray laboratory and staff, and proposed the experiment. Eventually, Laue was assisted by Sommerfeld's assistant W. Friedrich and the just graduated student from Röntgen's group, P. Knipping, both true experimentalists. The team agreed on irradiating a well-developed copper sulfate crystal and in June 1912, announced to the Bavarian Society the diffraction of X-rays, thus demonstrating its wave nature, and proving that crystals are indeed a spatially regular arrangement of atoms.

Soon after the publication of Laue's results Lawrence H. Bragg, then a student in Trinity College of the University of Cambridge, confirmed Laue's results and conducted experiments irradiating and solving the structure of zinc blende (ZnS) and publishing the results in November 1912 (*Proceedings of the Cambridge Philosophical Society*), Fig. 12.1. At the suggestion of Prof. Pope (Department of Chemistry, University of Cambridge), he also studied several crystals including NaCl, KCl, KBr, and KI solving the structures in June 1913 (*Proceedings of the Royal Society*). This seminal and productive stage led to the now famous Bragg's relationship

$$2d \cdot \sin\theta = n\lambda \tag{12.1}$$

although the original relationship was expressed as $2d{\cdot}\cos\Theta = n\lambda$, where $\Theta = 90^\circ - \theta$. L. H. Bragg also determined for the first time the absolute wavelength of X-rays [1].

In the amazingly short period of time (1912–1914), W. H. and W. L. Bragg, at Leeds and Cambridge respectively, established the fundamentals of structure determination by X-ray diffraction, developed recording techniques using the ionization chamber (the grandfather of modern detectors) in combination with photographic techniques, and the characterization of targets for X-ray production, fundamental aspects that today we take for granted. The first experimental setup for (wide-angle) X-ray diffraction in

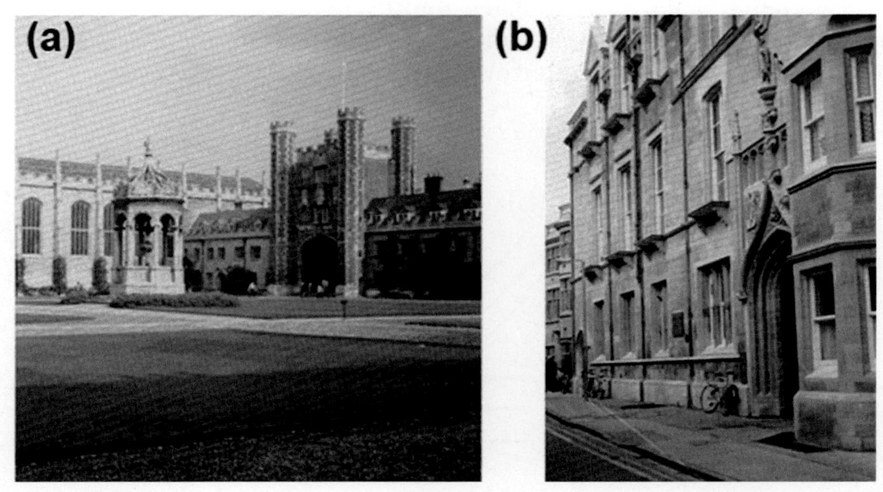

FIGURE 12.1 (A) Trinity College and (B) Cavendish Laboratory University of Cambridge (England) where the pioneering work on structure elucidation by X-ray diffraction was carried out by the then student, Lawrence H. Bragg, between 1912 and 1914. Years later, in 1953, with Bragg now as the director of the Cavendish Lab, Crick and Watson determined the structure of DNA.

symmetrical transmission mode and photographic recording, implemented by Friedrich and Knipping in 1912, is shown in Fig. 12.2A. Amazingly, the modern setups are, in essence, identical, as it will be shown in the next sections. Fig. 12.2B shows the principle adopted by Bragg to derive his relationship, Eq. (12.1). The sum of these two vectors denoting the incident and diffracted X-ray radiation, which closes the triangle in Fig. 12.2B, and the vector sum can only give elastically scattered radiation when, for given values of $1/\lambda$ and $1/d$, there is a particular angle θ between the incident beam and the normal to the planes. Therefore, a given set of standing waves, which we call crystal "planes," will only diffract when Bragg's condition is fulfilled [2,3]. Fig. 12.2C shows the X-ray diffraction pattern of *Fe* wire exhibiting a hexagonal crystal structure.

Many other developments followed at the same time, for instance, the work of H. G. J. Moseley in Prof. Rutherford's group in the University of Manchester and of L. de Broglie (in France) established the foundations of X-ray spectroscopy in the period from 1913 to 1914. At the same time, N. Bohr, a visiting scholar at Rutherford's lab, developed his atomic theory and set the foundations of quantum physics. Evidently, X-rays and its diffraction in the early 1900s were providing hard experimental evidence calling for new concepts in physics involving atomic-scale and high energies, which eventually gave rise to quantum physics.

After the outbreak of the Great War in Europe in 1914, all academic research was halted, and X-ray generators and instrumentation were mobilized to the battlefields for medical use. However, the foundations of (wide-angle) X-ray diffraction had been established. The following years saw improvements in instrumentation, including more powerful and stable X-ray sources such as the tube invented by W. D. Coolidge of the General Electric Co., cameras and detection and recording systems, filters, and monochromators, to name a few. Additionally, improvements in structure determination, where the complexity of structures studied rapidly increased, thanks to new methods such as Fourier analysis and Patterson functions and the eventual availability of

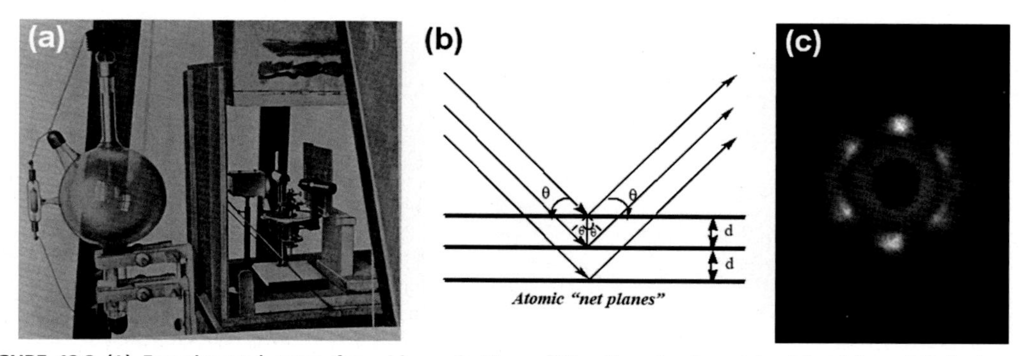

FIGURE 12.2 (A) Experimental setup for wide-angle X-ray diffraction developed by Friedrich and Knipping. (B) Lawrence H. Bragg envisioned X-ray diffraction as the reflection of X-ray from spatially periodic planes forming the crystal. (C) X-ray diffraction pattern of *Fe* wire denoting a hexagonal crystalline structure.

computers. For a detailed account of this amazing period in science and personal reminiscences of the major players involved, see [1].

Finally, the usefulness of X-ray diffraction in materials research is beyond question, from beginning was applied to study metals and organic materials, and today is also key in the pharmaceutical industry for drug development and to elucidate the structure of viruses.

12.2.2 Properties of X-rays

Nowadays, the most utilized radiation for laboratory X-ray generators is produced from the targets iron (Fe), copper (Cu) and molybdenum (Mo). The X-ray beam possesses a continuous spectrum of wavelengths (called the continuous or white radiation) with a broad maximum about 0.5 Å and a characteristic spectrum superposed on it consisting of two sharp and intense maxima, Kα and Kβ, where the Kα maximum is the most intense. Additionally, the Kα line is itself a doublet, but it is usually not separated for the study of polymers [3].

The white X-radiation is filtered by passing through an appropriate absorbing material (i.e., filter) to reduce the Kβ line to a negligible level at the cost of losing Kα intensity. A β-filter is the element whose K-absorption edge falls between the wavelengths of the Kα and the Kβ lines of the target, and the thickness of the filter is a compromise between reduction of the Kβ component at the cost of attenuating the Kα intensity. For Cu target, the Ni filter is about 15–20 μm thick; hence, Kα/Kβ ranges from 25 to 50 [2,3]. Some of the most common targets and respective filters are listed in Table 12.1.

12.2.3 Choosing the wavelength

The choice of X-ray wavelength is based on the material studied and experimental setup. For instance, in wide-angle mode *Cu* and *Mo* radiation of short wavelength is mostly utilized. On the other hand, in small-angle mode a longer wavelength is more suitable, thus *Cu* and *Fe* would be the choice.

That is, at constant sample-to-detector distance, the shorter wavelength *Mo* radiation will comprise more real space range in wide-angle, whereas the larger wavelength *Fe* radiation would increase real-space distance resolution, critical for small-angle scattering.

Table 12.1 X-ray targets, corresponding wavelengths, filters, and filters' thickness.[a]

Target	$\lambda_{K\alpha 1}$ (Å)	Filter	K-absorption edge (Å)	$K\beta_1/K\alpha_1 = 1/100$ (mm)	Loss $K\alpha_1$ (%)
Fe	1.9359	Mn	1.896	0.011	38
Cu	1.5405	Ni	1.488	0.015	45
Mo	0.7093	Zr	0.6888	0.081	57

[a]Alexander [3].

For instance, Fig. 12.3 shows wide-angle X-ray patterns of the same oriented amorphous polymer at constant sample-to-detector distance. The patterns were acquired using (A) *Mo* and (B) *Cu* targets, respectively. Note that the shorter *Mo* wavelength enables accessing higher-order reflections at constant angular range (indicated by the *arrow*), which corresponds to shorter real space distances. Likewise, note that the pattern acquired with *Cu* radiation exhibits larger resolution closer to the center of the pattern where the black dot in both patterns corresponds to a beam stop made of lead, a highly absorbing metal used for X-ray shielding.

12.2.4 Filters versus monochromators

It is noted that when a β-filter is employed, only the Kα X-rays produce sharp diffraction effects, while the residual continuous spectrum has the effect of generating diffuse background. Nearly monochromatic radiation can be obtained by using a crystal monochromator inserting it in the direct X-ray beam and reflecting the Kα line from a set of planes of the monochromator on the specimen. For instance, a silicon crystal is a narrow bandpass monochromator and pyrolytic graphite is a broad bandpass monochromator for Cu targets in X-ray generators. Double crystals are used as monochromators for synchrotron radiation [3,4]. Fig. 12.4 shows the X-ray diffraction patterns of an oriented cellulosic composite acquired with (A) filtered and (B) monochromatic X-radiation. Note the dramatic elimination of diffuse background, especially in the low angle range. Furthermore, the white tail of the radiation, not completely eliminated by a filter, can give fictitious signals that only experienced researchers can discern. The use of monochromators is much better choice; however, it is penalized by increasing the acquisition time due to the significant loss of radiation intensity, and this should be considered when designing experiments.

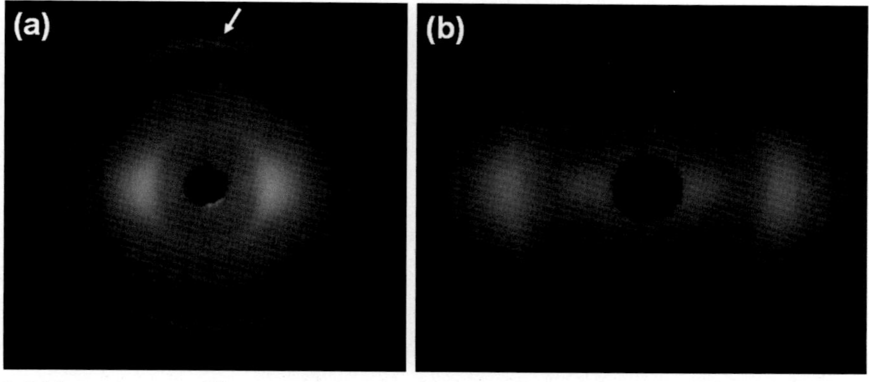

FIGURE 12.3 Wide-angle X-ray diffraction of an oriented amorphous polymer at constant sample-to-detector distance using (A) Zr-filtered MoKα and (B) Ni-filtered CuKα radiation. The shorter *Mo* wavelength enables accessing larger scattering range (i.e., shorter distances).

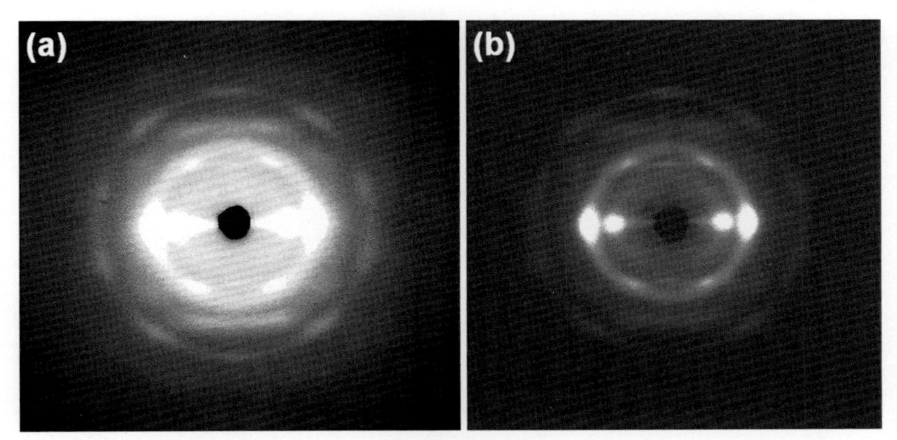

FIGURE 12.4 Wide-angle X-ray diffraction of an oriented cellulosic composite at constant sample-to-detector distance using CuKα radiation. (A) Ni-filter, and (B) graphite monochromator.

12.3 Methods in X-ray scattering

In general, the X-ray diffraction instrumentation is suitable for the study of the great majority of materials in practically any form, and hence, it is also applicable to polymers. The choice of the instrument is based on the specific kind of information investigated. Some properties of interest in polymer research include the presence or absence of crystalline structure, the determination of the size of crystalline regions, the mode of preferred orientation and its quantification, crystal structure and phase transitions due to, for instance, to changes in temperature or environment. Hence, instrumentation ideal for one goal may be quite unsuitable for a different goal.

As evidenced from Bragg's law, Eq. (12.1), the angular resolution of diffraction and the wavelength utilized define the real space distance accessible. Therefore, structure analysis is divided into wide-angle X-ray diffraction (WAXD) and small-angle X-ray scattering (SAXS), and the typical ranges accessible in a laboratory diffractometer equipped with Cu target are shown in Table 12.2.

12.3.1 X-ray scattering and polymers

Table 12.2 suggests that the information collected from WAXD and SAXS can, in principle, be merged into a single intensity trace covering all the angular range to get a complete mapping of the material structure. For instance, Fig. 12.5 shows the WAXD and SAXS patterns of an elastomer filled with nanoplatelets arranged in a lamellar superstructure. The patterns were recorded with the X-ray beam directed perpendicular to the plane of the specimen, and this will be discussed in detail below. Here the angular range 2θ has been converted to scattering angle q, which is independent of wavelength, and it is defined as

Table 12.2 Typical angular range and distance in X-ray diffractometers, based on Cu target of $\lambda = 1.5405$ Å wavelength.

Technique	2θ range (°)	Distance (Å)
WAXD	4–54	22.1–1.7
SAXS	0.1–4	882.6–22.1

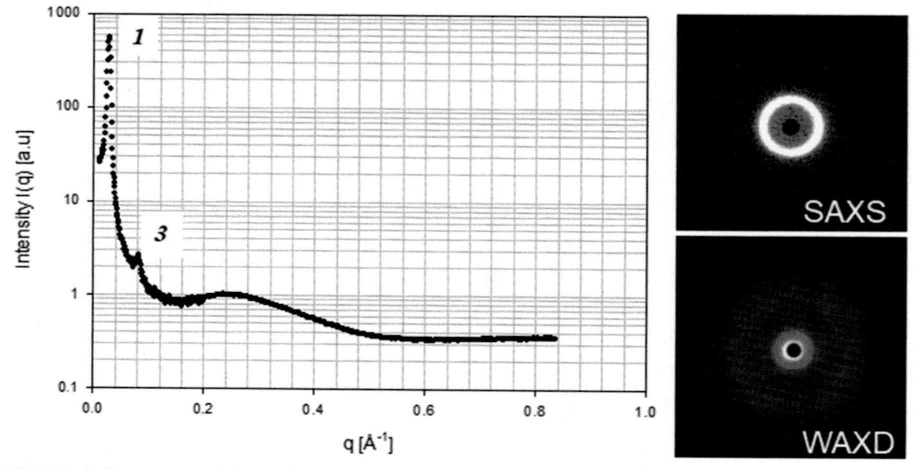

FIGURE 12.5 Radially averaged intensity trace obtained from the SAXS and WAXD patterns of an elastomer reinforced with nanoplatelets.

$$q = \left|\vec{q}\right| = \frac{4\pi\sin\theta}{\lambda} \tag{12.2}$$

The WAXD pattern exhibits only a halo (correspondingly the intensity trace shows a broad maximum at larger angular range or q), which demonstrates that the short-range structure, at monomers' length, is amorphous. This is typical of an elastomeric phase.

On the other hand, the SAXS pattern shows sharp rings, and these correspond to the intensity peaks in the small angular range (small q) labeled 1 and 3, and these peaks denote a layered spatially periodic superstructure, which is embedded in the elastomeric matrix. The d-spacing obtained from Bragg's law (or $d = 2\pi/q$) of the peak 1 is 339 Å and that of peak 3 is 113 Å, denoting first- and third-order maxima because $d_1 = 3d_3$. This will be discussed in more detail in Section 12.5 devoted to SAXS. Hence, Fig. 12.5 shows typical X-ray patterns of an elastomer composite reinforced with layered nanofillers.

Furthermore, note that the X-ray patterns of Fig. 12.5 are symmetrical, that is, exhibit a circular halo (WAXD) and rings (SAXS), which indicate that the short- and long-range structure in the elastomer composite is isotropic. This is another important piece of information readily available from X-ray diffraction. Therefore, it was possible to average the intensity around the azimuth and merge the intensities extracted from both techniques, as shown in Fig. 12.5. However, when the materials exhibit preferred orientation, i.e., anisotropy, this treatment is not possible.

12.4 Wide-angle X-ray diffraction, WAXD

As stated above, WAXD was first developed, and it is still the most utilized technique in materials research largely based on its apparent simplicity. Regarding polymeric materials, structural characteristics including crystallinity, macromolecular orientation, interchain distances, phase transitions, and crystal structure can be readily obtained.

12.4.1 WAXD configurations

The most common X-ray diffraction configurations for the study of (polymeric) materials are as follows:

(a) symmetrical transmission mode, Fig. 12.6A. This is the most popular configuration for 2D detection systems, and it is based on the original configuration developed by Friedrich and Knipping in 1912 (see [1]);
(b) symmetrical fiber diffraction with a cylindrical camera, which was developed by J. D. Bernal in 1924 when studying graphite at the Royal Institution, London;
(c) symmetrical reflection mode where the specimen is fixed, usually in horizontal position, and the incident X-ray beam and detector are rotated by θ and θ, respectively;
(d) asymmetrical reflection, parafocusing (Bragg-Brentano) mode developed in 1917, Fig. 12.6B, where the divergent and diffracted beams are focused on a fixed radius

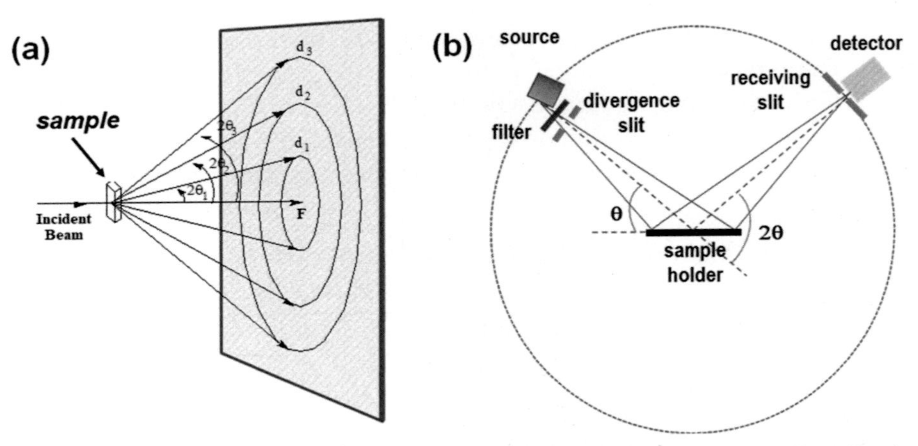

FIGURE 12.6 Typical configurations for WAXD. (A) symmetrical transmission mode usually utilized with 2D detection systems, and (B) reflection, Bragg-Brentano, mode typically utilized with 1D detection systems.

from the sample position. Here the incident X-ray beam is fixed, and the sample and detector are rotated by θ and 2θ, respectively.

Regardless of the configuration, the film-to-specimen distance D needs to be calibrated, and this is achieved using a material (standard) with known scattering angle 2θ (or d-spacing) and using the relationship

$$r = D \tan 2\theta \tag{12.3}$$

where r is the radial distance of a known reflection r, and 2θ is the corresponding Bragg angle.

In cylindrical (and Debye—Scherrer cameras), the scattering angle 2θ is proportional to the arc length L and the radius of the camera R,

$$\theta = \frac{L}{4R} \tag{12.4}$$

Some of the most utilized standards for polymer research are Si and quartz powders, the crystallographic data are listed in Tables 12.3 and 12.4.

12.4.2 X-ray patterns and preferred orientation

In general, natural and synthetic fibers exhibit some degree of preferred macromolecular orientation. However, macromolecular orientation can also be induced during processing and end-use applications, and this can be easily evaluated and quantified using WAXD and SAXS.

That is, common processing routes such as injection molding, extrusion, two-roll molding, or calendaring can induce preferred orientation in a finished part. Fiber production where a polymer melt or a concentrated polymer solution is forced through pin holes, named spinnerets, and then further stretched when solidifying in air (or when passing through a coagulating bath to extract the solvent) usually induce significant macromolecular alignment along the flow direction.

The X-ray patterns are commonly recorded with the fiber axis normal to the incident X-ray beam. Then, recording the X-ray pattern on a 2D detector the axis parallel to the fiber axis is designated the *meridian*. On the other hand, the axis at right angles is designated the *equator*. Any location on the pattern is defined by a radial distance r from the center of the pattern (which corresponds to the undeviated X-ray beam) and an

Table 12.3 Main reflections, *d*-spacing, and scattering angle (based Cu radiation $\lambda = 1.5405$ Å) of quartz powder.

Reflection	*d*-spacing (Å)	2θ (°)
100	4.25	20.88
101	3.34	26.66
110	2.46	36.57

Table 12.4 Main reflections, *d*-spacing, and scattering angle (based Cu radiation $\lambda = 1.5405$ Å) of Si powder [5].

Reflection	*d*-spacing (Å)	2θ (°)
111	3.135	28.44
220	1.920	47.30
311	1.637	56.12

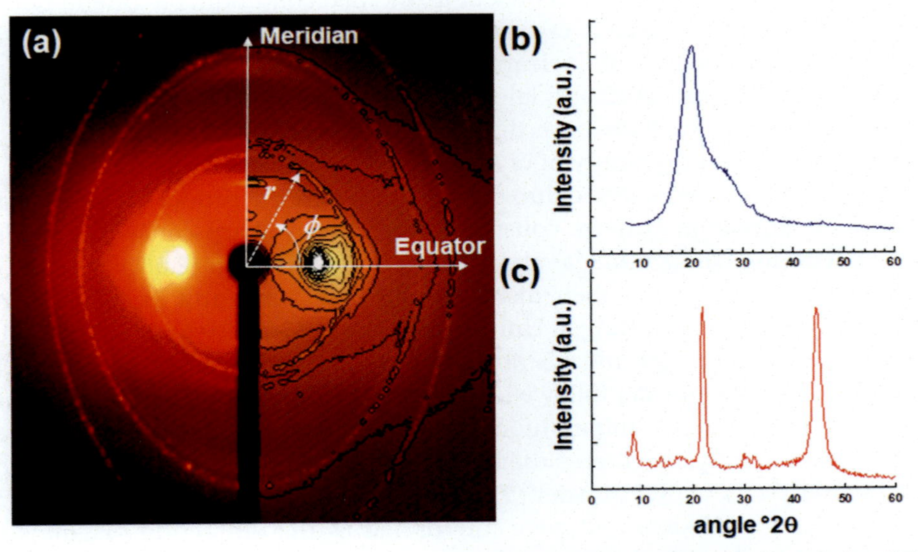

FIGURE 12.7 (A) WAXD pattern of a (semicrystalline) polymer fiber where the fiber axis is vertical, and the X-ray beam was directed orthogonal to the fiber. The sharp elliptical reflections correspond to quartz impregnated in the fibers to serve as internal standard, (B) intensity trace extracted along the equatorial axis, and (C) intensity trace extracted along the meridional axis. The scale in (C) is 1/3 of that in (B). Cylindrical camera and CuKα radiation were used.

azimuthal angle ϕ with respect to the equator, as shown in Fig. 12.7A. The radial distance r is, of course, proportional to the diffraction angle 2θ, and the conversion is carried out using Eqs. 12.3 or 12.4 and an appropriate calibration standard. Hence, when preferred macromolecular orientation is present, at least two axes on the patterns must be defined. Fig. 12.7B and C shows the intensity traces extracted along the equatorial and meridional axes.

The intensity maximum along the *equator* defines the interchain spacing, which is determined using Bragg's law. On the other hand, the intensity maxima along the *meridian* correspond to spatial regularity along the polymer chain.

We now go back to Fig. 12.3, which shows WAXD patterns of an amorphous polymer exhibiting preferred orientation. Using the notation defined in this section, we can see that the intensity is concentrated on the equatorial axis, and it is azimuthally (ϕ) and radially (2θ) spread, thus confirming that the polymer chains are on average aligned along the vertical axis but in amorphous state. Note the absence of meridional intensity maxima, consistent with the lack of spatial periodicity along the polymer chains.

The axes definition is not only applicable to WAXD patterns but also to SAXS patterns exhibiting anisotropy, and this will be discussed in Section 12.5.

12.4.3 Amorphous state and random microcrystallinity

In semicrystalline polymers, either in bulk or as-molded condition, the orientation of the crystalline domains (also named crystallites) is commonly statistically random. The randomness is a consequence of nucleation and crystal growth mechanisms in the melt or concentrated solutions. SAXS and later on transmission electron microscopy (TEM) have shown that the crystallites have dimensions around 50 nm, and therefore, a few milligrams contain a large quantity of crystallites, which results in all orientations being statistically present. Hence, a monochromatic X-ray beam will produce a WAXD pattern, named powder-diffraction pattern, consisting of relatively sharp concentric circles superposed on a diffuse background, as shown schematically in Fig. 12.6A. These are called Debye–Scherrer patterns after their discoverers, Peter Debye and Paul Scherrer, who conducted this research in Göttingen University in the 1913–14 period. Scherrer built an X-ray setup using a Cu target and his now famous cylindrical camera and successfully studied lithium fluoride powder followed by studies on the organic liquids, benzene, and cyclohexane. The crystalline lithium fluoride produced sharp concentric rings, and these bright scientists correctly interpreted the result as crystalline diffraction on the randomly oriented microcrystals of the powder. The organic liquids, however, produced broad interference rings (i.e., halos) and were interpreted as the diffraction diagrams of the carbon hexagons in these molecules.

Strikingly, the features observed by Debye and Scherrer are totally relevant to unoriented polymers, semicrystalline and amorphous, which exhibit the same features, as shown in Figs. 12.5 and 12.8. The sharp circular rings in Fig. 12.8A arise from the randomly oriented micro crystals in polyethylene (PE), whereas the diffuse scattering near the center of the pattern arises from the amorphous phase in PE (a distinction between metals and polymers). That is, the crystalline regions in a polymer coexist with the amorphous ones and, furthermore, the former regions contain a considerable number of irregularities. The concentric halos in Fig. 12.8B denote a liquid-like structure and the uniform intensity around the azimuth confirm that there is no preferred orientation in the elastomer. Another distinction between liquids and amorphous polymers is that liquids can not preserve orientation, but polymers can, as seen in Fig. 12.3. The presence or lack of order in amorphous polymers has received considerable attention, see Ref. [6].

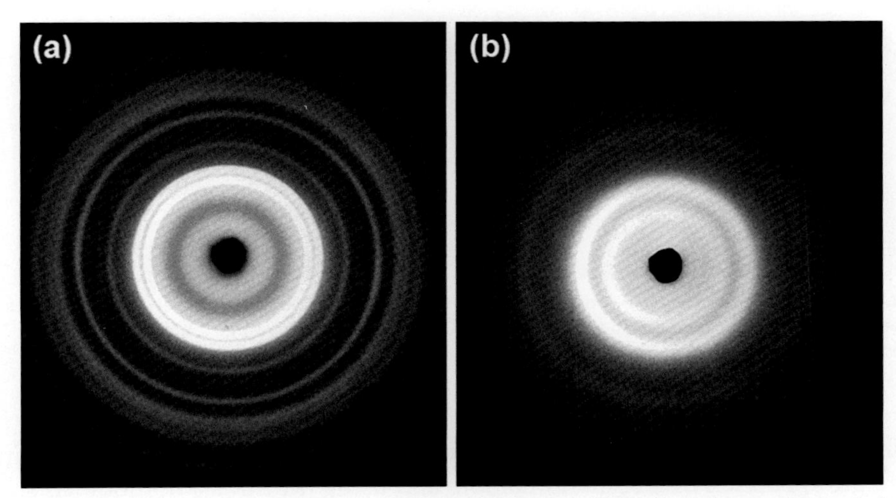

FIGURE 12.8 Debye–Scherrer WAXD patterns of unoriented polymers. (A) Molded plaque of polyethylene (PE), a semicrystalline polymer, and (B) a molded elastomer with halos denoting a liquid-like structure. The X-ray beam was directed orthogonal to the plane of the plaques. Flat plate camera and CuKα radiation were used.

12.4.4 Detection systems

The WAXD data is recorded with 1D and 2D detection systems.

(a) The 1D detectors consist of scintillation counters and position sensitive detectors (PSD), and here an intensity trace is recorded as a function of diffraction angle 2θ. This is usually applied to isotropic materials where there is no preferred orientation. Hence, the material can by studied in symmetrical transmission or reflection mode regardless of its orientation. The intensity traces recorded with scintillation counters can consume minutes to hours and require stable X-ray sources. Fast, time-resolved intensity traces can be obtained with PSDs and conventional sealed tube and rotating anode X-ray generators (an intensity trace can be recorded in a matter of few seconds). Fig. 12.9 shows X-ray diffractometers fitted with scintillation counters.

(b) The use of 2D detectors is highly advisable as these cover materials with and without preferred orientation. However, the cost can increase significantly and depends on their spatial resolution, intensity detection limit, and speed to store data. Photographic recording is by far the less expensive and quite effective although nowadays rarely used. Electronic detection is now common, but many of them do no attain the quality of data of photographic recording. The 2D detectors most common are (i) image plates, (ii) CCD cameras, and (iii) area detectors.

Because in θ/θ and $\theta/2\theta$ configuration, the specimen is held horizontal as shown in Fig. 12.8A, it is ideal to perform hot-stage X-ray diffraction and to study liquids (see Ref. [7]).

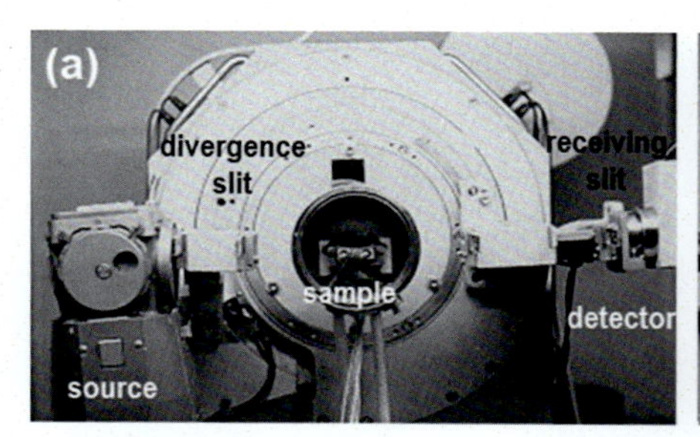

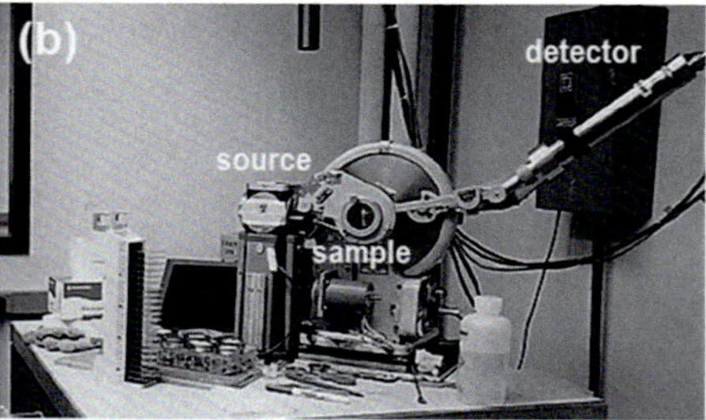

FIGURE 12.9 X-ray diffractometers working in reflection mode and equipped with scintillation counter. (A) Diffractometer working in θ/θ configuration (Siemens D500), and (B) diffractometer (Philips XRG) using Bragg–Brentano $\theta/2\theta$ configuration. In (A) the source, on the left, rotates along with the detector. In (B) the source is fixed to the X-ray generator.

It is noted that 2D patterns can still be recorded using scintillation counters by mounting the specimen in an Eulerian cradle and using transmission mode, as shown in Fig. 12.10A. In this case, the specimen can be rotated and intensity traces as a function of 2θ and as a function of the azimuth ϕ are recorded, as shown in Fig. 12.10B. This, however, is very time-consuming and several hours of recording are needed to acquire only one quadrant (varying ϕ from 0 to 90 degrees) of the WAXD pattern. However, as long as the X-ray source is stable (quite achievable nowadays), the quality of data is very good.

Typical 2D detection systems, working in transmission mode, are shown in Fig. 12.10. Note the amazing similarity with the original system developed by Friedrich and

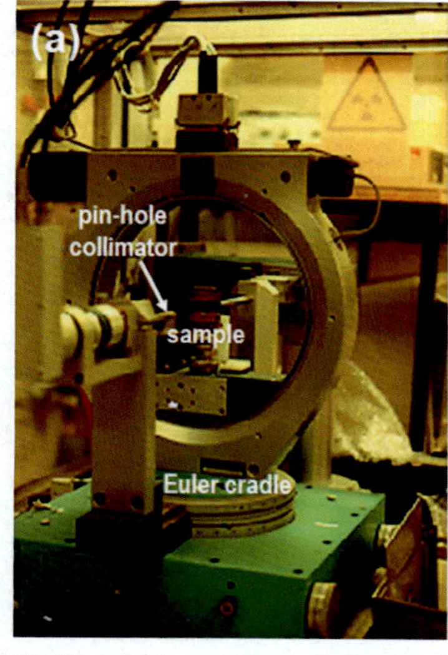

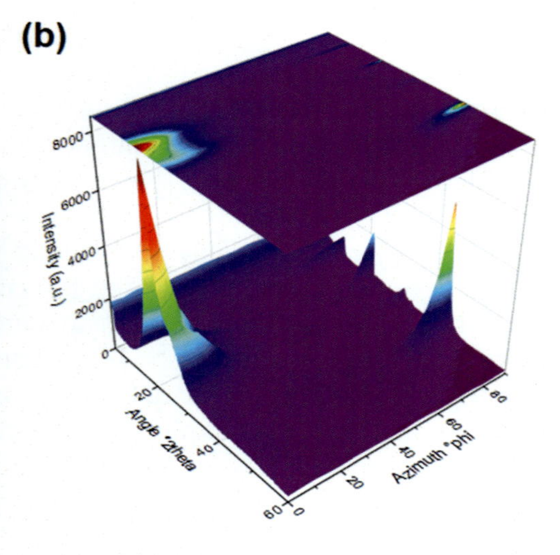

FIGURE 12.10 (A) Eulerian cradle and transmission mode, and (B) 2D WAXD pattern generated from 1D intensity traces varying 2θ in 0.2 degrees steps and varying ϕ in step mode in steps of 5 degrees. CuKα radiation was used.

Knipping in 1912 (see Fig. 12.2A). The modern systems consist of filters (or monochromators), pin-hole collimators, and the recording device. Photographic recording, image plates, CCD cameras, and area detectors can be used for the flat-plate camera, Fig. 12.11A. Image plates and photographic recording are used in cylindrical cameras, Fig. 12.11B. Fig. 12.11C shows a comprehensive system equipped with goniometer mounted on a semi-Eulerian cradle, pin-hole collimation, and the Hi–Star area detector (this system was originally developed by Siemens).

Interestingly enough, the recording time to get a decent WAXD pattern using any of these cameras is around 40 min for photographic recording or 20 min with the area detector, with typical sample-to-detector distance D of 40 mm, Cu target and generators powering from 1.2 to 2 kW. However, the newly developed electronic 2D cameras, such as the HyPix 3000 (manufactured by Rigaku) or the Vantec 2000 (resolution of 68×68 μm, manufactured by Brüker and suitable for Cu X-radiation) require as little as 10 min to get good-quality WAXD patterns.

12.4.5 Remarks

Finally, it can now be appreciated that before attempting any analysis of a material using WAXD, many decisions need to be made based on the specimen under study, the properties investigated, and the instrumentation available.

The commercially available X-ray diffractometers are equipped with a sealed tube or a rotating anode, the latter providing much higher power. Then, decisions need to be made on the filtering system, the geometry of the collimation (slits vs. pin-hole), the configuration (transmission vs. reflection), and the detection system. All these factors will impact the quality of data acquired and the time it takes to acquire. For instance, highly collimated, monochromatic radiation will provide a "clean" X-ray beam but at the cost of significantly reducing its intensity. Then, it is recommended to first survey the specimen with a high intensity beam and then, based on the results, refine the measurements.

Most polymers can be studied with little sample preparation if any, and this makes WAXD very versatile, thus enabling experiments as a function of temperature, atmosphere, humidity, and electric, magnetic or stress fields. High-intensity sources, such as

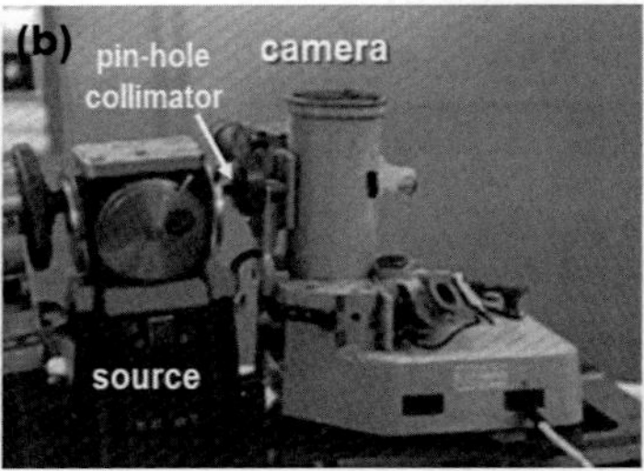

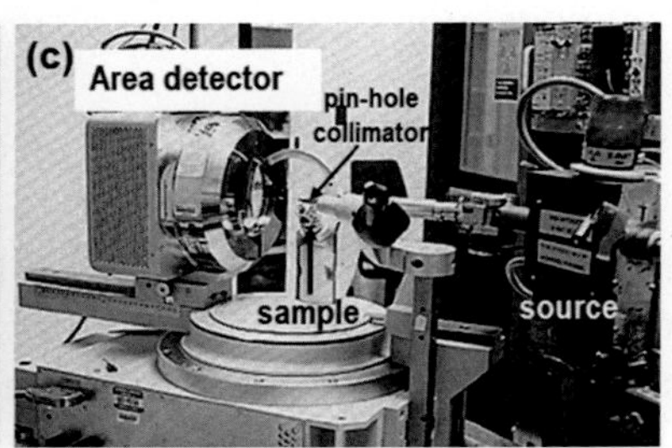

FIGURE 12.11 2D detection systems in transmission mode using pin-hole collimation. (A) flat-plate camera; (B) cylindrical camera, and (C) area detector.

synchrotron radiation, and modern detectors enable time-resolved structure studies [7–10].

12.5 Small-angle X-ray scattering (SAXS)

12.5.1 The beginnings of SAXS

Small-angle X-ray scattering was developed in the period 1937–1938 by Andre Guinier, then working under Ch. Mauguin. It was then recognized by Mauguin and Jean Laval some diffraction features quite close (within $1°2\theta$) to the primary X-ray beam and with intensity about 1/1000 of typical diffraction features in WAXD. Resolving such a low angle intensity was, and still is today, a challenge. For this, Guinier developed the basic principles of SAXS instrumentation: a monochromatic, well-collimated and highly intense X-ray beam and the *Guinier camera* were born. Furthermore, in a relatively short and productive period of time, Guinier established the *Guinier Law* to determine the size of a particle in dilute solution and then applied SAXS to study metal alloys and age-hardening and established the *Guinier–Preston zones*. Finally, in 1948 developed the *Guinier–Tennevin* method to study dislocations in crystals [1,11]. Thus, Guinier established methods to study particle size in colloidal dispersions and proteins in solution and fundamental results for solid-state physics. Most importantly, Guinier also established that small-angle scattering does not exist when the sample is homogeneous, i.e., there is no contrast in a homogeneous continuum.

12.5.2 SAXS and polymers

SAXS from polymers consist of either *diffuse* or *discrete* features [3,11,12], and both types of scattering contain important information regarding the long-range structure in the material. The properties that can be extracted when applying SAXS to polymers and complex fluids consist of (a) long-range structural orientation, (b) size of structures (e.g., crystalline regions in semicrystalline polymers, nanoparticles embedded in a matrix), (c) size and type of nanophases (e.g., block copolymers and polymer nanocomposites), (d) particle size in dilute polymer solutions and colloidal dispersions (i.e., radius of gyration), (e) phase transitions (e.g., block copolymers), (f) microvoid or micropore morphology and size, and (g) fractal dimension.

12.5.3 Diffuse small-angle scattering

Diffuse scattering is essentially particulate in origin but modulated to some extent by interparticle interferences. This is the basis for the concept of a dilute system, which produces pure particulate scattering [11]. The angular limit within which the scattering occurs is inversely proportional to the size of the inhomogeneities in the electron density distribution within the specimen. Hence, the diffuse scattering from a loose aggregate of oriented particles will have its largest extension in the direction of the smaller particle dimension and its smaller extension in the direction of the larger dimension. Therefore, the shape of the diffuse scattering depicts the particle shape in reverse (i.e., reciprocal space). Fig. 12.12 shows a typical SAXS pattern of an elastomer reinforced with Na-Montmorillonite nanoclay and the radially averaged intensity trace as a function of scattering vector q.

12.5.3.1 Guinier law

If the concentration of the particles is sufficiently dilute, the positions of individual particles, would be uncorrelated. Then, the waves scattered from different particles are incoherent among them, and the observed intensity simply becomes the sum of the individual scattering. If the shape of the particles is known or assumed based on independent information, the intensity of scattering from individual particles can be simulated and compared with the experimental results.

If, however, the particles are of irregular or unknown shape, the data may be analyzed according to the *Guinier Law* to determine the radius of gyration characterizing the size of the particles [3]. Thus, in the limit of small q,

$$I(q) = \rho_o^2 v^2 \exp\left(-\frac{1}{3}q^2 R_g^2\right) \tag{12.5}$$

where $I(q)$ is the intensity of independent scattering by a particle. Eq. (12.9) is called the *Guinier Law* and allows for the determination of the radius of gyration R_g of a particle from SAXS measurements. The Guinier Law is valid provided that:

(1) $q << 1/R_g$,
(2) the system is dilute, so that the particles in the system scatter independently of each other,

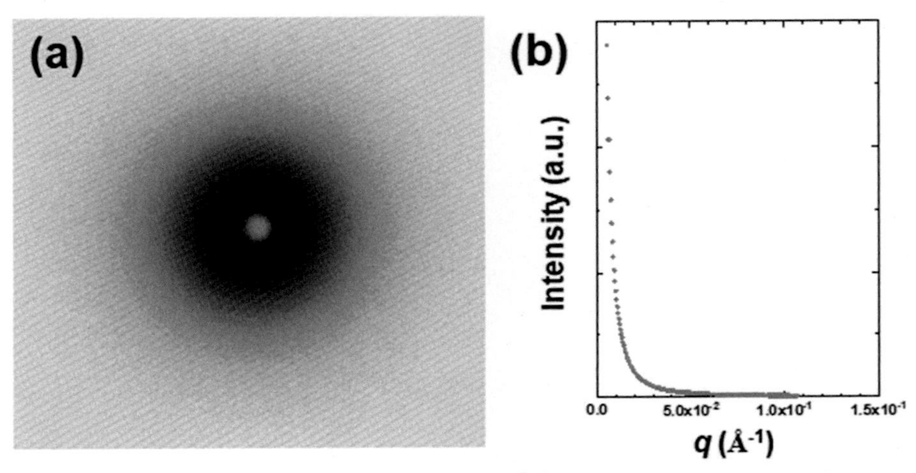

FIGURE 12.12 (A) SAXS pattern of an elastomer/nanoclay composite acquired using pin-hole collimation and area detector, and (B) azimuthally averaged intensity trace extracted from (A).

(3) the system is isotropic, i.e., the particles assume random orientations
(4) the matrix (or the solvent) in which the particles are dispersed is of constant density and devoid of any internal structure that can by itself give scattering in the q-range of interest.

12.5.3.2 Fractal structure

It is noted that in polymer nanocomposites with featureless SAXS patterns like that shown in Fig. 12.12, there may be important information. This is revealed in Fig. 12.13A where now log I versus log q has been plotted. These results show that the scattered intensity of the elastomeric nanocomposite obeys a *power-law scaling*, which is typical of *fractal* structures [14,15].

$$I(q) \sim \left|\vec{q}\right|^{-D} \tag{12.6}$$

The scaling power D is 2.05 ($\pm$0.07) and corresponds to that of randomly oriented lamellae or platelets and indicates that the montmorillonite nanoplatelets are aggregated into a self-similar, fractal, structure. Furthermore, the results show that the slope is constant over a broad size range, indicating that the self-similarity is conserved over nearly two orders of magnitude.

Fig. 12.13B shows the corresponding TEM micrograph, where the nanoclay particles are imaged, these appear aggregated, but nothing can be said about the fractal structure. This is just one example of many where SAXS proves much superior to electron microscopy techniques. Whereas SAXS did not require sample preparation and data

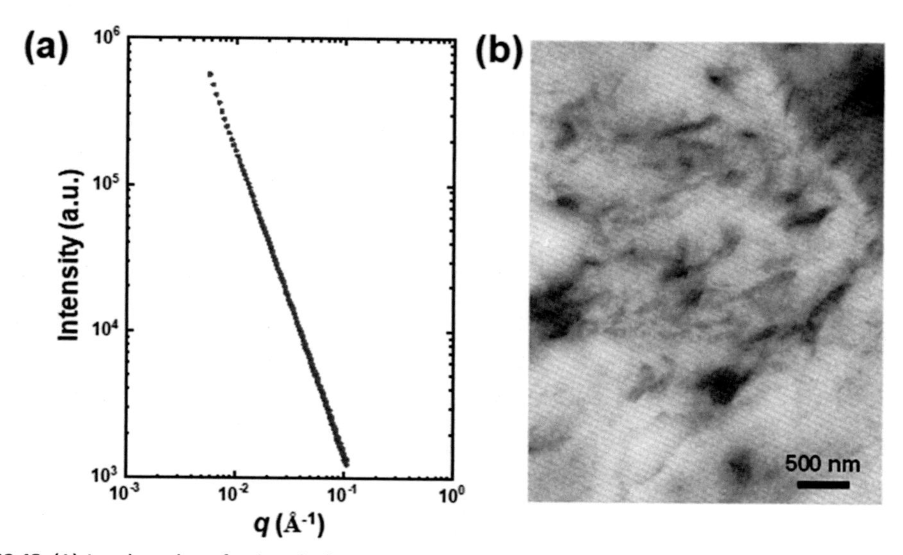

FIGURE 12.13 (A) Log-log plot of azimuthally averaged intensity extracted from SAXS pattern of Fig. 12.12A. The constant slope denotes that nanoclay is aggregated into a self-similar, fractal structure in the elastomeric continuum. (B) Corresponding TEM micrograph of the same specimen (see Ref. [13] for details).

acquisition required ca. 30 min (2 kW Cu radiation), hours of tedious sample preparation and data acquisition were required before a decent TEM micrograph could be obtained, and the results of TEM are not exempt of ambiguity.

Fig. 12.14 shows a TEM micrograph of an elastomer filled with nanosilica particles of average size 10–15 nm, per supplier specifications. Using image analysis, the average particle diameter was determined to be 20.2 nm (±2 nm).

Applying SAXS to the elastomer/nanosilica composite and simulating the scattering data, assuming spherical geometry of the nanoparticles (using NIST small-angle scattering packages and a commercial software such as Igor Pro), the particle size was determined. Fig. 12.14B shows the experimental SAXS trace (symbols) and the simulated curve (continuous line). The fit to the experimental data produced a particle average radius of 9.7 nm with a polydispersity index of 0.52, in agreement with the TEM results.

12.5.3.3 Scattering equivalents

Here is a good point to stress the fact that, as far as SAXS is concerned, the spherical nanoparticles embedded in the polymeric matrix of Fig. 12.14 behave like:

(a) particles suspended in a liquid,
(b) colloidal dispersions,
(c) polymer emulsions,
(d) pores in a hydrogel or pores in a catalyst.

These types of structures are called *scattering equivalents*. For instance, a hydrogel and a catalyst are scattering equivalents because they can be considered dilute particle systems where the particles are dispersed in an uniform matrix of a second material. In the hydrogel or in the catalyst, the "particles" consist of microvoids, and these are continuously dispersed throughout the material, the "matrix". Hence, the interpretation of diffuse small-angle scattering of Fig. 12.14B will be ambiguous unless we have extra information, such as the TEM micrograph of Fig. 12.14A.

Therefore, diffuse SAXS results always carry the following types of ambiguities [3,11]:

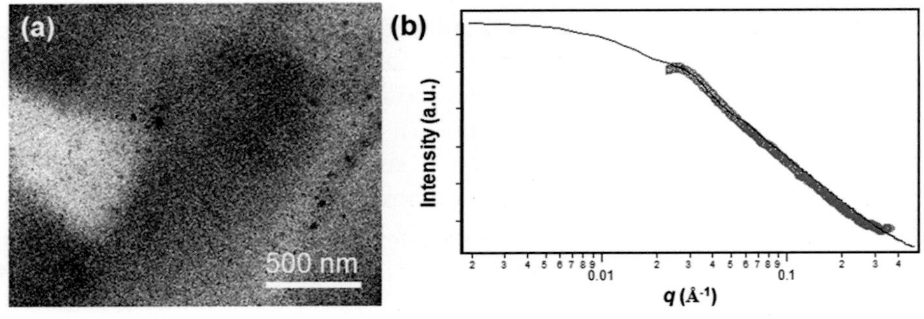

FIGURE 12.14 (A) TEM micrograph of silica nanoparticles embedded in an elastomer matrix, (B) SAXS intensity as a function of scattering vector q: experimental data (*symbols*) and simulated curve (*continuous curve*).

(a) It is not possible to distinguish the scattering by a system of particles in space from the scattering by a complementary system of micropores in a solid continuum. The choice between these alternatives is usually made on evidence bearing on the nature of the system concerned.

(b) It is not possible to differentiate, without some degree of uncertainty, the effects due to particle shape from those due to polydispersity. Hence, scattering equivalence may result from a special particulate shape and a certain distribution of particle sizes.

(c) In dense systems interparticle scattering cannot be neglected. Thus, it is difficult to discern to what extent the scattering data is determined by the isolated particles and to what extent by interparticle interferences.

12.5.4 Discrete small-angle scattering

Early in SAXS development, it was recognized that polymeric and biological materials exhibit spatially periodic structures having a period of the order of 10–1000 Å. However, the degree of periodic order in these systems is relatively poor and dealing with the "imperfections" is an important part of data analysis [3].

Some examples of spatially periodic systems that are subjected to small-angle scattering studies are folded-chain lamellar crystalline regions, membrane structures (e.g., liposomes) and lipidic bilayers, core–shell structures (e.g., latex particles), layered nanoparticles, block copolymers, and nanoparticle aggregates [16].

Discrete scattering is present in unoriented and preferentially oriented materials. We showed in Fig. 12.5 an example of discrete scattering in an elastomer filled with clay nanoparticles, On the other hand, oriented semicrystalline polymers (i.e., films and fibers) and natural fibrous systems exhibit discrete scattering, and the interferences appear in only one direction, most commonly along the orientation axis, denoted the Meridian (see Fig. 12.7). Meridional reflections in SAXS arise from some degree of large-scale periodic character in the structure parallel to the fiber axis. Analysis of these diffraction patterns enables, for instance, the determination of the lamellar or domain size in semicrystalline polymers and block copolymers, and the type of nanophase exhibited by narrow molecular weight distribution block copolymers.

An example of discrete small-angle scattering from synthetic spherical latex particles is shown in Fig. 12.15A. The original film displayed up to 17 circular intensity maxima. Like in WAXD (Fig. 12.8), the circular intensity maxima denote lack of preferred orientation. The pattern was photographically recorded using CuKα radiation, sample-to-film distance of 660 mm and 129 h exposure time [17].

On the other hand, an example of discrete small-angle scattering from nature is shown in Fig. 12.15B; this is the SAXS pattern obtained from kangaroo tail tendon using a CCD detector, sample-to-film distance of 400 mm, and 1 h exposure time. Note that the pattern shows only meridional reflections indicative of preferred macromolecular alignment and of the long-range periodic character of this system, which is 627 Å. Thus, biological materials display greater morphological perfection than synthetic polymeric

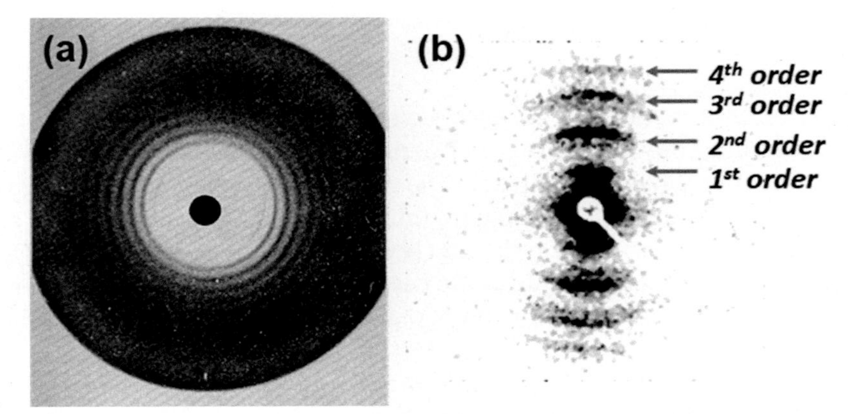

FIGURE 12.15 Discrete small-angle scattering can be found in synthetic and natural materials. SAXS patterns of (A) latex particles, and (B) kangaroo tail tendon. *(A) Adapted from W.E. Danielson, L. Shenfil, J.W.M. du Mond, Latex particle size determination using diffraction peaks obtained with the point focusing X-ray monochromator, J. Appl. Phys. 23 (1952) 860–865.*

ones, therefore, the pattern shows up to the fourth-order meridional reflection. Higher-ordered reflections would require longer exposure times.

Crystallizable and nanostructured polymers can exhibit limited spatial periodicity of their structures, and this would produce SAXS patterns with typically one and, more rarely, two or three intensity maxima. For instance, Fig. 12.16A shows the SAXS pattern of an extruded polyethylene (PE) film. The pattern shows three reflections along the meridional axis, which corresponds to the extrusion direction. The small-angle intensity maxima (labeled 1, 2, and 4) indicate that there is a large-scale periodic character in the microstructure parallel to the extrusion direction with relatively good spatial correlation. The secondary maxima are integer multiples of the first intensity maximum, and this is indicative of lamellar morphology. The analysis of the nanostructure in the oriented PE

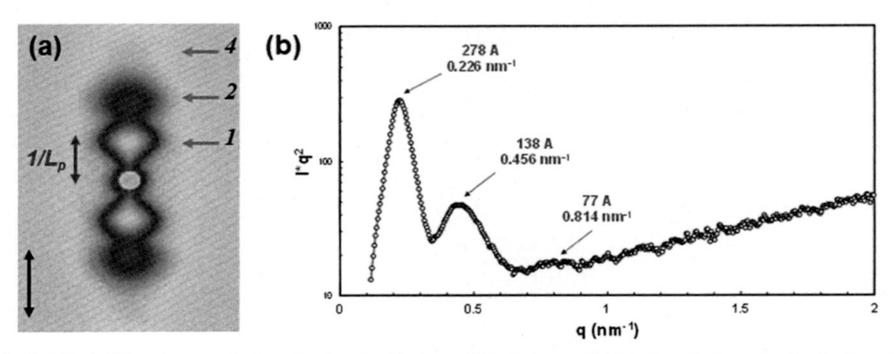

FIGURE 12.16 (A) SAXS pattern of stretched polyethylene film, where double-headed arrow indicates elongation axis. (B) Lorentz-corrected intensity trace.

can be carried out using two methods: (i) the two-phase model, and (ii) the invariant and radial correlation function.

12.5.4.1 Two-phase model and Lorentz correction

Information regarding the microstructural periodicity may be obtained from the SAXS pattern when the morphology is lamellar. The data can then be analyzed to give a one-dimensional Bragg spacing, L_p, by applying the Lorentz correction, q^2, to the scattering intensity, $I(q)$ [18]. If the first-order reflection is very strong, and additionally, if a second-order reflection is observed (as is indeed the case shown in Fig. 12.16), then the interface between the crystalline and the amorphous regions is sharp, and application of a two-phase model is valid.

For the calculation of the Bragg spacing, the first maximum in a plot of $I(q)\cdot q^2$ as a function of q is taken as q^*, so that

$$L_p = \frac{2\pi}{q^*} \tag{12.7}$$

Fig. 12.16B shows the Lorentz corrected meridional intensity trace as a function of scattering vector q. The first maximum occurs at $q^* = 0.0226/\text{Å}$, therefore the Bragg spacing, corresponding to the lamellar spacing, is determined to be $L_p = 278$ Å. Note that the other intensity maxima get broader and weaker as q increased due to the imperfect spatial periodicity of the crystalline structure. The distance associated to those maxima are 138 and 77 Å, thus $L_p \approx 2 \times 138$ Å and $L_p \approx 4 \times 77$ Å, therefore they are second- and fourth-order intensity maxima.

12.5.4.2 Invariant and radial correlation function

The scattering power, also designated as the *invariant*, Q, by Porod [19] is defined as

$$Q = \int_0^\infty I(q) q^2 dq \tag{12.8}$$

The invariant may be used to calculate the degree of crystallinity and lamellar thickness. However, the determination of the absolute value of Q requires absolute intensity measurements, subtraction of thermal background, and extrapolation of $I(q)\ q^2$ versus q data to $q = 0$ and $q = \infty$. The major contribution to the experimental invariant can be used to characterize structure development and is readily obtained from integration between experimentally accessible limits. In the absence of particulate clustering or a large degree of polydispersity, it is normally possible to determine the very low angle portion of the curve by extrapolating the plot of $I(q)\ q^2$ versus q to zero angle (by using, for instance, Guinier's Law). The high angle portion (tail end) of the intensity curve can be estimated by use of the theoretical finding of Porod [19] that the course of the tail end should conform to a constant limiting value of $q^4 \cdot I(q)$.

The SAXS data can then be analyzed using a one-dimensional correlation function in terms of an ideal lamellar morphology. The correlation function, designated $\gamma(r)$, is essentially a Fourier transform of a given one-dimensional SAXS intensity trace. $\gamma(r)$ is interpreted in terms of an imaginary rod moving through the structure of the material from which the SAXS pattern was obtained, so that $\gamma(r)$ may be considered as the probability that the rod is of length r, with equal electron densities at either end. Therefore, a frequently occurring spacing within a structure is manifested as a maximum in the one-dimensional correlation function. The interpretation of $\gamma(r)$ assumes that, within the SAXS length scale, spacings occur along one fixed axis, but that the axis assumes all possible directions throughout the material. The calculation and interpretation of the correlation function can be summarized as follows:

- extrapolation of the experimental SAXS curve to the limits $q = 0$ and $q = \infty$, which is required mathematically to apply the Fourier transform. As stated above, the tail-fitting process can be carried out using Porod's rule, for instance,

$$I(q) = I_b + \frac{K \cdot e^{\sigma^2 \cdot q^2}}{q^4} \tag{12.9}$$

  the parameter σ characterizes the form of the interface. The forward extrapolation can be achieved using Guinier's Law [11].
- Fourier transformation of the extrapolated data gives the correlation function $\gamma(r)$

$$\gamma(r) = \frac{1}{Q} \int_0^\infty I(q) \cdot q^2 \cdot \mathrm{Cos}(qr) \cdot dq \tag{12.10}$$

  where the invariant Q is used to normalize $\gamma(r)$ so that $\gamma(0) = 1$.
- The correlation function is interpreted based on an ideal lamellar morphology to yield the *long spacing* L_p, the *lamellar thickness* L_γ, and the *local degree of crystallinity* χ from the ratio

$$\chi = \frac{L_\gamma}{L_p} \tag{12.11}$$

  The long spacing L_p is determined from the position of the first maximum in the correlation function $\gamma(r)$; the lamellar thickness L_γ is obtained from the intersection of the ordinate to the first minimum of $\gamma(r)$ and the extrapolated tangent to the low end side of the correlation function curve.

  The method outlined above was applied to the oriented polyethylene film of Fig. 12.16. Fig. 12.17A shows the plot of the total meridional SAXS intensity trace I_{tot}. An additional scan was carried out but now with the sample holder being empty to correct the data for air scattering, this trace is labeled I_{backg}. Subtraction of the incoherent air background scattering was carried out and the correlation function was calculated from the corrected scattering curve as described above. The resulting correlation function is shown in Fig. 12.17B.

  Analysis of the correlation function yields the long spacing $L_p = 276$ Å. Note that this result is in excellent agreement with that obtained from the two-phase model. Furthermore, we can also determine the lamellar thickness, the results show that $L_\gamma \approx 80$ Å. Finally, using Eq. (12.11) the local degree of crystallinity was also

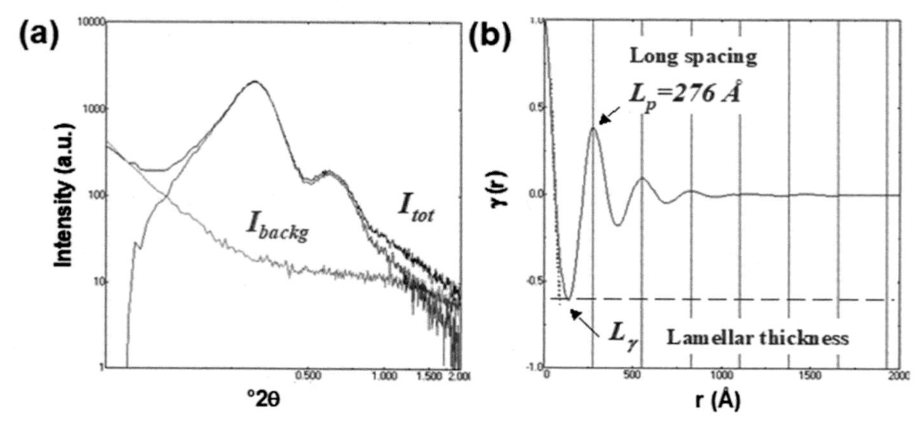

FIGURE 12.17 (A) Total, background, and corrected meridional SAXS intensity trace of stretched polyethylene film of Fig. 12.16. (B) Correlation function, $\gamma(r)$, as a function of distance r.

determined, $\chi = 29\%$. Hence, SAXS enables a complete characterization of the nanostructure of an oriented semicrystalline polymer.

Diffuse SAXS scattering is also found in microvoids and fibrous systems, and the analysis method has been described by Romo-Uribe et al. [7,20].

12.5.5 Instrumentation for small-angle X-ray scattering

The requirements for small-angle X-ray scattering are quite stringent compared with those met by WAXD instrumentation. Guinier clearly identified these hurdles:

(i) high-intensity X-ray source
(ii) focused, monochromatic intensity
(iii) highly collimated incident beam

In order to meet these requirements, a considerable loss of intensity must be paid. Then, small-angle scattering measurements are carried out in symmetrical transmission mode [3,11,19] and preferentially using rotating anode X-ray generators. The beta component of the X-ray radiation is first filtered. Then, the filtered X-rays are collimated using three collimators, either slits or pin-holes (e.g., s_1, s_2 and s_3 in Fig. 12.18A). In slit configurations, the diffracted X-rays are further collimated by a receiving slit (RS) to eliminate spurious scattering before being collected by a detector. D is the sample-to-detector distance. Schematic diagrams of typical SAXS systems and components are shown in Fig. 12.18, (a) slit collimation and (b) pin-hole collimation.

In a slit collimation system, the typical dimensions of the slits are $s_1 = 0.04$ mm, $s_2 = 0.03$ mm, $s_3 = 0.25$ mm, and RS $= 0.1$ mm, with sample-to-detector distance $D = 28.5$ cm. A typical instrument with this configuration, manufactured by Rigaku Inc. is shown in Fig. 12.19A. Note that the setup includes a vacuum chamber to reduce air scatter, and the scattered radiation is recorded using a scintillation counter. Due to the

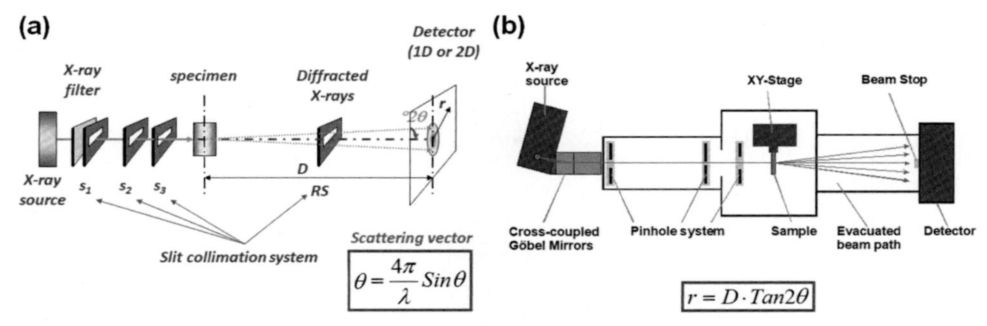

FIGURE 12.18 Typical SAXS configurations using: (A) slit collimators and (B) pin-hole collimators. *The latter diagram courtesy of Dr. K. Erlacher, Brüker AXS Inc.*

FIGURE 12.19 SAXS instrumentation implementing the configurations described in Fig. 12.18. (A) Dmax2500 with slit collimators (Rigaku USA), and (B) Nano-Star with pin-hole collimation (Brüker AXS Inc.).

very low scattered intensity a Ni-filtered is used thus maximizing the incident intensity. The small-angle scattering range achievable with this optics reaches a minimum angle of $0.08°2\theta$, which translates into a minimum scattering vector q of 0.0056/Å or a maximum real space distance of 1100 Å. This SAXS optics was installed on an 18 KW rotating anode X-ray generator, model Dmax 2500 (Rigaku Inc). The instrument was equipped with a copper (Cu) anode, giving X-ray radiation of wavelength 1.5405 Å and experiments are conducted typically at 50 KV and 150 mA. Using a scintillation counter typical 1D SAXS traces can be recorded in ca. 16 min using step scan mode of $0.006°2\theta$/step and 3s/step recording.

On the other hand, in a pin-hole collimation system, the pinhole diameters are typically 750, 400, and 1000 μm, and the distances source – first pinhole – second pinhole – third pinhole – sample – detector: 200 mm – 925 mm – 482 mm – 35 mm – 105 mm, respectively. This optics is typical of the Nanostar U system manufactured by Brüker AXS. The intensity is maximized by using cross-coupled Göbel mirrors, and therefore, a low intensity sealed tube X-ray generator with Cu target can be used instead of the significantly more expensive rotating anode generator. Experiments are conducted typically at 40 KV and 35 mA, and 2D SAXS patterns are recorded with an area detector. The scattering vector q range attained by this configuration is $0.01 \rightarrow 0.225$/Å (i.e., maximum real space distance of 628 Å). Depending on the scattering power of the specimen and its dimensions, good SAXS patterns can be recorded in as little as 10 s and typically 1000 s exposure time.

12.6 Applications

Well-controlled aluminosilicate nanostructures were obtained using block copolymers as templates by Prof. Wiesner's group [39] utilizing the natural phase separation in copolymers. The copolymers consisted of blocks of poly-isoprene (PI) and polyethylene oxide (PEO) with varying mol fraction of PEO, and the inorganic phase was synthesized via a sol–gel process. Fig. 12.20A shows the Lorentz-corrected SAXS intensity trace of the nanocomposite PI-b-PEO with $f_{PEO} = 0.13$ and inorganic-to-organic mass ratio of 0.51. The SAXS trace was recorded in step scan mode of $0.006°2\theta$/step and 9 s recording time for a total time of 48 min. The trace exhibits multiple scattering maxima that conform to the scaling $q_i = 1, 5^{1/2}, 7^{1/2}, 13^{1/2}$ relative to the first-order reflection. The maxima scaling implies the existence of hexagonally packed cylindrical microdomains, with long-range spatial order [40].

The *average interdomain distance D* can be determined from the SAXS results, i.e., the domain identity period for lamellae, or the nearest-neighbor distance between the microdomains for cylinders and spheres, or the *average radius R* of cylinders or spheres [40]. The interdomain distance D can be determined from the corresponding Bragg spacings [3]. That is, for lamellar microdomains $D = d_{001}$, for hexagonally packed cylindrical microdomains $D = \sqrt{4/3} \cdot d_{100}$, for the spherical microdomains, with simple cubic (sc) symmetry $D = d_{100}$ and for body-centered cubic (bcc) symmetry $D = \sqrt{3/2} \cdot d_{110}$. The Bragg spacing is determined from the multiple-order diffraction maxima (for instance, for lamellar microdomains $2d_{00l} \cdot \text{Sin}\theta_{00l} = \lambda$, where $l =$ integers and $d_{00l} = D/l$).

The results of the SAXS analysis are summarized in Table 12.5, giving and interdomain spacing of 39.2 nm. These results are compared with those obtained from atomic force microscopy (AFM). Fig. 12.20B shows the corresponding AFM micrograph,

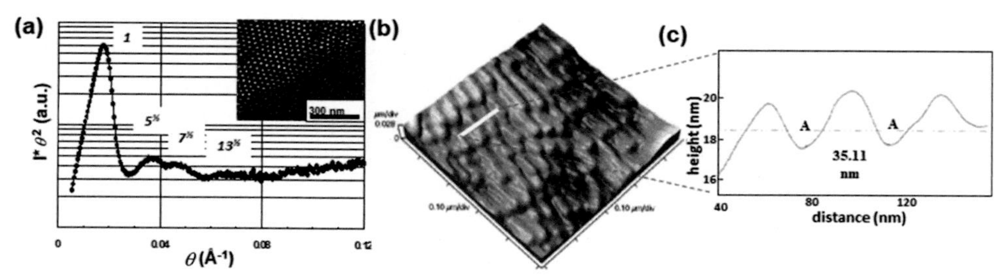

FIGURE 12.20 Hybrid nanocomposite PI-b-PEO with PEO fraction of 0.13 and inorganic-to-organic mass ratio is 0.51. (A) SAXS intensity trace. The periodic nanostructure corresponds to cylindrical morphology, as confirmed by TEM (inset) and AFM. (B) AFM micrograph, and (C) height profile extracted orthogonal to the cylinders' axes (see [21,22] for more details).

Table 12.5 Intensity maxima position (in nanometers) obtained from SAXS patterns.

Intensity maxima	1	$5^{1/2}$	$7^{1/2}$	$13^{1/2}$
Observed	33.9	16.6	12.9	9.2
Theoretical	33.9	15.1	12.8	9.4
First order q (nm^{-1})	d *(nm)*	D_{SAXS} (nm)	Morphology	D_{AFM} (nm)
0.185	$d_{100} = 33.9$	39.2	Cylindrical	35.1

Scaling corresponds to hexagonal cylindrical nanophase. The characteristic nanophase, the spacing *d*, and the interdomain spacing *D* determined by SAXS and by AFM.

which confirmed the nanocylinder morphology. A linear scan across the cylinders is shown in Fig. 12.20C, giving an average size of 35.1 nm cylinder diameter, in excellent agreement with the SAXS analysis. However, note that acquiring the TEM and AFM micrographs required many tedious hours of sample preparation, cryo-microtoming (for TEM) or fracturing in liquid nitrogen (AFM) and patiently scanning the specimen to find the "best" region, and being careful to avoid radiation damage, typical in TEM of polymers and other organic materials.

Varying the PEO fraction in the copolymer and the ratio of inorganic-to-organic phase, the morphology of the nanostructure can be tuned [39]. Fig. 12.21A shows the SAXS pattern and intensity trace of the cast film of PI-b-PEO with $f_{PEO} = 0.15$ and inorganic-to-organic mass ratio of 0.98. The SAXS pattern was acquired with an area detector and recording time of 1000 s (left side) and 10 s (right side) to reveal the three intensity maxima. Note that there is uniformity of the intensity around the azimuth indicating that there is no preferred orientation in the plane of the film. The intensity trace of Fig. 12.21A was extracted from the 2D pattern by averaging around the azimuth ϕ from 0 to 360 degrees. The position of the intensity maxima are integer multiples of the first-order reflection (labeled as 1), indicating the existence of alternating *lamellar* nanodomains with a long-range spatial order. A lamellar morphology results in relative q values of

$$q_i = 1,\ 2,\ 3,\ 4, \ldots. \tag{12.12}$$

The results of the SAXS analysis are summarized in Table 12.6, giving and interdomain spacing of $D = 33.9$ nm. These results are compared with those obtained from AFM. The AFM micrograph shown in Fig. 12.21B confirms the lamellar morphology and the linear scan across the lamellae shown in Fig. 12.21C, yields of $l = 63$ nm, that is $l \approx 2D$, thus confirming the striking regularity of the nanostructure.

Simultaneous WAXD and SAXS with pin-hole collimation system is ideal to investigate the short and long-range structure of nanostructured polymers and elastomers. Fig. 12.22A shows a typical configuration found in the S-Max3000 (Rigaku Inc.) system, which employs Cu Kα radiation typically at 45 kV and 0.88 mA. The WAXD patterns are recorded with flat-plate camera and (Fuji) image plates, with typical sample-to-detector distance of 6 cm, and the SAXS patterns are recorded with an area detector.

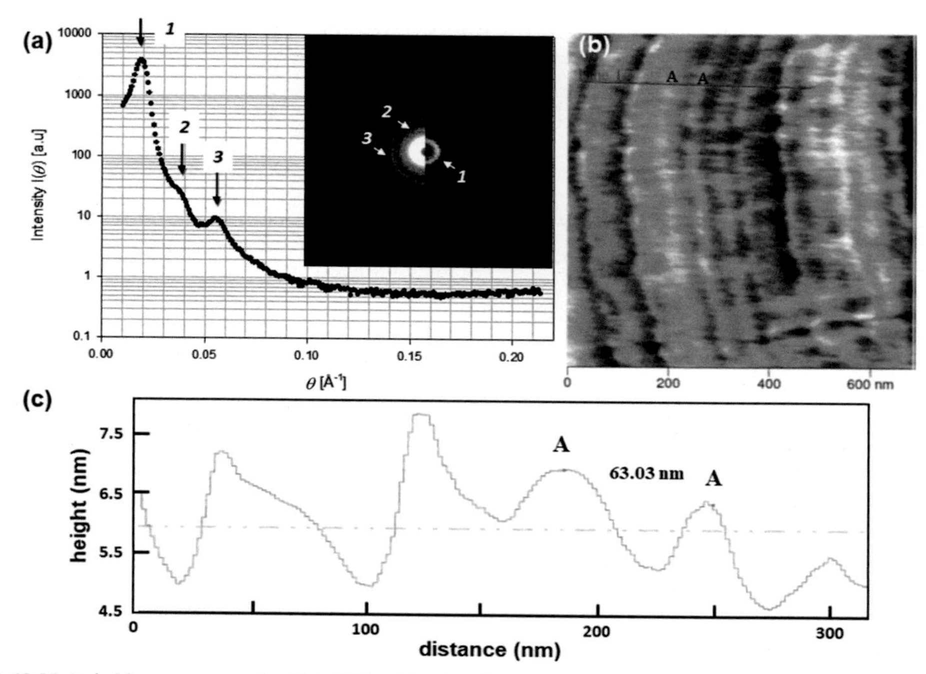

FIGURE 12.21 Hybrid nanocomposite PI-b-PEO with PEO fraction of 0.15 and inorganic-to-organic mass ratio is 0.98. (A) SAXS pattern and azimuthally averaged intensity trace. The periodic position of intensity maxima denotes a lamellar morphology. (B) AFM micrograph, and (C) height profile extracted across the layers, as indicated in (B) (see [21,22] for more details).

Table 12.6 Intensity maxima position (in nanometers) obtained from SAXS data of Fig. 12.19A.

Intensity maxima	**1**	**2**	**3**
Observed	33.9	16.8	11.3
Theoretical	33.9	16.9	11.3
First order q (nm^{-1})	d (nm)	D (nm)	Morphology
0.181	$d_{001} = 33.9$	33.9	Lamellar

Scaling corresponds to lamellar nanophase. The characteristic nanophase, the spacing *d*, and the interdomain spacing *D*.

Fig. 12.22B and C shows the WAXD and SAXS patterns of nanostructured elastomeric networks. These patterns revealed short- *and* long-range order in the elastomers. That is, the WAXD patterns show sharp crystalline rings indicating the presence of crystalline, unoriented, regions in the bulk. The amorphous halo shows the coexistence of the crystalline regions with an amorphous matrix. On the other hand, the SAXS patterns revealed the presence of spatially periodic super structures. Furthermore, the specimens were hot pressed and the processing induced anisotropy, which is exhibited by the SAXS

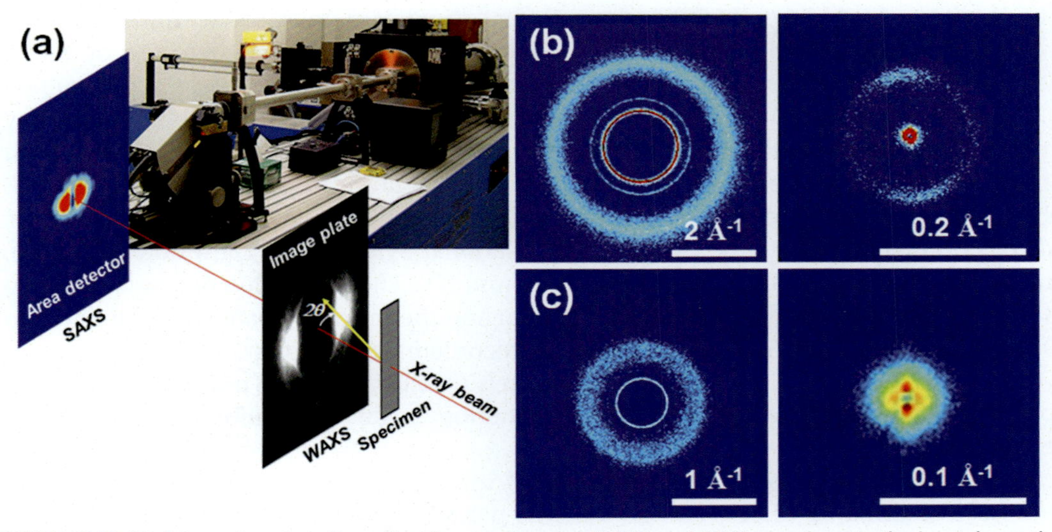

FIGURE 12.22 (A) Schematics of configuration for *simultaneous* WAXD and SAXS experiments, the inset shows the S-Max3000 system (Rigaku Inc.), (B) WAXD (left) and SAXS (right) patterns of elastomeric poly(lactic-co-glycolic acid) (PLGA) and polyhedral oligomeric silsesquioxane (POSS) molecular networks, and (C) WAXD (left) and SAXS (right) patterns of POSS/polyurethane thermoplastic. CuKα and 30 min exposure time.

FIGURE 12.23 (A) Floor at the Advanced Photon Source, Argonne National Laboratory (Lemont, IL, USA). (B) SAXS setting using pin-hole collimation and a mar 165 CCD detector. (C) Alignment system and vacuum chamber covered with polyimide Kapton film.

patterns. Therefore, X-ray diffraction is also ideal to study the microstructure of elastomeric networks with crystallizable components.

12.7 Synchrotron scattering

Synchrotron X-ray scattering is a powerful and unique technique that enables the study of polymeric systems in situ and in *real time*, see Fig. 12.23. Note that the combination of high flux and modern detectors has enabled the study of polymers under shear and extensional deformation [23–25], to elucidate the crystallization phenomena in semicrystalline polymers [26,27] and to characterize the shear-induced orientation in structured complex fluids [16,28–30]. For instance, studies on uniaxially stretched linear low-density polyethylene (LLDPE) showed that there is a martensitic transformation around the second yield point as well as interlamellar shearing, as detected by SAXS. Synchrotron radiation enabled the monitoring of nanostructural changes of strained LLDPE while simultaneously collecting the stress–strain data. This approach removes any ambiguity between bulk and microscopic measurements and enable to establish structure–property correlations. Fig. 12.24 shows synchrotron SAXS patterns of LLDPE while uniaxially deformed, at room temperature. These SAXS patterns were obtained at the Advanced Polymers Beamline (X27C) in the National Synchrotron Light Source (NSLS), Brookhaven National Laboratory (BNL), in Upton, NY. The X-ray radiation was tuned to $\lambda = 1.37$ Å, using a 3-pinhole collimation system and the incident beam was collimated to 600 μm diameter. The SAXS patterns were recorded using a mar 165 CCD detector (MARUSA) having a resolution of 512 × 512 pixels (pixel size = 257.6 μm). The sample-to-detector distance was 170 cm, and the scattering angle was calibrated using silver behenate [31].

Fig. 12.24 shows selected SAXS patterns of the as-molded and undeformed sample and those strained up to 22% and 42%. These patterns show the gradual splitting of the meridional reflections into four off meridional reflections (four-point pattern).

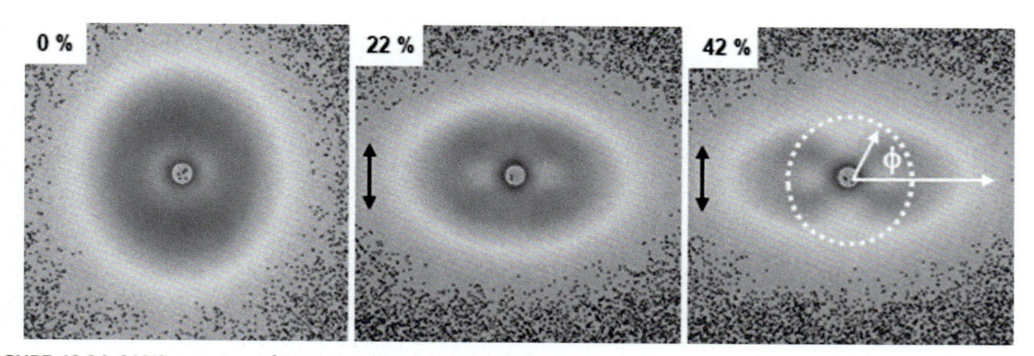

FIGURE 12.24 SAXS patterns of LLDPE strained at 5 mm/min and room temperature. Patterns recorded with a mar area detector, sample-to-detector distance of 1.70 m, 10 s recording time, and using synchrotron radiation (see Ref. [31] for further details).

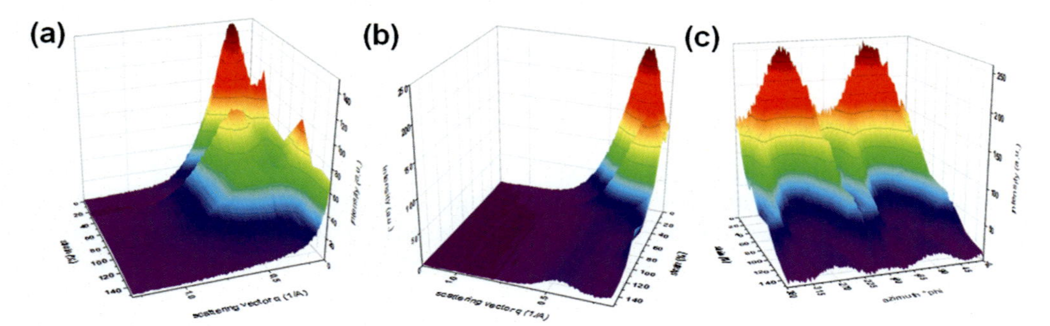

FIGURE 12.25 SAXS intensity maps as a function of strain and scattering vector: (A) equatorial traces, (B) meridional (parallel to extension axis) traces. (C) Intensity traces as a function of strain and azimuthal angle ϕ, at constant $q = 0.292$/nm. Uniaxial stretching at 5 mm/min and room temperature (see Ref. [31] for further details).

Furthermore, the tensile deformation induced a rotation of the off meridional reflections toward the equatorial axis. Hence, the azimuthal angle between the meridional reflections and the equatorial axis decreased. The influence of strain on the nanostructure evolution is clearly appreciated from the 3D SAXS maps shown in Fig. 12.25. The maps were constructed using the intensity traces extracted along the equatorial and meridional axes and around the azimuth, at constant q. It has been suggested that the transformation of the meridional reflection into off-meridional reflections originates from exhaustion, or "lock" of the interlamellar shear due to almost complete extension of the chain segments immobilized by adjacent crystals and by trapped entanglements. For a complete discussion of results, see Romo-Uribe et al. [31].

12.8 Conclusions

The hope to establish structure–property correlations and further the development and innovation of elastomeric (nano) composites requires detailed microstructure analysis and determination of the state of dispersion of fillers in the elastomeric matrix. That is, the fundamental understanding of the molecular mechanisms associated to changes in physical properties is a nonnegotiable requirement. It is of no use to report only a polymer-filler system and some changes in properties without understanding the fundamentals associated to the changes in those properties. Today is known that (nano) fillers induce confinement effects and the chain–filler interactions cannot be ruled out. Hence, the chain dynamics can be modified [13,32]. This chapter provided a practical overview of the utilization of X-ray scattering with broad spatial resolution from Å-scale to nm-scale and discussed structural features extracted from elastomeric materials reinforced with nanoparticles of different topology, spherical-like and plate-like. Tedious mathematical treatment of data was on purpose left out as those are well discussed in several textbooks listed in the Suggested Reading section. Here it was shown that rapid inspection of X-ray scattering data provides information on materials properties, for instance crystallinity, long-range order, or anisotropy, and these features can easily be correlated with the synthesis, processing, or end-use conditions. An unsurpassable advantage is that X-ray scattering is a nondestructive technique, involves negligible (if any) sample preparation and its applications in (polymeric) materials are very broad as it relies on electron density differences in materials. Furthermore, most organic materials are transparent to this radiation. The versatility of X-ray scattering on materials characterization was recognized since its discovery, over 100 years ago. Here it was shown that the fundamental principles of scattering and instrumentation have not changed over a century, rather more powerful and stable X-ray sources and detectors have been developed. Finally, the striking versatility of X-ray scattering is evident in the implementation of time-resolved measurements and its combination with other techniques to enable hyphenated techniques [8–10,21,22,33–38].

References

[1] P.P. Ewald (Ed.), Fifty Years of X-Ray Diffraction, International Union of Crystallography, The Netherlands, 1962.

[2] B.K. Vainshtein, Diffraction of X-Rays by Chain Molecules, Elsevier Publishing Company, Amsterdam, 1966.

[3] L.E. Alexander, X-ray Diffraction Methods in Polymer Science, John Wiley and Sons, London, 1969.

[4] J.A. Golovchenko, R.A. Levesque, P.L. Cowan, X-ray monochromator system for use with synchrotron radiation sources, Rev. Sci. Instrum. 52 (1981) 509–516, https://doi.org/10.1063/1.1136631.

[5] M.C. Morris, H.F. McMurdie, E.H. Evans, B. Paretzkin, J.H. de Groot, B.S. Weeks, N.J. Newberry, Standard X-Ray Diffraction Powder Patterns, NBS, Washington DC, 1978.

[6] S.E. Keinath, R.L. Miller, J.K. Rieke (Eds.), Order in the Amorphous "State" of Polymers, Plenum Press, New York, 1987.

[7] A. Romo-Uribe, Scattering and other miscellanies techniques for the characterization of shape memory polymers, in: P. Jyotishkumar (Ed.), Shape Memory Polymers, Blends and Composites - Advances and Applications, Springer, 2020, https://doi.org/10.1007/978-981-13-8574-2_12.

[8] P. Cebe, B.S. Hsiao, D.J. Lohse (Eds.), Scattering from Polymers, ACS Symp. Series, vol. 739, Oxford University Press, New York, 2000.

[9] A.I. Nakatani, M.D. Dadmun (Eds.), Flow-induced Structure in Polymers, ACS Symposium Series, vol. 597, 1995 (Washington DC).

[10] K. Sondegaard, J. Lyngaae-Jorgensen (Eds.), Rheo-physics of Multiphase Polymer Systems, Technomic, Lancaster, 1995.

[11] A. Guinier, G. Fournet, Small-angle X-Ray Scattering, John Wiley and Sons, New York, 1955.

[12] W.O. Statton, Small-angle X-ray studies of polymers, in: B. Ke (Ed.), Newer Methods of Polymer Characterization, Interscience Publishers, New York, 1964, p. 231.

[13] A. Romo-Uribe, Montmorillonite nanoclay aggregated in a fractal structure in acrylic-styrene coatings, slowed the chain dynamics and increased an order of magnitude the tensile modulus, Polym. Adv. Technol. (2021) 1–13. https://doi.org/10.1002/pat.5321.

[14] E.M. Anitas, Small-angle scattering from fractals: differentiating between various types of structures, Symmetry 12 (2020) 65. https://doi.org/10.3390/sym12010065.

[15] G. Beaucage, Small-angle scattering from polymeric mass fractals of arbitrary mass-fractal dimension, J. Appl. Crystallogr. 29 (1996) 134–146.

[16] I.W. Hamley, V. Castelletto, O.O. Mykhaylyk, A.J. Gleeson, Mesoscopic crystallography of shear-aligned soft materials, J. Appl. Crystallogr. 37 (2004) 341.

[17] W.E. Danielson, L. Shenfil, J.W.M. du Mond, Latex particle size determination using diffraction peaks obtained with the point focusing X-ray monochromator, J. Appl. Phys. 23 (1952) 860–865.

[18] F. Cser, About the Lorentz correction used in the interpretation of small angle X-ray scattering data of semicrystalline polymers, J. Appl. Polym. Sci. 80 (2001) 2300–2308.

[19] O. Glatter, O. Kratky (Eds.), Small Angle X-Ray Scattering, Academic Press, New York, 1982.

[20] A. Romo-Uribe, A. Reyes-Mayer, M. Calixto-Rodriguez, R. Benavente, M. Jaffe, Synchrotron scattering and thermo-mechanical properties of high performance thermotropic polymer. A multi-scale analysis and structure-property correlation, Polymer 153 (2018) 408–421.

[21] A. Romo-Uribe, On the molecular orientation and viscoelastic behaviour of liquid crystalline polymers. The influence of macromolecular architecture, Proc. Roy. Soc. Lond. A 457 (2001a) 207–229.

[22] A. Romo-Uribe, Smectic-like order in the log-rolling flow of thermotropic random copolymers. A time-resolved wide-angle X-ray scattering study, Proc. Roy. Soc. Lond. A 457 (2001b) 1327–1342.

[23] P.K. Agarwal, R.H. Somani, W. Weng, A. Mehta, L. Yang, S. Ran, L. Liu, B.S. Hsiao, Shear-Induced crystallization in novel long chain branched polypropylenes by in situ rheo-SAXS and −WAXD, Macromolecules 36 (2003) 5226.

[24] M.F. Butler, A.M. Donald, A.J. Ryan, Time resolved simultaneous small- and wide-angle X-ray scattering during polyethylene deformation- II. Cold drawing of linear polyethylene, Polymer 39 (1998) 39.

[25] S. Ran, D. Fang, X. Zong, B.S. Hsiao, B. Chu, P.M. Cunniff, Structural changes during deformation of Kevlar fibers via on-line synchrotron SAXS/WAXS techniques, Polymer 42 (2001) 1601−1612.

[26] L. Liu, B.S. Hsiao, X. Fu, X. Ran, S. Toki, B. Chu, A.H. Tsou, P. Agarwal, Structure changes during uniaxial deformation of ethylene-based semicrystalline ethylene-propylene copolymer. 1. SAXS study, Macromolecules 36 (2003) 1920.

[27] A.J. Ryan, Simultaneous small-angle X-ray scattering and wide-angle X-ray diffraction. A powerful new technique for thermal analysis, J. Therm. Anal. 40 (1993) 887−899.

[28] G.R. Mitchell, J.A. Pople, E.M. Andresen, P.G. Brownsey, Adv. X-ray Anal. (1997) 37.

[29] A. Romo-Uribe, P.T. Mather, K. Chaffee, C.D. Han, Molecular and textural ordering of thermotropic polymers in shear flow, MRS Symp Proc 461 (1997) 63−68.

[30] L. Soubiran, E. Staples, I. Tucker, J. Penfold, Effects of shear on the lamellar phase of diakyl cationic surfactant, Langmuir 17 (2001) 7988.

[31] A. Romo-Uribe, A. Manzur, R. Olayo, Synchrotron small-angle X-ray scattering study of linear low density polyethylene under uniaxial deformation, J. Mater. Res. 27 (2012) 1351−1359.

[32] A. Romo-Uribe, Dynamics and viscoelastic behavior of waterborne acrylic polymer/silica nano-composite coatings, Prog. Org. Coating 129 (2019) 125−132.

[33] A. Romo-Uribe, A.H. Windle, Flow-induced orientational transition in thermotropic random copolyesters, Macromolecules 26 (1993) 7100−7102.

[34] A. Romo-Uribe, A.H. Windle, Log-rolling alignment in main-chain thermotropic liquid crystalline polymers: an in-situ WAXS study, Macromolecules 29 (1996) 6246−6255.

[35] A. Romo-Uribe, A.H. Windle, A rheo-optical and dynamic X-ray scattering study of flow-induced textures in main-chain thermotropic polymers: the influence of molecular weight, Proc. Roy. Soc. Lond. A 455 (1999) 1175−1201.

[36] A. Romo-Uribe, Long-range orientation correlations and molecular alignment in sheared thermo-tropic copolyester. In-situ light and X-ray scattering, Polym. Adv. Technol. 18 (7) (2007a) 503−512.

[37] A. Romo-Uribe, Hybrid-block copolymer nanocomposites. Characterization of nanostructure by small-angle X-ray scattering (SAXS), Rev. Mexic. Fisica 53 (2007b) 171−178.

[38] W. Wilke, M. Bratrich, B. Heise, G. Peichel, The change of the superstructure of semicrystalline polymers during deformation: results from small-angle scattering with synchrotron radiation, Polym. Adv. Technol. 3 (1992) 179−190.

[39] M. Templin, A. Franck, A. Du Chesne, H. Leist, Y. Zhang, R. Ulrich, V. Schadler, U. Wiesner, Organically modified aluminosilicate mesostructures from block copolymer phases, Science 278 (1997) 1795.

[40] I.W. Hamley, Block Copolymers, Oxford University Press, England, 1998.

Further reading

[1] R. Hoseman, S.N. Bagchi, Direct Analysis of Diffraction by Matter, North-Holland, Amsterdam, 1962.

[2] M. Kakudo, N. Kasai, X-ray Diffraction by Polymers, Elsevier, Amsterdam, 1972.

[3] H.P. Klug, L.E. Alexander, X-ray Diffraction Procedures, second ed., Wiley, New York, 1974.

[4] H. Tadokoro, Structure of Crystalline Polymers, Wiley, New York, 1979.

[5] B. Alvarado-Tenorio, A. Romo-Uribe, P.T. Mather, Microstructure and phase behavior of POSS/PCL shape memory nanocomposites, Macromolecules 44 (2011) 5682–5692.

[6] B. Alvarado-Tenorio, A. Romo-Uribe, P.T. Mather, Stress-induced bimodal ordering in POSS/PCL biodegradable shape memory nanocomposites, MRS Symp Proc 1450 (2012). https://doi.org/10.1557/opl.2012.1327.

[7] B. Alvarado-Tenorio, A. Romo-Uribe, P.T. Mather, Nanoscale anisotropic orientation in shape memory random POSS/Polycaprolactone nanocomposites, MRS Symp Proc 1453 (2013). https://doi.org/10.1557/opl.2013.1117.

[8] B. Alvarado-Tenorio, A. Romo-Uribe, P.T. Mather, Nanoscale order and crystallization in POSS-PCL shape memory molecular networks, Macromolecules 48 (2015) 5770–5779.

[9] V. Castelleto, I.W. Hamley, Capillary flow behavior of worm-like micelles studied by small- angle X-ray scattering and small angle light scattering, Polym. Adv. Technol. 17 (2006) 137–144.

[10] B. Chu, B.S. Hsiao, Small-angle X-ray scattering of polymers, Chem. Rev. 101 (2001) 1727–1761.

[11] T. Chung, A. Romo-Uribe, P.T. Mather, Two-way reversible shape memory in a semicrystalline network, Macromolecules 41 (2008) 184–192.

[12] S. Hanna, A. Romo-Uribe, A.H. Windle, Sequence segregation in molten liquid-crystalline random copolymers, Nature 366 (1993) 546–549.

[13] E. Huitron-Rattinger, K. Ishida, A. Romo-Uribe, P.T. Mather, Thermally modulated nanostructure of poly(ε-caprolactone)–POSS multiblock thermoplastic polyurethanes, Polymer 54 (2013) 3350–3362.

[14] J. Janicki, Time-resolved small-angle X-ray scattering and wide-angle X-ray diffraction studies on the nanostructure of melt-processable molecular composites, J. Appl. Crystallogr. 36 (2003) 986–990.

[15] S. Kumar, S. Werner, D.T. Grubb, W. Adams, On the small-angle X-ray scattering of rigid-rod polymer fibres, Polymer 35 (1994) 5408–5412.

[16] X. Luo, P.T. Mather, Triple-shape polymeric composites (TSPCs), Adv. Funct. Mater. 20 (2010) 2649–2656.

[17] X. Luo, P.T. Mather, Shape memory assisted self-healing coating, ACS Macro Lett. 2 (2013) 152–156.

[18] X. Luo, R. Ou, D.E. Eberly, A. Singhal, W. Vyratyaporn, P.T. Mather, A thermoplastic/thermoset blend exhibiting thermal mending and reversible adhesion, ACS Appl. Mater. Interfaces 1 (2009) 612–620.

[19] U. Nochel, K. Kratz, M. Behl, A. Lendlein, Relation between nanostructural changes and macroscopic effects during reversible temperature-memory effect under stress-free conditions in semicrystalline polymer networks, MRS Symp Proc 1718 (2015). https://doi.org/10.1557/opl.2015.427.

[20] W. Ruland, X-ray determination of crystallinity and diffuse disorder scattering, Acta Crystallogr. 14 (1961) 1180–1185.

[21] A. Soto-Quintero, A. Meneses-Acosta, A. Romo-Uribe, Tailoring the viscoelastic, swelling kinetics and antibacterial behavior of poly(ethylene-glycol)-based hydrogels with polycaprolactone, Eur. Polym. J. 70 (2015) 1–17.

[22] D.L. Thomsen III, P. Keller, J. Daciri, R. Pink, H. Jeon, D. Shenoy, B.R. Ratna, Liquid crystal elastomers with mechanical properties of a muscle, Macromolecules 34 (2001) 5868–5875.

[23] A.H. Torbati, H. Birjandi Nejad, M. Ponce, J.P. Sutton, P.T. Mather, Properties of triple shape memory composites prepared via polymerization-induced phase separation, Soft Matter 10 (2014) 3112–3121.

Theoretical modeling and simulation of elastomer blends and nanocomposites

Jitha S. Jayan[1], B.D.S. Deeraj[2], Appukuttan Saritha[1], Kuruvilla Joseph[2]

[1]DEPARTMENT OF CHEMISTRY, SCHOOL OF ARTS AND SCIENCES, AMRITA VISHWA VIDYAPEETHAM, AMRITAPURI, KOLLAM, KERALA, INDIA; [2]DEPARTMENT OF CHEMISTRY, INDIAN INSTITUTE OF SPACE SCIENCE AND TECHNOLOGY, VALIAMALA, THIRUVANANTHAPURAM, KERALA, INDIA

13.1 Introduction

The computational approaches generally enable the predictions on the dynamics, microscopic and macroscopic material properties, and macromolecular structure thermodynamics [1]. Thus with the advancement of computer simulations, the polymer simulations can be carried out in determining various properties of elastomers. Elastomers are a special class of polymers having good elastic property [2–6]. They are loosely cross-linked polymers with advanced functional applications in various fields [7–11]. Rubbers, polyisoprene, styrene-butadiene block copolymers, ethylene propylene rubber, polybutadiene, silicone elastomers, and ethylene propylene diene rubber fall under the category of elastomers [12–16].

The complexity of the elastomers is considered as one of the major challenges faced by the computational scientists [14,17–20]. The improvement in computer hardware, the development of new algorithms and software helped in nullifying or addressing the few challenges in the field of elastomer simulation and modeling [21–25]. Nowadays, even the smart phones can be used for the modeling of polymers, and thus more advancements are observed in the field of elastomer modeling [26–28]. So the power of simulations is just in a distance of one touch or at our fingertips. The growing number of simulation studies in the field of elastomers help in giving new insights into the structure, property, and structure–property relations [29–35].

Simulations are considered as an easy and unreal tool for connecting the contrast to theory [1,36,37,37a]. It gives an idea about the microscopic changes associated with macroscopic problems [38–42]. In order to face the challenges in the polymer

Elastomer Blends and Composites. https://doi.org/10.1016/B978-0-323-85832-8.00009-2

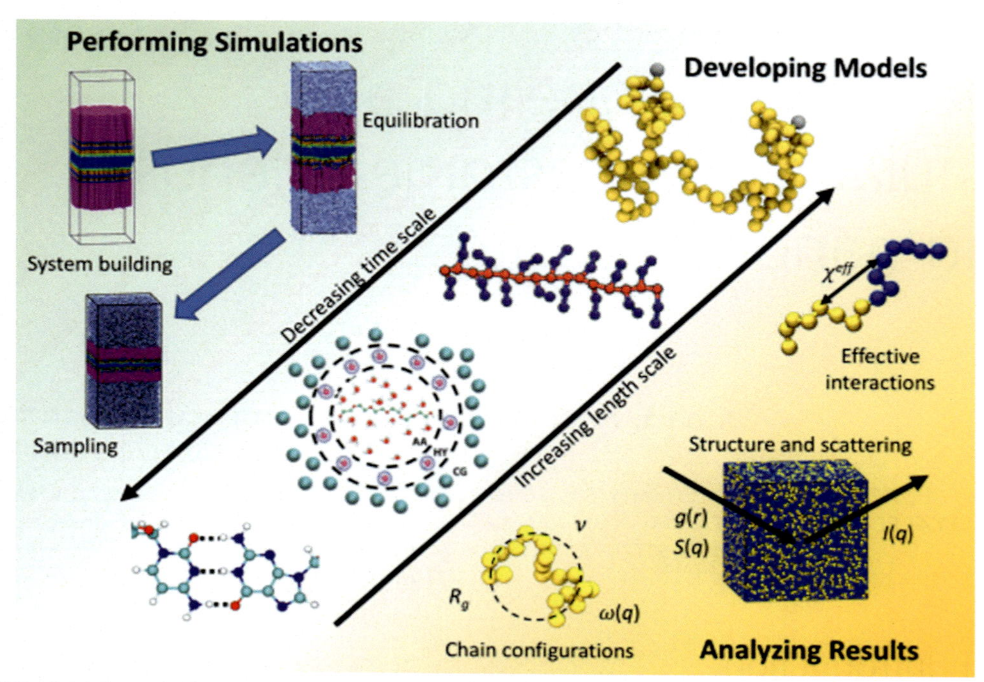

FIGURE 13.1 Schematic illustrating many of the key steps in performing a polymer simulation. *From T.E. Gartner, A. Jayaraman, Modeling and simulations of polymers: a roadmap, Macromolecules 52(3) (2019) 755–786. https://doi.org/10.1021/acs.macromol.8b01836.*

processing and property analysis, it is indeed very essential to have knowledge about the theoretical simulations and modeling [20,43–45]. The general steps in the simulation studies of polymer are represented in Fig. 13.1. Thus simulation and modeling are considered to be effective tools for understanding elastomer-related problems [46–53]. Generally, modeling or simulations belong to two major classes such as atomistic or coarse-grained types. An understanding of monomer-level arrangements, orientations, rearrangements, interactions with the polymer chains, and fluctuations are essential in carrying out the modeling and simulation studies in detail [54–57].

In elastomer-based composites and blends, the interaction of chains with the fillers, cross-link density, Poisson's ratio, the entanglement, and other properties of the fillers, etc., have to be considered [58–64]. This chapter tries to bring into light the different simulation studies and modeling that are generally used in determining the properties of elastomers-based blends and composites. This chapter also portrays different challenges associated and future scope of simulation and modeling studies of elastomers.

13.2 Simulations of elastomers

13.2.1 Thermoplastic elastomers

Thermoplastic elastomers are advanced materials having similar behavior as that of rubber at ambient conditions along with plastic such as processability behavior [65,66].

Thermoplastics derived from rubber blends can have the coexistence of rubber and thermoplastic domains [67,68]. The advancement in computational techniques helps in determining the properties, miscibility, and phase behavior of such complex elastomers [69–71]. Recent advances in the computational field help in predicting the thermoplastic properties with the help of Molecular Dynamics (MD) methods [72–74]. This method helps in connecting macro and microscopic properties by the utilization of statistical methods based on mathematical expressions [75–77].

Atomistic stimulation is a kind of simulation method used for the prediction of properties of elastomeric blends, which gives more priority to the interface [78]. New algorithms can be formulated aiming at the prediction of properties and can be conveniently utilized for comparing the experimental data [79]. Thus active research is taking place in the formulation and implementation of theoretical stimulation studies for thermoplastics. Dissipative particle dynamics (DPD) is another method in which atoms are grouped together up to a particular length of a polymer chain and can be extended by in length and timescale to several higher magnitudes compared to atomistic stimulation [80–82].

Recently, Sahu et al. [68] conducted atomistic stimulation study of thermoplastic blend and studied the compatibility and atomic-level interactions and later compared it with the experimental results and Flory–Huggins parameter. They have also utilized the DPD method for predicting the morphology. In DPD simulation method, atoms are considered as beads, and the simulation is carried out in a cubic simulation box. The immiscibility of the blended systems is clearly understood from the simulation as the domains of both the constituents are present. Fig. 13.2 represents the morphology obtained from the DPD analysis. Glass transition temperature was analyzed with the help of atomistic stimulation method, and the Tg of the thermoplastic elastomer (TPE) blends was in good agreement with the experimental value.

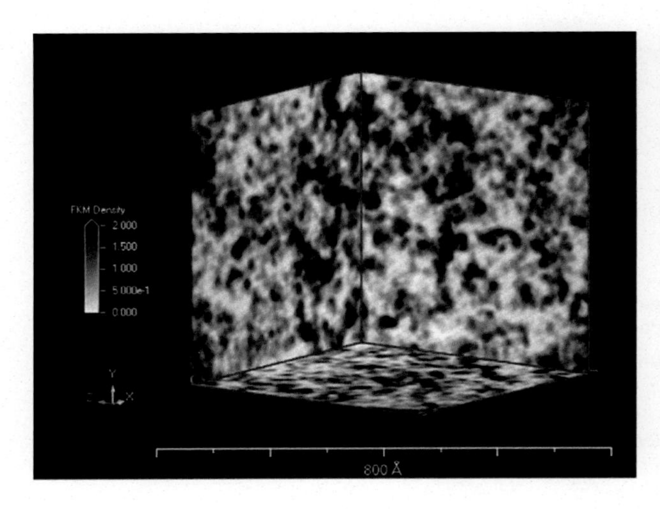

FIGURE 13.2 Morphology of 40/60 blend of PA6/FKM, FKM density profile after 50,000 steps DPD simulation; PA6: white, FKM: black. *From S. Saha, A.K. Bhowmick, Computer simulation of thermoplastic elastomers from rubber–plastic blends and comparison with experiments, Polymer 103 (2016) 233–242. https://doi.org/10.1016/j.polymer.2016.09.065.*

Thus atomistic MDs can be considered as a promising method for finding the miscibility of TPE-based blends. This gives a correlation between microscopic and macroscopic properties. Calculation of mean square displacement is possible from the MD method using Eq. (13.1), which helps in understanding the MD of the TPE blends.

$$MSD = < |r(t) - r(0)|^2 >= 6D \quad (13.1)$$

where position of center of mass of polymer chains at time t and 0 is represented by $r(t)$ and $r(0)$, and slope of the curve represents the diffusion coefficient D. Usually steeper curves are obtained for the pristine polymer indicating that the mobility is higher, whereas in case of blends, the presence of other constituents restricts the mobility of the chains. As shown in Fig. 13.3, the HNBR (Hydrogenated Nitrile Rubber) chains restrict the mobility of PVDF(Poly(vinylidene fluoride)) membranes [83].

Mechanistic understanding about the source of the deformation properties of thermoplastic Polyurethanes (TPU) is very essential because the molecular-level properties often remain mysterious due to their complex and heterogeneous structure. Hence the simulation studies can be utilized for the analysis of mechanical response under tensile deformation. Mostly, macroscopic continuum models such as numerical models are used to explain the deformation behaviors. These models suggest the softening mechanism behind the deformation process [84,85]. The softening effect called "Mullins effect" is considered as an outstanding property of polymers, but the molecular origin of this effect remains unclear. Qi et al. [84] proposed that the soft chains are getting activated during the deformation process. But the possibility of breakdown of hyper elastic polymer networks is also predicted, and these breakdowns are considered irreversible. This is further confirmed from the associated changes in the X−ray patterns during the deformation. But the "fuzzy interface" between the hard and soft segments is found responsible for the structural changes during the deformation process, and the molecular mechanism behind the Fuzzy interface remains unclear [86,87].

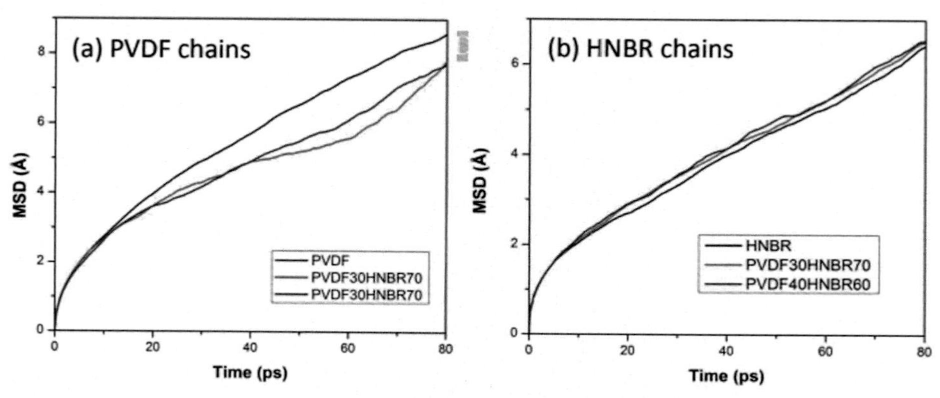

FIGURE 13.3 Mean square displacements of (A) PVDF chain and (B) HNBR chain in the pristine polymers as well as in the blends. *From S. Saha, A.K. Bhowmick, Computer aided simulation of thermoplastic elastomer from poly (vinylidene fluoride)/hydrogenated nitrile rubber blend and its experimental verification, Polymer 112 (2017) 402–413. https://doi.org/10.1016/j.polymer.2017.02.035.*

In spite of several theoretical and experimental works, the atomistic origin of the softening and the molecular deformation mechanisms are still unclear. Hence recently, Zhu et al. [88] tried to give insights into the molecular-level deformation of complex TPU systems using atomistic Interphase Monte Carlo (IMC) method and MD simulations. In the IMC method, both the connectivity and the molecular packaging of soft and hard domains are respected. In the IMC, they used unbiased sampling from the topology distribution, which is in equilibrium, and the configurations are then analyzed to quantify the topological features, and it is shown in Fig. 13.4, which shows the hard and soft domains in the form of crystalline lamellae and amorphous layer. Fig. 13.4D shows the possible bridges, and entanglements between loops and the other side without bridge are shown in Fig. 13.4C. They observed that during the large-scale deformation, different molecular-level mechanisms were observed, which are either recoverable or not. They observed that the destruction of hard domains is responsible for the shear band formation, block slip, localized melting and pulling out of hard segments. Adams et al. [89] established a novel method for characterizing the properties of elastomers and accurately implementing the finite element simulation for additive manufacturing.

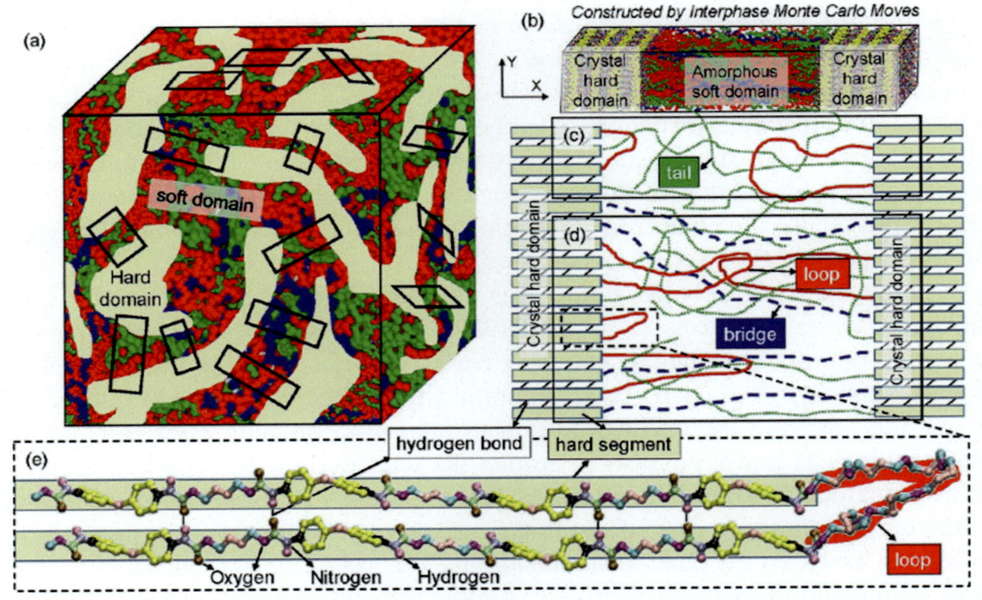

FIGURE 13.4 (A) Schematic showing a typical morphology of a TPU, (B) molecular configuration after Interphase Monte Carlo and molecular dynamics equilibration, (C) a system characterized without bridges, (D) a system characterized with bridges and (E) united atom model. *From S. Zhu, N. Lempesis, P. J. in't Veld, G.C. Rutledge, Molecular simulation of thermoplastic polyurethanes under large tensile deformation, Macromolecules 51(5) (2018) 1850–1864. https://doi.org/10.1021/acs.macromol.7b02367.*

13.2.2 Thermosetting elastomers

Thermosetting resins are having applications in different fields, and in order to reduce the experimental efforts in the synthesis and property optimization, computer simulation studies are very essential [90,91]. Monte-Carlo and MD simulations are widely used for thermosetting elastomers [91a,91b]. These methods are very helpful in predicting the behavior of polymer networks, and this method tries to correlate the network structure and physical properties [92,93]. The microscopic-level interpretation of properties can be comprehended with the help of these kinds of atomistic models. There are different simulation methods for predicting the properties of cross-linked networks. The first study on the polymer network was carried out in 1990 by representing the chains in the form of bead and spring [94]. Later in the mid-1990s, MD atomistic stimulations were carried out [95].

Using these models, they predicted both the elastic modulus and Tg of polymers containing about 200 atoms in its structure [95]. It is also possible to carry out the simulations at the time of polymerization as well. In this method, the monomers are packed and equilibrium is attained, and later it allows the repetitive cross-linking reactions to carry out until a definite conversion is attained, and the final step involves the energy minimization for the relaxation of the actual final structures [96]. This method is still in use, or it is being used widely by modifying the procedure or interaction descriptions. A cross-linking technique was utilized in which all the reactions related to cross-linking are considered to take place in one step, and the calculated conversion degrees are lower than the actual ones [97]. The relation between structure and elastic moduli can be studied in detail with the help of MD [98]. Later studies were conducted to understand the reaction involving between two different polymer networks using DREIDING force field; this ensures the systematic evaluation of structure–property relation as well as prediction of various structural factors [99]. The final network structure of the physically and chemically cross-linked polymer is shown in Fig. 13.5, which is a system-generated MD-level stick–ball model.

In a study, network polymerization was carried out by Coarse-Grained (CG) and atomistic MDs were used to map the model [100]. The possibility of formation of highly cross-linked thermoset polymer networks was a matter of study using molecular modeling. Varshney et al. [101] used different approaches for predicting cross-linked structure and observed the effect of multistep relaxation procedure for relaxing newly molded topology. It is also possible to predict the reaction between polymer cross-links using a simulated annealing algorithm, in this method after optimizing the parameters cross-linking reactions are carried out in a single step [102]. In certain studies, periodic cross-linking based on the chain distance followed relaxation is carried out [103,104]. In another studies, the breakage of existing bonds and the formation of new bonds are equally considered [97,105]. Evolution of atomic charges was involved during the polymerization reaction due to the electrostatic interactions. Hence in a precise simulation method, these charges should be accounted. Generally electronegativity

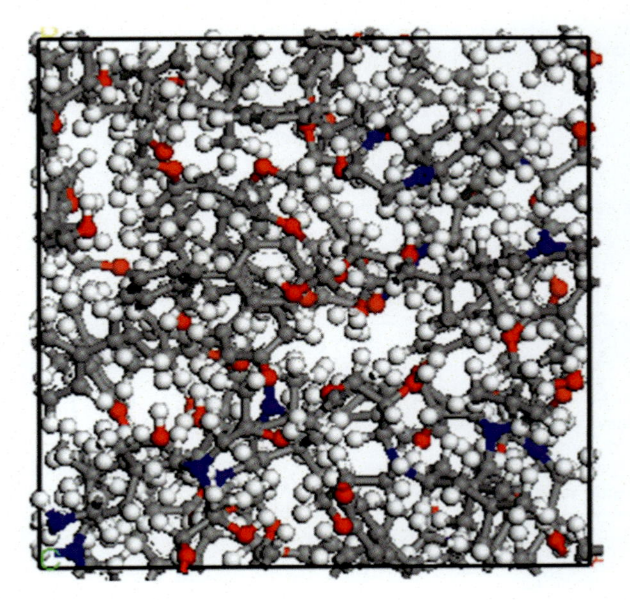

FIGURE 13.5 Molecular model system generated according to our methods: stick–ball model, balls with different colors represent different elemental atoms (red for oxygen, gray for carbon, white for hydrogen, blue for nitrogen). *From C. Wu, W. Xu, Atomistic molecular modeling of crosslinked epoxy resin, Polymer 47(16) (2006) 6004–6009. https://doi.org/10.1016/j.polymer.2006.06.025.*

equalization method (EEM) [106–108] or charge equilibration (QEq) [109] methods are followed. Charge update scheme based on ab initio methods is followed generally for small-molecule topology [101]. It is also possible to maintain the neutrality by assigning charges during the bond breakage and formation steps [105], and it is also possible to calculate these charges by EEM methods as described earlier [110,111]. Thus charge update procedure can be generated based on the computational techniques; hence the Materials Processes and Simulations (MAPS) were developed. The general steps in the cross-linking method are shown in Fig. 13.6.

Nowadays, thermosets for specially targeted coating applications are simulated using MD. In this study, the molecular interactions were parameterized by statistical associating fluid theory (SAFT) [112,113]. The kinetic data were devised using a theoretical model fit and cross-linking algorithms are implemented [114]. The Tg and volumetric thermal expansions of thermoset-based nanocomposites can be studied using MD. Condensed-phase optimized molecular potentials for atomistic simulation (COMPASS27) force field were used to build a matrix for representing the thermosets [115]. It is very important to note that the computational studies can be utilized to examine the properties of bio-based precursors, which are used to synthesize thermoset polymers. MD simulations were used for the generation and testing of the resin samples. Effect of the functionalities present on the bio-derived prepolymer on the formation of network was also investigated. After the calculation of partial charges, the monomer and hardener molecules were mixed in the stoichiometric ratios, and then PACKMOL software package was used for analysis. Then the mixture is subjected to geometry

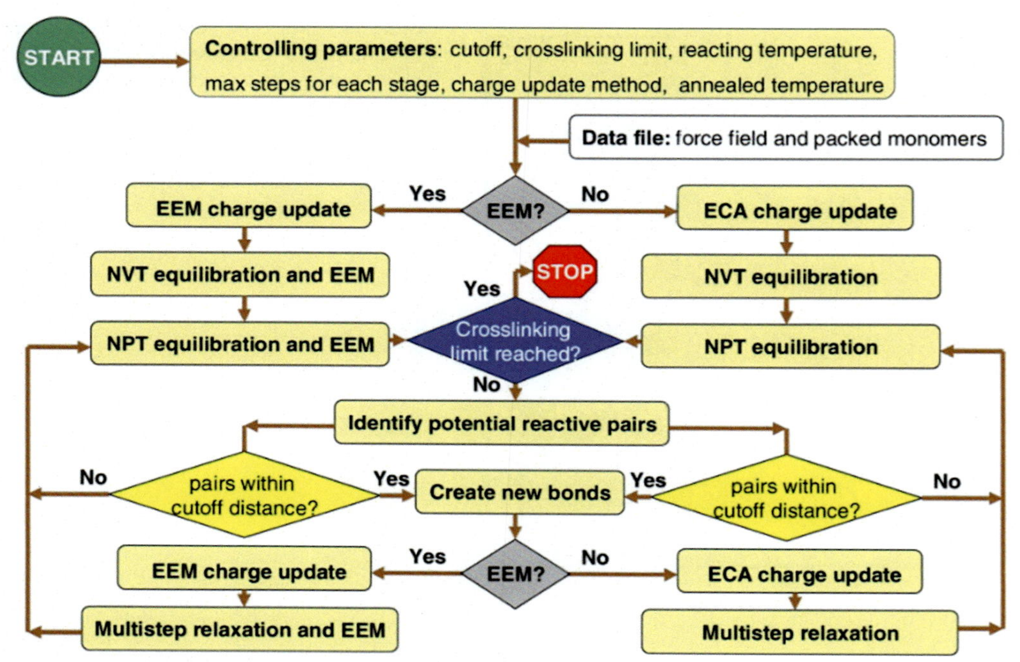

FIGURE 13.6 Flowchart of the MD cross-linking process. *From C. Li, A. Strachan, Molecular simulations of cross-linking process of thermosetting polymers, Polymer 51(25) (2010) 6058–6070. https://doi.org/10.1016/j.polymer.2010.10.033.*

optimization [116]. Thus the possible interactions between the monomer and hardener can be predicted [117]. Proximity-based MDs enable the curing simulation in a realistic way by incorporating cutoff distance, reaction probability, and cure temperature into the Arrhenius equation [118]. Recently, mechanical and energy storage properties of shape memory polymers are also analyzed using the MDs [119]. Simulation methodology was utilized for analyzing the properties of vitrimers by combining CG MDs and Monte Carlo method. This helps to capture the integrity of network and flowability of vitrimers at different temperatures. Thus rheological and volumetric properties of the vitrimers can be predicted by the simulation studies [120].

13.3 Modeling study of elastomer blends and composites

13.3.1 Thermal modeling

As described earlier, modeling and simulation correlate various structure–property relations. It is noted that the thermal response of elastomers is a critical area of study. Recently, in one of our studies we have utilized thermal modeling studies for the analysis of kinetics of glass transitions using two different models such as Kissinger and

Moynihan's model. Polyethylene Glycol (PEG)-based epoxy blends showed more proximity toward the Moynihan's model. This was the first reported study on the kinetics of glass transition of PEG-based epoxy blends [121,122]. The fitting curves are shown in Fig. 13.7.

Cai et al. [123] recently combined stimulation and experiment to study the thermal conductivity (TC) of graphene-oxide-based SBR composites. It was observed that the TC of the polymer generally depends on three main factors such as the nature of the matrix, the nature of filler, and compatibility and interaction between the matrix and the filler. So these factors have to be considered in the molecular modeling studies [124–126]. Generally the nonequilibrium molecular dynamic (NEMD) simulation methods are used for sandwich-like structures. From the simulation studies, Cai et al. [123] observed that the surface defects reduce the TC value, but the cover ratio helps in enhancing the same. Yang et al. [127] derived a relation between the Tg and cross-linking and degree of conversion. Later, Shenogina et al. [128,129] used Debenedetto equation for analyzing the Tg of thermoset-based elastomers using MD and COMPASS force field. They have used Accelrys program to build the network of the thermoset elastomers. Xie et al. [130] used simulation for the analysis of the cross-linked rubber networks using CG model, which ensures the presence of hard fillers in the network. Cross-linking density is calculated by dividing the number of particles forming cross-linking bond with the total number of particles where chain entanglements are not counted. From Fig. 13.8, it can be clearly understood that the Tg value is associated with the cross-linking density, and the heat buildup phenomenon is high, and it is decreasing linearly with the cross-linking density up to 13%, and later it increases linearly.

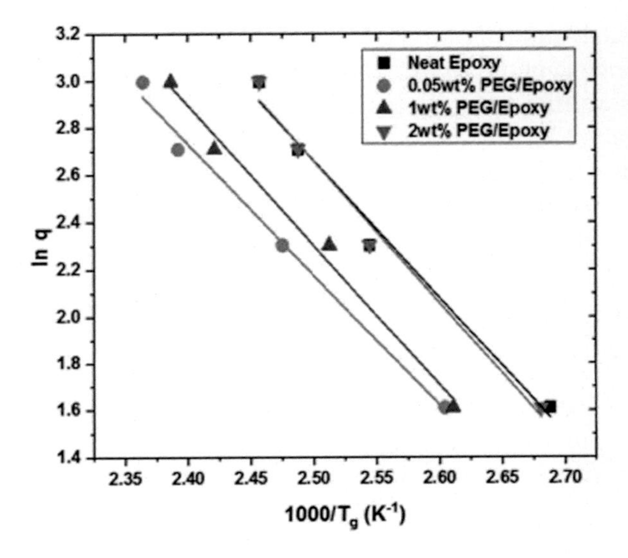

FIGURE 13.7 Moynihan plot obtained PEG/epoxy at different loadings. *From J.S. Jayan, B.D.S. Deeraj, A. Saritha, K. Joseph, Theoretical modeling of kinetics of glass transition temperature of PEG toughened epoxy, Plast. Rubber Compos. 49(6) (2020) 237–244. https://doi.org/10.1080/14658011.2020.1732124.*

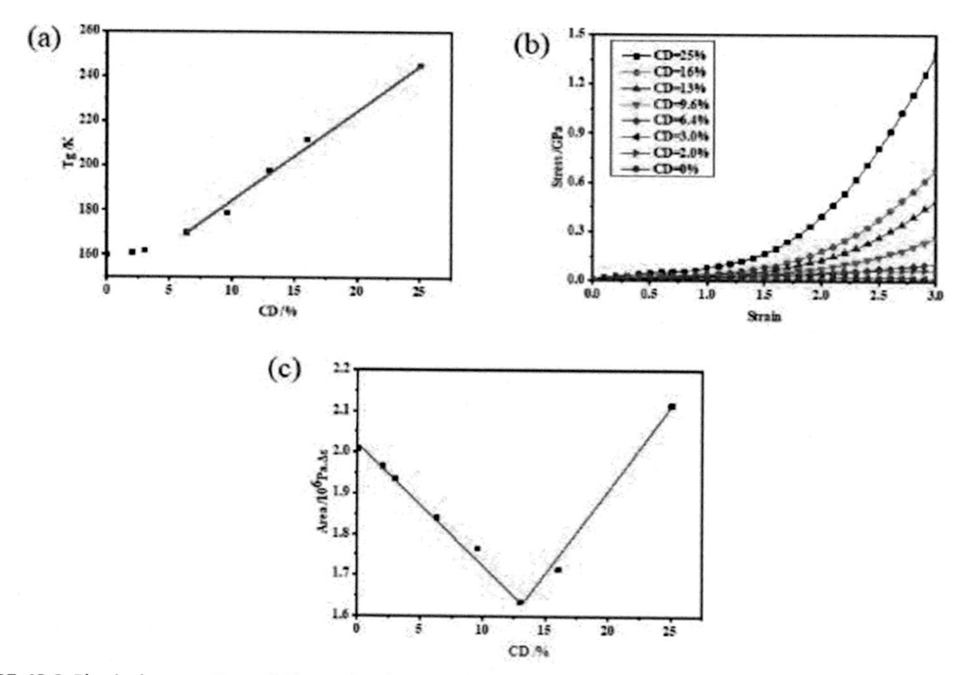

FIGURE 13.8 Physical properties of PI matrix along with the varied cross-linking degree. (A) Glass transition temperature; (B) Stress–strain curves; (C) heat buildup. *From X. Bin-Gen, W. Heng, L. Rong-Li, W. Hao, X. Ru, C. Peng, Q. Jia-Sheng, A combined simulation and experiment study on polyisoprene rubber composites, Compos. Sci. Technol. 200 (2020). https://doi.org/10.1016/j.compscitech.2020.108398.*

13.3.2 Mechanical modeling

Kunori model, parallel model, series model, and Halpin–Tsai models were used to study the tensile strength of ethylene-co-vinyl acetate (EVA) elastomers-based blends by Shafeeq et al. [131] The experimental quantities were showing better fit with the Kunori model as shown in Fig. 13.9. Abed et al. [132] carried out the modeling of reinforced concrete (RC) and compared with the experimental results. The axial behavior was studied by the help of nonlinear finite element models using Commercial software (ABAQUS). It was observed that the change in the parameters corresponds to number and the spacing among the wraps in the beams made by RC.

It was observed that the mechanical loading is capable of enhancing the thermal conductivity and elastic modulus of carbon-fiber-reinforced rubber composites. The representative volume elements (RVEs) have been simulated using the commercial FEA package ABAQUS [133]. Rule of Hybrid Mixture (RoHM) is a method used for the prediction of tensile strength and modulus of fiber-reinforced rubber composites [134]. Surface-induced and volume-induced effects have to be considered for the determination of the effect of filler on the mechanical properties of polymer composites. FE model is an in-built model for finding this relationship [135–142]. Axisymmetric model was put

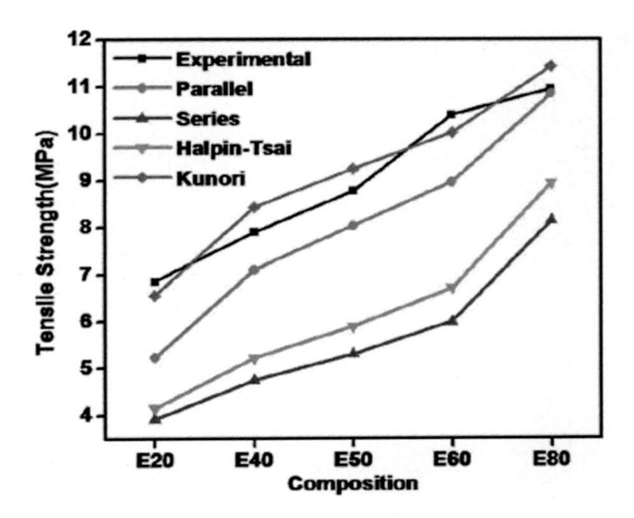

FIGURE 13.9 Comparison of experimental result with theoretical predictions of tensile strength. *From V.H. Shafeeq, G. Unnikrishnan, Experimental and theoretical evaluation of mechanical, thermal and morphological features of EVA-millable polyurethane blends, J. Polym. Res. 27(3) (2020). https://doi.org/10.1007/s10965-020-2027-7.*

forward by Fukahori and Seki for determining the properties of spherical filler-reinforced rubber matrix, and it can be considered as the pioneer in this group. By varying the stress fields in the secondary and the large primary crack, the correlation between the theories of elasticity and the numerical models was made possible [140,143–145]. Thus the mechanical and fracture-related properties can be determined by this method. Generally for obtaining this a unit cell is considered in which three phases of a molecule are represented and with a strong interface bounded with the filler and the elastomer matrix. The total volume fraction of the filler/reinforcement is generally represented by the particle and all surface related phenomena such as surface area of the filler, structure of the filler, filler–filler, and filler–elastomer interactions are symbolized by the interphase. The J-integral was calculated to estimate the energy release rate of a crack interrelating the interphase layers with hyperelastic or viscoelastic properties. The interphase layers are highly capable of reducing the crack driving force by affecting the local strain fields and thus inducing viscoelastic dissipations [146]. Parametric study reveals the effect of the shape of the filler, ratio of filler to matrix modulus, and mesh density on the stiffness of rubber composites in a two- or three-dimensional model by applying a compressive loading. The 2D micromechanical modeling is found having prominent effect in predicting the mechanical behavior of elastomer-based composites [147]. A sophisticated model was made by Peng et al. [148] for understanding the effect of surface properties and the clustering effect of filler in the determination of modulus of the elastomer-based composites. The Mori–Tanaka modified theory [149] and Eshelby's equivalent inclusion theory [150] are the theories generally used for predicting the modulus of fibers, which are used in elastomer matrix. Gao et al. [151] used the FE model for predicting the properties of aramid-fiber-reinforced elastomer composites. They have used certain random sequential adsorption algorithms, for generating fibers using the technique of

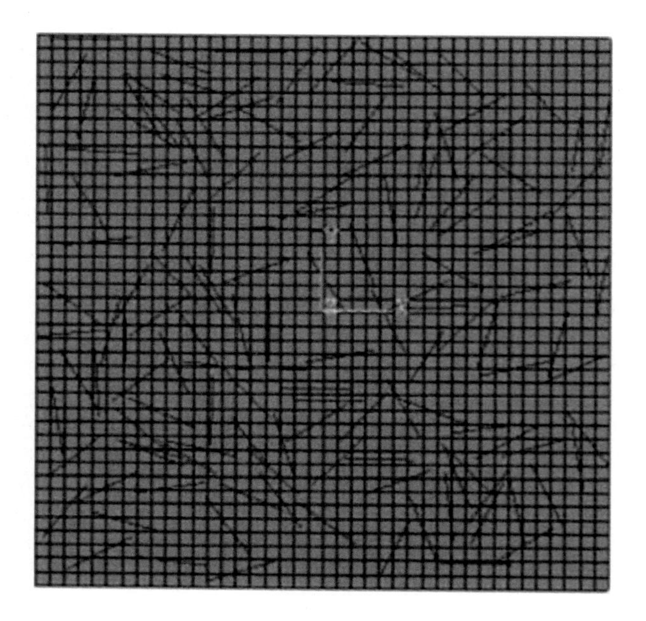

FIGURE 13.10 Mesh detail of RVE using embedded element technique. *From J. Gao, X. Yang, L.H. Huang, Numerical prediction of mechanical properties of rubber composites reinforced by aramid fiber under large deformation, Compos. Struct. 201 (2018) 29–37. https://doi.org/10.1016/j.compstruct.2018.05.132.*

embedded element, and this makes the modeling more appropriate and allows the investigation of two major issues such as large aspect ratio of fiber and deformation of elastomer. The results show that with increase in the FE model, the stress–strain responses coincide with the boundary conditions applied, and the results were also dependent on the mesh density. Mesh details obtained are shown in Fig. 13.10.

Halpin–Tsai is considered as a simple analytical model for describing the elastic response associated with fiber-reinforced elastomers [152]. Model associated with the multiscale progressive failure is currently used for predicting the damages in the cord–elastomer composite [153]. The mechanical properties of the rubber composites can be analyzed as a function of mechanical loading using representative volume element (RVE). RVE method is used for presenting the composite in the stimulation, and ABAQUS was used for the stimulation studies. The results suggest that the mechanical properties are associated with input loading and the properties also depend on the Poisson's ratio. More importantly, the alignment of fillers is affected by the applied load. Fig. 13.11 shows the RVE under 3.3% and 15% tensile strain [133].

13.3.3 Rheological modeling

Computer modeling and simulation play an important role in determining the rheological properties of elastomer-based blends and composites. The rheological and processing behavior can be thus analyzed by the help of theoretical models and simulations. The rheological behavior of polymer is of high demand as it is very essential in the processing of those polymers in functional applications. The processing and

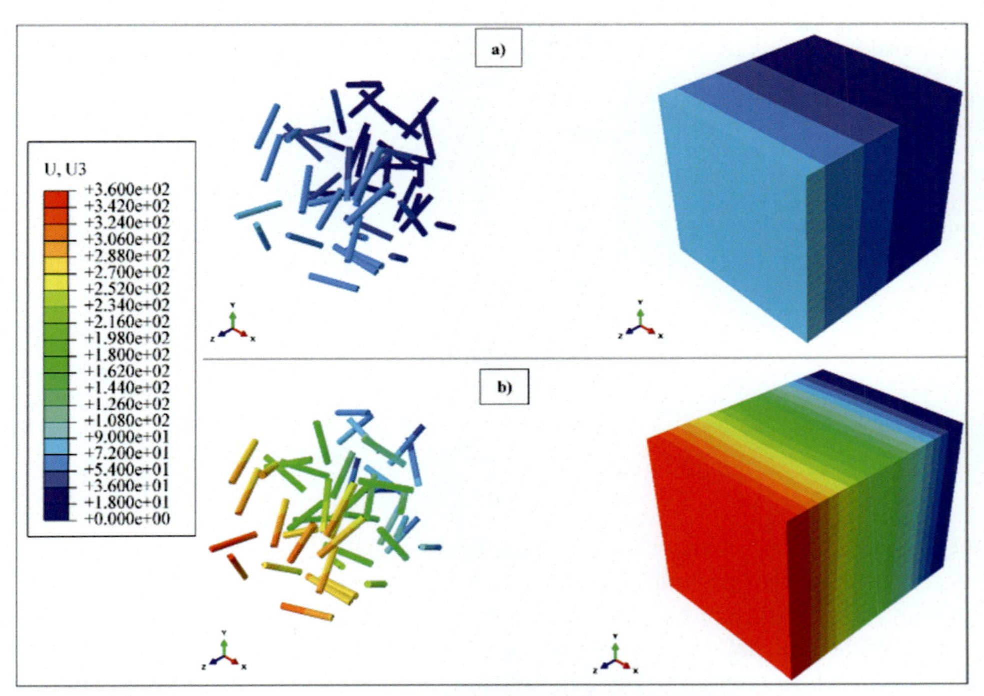

FIGURE 13.11 (A) Contours of displacement in deformed RVE with tensile strain 3.33% for the sample 1. (B) Deformed RVE with tensile strain 15% for the sample 1. *From M. Mahdavi, E. Yousefi, M. Baniassadi, M. Karimpour, M. Baghani, Effective thermal and mechanical properties of short carbon fiber/natural rubber composites as a function of mechanical loading, Appl. Thermal Eng. 117 (2017) 8–16. https://doi.org/10.1016/j.applthermaleng.2017.02.004.*

manufacturing methods can be triggered with the help of rheological analysis [154]. The storage and loss moduli are generally studied and give an idea about the nature of the polymer [121]. It is also possible to understand the effect of filler in the rheology of the system and thus the processing techniques can be chosen.

Recently, in one of our studies, the effect of PEG-grafted graphene oxide in the rheological properties of epoxy composites was analyzed, and it was found that Newtonian behavior was maintained by the polymer even after the incorporation of the filler [122]. The rheological data was showing a perfect fit with the Newtonian model as shown in Fig. 13.12.

Theoretical network model was used by Sarvestani et al. [155] and observed that the viscoelastic properties depend on the polymer–filler junctions and more importantly the solid-like characteristics are attributed by the networking of polymer–filler junctions. Unlike the Newtonian behavior, non-Newtonian behaviors such as shear thickening and thinning were also observed and can be modeled [156–160]. Several models such as Einstein, Quemada, Guth, etc., are used to predict the non-Newtonian rheological behavior.

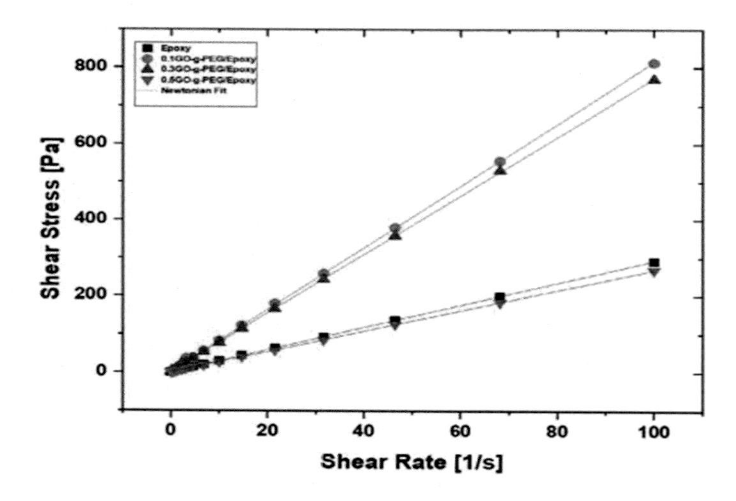

FIGURE 13.12 Newtonian fit of GO-g-PEG incorporated epoxy. *From J.S. Jayan, A. Saritha, B.D.S. Deeraj, K. Joseph, Graphene oxide as a prospective graft in polyethylene glycol for enhancing the toughness of epoxy nanocomposites, Polym. Eng. Sci. 60(4) (2020) 773–781. https://doi.org/10.1002/pen.25335.*

Petrone et al. [161], in their work, characterized rheological models, by varying the parameters under given preload and by simulating the results. They concluded that this methodology was found to be valid for identifying the elastomer specimens and their behavior simulations as well. Chen et al. [162] presented a rheological model in consideration with the dynamic behavior of magneto rheological elastomers. It modeled the viscoelastic nature, magnetic-field-induced properties, and interfacial slippage by analogy with a standard linear solid model, a stiffness variable spring, and a spring-Coulomb friction slider, respectively. The effectiveness of the model was verified by simulating the properties of MREs. The simulation results were consistent with the theories. The proposed rheological model is capable of predicting the dynamic behavior of MREs possessing any volume fraction of iron particles and properties of the matrix material. It offers a practical tool in the design and development of MREs. In particular, it can be employed as an effective input parameter model in the system control of MRE devices.

13.4 Major concern/challenges

The computer modeling and simulations can be used to predict the properties by adjusting the input parameters and thus can be effectively employed for the analysis of elastomers as well. This can be considered as an advantage of the modeling studies; at the same time it is also a major disadvantage. Thus if one fails to incorporate the accurate factors, the outcome may not be correct. The predictions of these parameters have to be more scientifically discussed to avoid all possible mistakes in the simulation and modeling as most of them are based on assumptions. Due to the highly cross-linking structures of elastomers, the simulation and modeling studies are highly challenging.

Moreover, in case of composites, the interfaces are more complex and hence the simulation studies are more difficult [163,164]. As very broad range of lengths as well as timescales are involved, the studies of elastomers are very difficult. The major challenge lies in the realistic systems with many nanoparticles. In order to avoid the challenges, better improved algorithms have to be developed, and there is urgent need for the development of systematic methods that are capable of working in wider length and timescale [165,166].

13.5 Conclusion and future scope

This chapter gives a detailed analysis of simulation and modeling studies of different elastomer-based blends and composites. It is clear that the simulation and modeling studies can bring light into the microscopic level of changes associated with the thermal, mechanical, and rheological properties. The effect of cross-link density of polymer, the shape and structure of filler, filler–matrix interactions, and Poisson's ratio are detailed in this chapter. Thus one can select suitable filler materials and incorporate it in the desired elastomer matrix and analyze the expected properties with the help of simulation studies. Prior to the experiment, one can analyze the properties, and if the desired properties are not enhanced, the experimental procedure can be neglected, thus avoiding the loss of chemicals. Once the challenges associated are rectified, the simulation and modeling studies can bring about tremendous changes in the elastomer applications. Thus simulation and modeling studies of elastomers and blends are dynamic and vibrant, and it can be further extended to ensure advanced applications.

References

[1] T.E. Gartner, A. Jayaraman, Modeling and simulations of polymers: a Roadmap, Macromolecules 52 (3) (2019) 755–786, https://doi.org/10.1021/acs.macromol.8b01836.

[2] S. Abdou-Sabet, R.C. Puydak, C.P. Rader, Dynamically vulcanized thermoplastic elastomers, Rubber Chem. Technol. 69 (3) (1996) 476–494, https://doi.org/10.5254/1.3538382.

[3] F. Carpi, D. Rossi, R. Kornbluh, R.E. Pelrine, P. Sommer-Larsen, Dielectric Elastomers as Electromechanical Transducers: Fundamentals, Materials, Devices, Models and Applications of an Emerging Electroactive Polymer Technology, 2011.

[4] Z.S. Petrović, J. Ferguson, Polyurethane elastomers, Prog. Polym. Sci. 16 (5) (1991) 695–836, https://doi.org/10.1016/0079-6700(91)90011-9.

[5] E.M. Terentjev, Liquid-crystalline elastomers, J. Phys. Condens. Matter 11 (24) (1999) R239–R257, https://doi.org/10.1088/0953-8984/11/24/201.

[6] G. Zanchin, G. Leone, Polyolefin thermoplastic elastomers from polymerization catalysis: advantages, pitfalls and future challenges, Prog. Polym. Sci. (2020).

[7] E. Filippidi, T.R. Cristiani, C.D. Eisenbach, J. Herbert Waite, J.N. Israelachvili, B. Kollbe Ahn, M.T. Valentine, Toughening elastomers using mussel-inspired iron-catechol complexes, Science 358 (6362) (2017) 502–505, https://doi.org/10.1126/science.aao0350.

[8] Q. He, Z. Wang, Y. Wang, A. Minori, M.T. Tolley, S. Cai, Electrically controlled liquid crystal elastomer-based soft tubular actuator with multimodal actuation, Sci. Adv. 5 (10) (2019), https://doi.org/10.1126/sciadv.aax5746.

[9] Z. Tang, J. Huang, B. Guo, L. Zhang, F. Liu, Bioinspired engineering of sacrificial metal-ligand bonds into elastomers with supramechanical performance and adaptive recovery, Macromolecules 49 (5) (2016) 1781–1789, https://doi.org/10.1021/acs.macromol.5b02756.

[10] Various Kinds of Self-Healing Elastomer Materials for 3D Printing, Video Proc. Adv. Mater. (2020). https://doi.org/10.5185/vpoam.2020.0835.

[11] Z. Zhao, H.J. Qi, D. Fang, A finite deformation theory of desolvation and swelling in partially photo-cross-linked polymer networks for 3D/4D printing applications, Soft Matter 15 (5) (2019) 1005–1016. https://doi.org/10.1039/c8sm02427h.

[12] I. Franta, Elastomers and Rubber Compounding Materials, 2012.

[13] G. Holden, Thermoplastic elastomers, in: Kirk-Othmer Encycl. Chem. Technol., 2000.

[14] G. Holden, Thermoplastic elastomers, in: Applied Plastics Engineering Handbook, Elsevier Inc., 2011, pp. 77–91. https://doi.org/10.1016/B978-1-4377-3514-7.10006-6.

[15] L. Imbernon, S. Norvez, From landfilling to vitrimer chemistry in rubber life cycle, Eur. Polym. J. 82 (2016) 347–376. https://doi.org/10.1016/j.eurpolymj.2016.03.016.

[16] A.O. Patil, T.S. Coolbaugh, Elastomers: a literature review with emphasis on oil resistance, Rubber Chem. Technol. 78 (3) (2005) 516–535. https://doi.org/10.5254/1.3547894.

[17] M. Alber, A.B. Tepole, W. Cannon, S. De, S. Dura-Bernal, K. Garikipati, G. Karniadakis, W.W. Lytton, P. Perdikaris, L. Petzold, E. Kuhl, Integrating machine learning and multiscale modeling: perspectives, challenges, and opportunities in the biological, biomedical, and behavioral sciences, ArXiv (2019). https://arxiv.org.

[18] E.V. Alexandrov, V.A. Blatov, A.V. Kochetkov, D.M. Proserpio, Underlying nets in three-periodic coordination polymers: topology, taxonomy and prediction from a computer-aided analysis of the Cambridge structural database, CrystEngComm 13 (12) (2011) 3947–3958. https://doi.org/10.1039/c0ce00636j.

[19] K. Binder, Monte Carlo and Molecular Dynamics Simulations in Polymer Science, 1995.

[20] G.H. Fredrickson, V. Ganesan, F. Drolet, Field-theoretic computer simulation methods for polymers and complex fluids, Macromolecules 35 (1) (2002) 16–39. https://doi.org/10.1021/ma011515t.

[21] A. Ferreira, S.S. Aphale, A survey of modeling and control techniques for micro- and nano-electromechanical systems, IEEE Trans. Syst. Man Cybern. C Appl. Rev. 41 (3) (2011) 350–364. https://doi.org/10.1109/TSMCC.2010.2072779.

[22] S.S. Jawalkar, T.M. Aminabhavi, Molecular modeling simulations and thermodynamic approaches to investigate compatibility/incompatibility of poly(l-lactide) and poly(vinyl alcohol) blends, Polymer 47 (23) (2006) 8061–8071. https://doi.org/10.1016/j.polymer.2006.09.030.

[23] S.S. Jawalkar, K.V.S.N. Raju, S.B. Halligudi, M. Sairam, T.M. Aminabhavi, Molecular modeling simulations to predict compatibility of poly(vinyl alcohol) and chitosan blends: a comparison with experiments, J. Phys. Chem. B 111 (10) (2007) 2431–2439. https://doi.org/10.1021/jp0668495.

[24] L. Levine, B. Lane, J. Heigel, K. Migler, M. Stoudt, T. Phan, R. Ricker, M. Strantza, M. Hill, F. Zhang, J. Seppala, E. Garboczi, E. Bain, D. Cole, A. Allen, J. Fox, C. Campbell, Outcomes and conclusions from the 2018 AM-bench measurements, challenge problems, modeling submissions, and conference, Integr. Mater. Manuf. Innovation 9 (1) (2020). https://doi.org/10.1007/s40192-019-00164-1.

[25] S. Thenozhi, W. Yu, Advances in modeling and vibration control of building structures, Annu. Rev. Control 37 (2) (2013) 346–364. https://doi.org/10.1016/j.arcontrol.2013.09.012.

[26] C. Jo, D. Pugal, I.K. Oh, K.J. Kim, K. Asaka, Recent advances in ionic polymer-metal composite actuators and their modeling and applications, Prog. Polym. Sci. 38 (7) (2013) 1037–1066. https://doi.org/10.1016/j.progpolymsci.2013.04.003.

[27] A. Roda, E. Michelini, L. Cevenini, D. Calabria, M.M. Calabretta, P. Simoni, Integrating biochemiluminescence detection on smartphones: mobile chemistry platform for point-of-need analysis, Anal. Chem. 86 (15) (2014) 7299–7304. https://doi.org/10.1021/ac502137s.

[28] S. Tarkoma, M. Siekkinen, E. Lagerspetz, Y. Xiao, Smartphone energy consumption: modeling and optimization, in: Smartphone Energy Consumption: Modeling and Optimization, Cambridge University Press, 2014, pp. 1–336. https://doi.org/10.1017/CBO9781107326279.

[29] G.A. Buxton, N. Clarke, Predicting structure and property relations in polymeric photovoltaic devices, Phys. Rev. B Condens. Matter 74 (8) (2006). https://doi.org/10.1103/PhysRevB.74.085207.

[30] J. Jancar, J.F. Douglas, F.W. Starr, S.K. Kumar, P. Cassagnau, A.J. Lesser, S.S. Sternstein, M.J. Buehler, Current issues in research on structure-property relationships in polymer nanocomposites, Polymer 51 (15) (2010) 3321–3343. https://doi.org/10.1016/j.polymer.2010.04.074.

[31] C. Li, A. Wu, W. Yu, Y. Hu, E. Li, C. Zhang, Q. Liu, Parameterizing starch chain-length distributions for structure-property relations, Carbohydr. Polym. 241 (2020). https://doi.org/10.1016/j.carbpol.2020.116390.

[32] X. Lu, J. Du, Quantitative structure-property relationship (QSPR) analysis of calcium aluminosilicate glasses based on molecular dynamics simulations, J. Non-Cryst. Solids 530 (2020). https://doi.org/10.1016/j.jnoncrysol.2019.119772.

[33] D.L. McDowell, S. Ghosh, S.R. Kalidindi, Representation and computational structure-property relations of random media, J. Occup. Med. 63 (3) (2011) 45–51. https://doi.org/10.1007/s11837-011-0045-y.

[34] D.L. McDowell, R.A. Lesar, The need for microstructure informatics in process-structure-property relations, MRS Bull. 41 (8) (2016) 587–593. https://doi.org/10.1557/mrs.2016.163.

[35] D.E. Yilmaz, A.C.T. van Duin, Investigating structure property relations of poly (p-phenylene terephthalamide) fibers via reactive molecular dynamics simulations, Polymer 154 (2018) 172–181. https://doi.org/10.1016/j.polymer.2018.09.001.

[36] O. Carsten, A.H. Jamson, Driving simulators as research tools in traffic psychology, in: Handbook of Traffic Psychology, Elsevier Inc, 2011, pp. 87–96. https://doi.org/10.1016/B978-0-12-381984-0.10007-4.

[37] D. Jia, J. Sun, A. Sharma, Z. Zheng, B. Liu, Integrated simulation platform for conventional, connected and automated driving: a design from cyber–physical systems perspective, Transport. Res. C Emerg. Technol. 124 (2021). https://doi.org/10.1016/j.trc.2021.102984.

[37a] D. Sumardani, A. Putri, R.R. Saraswati, D. Muliyati, F. Bakri, Virtual reality media: the simulation of relativity theory on smartphone, Formatif: Jurnal Ilmiah Pendidikan MIPA. 10 (1) (2020).

[38] J.L. Halary, F. Lauprêtre, L. Monnerie, Polymer Materials: Macroscopic Properties and Molecular Interpretations, 2011.

[39] B.J. Kirby, P. Jungwirth, Charge scaling manifesto: a way of reconciling the inherently macroscopic and microscopic natures of molecular simulations, J. Phys. Chem. Lett. 10 (23) (2019) 7531–7536. https://doi.org/10.1021/acs.jpclett.9b02652.

[40] V. Marry, P. Turq, Microscopic simulations of interlayer structure and dynamics in bihydrated heteroionic montmorillonites, J. Phys. Chem. B 107 (8) (2003) 1832–1839. https://doi.org/10.1021/jp022084z.

[41] D. Sandrin, D. Wagner, C.E. Sitta, R. Thoma, S. Felekyan, H.E. Hermes, C. Janiak, N. De Sousa Amadeu, R. Kühnemuth, H. Löwen, S.U. Egelhaaf, C.A.M. Seidel, Diffusion of macromolecules in a polymer hydrogel: from microscopic to macroscopic scales, Phys. Chem. Chem. Phys. 18 (18) (2016) 12860–12876. https://doi.org/10.1039/c5cp07781h.

[42] C. Svaneborg, G.S. Grest, R. Everaers, Strain-dependent localization, microscopic deformations, and macroscopic normal tensions in model polymer networks, Phys. Rev. Lett. 93 (25) (2004). https://doi.org/10.1103/PhysRevLett.93.257801.

[43] T.H. Epps, R.K. O'Reilly, Block copolymers: controlling nanostructure to generate functional materials - synthesis, characterization, and engineering, Chem. Sci. 7 (3) (2016) 1674–1689. https://doi.org/10.1039/c5sc03505h.

[44] M. Sahimi, Flow phenomena in rocks: from continuum models to fractals, percolation, cellular automata, and simulated annealing, Rev. Mod. Phys. 65 (4) (1993) 1393–1534. https://doi.org/10.1103/RevModPhys.65.1393.

[45] J.L. Suter, D. Groen, P.V. Coveney, Chemically specifi C multiscale modeling of clay-polymer nanocomposites reveals intercalation dynamics, tactoid self-assembly and emergent materials properties, Adv. Mater. 27 (6) (2015) 966–984. https://doi.org/10.1002/adma.201403361.

[46] T.V.M. Ndoro, M.C. Böhm, F. Müller-Plathe, Interface and interphase dynamics of polystyrene chains near grafted and ungrafted silica nanoparticles, Macromolecules 45 (1) (2012) 171–179. https://doi.org/10.1021/ma2020613.

[47] T.V.M. Ndoro, E. Voyiatzis, A. Ghanbari, D.N. Theodorou, M.C. Böhm, F. Müller-Plathe, Interface of grafted and ungrafted silica nanoparticles with a polystyrene matrix: atomistic molecular dynamics simulations, Macromolecules 44 (7) (2011) 2316–2327. https://doi.org/10.1021/ma102833u.

[48] Y.N. Pandey, M. Doxastakis, Detailed atomistic Monte Carlo simulations of a polymer melt on a solid surface and around a nanoparticle, J. Chem. Phys. 136 (9) (2012). https://doi.org/10.1063/1.3689316.

[49] Y.N. Pandey, G.J. Papakonstantopoulos, M. Doxastakis, Polymer/nanoparticle interactions: bridging the gap, Macromolecules 46 (13) (2013) 5097–5106. https://doi.org/10.1021/ma400444w.

[50] A.S. Pavlov, P.G. Khalatur, Filler reinforcement in cross-linked elastomer nanocomposites: insights from fully atomistic molecular dynamics simulation, Soft Matter 12 (24) (2016) 5402–5419. https://doi.org/10.1039/c6sm00543h.

[51] B.L. Peters, J.M.D. Lane, A.E. Ismail, G.S. Grest, Fully atomistic simulations of the response of silica nanoparticle coatings to alkane solvents, Langmuir 28 (50) (2012) 17443–17449. https://doi.org/10.1021/la3023166.

[52] J.S. Smith, O. Borodin, G.D. Smith, E.M. Kober, A molecular dynamics simulation and quantum chemistry study of poly(dimethylsiloxane)-silica nanoparticle interactions, J. Polym. Sci. B Polym. Phys. 45 (13) (2007) 1599–1615. https://doi.org/10.1002/polb.21119.

[53] G.G. Vogiatzis, D.N. Theodorou, Local segmental dynamics and stresses in polystyrene-C60 mixtures, Macromolecules 47 (1) (2014) 387–404. https://doi.org/10.1021/ma402214r.

[54] Z. Luo, J. Jiang, Molecular dynamics and dissipative particle dynamics simulations for the miscibility of poly(ethylene oxide)/poly(vinyl chloride) blends, Polymer 51 (1) (2010) 291–299. https://doi.org/10.1016/j.polymer.2009.11.024.

[55] Z. Luo, L. Zhang, J. Jiang, Atomistic insight into micro-phase separation and gas diffusion in PEO-PBT multiblock copolymers, Mol. Simulat. 39 (11) (2013) 902–907. https://doi.org/10.1080/08927022.2013.775441.

[56] V. Sethuraman, S. Mogurampelly, V. Ganesan, Multiscale simulations of lamellar PS-peo block copolymers doped with LiPF6 ions, Macromolecules 50 (11) (2017) 4542–4554. https://doi.org/10.1021/acs.macromol.7b00125.

[57] T. Spyriouni, C. Vergelati, A molecular modeling study of binary blend compatibility of polyamide 6 and poly(vinyl acetate) with different degrees of hydrolysis: an atomistic and mesoscopic approach, Macromolecules 34 (15) (2001) 5306–5316. https://doi.org/10.1021/ma001669t.

[58] D. Brown, P. Mélé, S. Marceau, N.D. Alberola, A molecular dynamics study of a model nanoparticle embedded in a polymer matrix, Macromolecules (2003) 1395–1406. https://doi.org/10.1021/ma020951s.

[59] H. Eslami, M. Rahimi, F. Müller-Plathe, Molecular dynamics simulation of a silica nanoparticle in oligomeric poly(methyl methacrylate): a model system for studying the interphase thickness in a polymer-nanocomposite via different properties, Macromolecules 46 (21) (2013) 8680–8692. https://doi.org/10.1021/ma401443v.

[60] V. Ganesan, A. Jayaraman, Theory and simulation studies of effective interactions, phase behavior and morphology in polymer nanocomposites, Soft Matter 10 (1) (2014) 13–38. https://doi.org/10.1039/c3sm51864g.

[61] A. Kyrychenko, O.M. Korsun, I.I. Gubin, S.M. Kovalenko, O.N. Kalugin, Atomistic simulations of coating of silver nanoparticles with poly(vinylpyrrolidone) oligomers: effect of oligomer chain length, J. Phys. Chem. C 119 (14) (2015) 7888–7899. https://doi.org/10.1021/jp510369a.

[62] A. Kyrychenko, D.A. Pasko, O.N. Kalugin, Poly(vinyl alcohol) as a water protecting agent for silver nanoparticles: the role of polymer size and structure, Phys. Chem. Chem. Phys. 19 (13) (2017) 8742–8756. https://doi.org/10.1039/c6cp05562a.

[63] J.M.D. Lane, A.E. Ismail, M. Chandross, G.S. Grest, Forces Between Functionalized Silica Nanoparticles, 2008.

[64] J.M.D. Lane, A.E. Ismail, M. Chandross, C.D. Lorenz, G.S. Grest, Forces between functionalized silica nanoparticles in solution, Phys. Rev. E - Stat. Nonlinear Soft Matter Phys. 79 (5) (2009). https://doi.org/10.1103/PhysRevE.79.050501.

[65] A.Y. Coran, R. Patel, Rubber-thermoplastic compositions - 1. EPDM-polypropylene thermoplastic vulcanizates, Rubber Chem. Technol. 53 (1) (1980) 141–150. https://doi.org/10.5254/1.3535023.

[66] S. Fakirov, Handbook of Thermoplastics (By Olabisi, O), 1994.

[67] R. Asaletha, M.G. Kumaran, S. Thomas, Thermoplastic elastomers from blends of polystyrene and natural rubber: morphology and mechanical properties, Eur. Polym. J. 35 (2) (1999) 253–271. https://doi.org/10.1016/S0014-3057(98)00115-3.

[68] S. Saha, A.K. Bhowmick, Computer simulation of thermoplastic elastomers from rubber-plastic blends and comparison with experiments, Polymer 103 (2016) 233–242. https://doi.org/10.1016/j.polymer.2016.09.065.

[69] M. A, I. U.S, M.I. Z.A, Rheological properties of dynamically vulcanized poly(vinyl chloride)/epoxidized natural rubber thermoplastic elastomers: effect of processing variables, Polym. Test. (2000) 193–204. https://doi.org/10.1016/s0142-9418(98)00093-2.

[70] S.S. Banerjee, K.D. Kumar, A.K. Bhowmick, Distinct melt viscoelastic properties of novel nanostructured and microstructured thermoplastic elastomeric blends from polyamide 6 and fluoroelastomer, Macromol. Mater. Eng. 300 (3) (2015) 283–290. https://doi.org/10.1002/mame.201400264.

[71] W.G.F. Sengers, P. Sengupta, J.W.M. Noordermeer, S.J. Picken, A.D. Gotsis, Linear viscoelastic properties of olefinic thermoplastic elastomer blends: melt state properties, Polymer 45 (26) (2004) 8881–8891. https://doi.org/10.1016/j.polymer.2004.10.030.

[72] M.A. Ditzler, M. Otyepka, J. Šponer, N.G. Walter, Molecular dynamics and quantum mechanics of RNA: conformational and chemical change we can believe in, Acc. Chem. Res. 43 (1) (2010) 40–47. https://doi.org/10.1021/ar900093g.

[73] A. Farazin, M. Mohammadimehr, Nano research for investigating the effect of SWCNTs dimensions on the properties of the simulated nanocomposites: a molecular dynamics simulation, Adv. Nano Res. 9 (2) (2020) 83–90. https://doi.org/10.12989/anr.2020.9.2.083.

[74] S.R.A. NAIR, A concurrently coupled multi-scale model for predicting properties of thermoplastic N anocomposites, Nanocomposites (2012).

[75] S.A. Adcock, J.A. McCammon, Molecular dynamics: survey of methods for simulating the activity of proteins, Chem. Rev. 106 (5) (2006) 1589–1615. https://doi.org/10.1021/cr040426m.

[76] G.E. Norman, V.V. Stegailov, Stochastic theory of the classical molecular dynamics method, Math. Models Comput. Simul. 5 (4) (2013) 305–333. https://doi.org/10.1134/S2070048213040108.

[77] W.F. van Gunsteren, H.J.C. Berendsen, Computer simulation of molecular dynamics: methodology, applications, and perspectives in chemistry, Angew Chem. Int. Ed. Engl. 29 (9) (1990) 992–1023. https://doi.org/10.1002/anie.199009921.

[78] H. Eslami, M. Behrouz, Molecular dynamics simulation of a polyamide-66/carbon nanotube nanocomposite, J. Phys. Chem. C 118 (18) (2014) 9841–9851. https://doi.org/10.1021/jp501672t.

[79] A.R. Tiller, B. Gorella, Estimation of polymer compatibility from molecular mechanics calculations, Polymer 35 (15) (1994) 3251–3259. https://doi.org/10.1016/0032-3861(94)90130-9.

[80] J.G. Gai, H.L. Li, C. Schrauwen, G.H. Hu, Dissipative particle dynamics study on the phase morphologies of the ultrahigh molecular weight polyethylene/polypropylene/poly(ethylene glycol) blends, Polymer 50 (1) (2009) 336–346. https://doi.org/10.1016/j.polymer.2008.10.020.

[81] R.D. Groot, T.J. Madden, Dynamic simulation of diblock copolymer microphase separation, J. Chem. Phys. 108 (20) (1998) 8713–8724. https://doi.org/10.1063/1.476300.

[82] S.S. Jawalkar, S.G. Adoor, M. Sairam, M.N. Nadagouda, T.M. Aminabhavi, Molecular modeling on the binary blend compatibility of poly (vinyl alcohol) and poly (methyl methacrylate): an atomistic simulation and thermodynamic approach, J. Phys. Chem. B 109 (32) (2005) 15611–15620. https://doi.org/10.1021/jp051206v.

[83] S. Saha, A.K. Bhowmick, Computer aided simulation of thermoplastic elastomer from poly (vinylidene fluoride)/hydrogenated nitrile rubber blend and its experimental verification, Polymer 112 (2017) 402–413. https://doi.org/10.1016/j.polymer.2017.02.035.

[84] H.J. Qi, M.C. Boyce, Stress-strain behavior of thermoplastic polyurethanes, Mech. Mater. 37 (8) (2005) 817–839. https://doi.org/10.1016/j.mechmat.2004.08.001.

[85] J. Yi, M.C. Boyce, G.F. Lee, E. Balizer, Large deformation rate-dependent stress-strain behavior of polyurea and polyurethanes, Polymer 47 (1) (2006) 319–329. https://doi.org/10.1016/j.polymer.2005.10.107.

[86] H. Cho, S. Mayer, E. Pöselt, M. Susoff, P.J. Veld, G.C. Rutledge, M.C. Boyce, Deformation mechanisms of thermoplastic elastomers: stress-strain behavior and constitutive modeling, Polymer 128 (2017) 87–99. https://doi.org/10.1016/j.polymer.2017.08.065.

[87] R.G. Rinaldi, M.C. Boyce, S.J. Weigand, D.J. Londono, M.W. Guise, Microstructure evolution during tensile loading histories of a polyurea, J. Polym. Sci. B Polym. Phys. 49 (23) (2011) 1660–1671. https://doi.org/10.1002/polb.22352.

[88] S. Zhu, N. Lempesis, P.J. In 'T Veld, G.C. Rutledge, Molecular simulation of thermoplastic polyurethanes under large tensile deformation, Macromolecules 51 (5) (2018) 1850–1864. https://doi.org/10.1021/acs.macromol.7b02367.

[89] R. Adams, S.P. Soe, R. Santiago, M. Robinson, B. Hanna, G. McShane, M. Alves, R. Burek, P. Theobald, A novel pathway for efficient characterisation of additively manufactured thermoplastic elastomers, Mater. Des. 180 (2019). https://doi.org/10.1016/j.matdes.2019.107917.

[90] M. Jenei, R.L.C. Akkermans, S. Robertson, J.A. Elliott, Molecular simulation of thermoset curing: application to 3D printing materials, Mol. Simulat. (2020). https://doi.org/10.1080/08927022.2020.1829613.

[91] R. Matsuzaki, T. Kobara, R. Yokoyama, Efficient estimation of thermal conductivity distribution during curing of thermoset composites, Adv. Compos. Mater. (2020). https://doi.org/10.1080/09243046.2020.1801337.

[91a] W.H. Jo, M.B. Ko, Structure development in epoxy resin modified with thermoplastic polymer: a Monte Carlo simulation approach, Macromolecules 26 (20) (1993) 5473–5478. https://doi.org/10.1021/ma00072a027.

[91b] M.J. Stevens, Manipulating connectivity to control fracture in network polymer adhesives, Macromolecules 34 (5) (2001) 1411–1415. https://doi.org/10.1021/ma0009505.

[92] E. Jankowski, N. Ellyson, J.W. Fothergill, M.M. Henry, M.H. Leibowitz, E.D. Miller, M. Alberts, S. Chesser, J.D. Guevara, C.D. Jones, M. Klopfenstein, K.K. Noneman, R. Singleton, R.A. Uriarte-Mendoza, S. Thomas, C.E. Estridge, M.L. Jones, Perspective on coarse-graining, cognitive load, and materials simulation, Comput. Mater. Sci. 171 (2020). https://doi.org/10.1016/j.commatsci.2019.109129.

[93] Z. Marie, V. Nicolas, A. Celzard, V. Fierro, First approach for modelling the physical foaming of tannin-based thermoset foams, Int. J. Therm. Sci. 149 (2020). https://doi.org/10.1016/j.ijthermalsci.2019.106212.

[94] G.S. Grest, K. Kremer, Statistical properties of random cross-linked rubbers, Macromolecules 23 (23) (1990) 4994–5000. https://doi.org/10.1021/ma00225a020.

[95] I. Hamerton, C.R. Heald, B.J. Howlin, Molecular modelling of the physical and mechanical properties of two polycyanurate network polymers, J. Mater. Chem. 6 (3) (1996) 311–314. https://doi.org/10.1039/jm9960600311.

[96] D.C. Doherty, P. Leung, B.N. Holmes, R.B. Ross, Polymerization molecular dynamics simulations. I. Cross-linked atomistic models for poly(methacrylate) networks, Comput. Theor. Polym. Sci. (1998) 169–178. https://doi.org/10.1016/s1089-3156(98)00030-0.

[97] Y. I, Computer simulation of structure and properties of crosslinked polymers: application to epoxy resins, Polymer (2002) 963–969. https://doi.org/10.1016/s0032-3861(01)00634-6.

[98] D.R. Heine, G.S. Grest, C.D. Lorenz, M. Tsige, M.J. Stevens, Atomistic simulations of end-linked poly(dimethylsiloxane) networks: structure and relaxation, Macromolecules 37 (10) (2004) 3857–3864. https://doi.org/10.1021/ma035760j.

[99] C. Wu, W. Xu, Atomistic molecular modelling of crosslinked epoxy resin, Polymer 47 (16) (2006) 6004–6009. https://doi.org/10.1016/j.polymer.2006.06.025.

[100] P.V. Komarov, Y.T. Chiu, S.M. Chen, P.G. Khalatur, P. Reineker, Highly cross-linked epoxy resins: an atomistic molecular dynamics simulation combined with a mapping/reverse mapping procedure, Macromolecules 40 (22) (2007) 8104–8113. https://doi.org/10.1021/ma070702+.

[101] V. Vikas, S.S. Patnaik, A.K. Roy, B.L. Farmer, A molecular dynamics study of epoxy-based networks: cross-linking procedure and prediction of molecular and material properties, Macromolecules (2008) 6837–6842. https://doi.org/10.1021/ma801153e.

[102] P.H. Lin, R. Khare, Molecular simulation of cross-linked epoxy and epoxy-POSS nanocomposite, Macromolecules 42 (12) (2009) 4319–4327. https://doi.org/10.1021/ma9004007.

[103] J.S. Bermejo, C.M. Ugarte, Chemical crosslinking of PVA and prediction of material properties by means of fully atomistic MD simulations, Macromol. Theory Simul. 18 (4–5) (2009) 259–267. https://doi.org/10.1002/mats.200800099.

[104] J.S. Bermejo, C.M. Ugarte, Influence of cross-linking density on the glass transition and structure of chemically cross-linked PVA: a molecular dynamics study, Macromol. Theory Simul. 18 (6) (2009) 317–327. https://doi.org/10.1002/mats.200900032.

[105] A. Bandyopadhyay, P.K. Valavala, T.C. Clancy, K.E. Wise, G.M. Odegard, Molecular modeling of crosslinked epoxy polymers: the effect of crosslink density on thermomechanical properties, Polymer 52 (11) (2011) 2445–2452. https://doi.org/10.1016/j.polymer.2011.03.052.

[106] P. Bultinck, W. Langenaeker, P. Lahorte, F. De Proft, P. Geerlings, M. Waroquier, J.P. Tollenaere, The electronegativity equalization method I: parametrization and validation for atomic charge calculations, J. Phys. Chem. 106 (34) (2002) 7887–7894. https://doi.org/10.1021/jp0205463.

[107] W.J. Mortier, K.V. Genechten, J. Gasteiger, Electronegativity equalization: application and parametrization, J. Am. Chem. Soc. 107 (4) (1985) 829–835. https://doi.org/10.1021/ja00290a017.

[108] S.L. Njo, J. Fan, B. Van De Graaf, Extending and simplifying the electronegativity equalization method, J. Mol. Catal. Chem. 134 (1–3) (1998) 79–88. https://doi.org/10.1016/S1381-1169(98)00024-7.

[109] A.K. Rappe, W.A. Goddard, Charge equilibration for molecular dynamics simulations, J. Phys. Chem. (1991) 3358–3363. https://doi.org/10.1021/j100161a070.

[110] C. Li, A. Strachan, Molecular simulations of crosslinking process of thermosetting polymers, Polymer 51 (25) (2010) 6058–6070. https://doi.org/10.1016/j.polymer.2010.10.033.

[111] C. Li, A. Strachan, Molecular dynamics predictions of thermal and mechanical properties of thermoset polymer EPON862/DETDA, Polymer 52 (13) (2011) 2920–2928. https://doi.org/10.1016/j.polymer.2011.04.041.

[112] A. Mejía, E.A. Müller, G. Chaparro Maldonado, SGTPy: a Python code for calculating the interfacial properties of fluids based on the square gradient theory using the SAFT-VR mie equation of state, J. Chem. Inf. Model. (2021). https://doi.org/10.1021/acs.jcim.0c01324.

[113] E.A. Müller, A. Mejía, Extension of the SAFT-VR mie EoS to model homonuclear rings and its parametrization based on the principle of corresponding states, Langmuir 33 (42) (2017) 11518–11529. https://doi.org/10.1021/acs.langmuir.7b00976.

[114] A.K. Pervaje, J.C. Tilly, A.T. Detwiler, R.J. Spontak, S.A. Khan, E.E. Santiso, Molecular simulations of thermoset polymers implementing theoretical kinetics with top-down coarse-grained models, Macromolecules 53 (7) (2020) 2310–2322. https://doi.org/10.1021/acs.macromol.9b02255.

[115] M. Hadipeykani, F. Aghadavoudi, D. Toghraie, A molecular dynamics simulation of the glass transition temperature and volumetric thermal expansion coefficient of thermoset polymer based epoxy nanocomposite reinforced by CNT: a statistical study, Phys. Stat. Mech. Appl. (2020) 546. https://doi.org/10.1016/j.physa.2019.123995.

[116] B. Demir, In silico study of bio-based epoxy precursors for sustainable and renewable thermosets, Polymer 191 (2020). https://doi.org/10.1016/j.polymer.2020.122253.

[117] A.V. Radchenko, H. Chabane, B. Demir, D.J. Searles, J. Duchet-Rumeau, J.F. Gérard, J. Baudoux, S. Livi, New epoxy thermosets derived from a bisimidazolium ionic liquid monomer: an experimental and modeling investigation, ACS Sustain. Chem. Eng. 8 (32) (2020) 12208–12221. https://doi.org/10.1021/acssuschemeng.0c03832.

[118] J.J. Schichtel, A. Chattopadhyay, Modeling thermoset polymers using an improved molecular dynamics crosslinking methodology, Comput. Mater. Sci. 174 (2020). https://doi.org/10.1016/j.commatsci.2019.109469.

[119] C.D. Wick, A.J. Peters, G. Li, Quantifying the contributions of energy storage in a thermoset shape memory polymer with high stress recovery: a molecular dynamics study, Polymer 213 (2021). https://doi.org/10.1016/j.polymer.2020.123319.

[120] A. Perego, F. Khabaz, Volumetric and rheological properties of vitrimers: a hybrid molecular dynamics and Monte Carlo simulation study, Macromolecules 53 (19) (2020) 8406–8416. https://doi.org/10.1021/acs.macromol.0c01423.

[121] J.S. Jayan, B.D.S. Deeraj, A. Saritha, K. Joseph, Theoretical modelling of kinetics of glass transition temperature of PEG toughened epoxy, Plast., Rubber Compos. 49 (6) (2020) 237–244. https://doi.org/10.1080/14658011.2020.1732124.

[122] J.S. Jayan, A. Saritha, B.D.S. Deeraj, K. Joseph, Graphene oxide as a prospective graft in polyethylene glycol for enhancing the toughness of epoxy nanocomposites, Polym. Eng. Sci. 60 (4) (2020) 773–781. https://doi.org/10.1002/pen.25335.

[123] F. Cai, G. You, K. Luo, H. Zhang, X. Zhao, S. Wu, Click chemistry modified graphene oxide/styrene-butadiene rubber composites and molecular simulation study, Compos. Sci. Technol. 190 (2020). https://doi.org/10.1016/j.compscitech.2020.108061.

[124] N.J. Vickers, Animal communication: when i'm calling you, will you answer too? Curr. Biol. 27 (2017).

[125] X. Xiangfan, C. Jie, Z. Jun, L. Baowen, Thermal conductivity of polymers and their nanocomposites, Adv. Mater. (2018) 1705544. https://doi.org/10.1002/adma.201705544.

[126] P. Yuan, P. Zhang, T. Liang, S. Zhai, Effects of surface functionalization on thermal and mechanical properties of graphene/polyethylene glycol composite phase change materials, Appl. Surf. Sci. 485 (2019) 402–412. https://doi.org/10.1016/j.apsusc.2019.04.011.

[127] S. Yang, J. Qu, Computing thermomechanical properties of crosslinked epoxy by molecular dynamic simulations, Polymer 53 (21) (2012) 4806–4817. https://doi.org/10.1016/j.polymer.2012.08.045.

[128] N.B. Shenogina, M. Tsige, S.S. Patnaik, S.M. Mukhopadhyay, Molecular modeling approach to prediction of thermo-mechanical behavior of thermoset polymer networks, Macromolecules 45 (12) (2012) 5307–5315. https://doi.org/10.1021/ma3007587.

[129] N.B. Shenogina, M. Tsige, S.S. Patnaik, S.M. Mukhopadhyay, Molecular modeling of elastic properties of thermosetting polymers using a dynamic deformation approach, Polymer 54 (13) (2013) 3370–3376. https://doi.org/10.1016/j.polymer.2013.04.034.

[130] X. Bin-Gen, W. Heng, L. Rong-Li, W. Hao, X. Ru, C. Peng, Q. Jia-Sheng, A combined simulation and experiment study on polyisoprene rubber composites, Compos. Sci. Technol. (2020) 108398. https://doi.org/10.1016/j.compscitech.2020.108398.

[131] V.H. Shafeeq, G. Unnikrishnan, Experimental and theoretical evaluation of mechanical, thermal and morphological features of EVA-millable polyurethane blends, J. Polym. Res. 27 (3) (2020). https://doi.org/10.1007/s10965-020-2027-7.

[132] F. Abed, C. Oucif, Y. Awera, H.H. Mhanna, H. Alkhraisha, FE modeling of concrete beams and columns reinforced with FRP composites, Defence Technol. 17 (1) (2021) 1–14. https://doi.org/10.1016/j.dt.2020.02.015.

[133] M. Mahdavi, E. Yousefi, M. Baniassadi, M. Karimpour, M. Baghani, Effective thermal and mechanical properties of short carbon fiber/natural rubber composites as a function of mechanical loading, Appl. Therm. Eng. 117 (2017) 8–16. https://doi.org/10.1016/j.applthermaleng.2017.02.004.

[134] N. Venkateshwaran, A. Elayaperumal, G.K. Sathiya, Prediction of tensile properties of hybrid-natural fiber composites, Compos. B Eng. 43 (2) (2012) 793–796. https://doi.org/10.1016/j.compositesb.2011.08.023.

[135] R.D. Bradshaw, F.T. Fisher, L.C. Brinson, Fiber waviness in nanotube-reinforced polymer composites-II: modeling via numerical approximation of the dilute strain concentration tensor, Compos. Sci. Technol. 63 (11) (2003) 1705–1722. https://doi.org/10.1016/S0266-3538(03)00070-8.

[136] N. Esmaeili, Y. Tomita, Micro- to macroscopic responses of a glass particle-blended polymer in the presence of an interphase layer, Int. J. Mech. Sci. 48 (10) (2006) 1186–1195. https://doi.org/10.1016/j.ijmecsci.2006.03.011.

[137] R.S. Fertig, M.R. Garnich, Influence of constituent properties and microstructural parameters on the tensile modulus of a polymer/clay nanocomposite, Compos. Sci. Technol. 64 (16) (2004) 2577–2588. https://doi.org/10.1016/j.compscitech.2004.06.002.

[138] F.T. Fisher, R.D. Bradshaw, L.C. Brinson, Fiber waviness in nanotube-reinforced polymer composites-I: modulus predictions using effective nanotube properties, Compos. Sci. Technol. 63 (11) (2003) 1689–1703. https://doi.org/10.1016/S0266-3538(03)00069-1.

[139] K. Hbaieb, Q.X. Wang, Y.H.J. Chia, B. Cotterell, Modelling stiffness of polymer/clay nanocomposites, Polymer 48 (3) (2007) 901–909. https://doi.org/10.1016/j.polymer.2006.11.062.

[140] W.T. Kern, W. Kim, A. Argento, E.C. Lee, D.F. Mielewski, Finite element analysis and microscopy of natural fiber composites containing microcellular voids, Mater. Des. 106 (2016) 285–294. https://doi.org/10.1016/j.matdes.2016.05.094.

[141] A. Mesbah, F. Zaïri, S. Boutaleb, J.M. Gloaguen, M. Naït-Abdelaziz, S. Xie, T. Boukharouba, J.M. Lefebvre, Experimental characterization and modeling stiffness of polymer/clay nanocomposites within a hierarchical multiscale framework, J. Appl. Polym. Sci. 114 (5) (2009) 3274–3291. https://doi.org/10.1002/app.30547.

[142] H.W. Wang, H.W. Zhou, R.D. Peng, L. Mishnaevsky, Nanoreinforced polymer composites: 3D FEM modeling with effective interface concept, Compos. Sci. Technol. 71 (7) (2011) 980–988. https://doi.org/10.1016/j.compscitech.2011.03.003.

[143] Y. Fukahori, New progress in the theory and model of carbon black reinforcement of elastomers, J. Appl. Polym. Sci. 95 (1) (2005) 60–67. https://doi.org/10.1002/app.20802.

[144] M. Sabzevari, R.J. Teymoori, S.A. Sajjadi, FE modeling of the compressive behavior of porous copper-matrix nanocomposites, Mater. Des. 86 (2015) 178–183. https://doi.org/10.1016/j.matdes.2015.07.080.

[145] Y. Fukahori, W. Seki, Stress analysis of elastomeric materials at large extensions using the finite element method, J. Mater. Sci. (1993) 4471–4482. https://doi.org/10.1007/bf01154959.

[146] M. Alimardani, M. Razzaghi-Kashani, M.H.R. Ghoreishy, Prediction of mechanical and fracture properties of rubber composites by microstructural modeling of polymer-filler interfacial effects, Mater. Des. 115 (2017) 348–354. https://doi.org/10.1016/j.matdes.2016.11.061.

[147] J.S. Bergström, M.C. Boyce, Mechanical behavior of particle filled elastomers, Rubber Chem. Technol. 72 (4) (1999) 633–656. https://doi.org/10.5254/1.3538823.

[148] R.D. Peng, H.W. Zhou, H.W. Wang, L. Mishnaevsky, Modeling of nano-reinforced polymer composites: microstructure effect on Young's modulus, Comput. Mater. Sci. 60 (2012) 19–31. https://doi.org/10.1016/j.commatsci.2012.03.010.

[149] T. Mori, K. Tanaka, Average stress in matrix and average elastic energy of materials with misfitting inclusions, Acta Metall. 21 (5) (1973) 571–574. https://doi.org/10.1016/0001-6160(73)90064-3.

[150] The determination of the elastic field of an ellipsoidal inclusion, and related problems, Proc. Roy. Soc. Lond. Math. Phys. Sci. 241 (1226) (1957) 376–396. https://doi.org/10.1098/rspa.1957.0133.

[151] J. Gao, X. Yang, L.H. Huang, Numerical prediction of mechanical properties of rubber composites reinforced by aramid fiber under large deformation, Compos. Struct. 201 (2018) 29–37. https://doi.org/10.1016/j.compstruct.2018.05.132.

[152] J.C.H. Affdl, J.L. Kardos, The Halpin-Tsai equations: a review, Polym. Eng. Sci. 16 (5) (1976) 344–352. https://doi.org/10.1002/pen.760160512.

[153] P. Behroozinia, S. Taheri, R. Mirzaeifar, An investigation of intelligent tires using multiscale modeling of cord-rubber composites, Mech. Base. Des. Struct. Mach. 46 (2) (2018) 168–183. https://doi.org/10.1080/15397734.2017.1321488.

[154] B.D.S. Deeraj, R. Harikrishnan, J.S. Jayan, A. Saritha, K. Joseph, Enhanced visco-elastic and rheological behavior of epoxy composites reinforced with polyimide nanofiber, Nano-Struct. Nano-Objects 21 (2020). https://doi.org/10.1016/j.nanoso.2019.100421.

[155] A.S. Sarvestani, C.R. Picu, Network model for the viscoelastic behavior of polymer nanocomposites, Polymer 45 (22) (2004) 7779–7790. https://doi.org/10.1016/j.polymer.2004.08.060.

[156] R. Krishnamoorti, J. Ren, A.S. Silva, Shear response of layered silicate nanocomposites, J. Chem. Phys. 114 (11) (2001) 4968–4973. https://doi.org/10.1063/1.1345908.

[157] J. Ren, R. Krishnamoorti, Nonlinear viscoelastic properties of layered-silicate-based intercalated nanocomposites, Macromolecules 36 (12) (2003) 4443–4451. https://doi.org/10.1021/ma020412n.

[158] L. Xu, S. Reeder, M. Thopasridharan, J. Ren, D.A. Shipp, R. Krishnamoorti, Structure and melt rheology of polystyrene-based layered silicate nanocomposites, Nanotechnology 16 (7) (2005) S514–S521. https://doi.org/10.1088/0957-4484/16/7/028.

[159] Q. Zhang, L.A. Archer, Poly(ethylene oxide)/silica nanocomposites: structure and rheology, Langmuir 18 (26) (2002) 10435–10442. https://doi.org/10.1021/la026338j.

[160] Q. Zhang, L.A. Archer, Structure and Rheology of Polyethylene Oxide/Silica Nanocomposites, APS March Meet. Abstr, 2003.

[161] F. Petrone, M. Lacagnina, M. Scionti, Dynamic characterization of elastomers and identification with rheological models, J. Sound Vib. 271 (1–2) (2004) 339–363. https://doi.org/10.1016/j.jsv.2003.02.001.

[162] L. Chen, S. Jerrams, A rheological model of the dynamic behavior of magnetorheological elastomers, J. Appl. Phys. 110 (1) (2011). https://doi.org/10.1063/1.3603052.

[163] L. Delle Site, C.F. Abrams, A. Alavi, K. Kremer, Polymers near metal surfaces: selective adsorption and global conformations, Phys. Rev. Lett. 89 (15) (2002) 156103. https://doi.org/10.1103/PhysRevLett.89.156103.

[164] K. Johnston, V. Harmandaris, Hierarchical simulations of hybrid polymer-solid materials, Soft Matter 9 (29) (2013) 6696–6710. https://doi.org/10.1039/c3sm50330e.

[165] S. Sinha Ray, M. Okamoto, Polymer/layered silicate nanocomposites: a review from preparation to processing, Prog. Polym. Sci. 28 (11) (2003) 1539–1641. https://doi.org/10.1016/j.progpolymsci.2003.08.002.

[166] Q.H. Zeng, A.B. Yu, G.Q. Lu, Multiscale modeling and simulation of polymer nanocomposites, Prog. Polym. Sci. 33 (2) (2008) 191–269. https://doi.org/10.1016/j.progpolymsci.2007.09.002.

14

Recycling of elastomer blends and composites

Jitha S. Jayan[1], A.S. Sethulekshmi[1], Gopika Venu[1], B.D.S. Deeraj[2], Appukuttan Saritha[1], Kuruvilla Joseph[2]

[1]DEPARTMENT OF CHEMISTRY, SCHOOL OF ARTS AND SCIENCES, AMRITA VISHWA VIDYAPEETHAM, AMRITAPURI, KOLLAM, KERALA, INDIA; [2]DEPARTMENT OF CHEMISTRY, INDIAN INSTITUTE OF SPACE SCIENCE AND TECHNOLOGY, VALIAMALA, THIRUVANANTHAPURAM, KERALA, INDIA

14.1 Introduction

Waste disposal and management are a serious problem of the 21st century, especially synthetic polymeric materials are very difficult to degrade or recycle [1,2]. The highly cross-linked networks in the polymer are good enough to enhance the mechanical properties and performance, whereas the cross-links make it difficult to degrade [3–5]. As the use of automobiles is increasing day by day, the scrap tire amount is increasing in the range of millions of tons [3–5]. The annual release of elastomeric waste is tremendously increasing day by day due to the increase in the usage of rubber products [6–8]. Mainly the waste rubbers are in the form of hoses, automobile tires and tubes, gloves, and conveyor belts, etc. [9,10]. These waste rubbers are usually discarded into the land or otherwise burnt. These methods are responsible for different environmental problems as they promote severe pollution. These materials never return to the ecological system by simple degradation or decomposition methods and are a serious factor for soil pollution since they are a kind of nonenvironmental material [11,12]. The piles of scrap tires will act as breeding place for mosquito [13], and the burning leads to the production of CO_2 gases and acts as a major source of air pollution [14,15]. This demands the need of sustainable methods of utilization of scrap rubber/other elastomeric materials. Land filling is a method of disposal of scrap rubber by directly dumping it into the land, but the scarcity of available site and the other environmental issues have led to the rejection of this method [16–19]. Hence the utilization of discarded rubber products is more focused by different nations, and the recycling of elastomeric waste materials is developing as a prominent research field. People are paying much interest toward the sustainable elastomer/rubber technology. The recycling of scrap elastomer is a major solution for all the environmental problems caused by it, the recycling of elastomeric

Elastomer Blends and Composites. https://doi.org/10.1016/B978-0-323-85832-8.00001-8

products is being analyzed by various researchers. Several recycling methods are followed, such as the elastomeric wastes are used as a fuel in the cement industry and as a raw material in rubber industry. Modification, modification asphalt, regenerate rubber, powdered rubber, etc., are practiced as recycling methods for elastomeric waste materials [2,20–27].

There are mainly three methods of recycling of scrap rubber; they are recycling, reclaiming, and devulcanization [28–30]. The reutilization of scrap elastomer in its original form is a much better option than the recycling process, and the method of reutilization of scrap elastomer is generally known as devulcanization or reclaiming process [31–34]. Reclaiming is a process in which uncontrolled breakdown of the networks occurs, even the main chain can be broken [35,36]. A devulcanization is a combination of processes such as oxidation, depolymerization, and increasing plasticity. Simply as the name indicates, it is the reverse of vulcanization, i.e., the breaking of cross-links without affecting the main chain [37,38]. The breakage of main chain adversely affects the properties; hence extra care must be taken during the devulcanization process [39]. Devulcanization process only aims at the S–S and C–S linkages, which are responsible for the three-dimensional cross-linking in rubber [33,40]. An ideal devulcanization is the one in which no main chain breakage occurs. The breakage of the main chain declines the mechanical properties of the revulcanized rubber. In the devulcanization process, the di- and polysulfide bonds are converted to monosulfide and converted to recycled uncured rubber, and later it is revulcanized, i.e., the devulcanizates are expected to behave like fresh rubber [41,42]. The devulcanization process is schematically represented in Fig. 14.1. Depending on the energy applied for the breaking of the bonds, there are several methods of devulcanization. There are generally three methods of devulcanization, i.e., physical, chemical, and biological [1,43,44]. The different devulcanization methods and their submethods are shown in Fig. 14.2. In

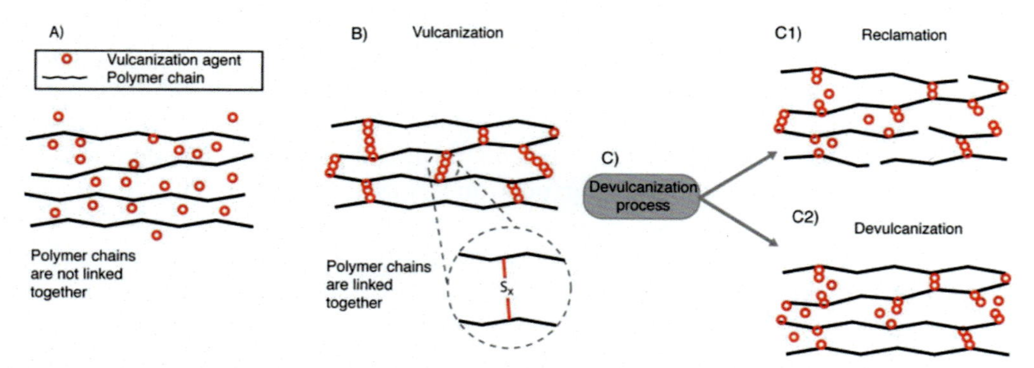

FIGURE 14.1 (A) Polymer chains before vulcanization, (B) rubber network after vulcanization, (C) possible results of a devulcanization process, (C1) reclamation and (C2) devulcanization [45]. *Reproduced with permission from Elsevier. From R. Diaz, G. Colomines, E. Peuvrel-Disdier, R. Deterre, Thermo-mechanical recycling of rubber: Relationship between material properties and specific mechanical energy, J. Mater. Process. Technol. 252 (2018) 454–468. https://doi.org/10.1016/j.jmatprotec.2017.10.014.*

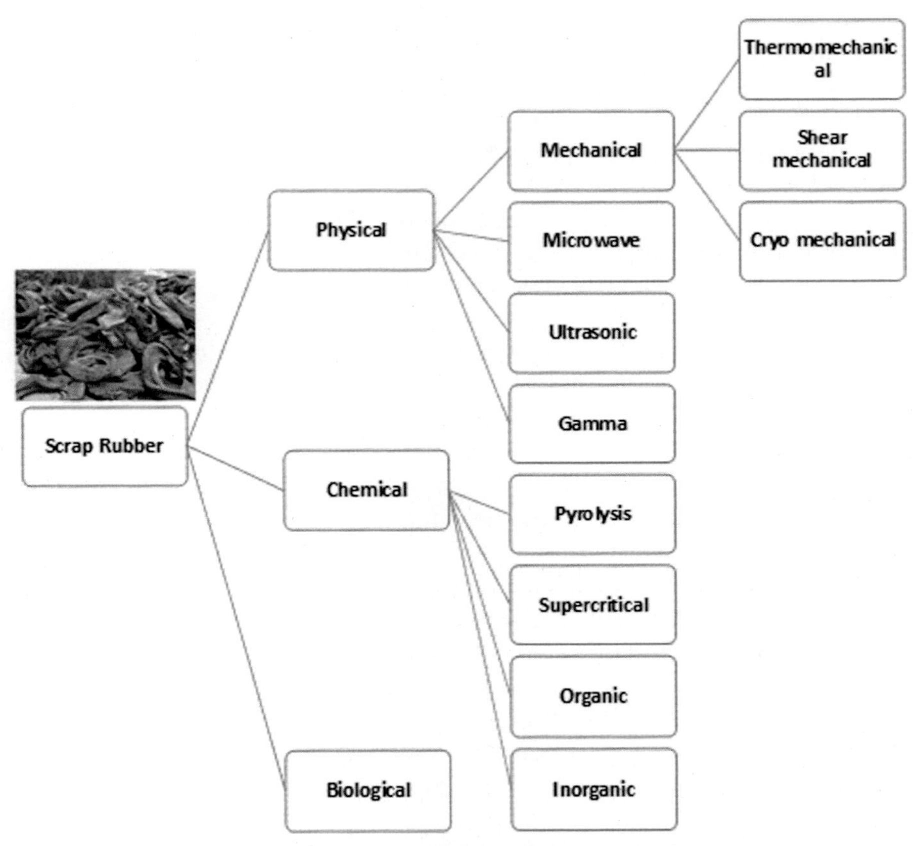

FIGURE 14.2 Different devulcanization methods.

physical method, physical forces are usually used for the devulcanization process, whereas biological devulcanization is carried out by the help of microorganisms, and in chemical methods the process is carried out with the help of chemical reagents. This chapter tries to portray different devulcanization methods such as chemical, ultrasound, microwave, thermomechanical, biological, and supercritical methods in detail. This chapter also tries to bring light into the products derived from scrap rubber and its application. Hence the chapter gives an idea about the different devulcanization methods and its applications in a comprehensive way [32].

14.2 Devulcanization methods

14.2.1 Chemical method

Chemical devulcanization is the process of breaking chemical bonds such as C–S and S–S in rubber materials by the action of some chemical reagents. Usually this method takes place with the help of mechanical and thermal energy for accelerating the devulcanization process [35]. Different chemical regents such as inorganic compounds [46], organic disulfides (diphenyl disulfide, thiophenols, and their zinc salts), mercaptans [47,48], and hydroxide or chlorinated hydrocarbons can be used for devulcanizing rubber. The concentrations of these reagents are usually in the range of 0.5–10 wt%. Excess usage of solvents in these methods causes many economic and environmental problems. Super critical fluids (cheap, nontoxic, nonflammable, and easily removal) can be considered as the best substitute in industries for these solvents. Even though the reaction mechanism of chemical agent assisted devulcanization is not fully known, but it ensures a better devulcanization to the waste rubber materials. However, we should determine the toxicity of chemical reagents involved in the reaction before carrying out the process.

Sadaka et al. [49] used periodic acid (cheap, nontoxic, and efficient degradation agent) for one-step oxidative method for degradation of carbonyl telechelic cis-1,4-oligoisoprenes and natural rubber (NR). Degradation of ground waste tire rubber was also carried out by following the same method. Reaction time and quantity of periodic acid are able to control the degree of degradation; this can be analyzed from average molecular weight analysis of degraded materials. This method can be also applied to degradation of virgin rubber and recycling elastomers containing compounding ingredients. Periodic acid degradation of waste tire rubber was independent of particle size. Degradation of waste tire was carried out in a controlled manner. Degradation product obtained was a viscous liquid with light color and a sticky elastomeric character. Ketone and aldehyde group at the chain end of the product were evaluated by proton NMR.

Cavalieri et al. [50] published a work related to devulcanization of cured rubber powder with the help of high-energy ball milling. They tried to increase the efficiency of this method by using 2,6-di-tert-butyl-4-methyl-phenol (BHT) as an antioxidant. Surface of rubber powder was grafted by devulcanized NR under an inert atmosphere. In the presence of BHT, devulcanization of sulfur cross-links and degradation of carbon backbone of rubber take place. Addition of 15 phr devulcanized GR into virgin NR enhanced its plasticity. The vulcanized product maintained their mechanical characteristics; however, degradation of matrix induced by rubber particle led to poor aging properties. Experimental milling conditions can strongly influence the mechanical and chemical characteristics; therefore controlling is very important for enhancing the efficiency of reclaimed rubber.

Gupta and coworkers [51] developed a carbonaceous adsorbent for removing pesticides from waste water by chemical and thermal treatment of waste rubber. Chemical treatment of tire helped to introduce surface porosity and carbonyl and hydroxyl functional groups on surface of activated carbon. After carbonization of ground tire granules, it was treated with KOH and heated at 900°C for 2 h. For removing ash content, they treated it with HCl, washed, dried, and the obtained product was named as CTRTAC. The SEM images of unactivated rubber and pesticides adsorbed rubber are shown in Fig. 14.3.

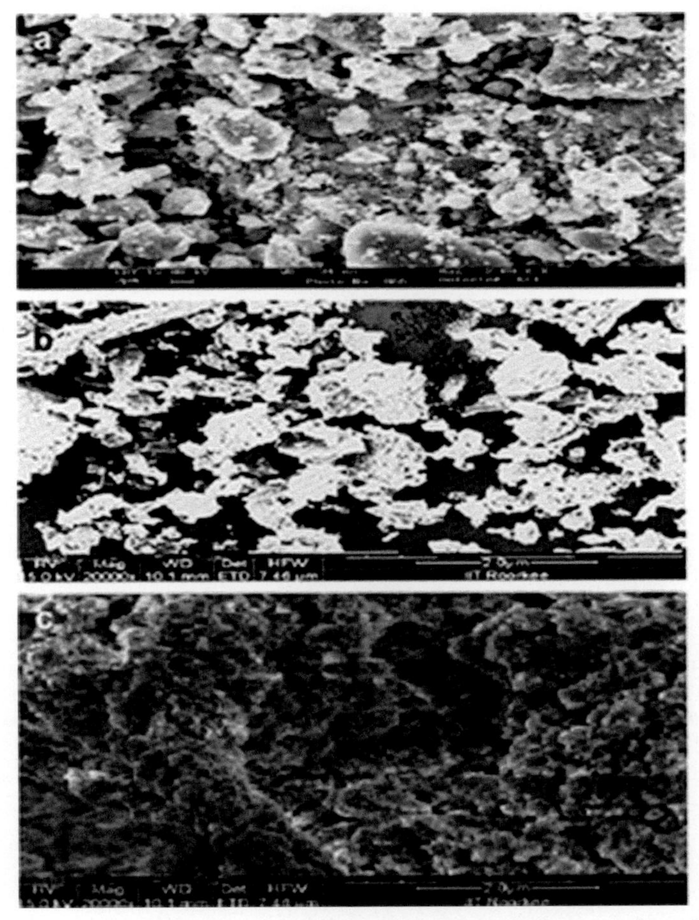

FIGURE 14.3 SEM images of (A) unactivated tire rubber (B and C) CTRTAC before and after adsorption of methoxy pesticide [51]. *Reproduced with permission from Elsevier. From V.K. Gupta, B. Gupta, A. Rastogi, S. Agarwal, A. Nayak, Pesticides removal from waste water by activated carbon prepared from waste rubber tire, Water Res. 45(13) (2011) 4047–4055. https://doi.org/10.1016/j.watres.2011.05.016.*

Influence of reaction time, temperature, and concentration of devulcanizing agent on the effectiveness of devulcanization method and the mechanical properties of the derived products were carried out by Sabzekar and coworkers [52]. They devulcanized sulfur-cured NR by using the devulcanizing agent known as benzoyl peroxide (BPO), and they used xylene as solvent. The selective devulcanization of cross-links in nNR was carried out at a reaction time of 2 h with less concentration of devulcanization agents such as 2, 4, 6, and 8 phr. But increased amount of BPO concentration showed a nonselective cross-link and backbone cleavage. As a result of cage effect (i.e., the recombination of some of the macro radicals formed after the scission), macro radicals from devulcanization reaction were recross-linked. At low concentration of BPO (4phr) and at higher temperature, cage effect is observed well, and hence this concentration and high temperature are considered as the optimum conditions. Mechanical properties of all the samples were analyzed by varying the time and temperature. Tensile strength and elongation at break were decreased due to lesser cross-link density as a result of longer reaction time (Figs. 14.4 and 14.5). The scission of cross-links happened at very high temperature, which led to lowering of tensile properties.

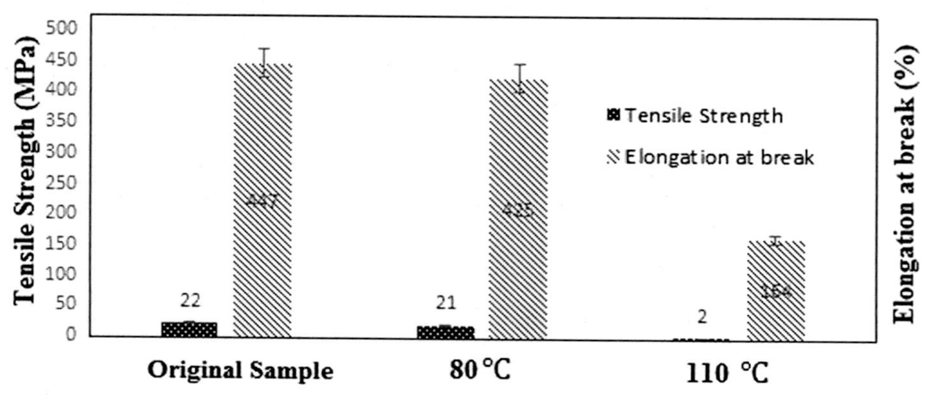

FIGURE 14.4 Influence of processing temperature on tensile strength and elongation at break [52]. *Reproduced with permission from Elsevier. From M. Sabzekar, M.P. Chenar, S.M. Mortazavi, M. Kariminejad, S. Asadi, G. Zohuri, Influence of process variables on chemical devulcanization of sulfur-cured natural rubber, Polym. Degrad. Stabil. 118 (2015) 88–95. https://doi.org/10.1016/j.polymdegradstab.2015.04.013.*

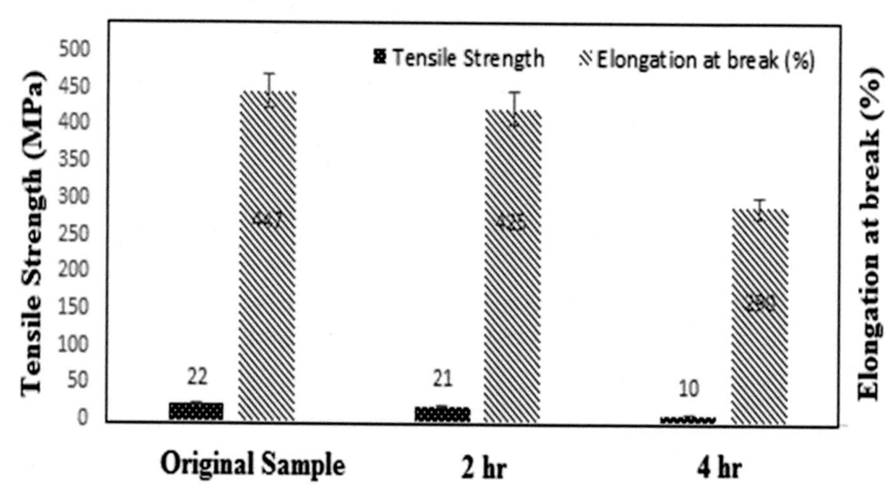

FIGURE 14.5 Influence of processing time on tensile strength and elongation at break [52]. *Reproduced with permission from Elsevier. From M. Sabzekar, M.P. Chenar, S.M. Mortazavi, M. Kariminejad, S. Asadi, G. Zohuri, Influence of process variables on chemical devulcanization of sulfur-cured natural rubber, Polym. Degrad. Stabil. 118 (2015) 88–95. https://doi.org/10.1016/j.polymdegradstab.2015.04.013.*

Rooj et al. [53] conducted devulcanization of NR in 2011 through chemical method and mechanochemical treatment using BPO. They determined the influence of concentration of devulcanizing agent and time on the rate of devulcanization. The devulcanized NR(DNR) from the mechanochemical treatment was added into the virgin NR with varying ratio. Morphological and mechanical results revealed that waste rubber recycling can be fulfilled by this method. Higher amount of DNR led to poor properties as well as poor morphology. Devulcanized rubber (DVR) can act as filler in composites only at a particulate level. Thaicharoen and coworkers [54] performed a mechanochemical devulcanization of vulcanized natural rubber (NV) using Thiosalicylic acid as devulcanizing agent. They optimized the conditions for mechanochemical method for NR vulcanizate, and it was found to be 1 phr thiosalicylic acid for 30 min at 140°C. Revulcanized rubber was also synthesized by adding different proportions of DVR into virgin NR. Mechanical studies revealed that DVR/NV composite shows 5%–10% reduction in tensile strength and 5%–10% increment in elongation at break of 5%–15% DVR content. Application of this method includes devulcanization of rubber tire of truck and partial recycling of waste tires in order to use in different rubber industries. This helps in reducing the need for new NR.

14.2.2 Ultrasound method

Ultrasound method can be considered as one of the promising methods for recycling of elastomeric products [1,55–57]. Ultrasonic waves selectively break chemical bonds such as C–S and S–S without any degradation of the C–C bond. But, after a particular level, these waves are able to cut carbon backbone of the rubber. Therefore, optimization of reaction conditions is very important in this method. This treatment, which is easily and quickly applied to materials, can provide high degree of devulcanization and control toward properties of devulcanizates. Ultrasound devulcanization of Ground Rubber Tire (GRT), Styrene Butadiene Styrene rubber (SBR), NR, and silicone rubber showed that three-dimensional cross-link breaking in vulcanized rubber materials can take place only at a particular temperature and pressure [58–63]. Devulcanization of rubber by means of ultrasonic method can be considered as a continuous process in the absence of chemicals [64]. The interesting factor is that the ultrasonically devulcanized rubber can be revulcanized in the same manner to the unvulcanized rubber.

In 1973, Pelofsky [65] carried out the first work in devulcanization of rubber through ultrasonic energy. Jushik et al. [66] developed a new grooved barrel ultrasonic reactor for recycling Ethylene Propylene Diene Terpolymer (EPDM) rubber (a roofing membrane). Mechanical properties, dynamic properties, cross-link density, gel fraction, and cure characteristics of virgin vulcanizate, DVR, and revulcanized rubber were evaluated. These results showed a strong dependency toward devulcanization processing conditions. Revulcanized rubber showed the same tensile strength as that of virgin vulcanizate. From TGA results, it is seen that the thermal stability of both the samples in air and nitrogen conditions remained same. However, thermal stability of devulcanized rubber depends on processing conditions in air environment. Revulcanized roofing membrane elasticity is found smaller than virgin roofing membrane, while elasticity of devulcanized roofing membrane is much higher than uncured virgin membrane (from DMA). Effect of particle size on ultrasonic devulcanization of tire rubber was investigated by Isayev et al. [35,36]. They compared two different particle-sized (10 and 30 mesh) tire rubbers manufactured by a latest ultrasonic twin-screw devulcanization extruder. The cross-link density, revulcanization behavior, and gel fraction were determined for understanding the degree of devulcanization. For understanding the variation between two different meshed tire rubbers after devulcanization and revulcanization, they carried out mechanical tests, rheological and swelling studies. The obtained properties of devulcanized and revulcanized rubbers of 10 and 30 mesh at various amplitudes can be grouped into two—smooth and sticky extrudates and rough extrudates and cohesive powder. The tensile strength and modulus at 100% elongation of the rough extrudates and cohesive powder were higher and the elongation at break was lower than that of smooth and sticky extrudates.

Feng et al. [67] conducted ^{1}H transverse relaxation for analyzing structural variations in ultrasonically treated devulcanized butyl rubber and butyl rubber gum after and before ultrasonic treatment. They expressed relaxation decay with the help of two-component model. A destructive effect was investigated in physical and chemical network and along the chain on gum and cross-linked butyl rubber as a result of ultrasound treatment. Seok and Isayev [68] carried out sulfur-cured unfilled butadiene rubber (BR) under various reaction conditions. They observed that BR has a narrow devulcanization window, and there is no notable deterioration in mechanical properties and has more elasticity compared to virgin BR. Gel permeation chromatography data indicated the expansion of molecular weight distribution and decrease in molecular weight during devulcanization and concluded that degradation and devulcanization of rubber happened at the same time. From rheological data, it can be understood that the uncured virgin BR was less elastic than devulcanized due to the weakening of bonds by the application of ultrasonic waves. Thermal degradation of devulcanized BR started earlier when compared to virgin vulcanizate and gum due to the same reasons [67].

14.2.3 Microwave methods

Microwave energy can be used for vulcanization, bonding, foaming, foam curing, and preheating due to their better efficiency, energy regulation, temperature control, and easiness of handling [6–8]. These radiations can also be used in devulcanization process in order to break S–S and C–S bond in rubbers. In this method, material absorbs microwave radiation through molecular interactions and change into heat, thereby temperature increases to 260–350°C, and the material can maintain their C–C bond in the main chain [10]. Comparing all other methods, microwave devulcanization is one of the most favorable methods for recycling of rubber products [1]. This method can be considered as an eco-friendly process because it never uses any chemicals. Moreover, uniform energy distribution can be achieved by following this technique [69]. Energy consumption in this method is very less due to the reduction in reaction time and consuming gases. Compared to conventional furnaces, microwave furnaces are smaller. Process parameters such as reaction time and power source can be easily controlled, thereby it is applicable to various rubbers with different characters for producing particular degree of devulcanization [70]. Microwave devulcanization helps to regain fluidity of rubber, and the products obtained can be recycled into various products. Distribution of temperature inside the heating material happens very quickly and accurately. Microwave-based devulcanization of waste rubber is familiar for 40 years; however, its application in industry is very less [3–5]. Microwave methods were first discovered by Goodyear [8] and Novotny [6–8]. In this method, waste rubber is exposed to microwave irradiation for a short time. This method is only applicable to polar group containing rubber, which is vulcanized by sulfur. There is an interaction between electric field of microwaves and charged ion and/or molecular dipoles in the material. Microwave method is based on three mechanisms such as interfacial polarization, ionic conduction, and dipolar polarization. Seghar et al. [28,31,67,68,71] studied the effect of ionic liquid (IL), pyrrolidinium hydrogen sulfate[Pyrr][H_2SO_4] on devulcanization of SBR by microwave method. They found that IL has a strong affinity for devulcanization of rubbers and decreases the energy needed. There is a decrease in cross-linking density. Zanchet et al. [72] used microwave method for devulcanizing SBR waste, and the obtained products are used in automobile applications. They prepared SBR composites by incorporating SBR industrial scraps (SBR-r). Initially, ground SBR-r is irradiated with microwave radiations for 1, 2, and 3 min, and then they added 80 phr SBR-r into SBR matrix. Mechanical and rheometric properties were investigated. Better results were obtained in composite with SBR-r devulcanized by microwave for 2 min. Enhancement in elongation at break and tensile properties are due to the increased interaction between virgin rubber matrix and devulcanized products. ATR-FTIR and cross-link density tests revealed some variations in chemical characteristics as a result of aging. Thus it could be understood that SBR-r and time can influence the aging process.

Liu et al. [73] developed a magnetic rubber composite from waste rubber by microwave devulcanization process. Powder of waste rubber and barium ferrite can be used as a starting material for the preparation of composite without any other additives. Microwave radiation was used for waste rubber's semidevulcanization and barium ferrite powder's surface modification. Enhanced compatibilities of barium ferrite and rubber matrix led to the synthesis of cost-effective magnetic composites having better ferromagnetic characters. This can be considered as a novel method to fabricate inorganic–organic nanocomposites. The TEM images of the waste rubber particles are shown in Fig. 14.6.

Colom et al. [74] conducted an experiment for analyzing the phenomenon behind devulcanization of GTR. For that they choose three different types of GTR containing carbon black and other inorganic compounds, and its structure after and before the process was investigated by FTIR, TGA, and SEM. As a result of thermo oxidation,

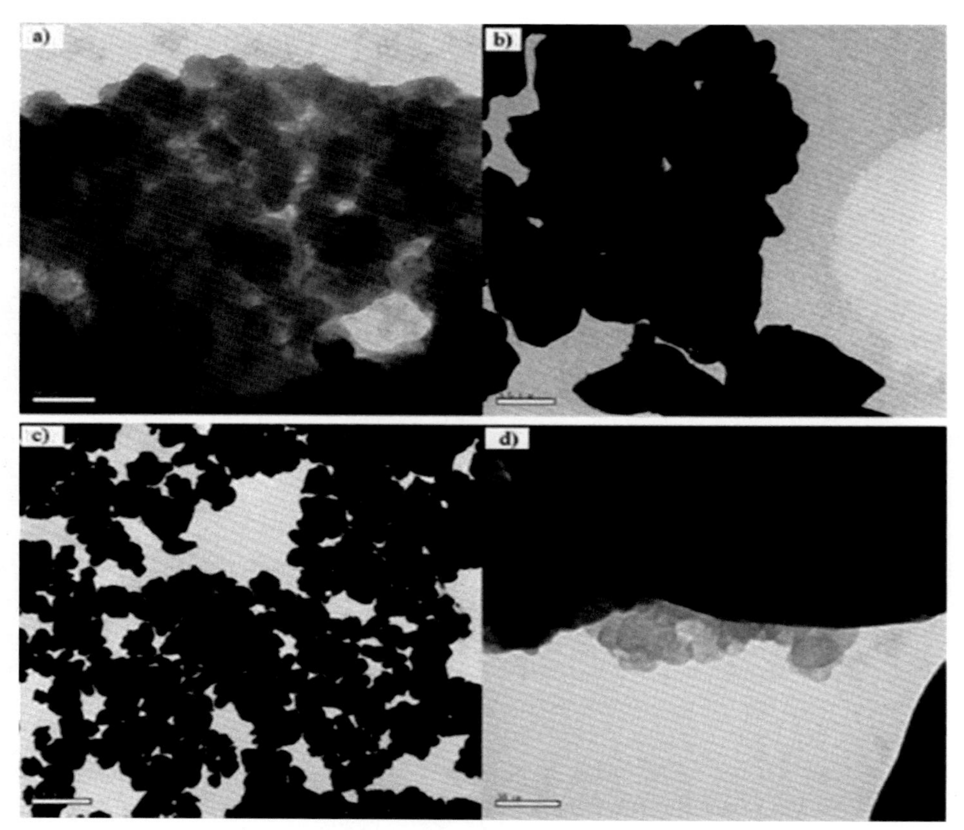

FIGURE 14.6 TEM images (A) WRP, (B) BaM powders, (C and D) mixtures of WRP/BaM after microwave treatment [73]. *Reproduced with permission from Elsevier. From J. Liu, P. Liu, X. Zhang, P. Lu, X. Zhang, M. Zhang, Fabrication of magnetic rubber composites by recycling waste rubber powders via a microwave-assisted in situ surface modification and semi-devulcanization process, Chem. Eng. J. 295 (2016) 73–79. https://doi.org/10.1016/j.cej.2016.03.025.*

reduction of black carbon occurs with the evolution of CO_2 during GTR devulcanization (from FTIR). This is also verified with the help of TGA results. Reduction of structural groups present on the double bonds as well as branching occurs due to microwave treatment. This method can also break C–H, C–S, and S–S bonds. In addition to this, holes, pores, and roughness can be introduced on the sample surface by this process. They concluded that SiO_2 has a major role for determining the degree of devulcanization. That means high SiO_2 content in GTR can be easily devulcanized than GTR with lower SiO_2, and SiO_2 can be considered as a catalyst in devulcanization process. Fig. 14.7 schematically represents the devulcanization process in detail.

Garcia et al. [75] tried to understand changes associated with microwave treatment in ground tire rubber, and they carried out thermal, chemical, morphological, and rheological studies on two types of rubber tires. Microwave treatment helped to enhance

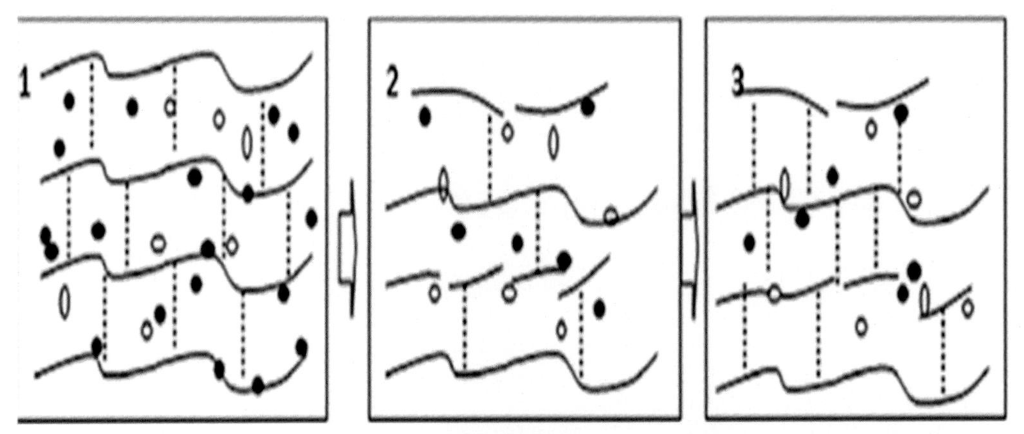

FIGURE 14.7 Evolution: Devulcanized and Revulcanized by MW method (1) previous devulcanization. (2) MW method: breakage of some backbone and S—S bond, reducing CB particles (3) Revulcanization process (*white dot,* SiO_2; *Black dot,* CB; *dotted line,* sulfur bridges [11]. *Reproduced with permission from Elsevier. From X. Colom, A. Faliq, K. Formela, J. Cañavate, FTIR spectroscopic and thermogravimetric characterization of ground tyre rubber devulcanized by microwave treatment, Polym. Testing 52 (2016) 200–208. https://doi.org/10.1016/j.polymertesting.2016.04.020.*

rubber fluidity through S—S bond breaking. But long-term irradiation caused degradation and changes in material properties in NR and SBR. Aoudia and coworkers [76] conducted a study for reusing waste tire rubber in polymer composite field. Grinded Waste Tire Rubber (WTR) was treated with microwave electromagnetic energy for devulcanization and the resultant powder inserted into a thermoset resin to generate new composite. S—S and C—S bond degradation and degree of devulcanization were confirmed from FTIR and swelling studies, respectively. They used these GTR and devulcanized ground tire rubber (DGTR) in epoxy resins separately for the preparation of composites. Mechanical properties of DGTR-filled epoxy composites are higher than that of epoxy filled with GTR. SEM results were also in agreement with the above observation. Microwave energy can influence the degree of devulcanization (maximum energy is 1.389 $kJ\ kg^{-1}$) but still high energy is needed. The relation between devulcanization and swelling degree is plotted in Fig. 14.8.

Molanorouzi and coworkers [77] also carried out devulcanization of waste rubber in a microwave oven in the presence of a devulcanizing agent and an oil. They found that particle size distribution of waste is a major factor for determining effective devulcanization, and effective devulcanization occurred at a particle size of 319 μm. Among various devulcanizing agents, diphenyl disulfide (DPDS) was found to be the better one. A part of total cross-links of polysulfide undergone breakage and remaining part exists as a mono sulfide bond. The free radical mechanism of tire rubber and polymer chain growth and sulfur cross-link are represented in Figs. 14.9 and 14.10.

Sousa et al. [78] studied the structural changes of GTR by the application of microwaves. They observed that final temperature plays a vital role in the success of the

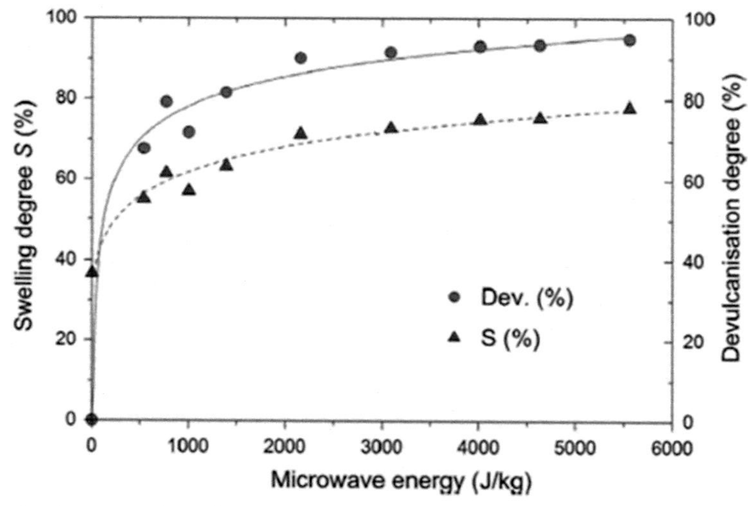

FIGURE 14.8 Devulcanization and swelling degree as a function of energy of treatment [76]. *Reproduced with permission from Elsevier. From K. Aoudia, S. Azem, N. Aït Hocine, M. Gratton, V. Pettarin, S. Seghar, Recycling of waste tire rubber: Microwave devulcanization and incorporation in a thermoset resin, Waste Manag. 60 (2017) 471–481. https://doi.org/10.1016/j.wasman.2016.10.051.*

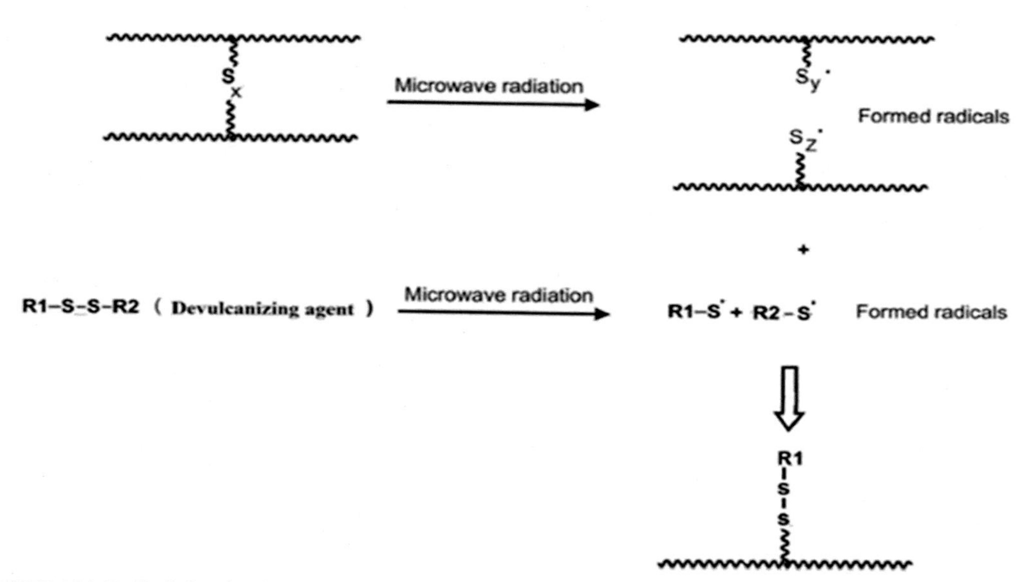

FIGURE 14.9 Radical devulcanization mechanism of tire rubber [77]. *Reproduced with permission from Elsevier. From M. Molanorouzi, S.O. Mohaved, Reclaiming waste tire rubber by an irradiation technique, Polym. Degrad. Stabil. 128 (2016) 115–125. https://doi.org/10.1016/j.polymdegradstab.2016.03.009.*

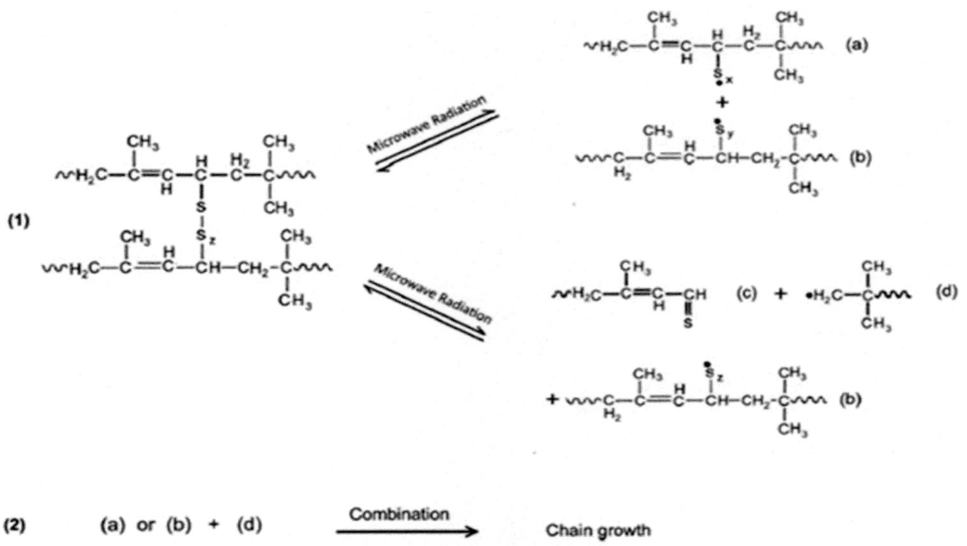

FIGURE 14.10 Polymer chain growth and sulfur cross-link mechanism [77]. *Reproduced with permission from Elsevier. From M. Molanorouzi, S.O. Mohaved, Reclaiming waste tire rubber by an irradiation technique, Polym. Degrad. Stabil. 128 (2016) 115–125. https://doi.org/10.1016/j.polymdegradstab.2016.03.009.*

process. Various kinds of sulfur bonds were broken depending upon the final sample temperature and the formation and breakdown of bonds balanced each other as shown in Fig. 14.11. Solvent extraction and swelling results showed that degree of devulcanization increased with increase in time of microwave exposure due the degradation of cross-linking. Presence of carbon black led to the absorption of more quantity of energy during treatment.

14.2.4 Thermomechanical methods

During the last few decades, waste rubbers were devulcanized using different thermomechanical methods. The thermomechanical method is a combination of mechanical shearing and high-temperature heat treatment of the material. Nowadays, the recycling of rubber materials by thermomechanical devulcanization methods uses chemical agents such as thiols, amines, unsaturated compounds, etc. Usually these chemicals act as softeners by reducing the thermal degradation of the vulcanizate by weakening the interaction between the filler and the rubber chains. The main advantage of this method is that this method can regain about 70%–80% of the material properties when compared with the original gum vulcanizate [79]. Using an open roll mixing mill and (3-triethoxysilyl propyl) tetra-sulfide as a devulcanizing agent, the NR devulcanization has been done by Ghorai et al. [80], and Diaz et al. [41] analyzed the devulcanization of EPDM rubber by shear mixing.

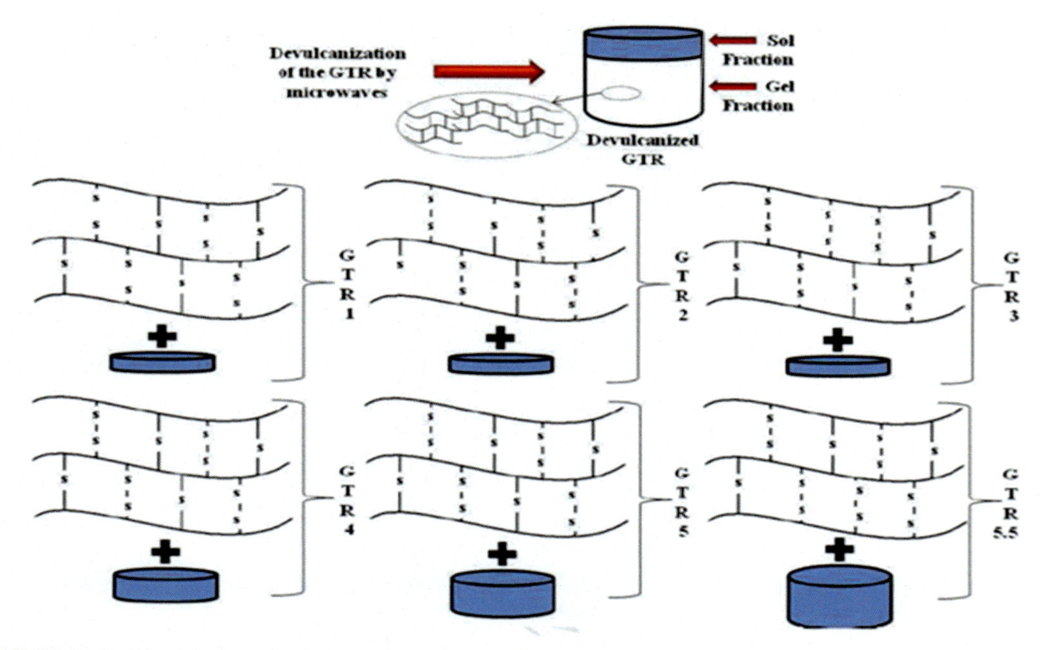

FIGURE 14.11 Scheme of C–S and S–S bond changes and gel fraction as a function of microwave irradiation time [78]. *Reproduced with permission from Elsevier. From F.D.B. de Sousa, C.H. Scuracchio, G.H. Hu, S. Hoppe, Devulcanization of waste tire rubber by microwaves, Polym. Degrad. Stabil. 138 (2017) 169–181. https://doi.org/10.1016/j.polymdegradstab.2017.03.008.*

An investigation about the continuous recycling of the EPDM in an extruder was reported by Suzuki et al. [81] and Mauri et al. [82] in 1998 and 2000, respectively. During the process, the selective C–S bond breakage leads to the formation of sol and gel components, which can be treated like a new elastomer. The use of the devulcanized rubber and ground rubber powder as fillers for NR has been studied by Li et al. [83]. They carried out the devulcanization of the rubber using shear flow extruder, and the addition of waste rubber did not alter the processability, whereas the abrasion resistance was improved. Zhang et al. [84] in 2009 carried out the devulcanization of GTR and a mixture of devulcanized rubber with virgin NR material in a pan-mill mechanochemical reactor at different ratios. The incorporation of waste tire powder was capable of enhancing the mechanical properties to a greater extent. Si et al. [85] analyzed the devulcanization of different GTR and thermoplastic elastomer materials in a twin-screw extruder. Higher shear rate was capable of bringing the chain scission and cross-link breaking, which leads to a reduction in gel content. The devulcanized elastomer blends were capable of showing enhanced mechanical properties. Wang et al. [86] carried out an experiment based on the devulcanization reaction of SBR-based ground rubber, which has been blended with EPDM using subcritical water as reaction medium in a twin-screw extruder. Swelling agent and the reaction medium promoted the selective bond breakage of cross-links and thereby reduced the gel particle and enhanced the mechanical properties in the blends made out of these devulcanized material. Using the thermomechanical shearing devulcanization, the processing ability of recycled rubber was studied by Tao et al. [87]. It was observed that the materials were capable of showing plasticity even after the devulcanization and observed distinct structures more specifically vulcanized structures among the revulcanizates. The devulcanization of rubber using a special screw design in a twin-screw extruder was studied by Formela et al. [88]. They observed that the efficiency of the process relay on the shear force and the pulverization helps in the processing in the screw extruder. Shi et al. [89] produced liquid reclaimed rubber (LRR) from GTR in a twin-screw extruder. This LRR was effectively used as a reactive polymer plasticizer in NR instead of conventional leachable oils, which helped in exhibiting enhanced mechanical properties.

14.2.5 Biological methods

Several microorganisms show biological activity toward sulfur, and hence, in the recent years microorganisms have been employed as a devulvanizing agent for scarp elastomers. The biological devulcanization ensures the selective breaking of cross-links without affecting the main chain of the polymer. Moreover, this is environmentally friendly as well as cost-effective way toward the recycling of elastomer waste. In other techniques, hazardous chemicals or high energy is required for the devulcanization; hence biological method can be considered as effective over other techniques. Li et al. [83]. devulcanized GTR by the biological agent Thiobacillus spp., which is having a sulfur oxidizing ability. Kim et al. [90] found an economical way of recycling scrap car tires by comparing the two different methods such as chemical treatment and microbial. In chemical they have used di-(cobenzanidophenyl)disulfide and in microbial they used *T. peromatabolis as* the devulcanizing agent. They concluded that both the treatments are efficient for devulcanizing the scrap rubber, but the microbial treatment was economically feasible process. In another study, Yao et al. [91] utilized desulfurizing capability of Alicyclobacillus spp. for the devulcanization of waste latex rubber (WLR). They have conducted a detailed study on the relation between the growth characteristics of the microorganism and the coculture desulfurization conditions as shown in Fig. 14.12.

The study reveals that Tween 80 has a great toxic effect on the microorganism Alicyclobacillus spp. and hence it should not be added individually to the culture medium. Romine et al. [92] compared the devulcanizing effect of bacterial strains such as Thiobacillus thiooxidans, Thiobacillus ferrooxidans, Sulfolobus acidocaldarius, Thiobacillus thioparus, Rhodococcus rhodochrous, and ATCC 39,327 on the GTR and concluded that among the bacteria, S. acidocaldarius had the higher efficiency in GTR. They also suggested a pathway known as “4S,” which ensures metabolization of sulfur into sulfate sulfoxide/sulfonate/sulfone by S. acidocaldarius. Bredberg et al. [93] used Pyrococcus furiosus as devulcanizing agent for GTR, but one of the drawbacks is that the

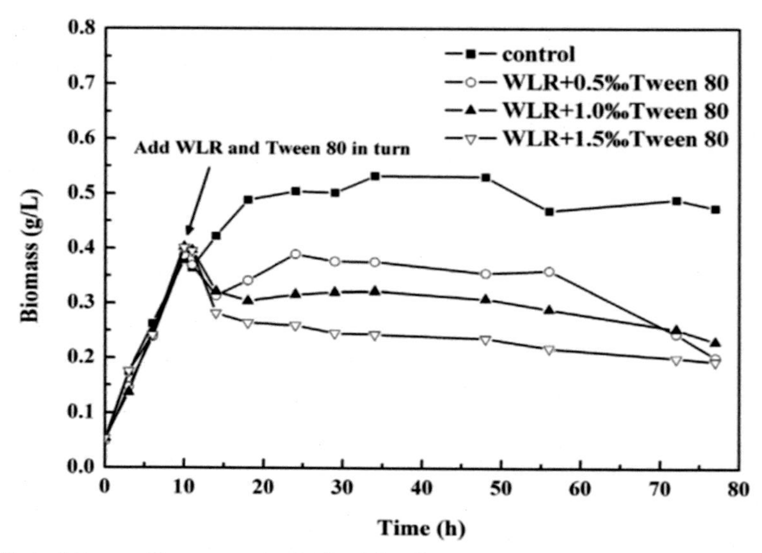

FIGURE 14.12 Effect of Tween 80 on growth of Alicyclobacillus spp. (WLR: 5%) [91]. *Reproduced with permission from Elsevier. From C. Yao, S. Zhao, Y. Wang, B. Wang, M. Wei, M. Hu, Microbial desulfurization of waste latex rubber with Alicyclobacillus sp, Polym. Degrad. Stabil. 98(9) (2013) 1724–1730. https://doi.org/10.1016/j.polymdegradstab.2013.06.002.*

rubber additives can adversely affect the bacterium. Li et al. [94] compared the desulfurization property of different microbes such as Sphingomonas, Thiobacillus spp., and T. ferrooxidans on GTR and found that Sphingomonas had great desulfurization effect because of its wide range of growth conditions. Löffler et al. [95] evaluated the effects of T. thioparus, T. ferrooxidans, and T. thiooxidans on the devulcanization of GTR. The obtained results were in favor of the T. thioparus bacteria, which are capable of devulcanizing rubber effectively than other bacteria. The study also revealed that the desulfurization and the particle size of GTR are associated with each other. Sato et al. [96] cultivated two types of microbes such as Dichomitus squalens and Ceriporiopsis subvermispora and kept in NR matrix for about 250 days, and the results show that the C. subvermispora decreased the intensity of SeC bonds; however, D. squalens had no effect on these bonds. Cui et al. [97] used both the bacteria Sphingomonas spp. and Gordonia spp. for the devulcanization of rubber. They have also used some other microorganisms such as Gordonia, Nocardia, Amycolatopsis, and Pseudomonas for the degradation of rubber. The effect of Nocardia on the devulcanization of rubber was carried out by Tsuchii et al. [98]. Shah et al. [99] also used the microbes such as genera, Gordonia (Gordonia westfalica and Gordonia polyisoprenivorans), Mycobacterium, and Nocardia to analyze its effect on the degradability of synthetic as well as NRs. Li et al. [100] used Acidithiobacillus ferrooxidans for the microbial reclaiming GTR by YT-1 cultured in media having varying Fe^{2+} concentrations. They observed that these bacteria have the ability to oxidize sulfur and subsequently devulcanize scrap rubber. Torma et al. [101]

used pure as well as mixed cultures of bacteria such as Thiobacillus thiooxidans and Thiobacillus ferrooxidans for the devulcanization of rubber. By measuring the amount of liberated SO_2 in the devulcanization process, they concluded that both the bacteria are capable of devulcanizing the rubber to the same extent in the pure as well as mixed proportions. This is because of the synergistic growth effect of the both the microorganisms.

14.2.6 Supercritical methods

With the advent of green chemistry, green solvents such as supercritical carbon dioxide and water started gaining importance in the field of devulcanization. Supercritical CO_2 ($scCO_2$) has the ability to exhibit diffusion behavior, which diffuses into polymers and is hence regarded as one of the best solvents for these matrices. Supercritical CO_2 possesses properties such as chemical inactiveness, nontoxicity, nonflammability, and cost-effectiveness. Additionally, it has an easily accessible critical point. Moreover, its removal is not a problem as it is gaseous at room temperature. Owing to this amazing property of supercritical fluids, researchers were in the process of exploring the various possible applications using polymers such as preparation of composites, blends, and foams. Thus the process of devulcanization started gaining momentum as a green technique or a sustainable process of waste management. Supercritical fluids (SCFs), especially supercritical water serve as a potential media for the recycling of fibers and resins owing to the cost-effectiveness and nontoxic nature of the medium of reaction. This can be attributed to the properties such as low viscosity, high mass transport coefficients, high diffusivity, and solvation power, which enable the effective recycling of rubber.

Mangili et al. [102] have carried out a statistical approach for evaluating the GTR devulcanization by making use of a green solvent such as $scCO_2$ and diphenyl disulfide (DD) as a devulcanizing reagent. An exhaustible portion of the previous literature focuses on the reclaiming of NR. Nevertheless, there are some works that focus on other matrices also. Raul et al. [103] conducted a detailed investigation on the recycling of carbon fiber epoxy composites using near-critical and supercritical water. They also conducted a study on the effect of reaction conditions on the process and suggested the potential chemical additives that can cause an increase in the rate of the reaction. Asaro et al. [35] have done a detailed study on the thermomechanical properties in connection with GTR devulcanization. The study was carried out in a twin-screw extruder under diverse operation conditions. This study also exploited the use of supercritical carbon dioxide in the process of devulcanization by varying the parameters such as temperature, screw speed, and quantity of CO_2. They have also analyzed the cross-link density and the percentage of devulcanization in the course of the study. Various characterization techniques such as infrared spectroscopy (FTIR) and gel permeation chromatography (GPC) were carried out and correlated with the viscosity of the substrate. The results indicated that the extruder temperature and speed of extrusion have an effect on the cross-link density of the material.

Zhang et al. [104] also utilized green solvents such as supercritical carbon dioxide in the devulcanization process of recycled tire rubber crumb via a twin-screw extruder. The primary aim of their study was to evaluate whether recycled rubber powder of varying particle sizes could be extruded in a twin-screw extruder in the absence of supercritical CO_2 and to observe the chemical changes proceeding the devulcanization process. They also probed into the mechanical properties of the devulcanized system after compounding with curing agents. The effective use of green solvents marked a change in the concept of devulcanization and marked a change in the concept that vulcanized rubbers cannot be reextruded due to cross-linking. Park et al. [94] investigated on the conversion of waste plastics and rubbers into low-molecular-weight compounds and its effective utilization as a fuel using supercritical carbon dioxide as the medium. Kojima et al. [105] found that green solvents are effective in penetrating into a polymer matrix such as rubber (NR) and synthetic rubbers leading to the swelling of rubber vulcanizates. Tzoganakis [81,104] carried out an investigation on the devulcanization of waste rubber and found that the process could be carried out inside an extruder itself by utilizing a supercritical solvent such as carbon dioxide.

14.3 Value-added products from revulcanized elastomeric blends and composites

As discussed earlier, it is possible to recycle elastomers, their blends and composites by various methods. But the process will be worth if it could find application in different fields. Recycled elastomeric materials thus obtained should be tested/analyzed for ensuring their application in different fields and for the development of value-added products. This section discusses various products or applications derived from recycled elastomeric blends and composites. Zanchet et al. [72] devulcanized styrene BR by means of microwave treatment and incorporated into the rubber matrix followed by the revulcanization; they have made rubber composites for the automotive applications. The study shows the effect of microwave-assisted DVR as a filler or a reinforcing agent. The increased tensile strength and reduction in modulus were in support of the suggested application of the DVR. Zhang et al. [104] followed mechanochemical treatment of devulcanization for GTR for the enhancement of acoustic absorption polyurethane foam. The devulcanization of GTR increases the flexible chain content and subsequent adsorption of sound waves. Aoudia et al. [76] used recycled waste rubber material as a filler for epoxy matrix for enhancing the mechanical properties of epoxy by the microwave treatment. Luo et al. [106] devulcanized GRT by means of ultrasonic and dynamic methods, and then revulcanized (RGRT) forms have used this for the synthesis of Polypropylene (PP) blends. The RGRT/PP blends were showing two-phase morphology and have shown better enhancement in mechanical properties. Joseph et al. [107] devulcanized carbon-black-filled rubber by means of mechanical methods and made blends of devulcanized and virgin rubber. Blends were showing comparable vulcanizate properties, aging resistance, and improved tear strength. Lapkovskis et al. [108] used

DVR tires for the for oil spill remediation. The DVR contains coarse particles, the end product may be in the form of powder or sponge as shown in Fig. 14.13.

Thus, the obtained particles can be used as adsorbent materials since they are oleophilic in nature, i.e., they are capable of absorbing oil and petroleum products. Compared to conventional crumb rubber (CCR), devulcanized crumb rubber (DCR) is capable of showing the maximum adsorption and oil-retention capacity (Fig. 14.14). Allouch et al. [109] in their work employed easy and cost-efficient way to prepare new composite from scrap elastomers (NR) by combining different composition of aluminum powder. The main objective of their work was to investigate the improvement in the

FIGURE 14.13 (A) DCR powder; (B) DCR sponge [108]. *Reproduced with permission from Elsevier. From V. Lapkovskis, V. Mironovs, D. Goljandin, Suitability of devulcanized crumb rubber for oil spills remediation, Energy Procedia 147 (2018) 351–357. https://doi.org/10.1016/j.egypro.2018.07.103.*

FIGURE 14.14 Wettability crumb rubber by motor oil (motor oil density 0.89 g/cm^3, test duration 30 s): (A) CCR; (B) DCR. [108] *Reproduced with permission from Elsevier. From V. Lapkovskis, V. Mironovs, D. Goljandin, Suitability of devulcanized crumb rubber for oil spills remediation, Energy Procedia 147 (2018) 351–357. https://doi.org/10.1016/j.egypro.2018.07.103.*

mechanical properties of recycled composites, which are needed for specific applications. The results showed aluminum powder incorporation lead to a positive and specific effect on rubber performance. The micrographs obtained from SEM highlight homogeneous distribution of aluminum powder in the matrix, which ensures better performance of prepared composites. They concluded that the prepared composites are "environmentally friendly" materials with specific properties.

Gupta et al. [110] developed adsorbent for the selective removal of lead and nickel from the aqueous solutions by the physical treatment of waste tire materials. They have developed novel carbon having porous structure from the waste tire, and these materials were capable of showing desorption as well as reusability. SEM and EDAX of the adsorbent material before and after adsorption are given in Fig. 14.15.

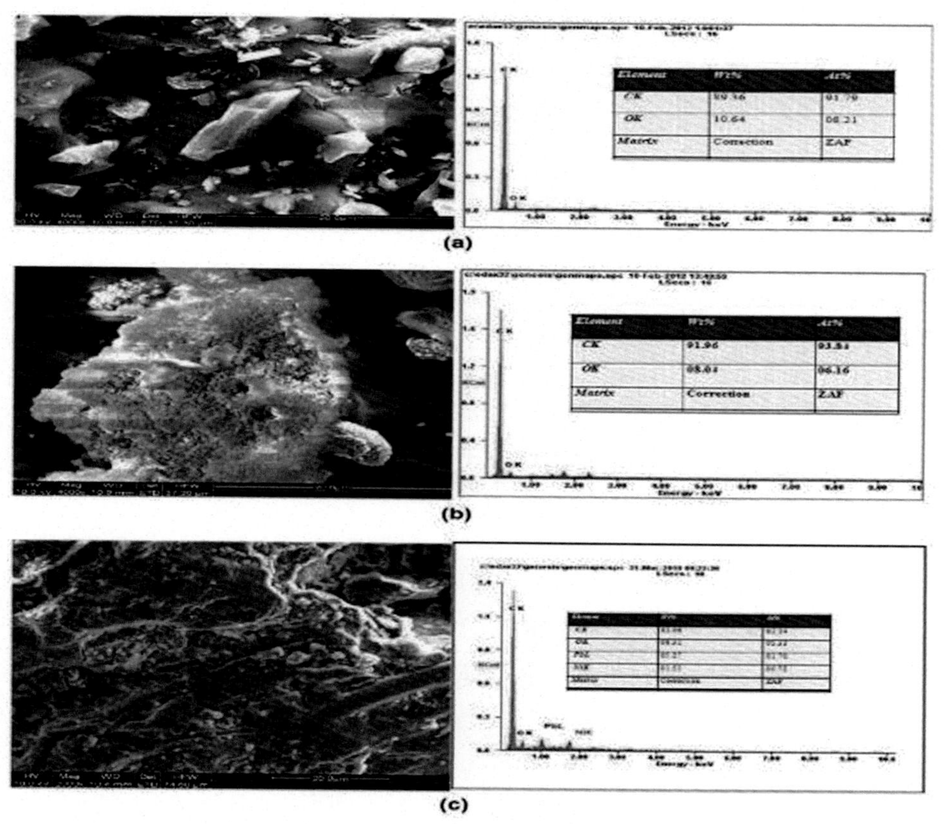

FIGURE 14.15 SEM micrograph and EDAX of (A) CAC, (B) RTAC before adsorption, (C) RTAC after adsorption of Pb^{2+} and Ni^{2+} [110]. *Reproduced with permission from Elsevier. From V.K. Gupta, M.R. Ganjali, A. Nayak, B. Bhushan, S. Agarwal, Enhanced heavy metals removal and recovery by mesoporous adsorbent prepared from waste rubber tire, Chem. Eng. J. 197 (2012) 330–342. https://doi.org/10.1016/j.cej.2012.04.104.*

Vitrimers possess exchangeable links in the cross-linking networks, giving them abilities of reprocessing and recycling. But most of vitrimers are usually fabricated by complex synthesis and polymerization. Toxic catalysts are applied to activate the reaction to rearrange the cross-linking networks. These drawbacks limit the widespread applications of vitrimers. Moreover, most reported vitrimers could only partially maintain or severely deteriorate their mechanical properties after recycling. Wang et al. [86] for the first time revealed a catalyst-free and recycle-reinforcing elastomer vitrimer. The elastomer vitrimer associated with exchangeable b-hydroxyl ester bonds was obtained by reactive blending of epoxidized NR and carboxylated nitrile rubber. It is observed that the vitrimer exhibits an exceptional recycle-reinforcing property. Thus they developed a feasible way to fabricate elastomer vitrimers, which promote the recycling of cross-linking commercial available elastomers.

The present use of cross-linking methods in rubber industry is lacking variable persistent issues, such as the use of toxic curing packages, release of volatile organic compounds (VOCs). Zhang et al. [111] reported a new "green" strategy for cross-linking diene-type elastomers. They efficiently cross-linked carboxylated nitrile rubber (XNBR), by a bio-based agent, epoxidized soybean oil (ESO), without any toxic additives. They observed that ESO exhibits good plasticization effect and very high scorch safety for XNBR. They also noticed that by simply varying the ESO content, the cross-linking density and mechanical performance of ESO-cured XNBR can be varied. Further, zinc oxide was added as a catalyst to accelerate the reaction and enhance the cross-linking. This zinc oxide also acted as reinforcement to improve the overall mechanical performance of the ESO-cured XNBR. The end-of-life elastomer materials exhibit a closed-loop recovery by selectively cleaving the ester bonds, resulting in very high recovery of the mechanical performance of the recycled composites. They concluded that this strategy gives a green platform to cross-link diene elastomers.

Karabork et al. [112] used microwave-assisted devulcanized into matrix and analyzed the friction as well as wear resistance. They have mixed unreacted and devulcanized (DVR), and then it is mixed with virgin and compared the properties. It is observed that the DVR added shows improved mechanical and tribological property. Tripathy et al. [113] devulcanized NR as well as and analyzed its polymeric behavior by the differential scanning calorimetry (DSC) in order to use it as a plasticizer instead of oil. The presence of particulate carbon content in the DVR showed better mechanical property retention after aging, moreover showed lesser extractable content in acetone. Gugliemotti et al. [114] derived a new technique of rubber recycling called direct powder molding, in which rubber pads were made from rubber powders by mechanical molding without virgin rubber or linking agents. They have found that the rubber powder distribution strongly affects the mechanical performance of the molded products. Smaller rubber particles showed better property than the fine and medium-sized particles. The absence of virgin rubber and the complete reuse of scrap rubber make this a better and advanced choice of recycling.

De et al. [31] devulcanized GTR by mechanical method in the presence of a multifunctional reclaiming agent, this was revulcanized with NR, and the blends were capable of good aging performance than the control. Lehmann et al. [115] made activated carbon from the rubber waste by following the method of pyrolysis and employed it as a good adsorbent to solve the air quality problems. The mesoporous structure of these activated carbons is capable of enhancing the adsorbing properties and has the ability to absorb the poisonous gases. Sasson et al. [116] derived a technique for the development of adsorbents from the waste rubber by the process of devulcanization. Miguel et al. [117] pyrolyzed rubber waste and have used it for the adsorption of nitrogen gas. The synthesized chars were showing better adsorption capacity due to the higher mesoporous volume of 0.1 mL/g. In order to avoid noise and environmental pollution, Gandoman et al. [118] developed a new class of concrete composite using DVR (Fig. 14.16). In the study, instead of Portland cement, geopolymeric binders were used as the concrete material and mixed with DVR for the production of noise screens. It is observed that as the loading of the rubber content increases, high transmission loss occurs showing better sound barrier properties (Fig. 14.17).

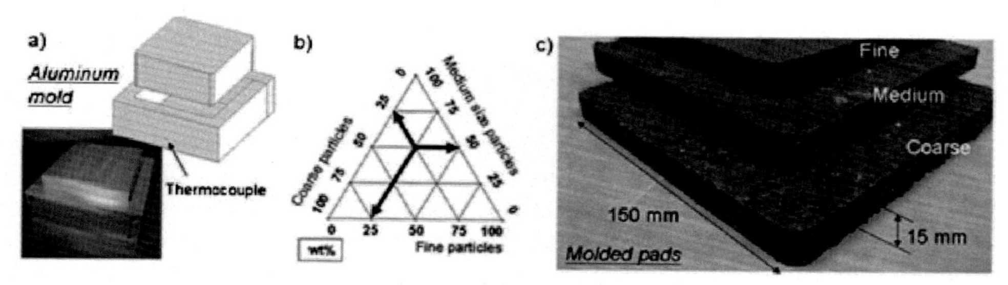

FIGURE 14.16 (A) mold; (B) molded blends of rubber powder; (C) molded plates with 100% of initial powders [118]. *Reproduced with permission from Elsevier. From M. Gandoman, M. Kokabi, Sound barrier properties of sustainable waste rubber/geopolymer concretes, Iranian Polym. J. 24(2) (2015) 105–112. https://doi.org/10.1007/s13726-014-0304-1.*

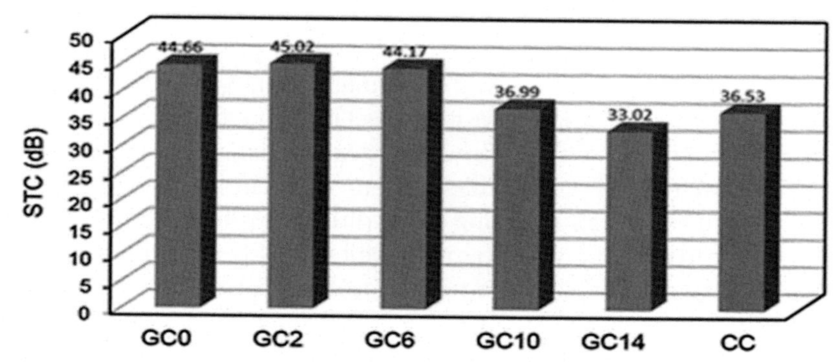

FIGURE 14.17 Variation of sound transmission classes for conventional and waste rubber geopolymer concrete samples [118]. *Reproduced with permission from Elsevier. From M. Gandoman, M. Kokabi, Sound barrier properties of sustainable waste rubber/geopolymer concretes, Iranian Polym. J. 24(2) (2015) 105–112. https://doi.org/10.1007/s13726-014-0304-1.*

Jaimson et al. [119] derived a method of developing particle board products from DVR materials with desirable acoustical values. Seok et al. [68] developed floor noise isolation system using the DVR material. Irez et al. [45] developed DVR modified epoxy-based composites reinforced with nano-magnetic iron oxide. For the synthesis of the composites, the matrix was prepared by treatment of epoxy with 10% recycled rubber. In order to make the rubber compatible with the epoxy matrix, devulcanization was carried out to break the existing bonds and other links were generated. After the chemical as well as microwave treatment, the epoxy is mixed with it for the synthesis of matrix, and subsequent addition of the iron oxide was done for composite synthesis. These composites were good enough for the development of electronic devices for the field of aeronautical engineering. Shi et al. [89] produced liquid reclaimed rubber (LRR) from GTR and used it as a polymeric plasticizer instead of environmental aromatic oil (EAO) in NR and compared the properties of LRR/NR and EAO/NR. Plasticized LRR has advanced tensile strength, modulus at 100% and 300% elongation, and it showed better hardness. The extraction resistance as well as thermostability was higher than EAO/NR (Fig. 14.18). The extraction resistance of NR increased with higher loadings of LRR, irrespective of EAO. In the case of EAO-modified systems, the resistance is not much pronounced. In both the cases, the tensile strength decreases and the elongation at break increases, showing the plasticizing effect of both. But for a particular content, higher tensile strength is shown by LRR than EAO. Stress–strain curve of both the plasticizer added NR samples is shown in Fig. 14.19.

Many investigations were done to develop cost-effective seismic protection systems with large-scale utilization range. In this view, researchers attempted to replace steel sheets of conventional steel reinforced elastomeric isolators with flexible

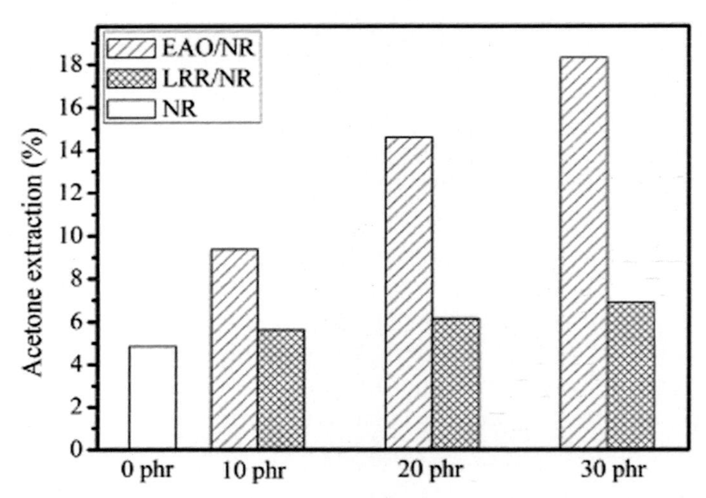

FIGURE 14.18 Comparison of acetone extraction of NR vulcanizates plasticized by LRR and EAO [34]. *Reproduced with permission from Elsevier. From J. Shi, H. Zou, L. Ding, X. Li, K. Jiang, T. Chen, X. Zhang, L. Zhang, D. Ren, Continuous production of liquid reclaimed rubber from ground tire rubber and its application as reactive polymeric plasticizer, Polym. Degrad. Stabil. 99(1) (2014) 166–175. https://doi.org/10.1016/j.polymdegradstab.2013.11.010.*

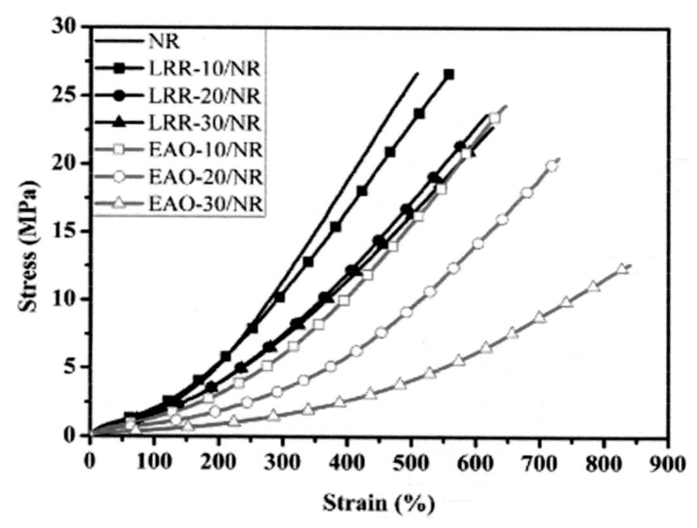

FIGURE 14.19 Stress–strain curves of NR vulcanizates plasticized by different amounts of EAO and LRR [34]. *Reproduced with permission from Elsevier. From J. Shi, H. Zou, L. Ding, X. Li, K. Jiang, T. Chen, X. Zhang, L. Zhang, D. Ren, Continuous production of liquid reclaimed rubber from ground tire rubber and its application as reactive polymeric plasticizer, Polym. Degrad. Stabil. 99(1) (2014) 166–175. https://doi.org/10.1016/j.polymdegradstab.2013.11.010.*

reinforcements. The performance of fiber-reinforced elastomeric isolators in unbounded configuration employing recycled elastomers was verified by previous studies. The use of recycled rubber for such isolaters is very interesting as they are very economic and have less environment impact. Losanno et al. [101] in their study compared the performance of such recycled rubber bearings and NR fiber reinforced elastomeric isolators. They discussed the manufacturing process and the features of these two kinds of rubber. They concluded that the performance of NR isolators was found to be better than that of recycled isolators. But recycled rubber devices are very promising when it comes to protecting nonengineered constructions from moderate to high seismic inputs at a fraction of the cost of traditional technologies.

Liu et al. [73] recycled rubber particle waste by microwave assisted in-situ surface modification as well as semidevulcanization process. The semidevulcanization and the modification of the filler barium ferrite were carried out simultaneously for the synthesis of magnetic rubber composites. The schematic representation of low cost fabrication method is given in Fig. 14.20. The magnetic behavior of the composites was determined by the help of vibrating sample magnetometer. The obtained composite results are shown in Fig. 14.21. From the figure it is clear that the composites are showing characteristic ferromagnetic behavior. The broad hysteresis loop shows the effective application of the composite as flexible magnets.

Jia et al. [120] made flexible electromagnetic interference shielding (EMI shielding) material from GTR waste in a low cost way. These flexible materials were having high

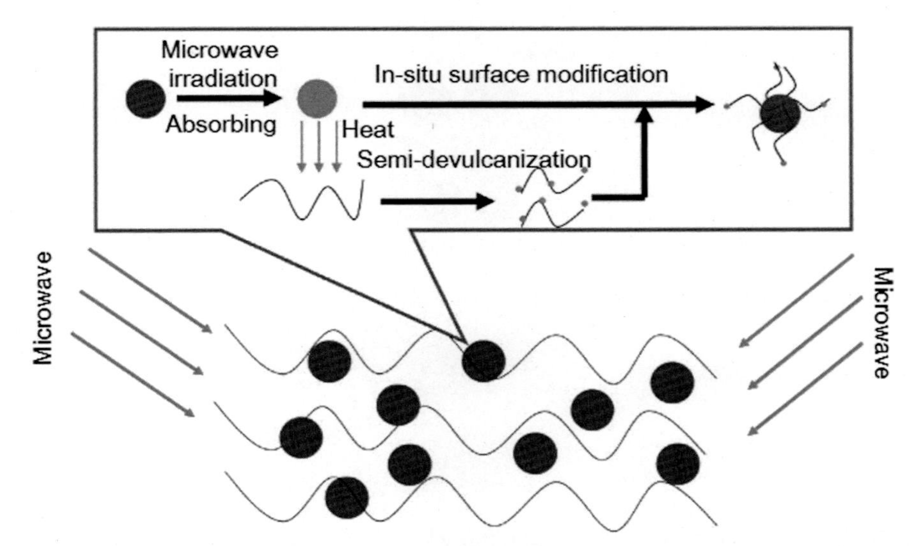

FIGURE 14.20 Schematic representation of fabrication of the magnetic rubber based on barium ferrite and waste rubber via a microwave-assisted in situ-surface modification and semidevulcanization process [73]. *Reproduced with permission from Elsevier. From J. Liu, P. Liu, X. Zhang, P. Lu, X. Zhang, M. Zhang, Fabrication of magnetic rubber composites by recycling waste rubber powders via a microwave-assisted in situ surface modification and semi-devulcanization process, Chem. Eng. J. 295 (2016) 73–79. https://doi.org/10.1016/j.cej.2016.03.025.*

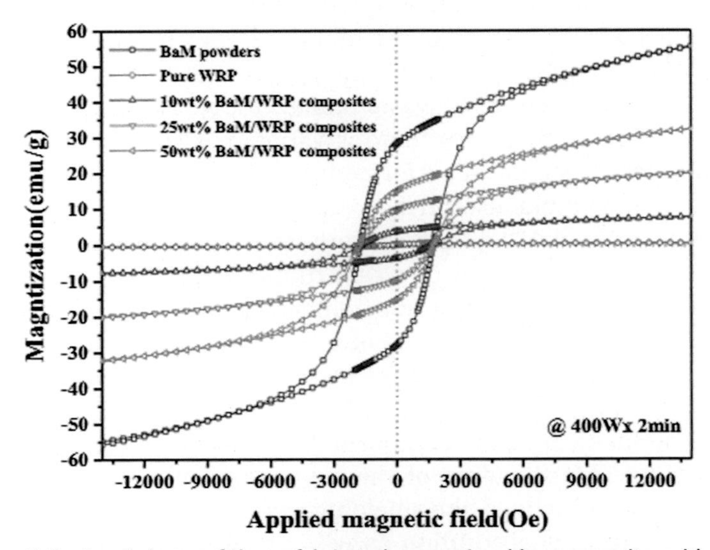

FIGURE 14.21 Magnetic hysteresis loops of the as-fabricated magnetic rubber composites with different BaM contents after microwave irradiation at 400 W for 2 min (the loops of both BaM and pure WRP are included for comparison) [73]. *Reproduced with permission from Elsevier. From J. Liu, P. Liu, X. Zhang, P. Lu, X. Zhang, M. Zhang, Fabrication of magnetic rubber composites by recycling waste rubber powders via a microwave-assisted in situ surface modification and semi-devulcanization process, Chem. Eng. J. 295 (2016) 73–79. https://doi.org/10.1016/j.cej.2016.03.025.*

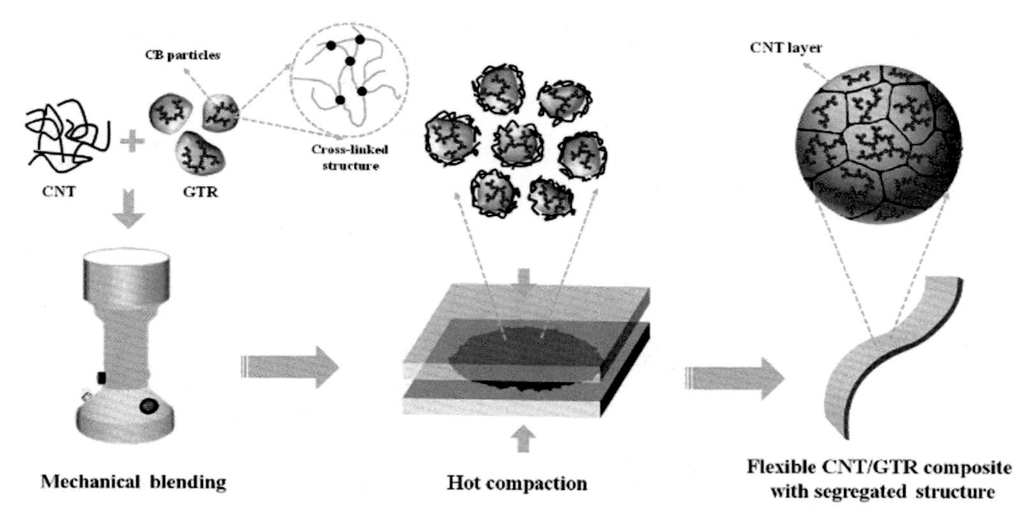

FIGURE 14.22 Schematic for the fabrication of flexible CNT/GTR composite with segregated structure [121]. *Reproduced with permission from Elsevier. From L.C. Jia, Y.K. Li, D.X. Yan, Flexible and efficient electromagnetic interference shielding materials from ground tire rubber, Carbon 121 (2017) 267–273. https://doi.org/10.1016/j.carbon.2017.05.100.*

potential application as flexible electronics in the field of flexible displays, health monitoring systems, and flexible power supply systems. In the study they have converted highly cross-linked 3D GTR into EMI shielding material by making composite with carbon nanotube. The composite synthesis method is shown in Fig. 14.22. Recycling of waste rubber is always a challenge in waste management. The use of mechanical recycling leads to processing difficulty as there is variation of raw materials, and the end products often have less mechanical performance. This drawback of recycling of plastics blends and plastic–rubber blends can be rectified by application of compatibilizing agents. L. Simon-Stőger et al. [121] in their work investigated the role of compatiblizing agents for enhancing the properties of waste rubber containing polypropylene. They studied the properties on effective elastomer concentrations, on elastomer ratios, and on proper structures of compatibilizers. It is observed that the mechanical performance could be improved by almost 44% in additive blends compared to neat uncompatibilized blends. They confirmed the test results by SEM, FTIR, and rheology [120].

The electrical conductivity of the composite increased with the increase in the CNT concentration and the EMI shielding of the composites also increased with the CNT loading. The electrical conductivity and shielding were showing the same trend as shown in Fig. 14.23. The microwave absorption as well as reflection was monitored, the absorption increases noticeably, whereas reflection varies slightly. Tanaka et al. [122] developed a technology to develop thermoplastic elastomer rubber scrap. Ethylene-propylene diene rubber wastes were devulcanized and blended with PP for the synthesis of elastomers. These materials are capable of showing elastic and mechanical

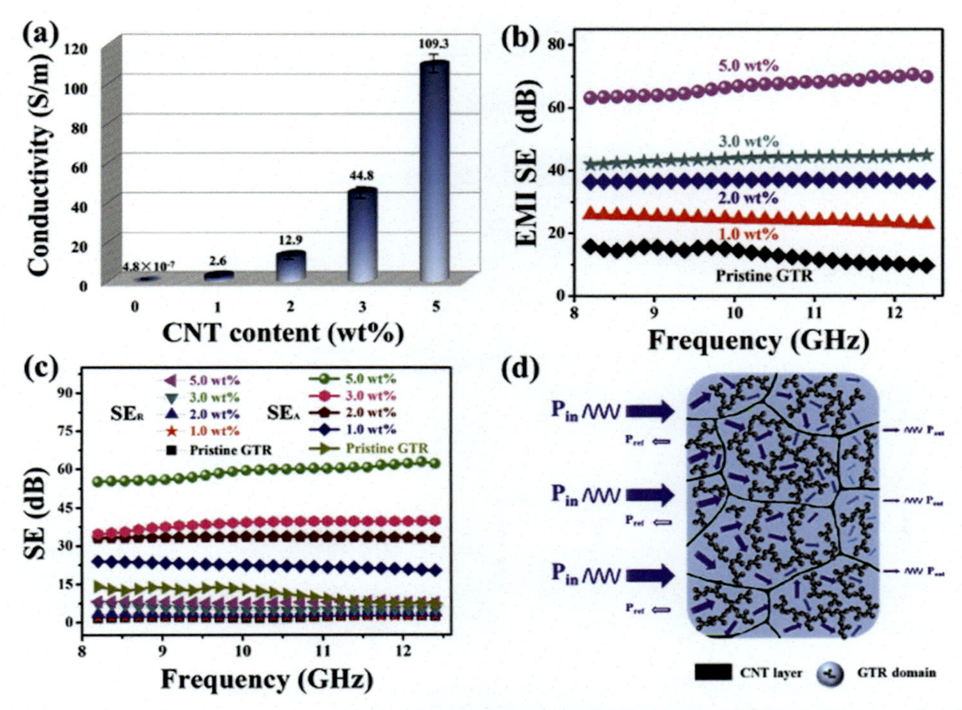

FIGURE 14.23 Electrical conductivity (A) and EMI SE (B) of the pristine GTR and the s-GTR composites as a function of CNT loading; (C) Microwave reflection (SER) and microwave absorption (SEA) versus CNT loading for the s-GTR composite and the pristine GTR; (D) Schematic representation of the EMI shielding mechanism for the s-GTR composites [124]. *Reproduced with permission from Elsevier. From L.C. Jia, Y.K. Li, D.X. Yan, Flexible and efficient electromagnetic interference shielding materials from ground tire rubber, Carbon 121 (2017) 267–273. https://doi.org/10.1016/j.carbon.2017.05.100.*

properties similar to that of thermoplastic olefins due to the presence of phase structure. Sarath et al. [71] revulcanized silicon rubber and plastics obtained from mobile phones to enhance the inflammability and mechanical strength of rubber. Gupta et al. [90] developed mesoporous material from scrap rubber by means of chemical and thermal treatment. Highly carbonaceous adsorbent thus formed is used for pesticide removal from water. Gupta et al. [123] developed carbonaceous materials from the waste rubber for the removal of phenolics from aqueous bodies.

Wölfel [125] worked on Thermoplastic Polyurethane (TPU), a unique engineering material due to the interactions of hard and soft segments within the chain. Since TPU is highly expensive, its recycling is of having high research interest. Hence the recycling of TPU and its utilization as a functional product are very much needed for a circular economy. In this study, they made new insights into the change in properties of TPU after recycling and then the processing of recycled TPU into nonwoven fabrics. The raw material reveals an exponential decrease in molar mass with recycling runs. The most commonly used materials become brittle as a result of recycling, thus TPU gets weaker

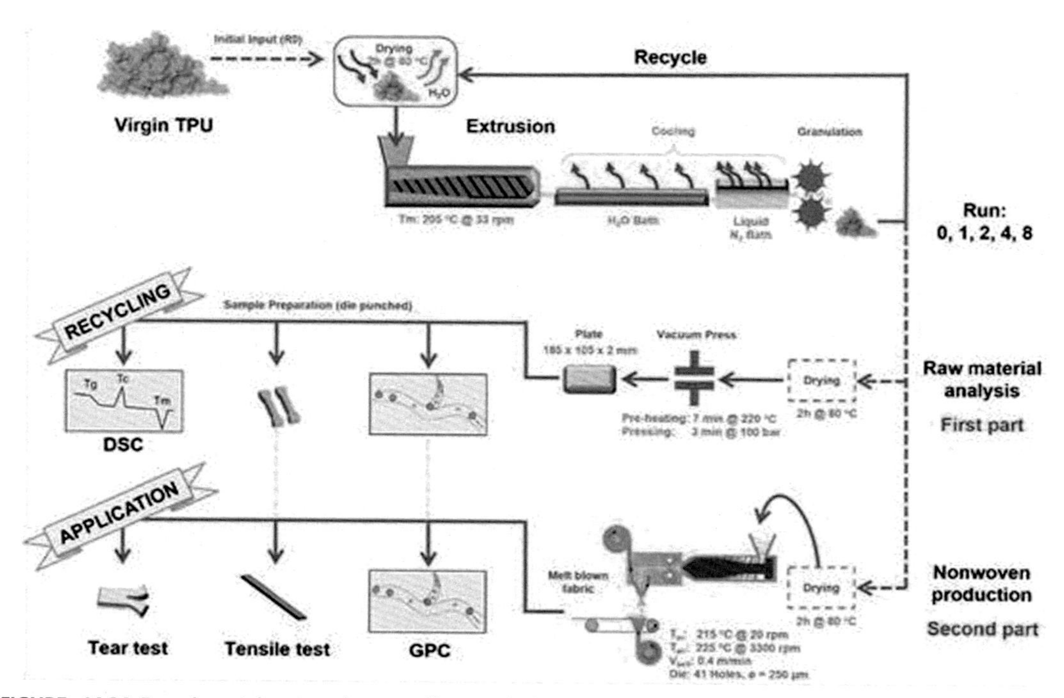

FIGURE 14.24 Experimental setup for recycling and development of value-added product from TPU [125]. *Reproduced with permission from Multidisciplinary Digital Publishing Institute. From L.C. Jia, Y.K. Li, D.X. Yan, Flexible and efficient electromagnetic interference shielding materials from ground tire rubber, Carbon 121 (2017) 267–273. https://doi.org/10.1016/j.carbon.2017.05.100.*

for deformations larger than 100% and stays identical. DSC measurements indicate that the crystallinity of virgin and recycled material (R8) is comparable even after eight times of recycling. Nonwoven fabrics were produced from various recycled grades by following melt blowing (Fig. 14.24) at constant extrusion conditions, and these fabrics hold the possibility to tailor the aspired material performance. Thus it is possible to develop advanced functional materials from recycled elastomer blends and composites.

14.4 Conclusion

From the detailed study, it can be understood that devulcanization methods are very effective ways of recycling of scrap elastomer than other conventional methods. The conversion of scrap elastomer into devulcanized form having the similar properties of virgin elastomer offers possibility of revulcanization and the development of advanced materials from it. Being a sustainable method, the devulcanization of scrap rubber and subsequent revulcanization can lower environmental pollution. The application of the DVR in sound adsorption and heavy metal adsorption is effective in combating sound and water pollution. The application of DVR as filler as well as matrix reduces the uses of virgin rubber and their application in automobiles, electronics, and EMI shielding. Hence the development of advanced functional materials from the devlcanized rubber is to be treated with immense research interest.

14.5 Future perspectives

As the environmental impact of elastomers is considered as a crucial problem, the devulcanization and reutilization of elastomer materials are very essential. The development of value-added products from devulcanized plastics and elastomers is indispensable for a sustainable world; hence the development and tailoring of applications of recycled material are highly vibrant field of research. It can avoid many of the environmental crisis in future. The recycling of elastomers is still in its initial stage; hence the development of new methods for recycling is highly needed for a sustainable future. The development of value-added products with high mechanical strength and physical properties is of having high research interest. It would be highly beneficial if one could develop advanced products with better stability and lifetime from recycled rubber. Hence the recycling of elastomers can be considered as an active area of research.

References

[1] B. Adhikari, D. De, S. Maiti, Reclamation and recycling of waste rubber, Prog. Polym. Sci. 25 (7) (2000) 909–948, https://doi.org/10.1016/S0079-6700(00)00020-4.

[2] K. Fukumori, M. Matsushita, H. Okamoto, N. Sato, Y. Suzuki, K. Takeuchi, Recycling technology of tire rubber, JSAE Rev. 23 (2) (2002) 259–264, https://doi.org/10.1016/S0389-4304(02)00173-X.

[3] K. Formela, A. Hejna, Zedler, X. Colom, J. Cañavate, Microwave treatment in waste rubber recycling – recent advances and limitations, Express Polym. Lett. 13 (6) (2019) 565–588, https://doi.org/10.3144/expresspolymlett.2019.48.

[4] K. Stevenson, B. Stallwood, A.G. Hart, Tire rubber recycling and bioremediation: a review, Ann. Finance 12 (1) (2008) 1–11, https://doi.org/10.1080/10889860701866263.

[5] F. Yi, Z. Maosheng, W. Ying, The status of recycling of waste rubber, Mater. Des. (2001) 123–128, https://doi.org/10.1016/s0261-3069(00) 00052–2.

[6] R. Chhabra, S. Marik, Review literature on the use of waste plastics and waste rubber tyres in pavement, Int. J. Core Eng. Manag. 1 (2014).

[7] G. Marković, O. Veljković, M. Marinović-Cincović, V. Jovanović, S. Samaržija-Jovanović, J. Budinski-Simendić, Composites based on waste rubber powder and rubber blends: BR/CSM, Compos. B Eng. 45 (1) (2013) 178–184, https://doi.org/10.1016/j.compositesb.2012.08.013.

[8] Microwave Devulcanization of Rubber, 1978.

[9] E. Abraham, B. Cherian, P. Elbi, L. Pothen, S. Thomas, Recent advances in the recycling of rubber waste, Recent Dev. Polym. Recycl. 47 (2011).

[10] V.V. Rajan, W.K. Dierkes, R. Joseph, J.W.M. Noordermeer, Science and technology of rubber reclamation with special attention to NR-based waste latex products, Prog. Polym. Sci. 31 (9) (2006) 811–834. https://doi.org/10.1016/j.progpolymsci.2006.08.003.

[11] H.Z. Mousavi, A. Hosseynifar, V. Jahed, S.A.M. Dehghani, Removal of lead from aqueous solution using waste tire rubber ash as an adsorbent, Braz. J. Chem. Eng. 27 (1) (2010) 79–87. https://doi.org/10.1590/S0104-66322010000100007.

[12] B.S. Thomas, R.C. Gupta, A comprehensive review on the applications of waste tire rubber in cement concrete, Renew. Sustain. Energy Rev. 54 (2016) 1323–1333. https://doi.org/10.1016/j.rser.2015.10.092.

[13] E. Masad, R. Taha, C. Ho, T. Papagiannakis, Engineering properties of tire/soil mixtures as a lightweight fill material, Geotech. Test J. 19 (3) (1996) 297–304. https://doi.org/10.1520/gtj10355j.

[14] J.G. Bryson, Reclaim Oil for Digester Process for Rubber Reclaiming, 1979.

[15] S. Rokade, Use of waste plastic and waste rubber tyres in flexible highway pavements, in: Int. Conf. Futur. Environ. Energy, vol. 28, IPCBEE, 2012.

[16] M. Juma, Z. Koreňová, J. Markoš, J. Annus, Ľ. Jelemenský, Pyrolysis and combustion of scrap tire, Pet. Coal 48 (2006) 15–26.

[17] J.R. Kershaw, Supercritical fluid extraction of scrap tyres, Fuel 77 (9–10) (1998) 1113–1115. https://doi.org/10.1016/S0016-2361(97)00280-9.

[18] K. Reschner, Scrap Tire Recycling, A Summ Prevalent Dispos Recycl Methods Entire-Engineering, 2008.

[19] H.H. Tsang, Uses of scrap rubber tires, in: Rubber: Types, Properties and Uses, Nova Science Publishers, Inc., 2013, pp. 477–491. https://www.novapublishers.com/catalog/product_info.php?products_id=13225.

[20] A. Acetta, J.M. Vergnaud, Upgrading of scrap rubber powder by vulcanization without new rubber, Rubber Chem. Technol. 54 (2) (1981) 302–310. https://doi.org/10.5254/1.3535807.

[21] F.A. Aisien, F.K. Hymore, R.O. Ebewele, Potential application of recycled rubber in oil pollution control, Environ. Monit. Assess. 85 (2) (2003) 175–190. https://doi.org/10.1023/A:1023690029575.

[22] S. Akbulut, S. Arasan, E. Kalkan, Modification of clayey soils using scrap tire rubber and synthetic fibers, Appl. Clay Sci. 38 (1–2) (2007) 23–32. https://doi.org/10.1016/j.clay.2007.02.001.

[23] T. Amari, N.J. Themelis, I.K. Wernick, Resource recovery from used rubber tires, Resour. Pol. 25 (3) (1999) 179–188. https://doi.org/10.1016/S0301-4207(99)00025-2.

[24] J. Fisher, J. Jury, M.D. Burgoyne, Process for Regeneration of Rubber, 1999.

[25] J. Jarrell, Apparatus for Producing Fuel and Carbon Black from Rubber Tires, 1993.

[26] J. Oliver, Modification of Paving Asphalts by Digestion with Scrap Rubber, ARRB, 1981.

[27] R. Siddique, T.R. Naik, Properties of concrete containing scrap-tire rubber - an overview, Waste Manag. 24 (6) (2004) 563–569. https://doi.org/10.1016/j.wasman.2004.01.006.

[28] D. De, S. Maiti, B. Adhikari, Reclaiming of rubber by a renewable resource material (RRM). III. Evaluation of properties of NR reclaim, J. Appl. Polym. Sci. 75 (12) (2000) 1493–1502. https://doi.org/10.1002/(SICI)1097-4628(20000321)75:12<1493::AID-APP8>3.0.CO;2-U.

[29] D.F. Martinez, Waste Tire Disposal and Recycling, 1994.

[30] P.P. Nicholas, Devulcanized Rubber Composition and Process for Preparing Same, 1979.

[31] D. De, D. De, G.M. Singharoy, Reclaiming of ground rubber tire by a novel reclaiming agent. I. Virgin natural rubber/reclaimed GRT vulcanizates, Polym. Eng. Sci. 47 (7) (2007) 1091–1100. https://doi.org/10.1002/pen.20790.

[32] D. Dobrotă, G. Dobrotă, An innovative method in the regeneration of waste rubber and the sustainable development, J. Clean. Prod. 172 (2018) 3591–3599. https://doi.org/10.1016/j.jclepro.2017.03.022.

[33] S.O. Movahed, A. Ansarifar, S. Estagy, Review of the reclaiming of rubber waste and recent work on the recycling of ethylene-propylene-diene rubber waste, Rubber Chem. Technol. 89 (1) (2016) 54–78. https://doi.org/10.5254/rct.15.84850.

[34] D.S. le Beau, Science and technology of reclaimed rubber, Rubber Chem. Technol. (1967) 217–237. https://doi.org/10.5254/1.3539043.

[35] L. Asaro, M. Gratton, S. Seghar, N. Aït Hocine, Recycling of rubber wastes by devulcanization, Resour., Conserv. Recycl. 133 (2018) 250–262. https://doi.org/10.1016/j.resconrec.2018.02.016.

[36] A.I. Isayev, Elastomers, Recycling of Rubbers, Science and Technology of Rubber, Third Edition, Elsevier, 2001.

[37] S. Maduwage, A. Amarasinghe, D. Munindradasa, Devulcanization of Sulfur Vulcanized Natural Rubber Using Ultrasonic Technology, 2014.

[38] W.C. Warner, Methods of devulcanization, Rubber Chem. Technol. 67 (3) (1994) 559–566. https://doi.org/10.5254/1.3538692.

[39] S. Saiwari, W.K. Dierkes, J.W.M. Noordermeer, Comparative investigation of the devulcanization parameters of tire rubbers, Rubber Chem. Technol. 87 (1) (2014) 31–42. https://doi.org/10.5254/rct.13.87933.

[40] M.G. Aboelkheir, L.Y. Visconte, G.E. Oliveira, R.D. Toledo Filho, F.G. Souza, The biodegradative effect of *Tenebrio molitor Linnaeus larvae* on vulcanized SBR and tire crumb, Sci. Total Environ. 649 (2019) 1075–1082. https://doi.org/10.1016/j.scitotenv.2018.08.228.

[41] R. Diaz, G. Colomines, E. Peuvrel-Disdier, R. Deterre, Thermo-mechanical recycling of rubber: relationship between material properties and specific mechanical energy, J. Mater. Process. Technol. 252 (2018) 454–468. https://doi.org/10.1016/j.jmatprotec.2017.10.014.

[42] M. Mouri, N. Sato, H. Okamoto, M. Matsushita, K. Fukumori, H. Honda, et al., New Continuous Recycling Technology for Vulcanized Rubbers, Pap Chem Soc Div Rubber Chem., 1999.

[43] C. Carné, M. del, Study of Compatibilization Methods for High Density Polyethylene and Ground Tyre Rubber: Exploring New Route to Recycle Scrap Tyres, 2010.

[44] B.A. Rahman, M. S, A Comparative Study of Natural Rubber Modified with Devulcanized Ground Tire Rubber (CAR), 2016.

[45] A.B. Irez, E. Bayraktar, I. Miskioglu, Recycled and devulcanized rubber modified epoxy-based composites reinforced with nano-magnetic iron oxide, Fe_3O_4, Compos. B Eng. 148 (2018) 1–13. https://doi.org/10.1016/j.compositesb.2018.04.047.

[46] R.D. Myers, P.C. Nicholson, J.B. Macleod, M.E. Moir, Rubber devulcanization process, Patente Norteamericana US (1998 Aug) 5798394.

[47] G.K. Jana, C.K. Das, Recycling natural rubber vulcanizates through mechanochemical devulcanization, Macromol. Res. 13 (1) (2005) 30–38. https://doi.org/10.1007/BF03219012.

[48] G.K. Jana, R.N. Mahaling, T. Rath, A. Kozlowska, M. Kozlowski, C.K. Das, Mechano-chemical recycling of sulfur cured natural rubber, Polimery/Polymers 52 (2) (2007) 131–136. https://doi.org/10.14314/polimery.2007.131.

[49] F. Sadaka, I. Campistron, A. Laguerre, J.F. Pilard, Controlled chemical degradation of natural rubber using periodic acid: application for recycling waste tyre rubber, Polym. Degrad. Stabil. 97 (5) (2012) 816–828. https://doi.org/10.1016/j.polymdegradstab.2012.01.019.

[50] F. Cavalieri, F. Padella, F. Cataldo, Mechanochemical surface activation of ground tire rubber by solid-state devulcanization and grafting, J. Appl. Polym. Sci. 90 (6) (2003) 1631–1638. https://doi.org/10.1002/app.12829.

[51] V.K. Gupta, B. Gupta, A. Rastogi, S. Agarwal, A. Nayak, Pesticides removal from waste water by activated carbon prepared from waste rubber tire, Water Res. 45 (13) (2011) 4047–4055. https://doi.org/10.1016/j.watres.2011.05.016.

[52] M. Sabzekar, M.P. Chenar, S.M. Mortazavi, M. Kariminejad, S. Asadi, G. Zohuri, Influence of process variables on chemical devulcanization of sulfur-cured natural rubber, Polym. Degrad. Stabil. 118 (2015) 88–95. https://doi.org/10.1016/j.polymdegradstab.2015.04.013.

[53] S. Rooj, G.C. Basak, P.K. Maji, A.K. Bhowmick, New route for devulcanization of natural rubber and the properties of devulcanized rubber, J. Polym. Environ. 19 (2) (2011) 382–390. https://doi.org/10.1007/s10924-011-0293-5.

[54] P. Thaicharoen, P. Thamyongkit, S. Poompradub, Thiosalicylic acid as a devulcanizing agent for mechano-chemical devulcanization, Kor. J. Chem. Eng. 27 (4) (2010) 1177–1183. https://doi.org/10.1007/s11814-010-0168-9.

[55] De, J.R. White, Rubber Technologist's Handbook, vol. 1, 2001.

[56] Ghose, A.I. Isayev, Ultrasonic Devulcanization of Used Tires and Waste Rubbers, 2005, pp. 353–445.

[57] Tire Technol. Int. '96, 1996, pp. 82–84.

[58] B. Diao, A.I. Isayev, V.Y. Levin, Basic study of continuous ultrasonic devulcanization of unfilled silicone rubber, Rubber Chem. Technol. 72 (1) (1999) 152–164. https://doi.org/10.5254/1.3538784.

[59] A.I. Isayev, J. Chen, A. Tukachinsky, Novel ultrasonic technology for devulcanization of waste rubbers, Rubber Chem. Technol. (1995) 267–280. https://doi.org/10.5254/1.3538741.

[60] V.Y. Levin, S.H. Kim, A.I. Isayev, J. Massey, E. Von Meerwall, Ultrasound devulcanization of sulfur vulcanized SBR: crosslink density and molecular mobility, Rubber Chem. Technol. 69 (1) (1996) 104–114. https://doi.org/10.5254/1.3538350.

[61] M. Tapale, A.I. Isayev, Continuous ultrasonic devulcanization of unfilled NR vulcanizates, J. Appl. Polym. Sci. 70 (10) (1998) 2007–2019. https://doi.org/10.1002/(SICI)1097-4628(19981205)70:10<2007::AID-APP17>3.0.CO;2-D.

[62] A. Tukachinsky, D. Schworm, A.I. Isayev, Devulcanization of waste tire rubber by powerful ultrasound, Rubber Chem. Technol. 69 (1) (1996) 92–103. https://doi.org/10.5254/1.3538362.

[63] J. Yun, J.S. Oh, A.I. Isayev, Ultrasonic devulcanization reactors for recycling of GRT: comparative study, Rubber Chem. Technol. 74 (2) (2001) 317–330. https://doi.org/10.5254/1.3544953.

[64] S.E. Shim, A.I. Isayev, Ultrasonic devulcanization of precipitated silica-filled silicone rubber, Rubber Chem. Technol. 74 (2) (2001) 303–316. https://doi.org/10.5254/1.3544952.

[65] A. Pelofsky, Rubber Reclamation Using Ultrasonic Energy, 1973.

[66] J. Yun, A.I. Isayev, Recycling of roofing membrane rubber by ultrasonic devulcanization, Polym. Eng. Sci. 43 (4) (2003) 809–821. https://doi.org/10.1002/pen.10067.

[67] W. Feng, A.I. Isayev, E. Von Meerwall, Molecular mobility in ultrasonically treated butyl gum and devulcanized butyl rubber, Polymer 45 (25) (2004) 8459–8467. https://doi.org/10.1016/j.polymer.2004.09.072.

[68] J.S. Oh, A.I. Isayev, Continuous ultrasonic devulcanization of unfilled butadiene rubber, J. Appl. Polym. Sci. 93 (3) (2004) 1166–1174. https://doi.org/10.1002/app.20508.

[69] Y.J. Hong, K.M. Jeong, P. Saha, J. Suh, J.K. Kim, Processing and characterization of microwave and ultrasonically treated waste-EPDM/LDPE polymer composites, Polym. Eng. Sci. 55 (3) (2015) 533–540. https://doi.org/10.1002/pen.23916.

[70] C.H. Scuracchio, D.A. Waki, M.L.C.P. Da Silva, Thermal analysis of ground tire rubber devulcanized by microwaves, J. Therm. Anal. Calorim. 87 (3) (2007) 893–897. https://doi.org/10.1007/s10973-005-7419-8.

[71] P. Sarath, M. Biswal, S. Mohanty, S.K. Nayak, Effect of silicone rubber based impact modifier on mechanical and flammability properties of plastics recovered from waste mobile phones, J. Clean. Prod. 171 (2018) 209–219. https://doi.org/10.1016/j.jclepro.2017.10.024.

[72] A. Zanchet, L.N. Carli, M. Giovanela, R.N. Brandalise, J.S. Crespo, Use of styrene butadiene rubber industrial waste devulcanized by microwave in rubber composites for automotive application, Mater. Des. 39 (2012) 437–443. https://doi.org/10.1016/j.matdes.2012.03.014.

[73] J. Liu, P. Liu, X. Zhang, P. Lu, X. Zhang, M. Zhang, Fabrication of magnetic rubber composites by recycling waste rubber powders via a microwave-assisted in situ surface modification and semi-devulcanization process, Chem. Eng. J. 295 (2016) 73–79. https://doi.org/10.1016/j.cej.2016.03.025.

[74] X. Colom, A. Faliq, K. Formela, J. Cañavate, FTIR spectroscopic and thermogravimetric characterization of ground tyre rubber devulcanized by microwave treatment, Polym. Test. 52 (2016) 200–208. https://doi.org/10.1016/j.polymertesting.2016.04.020.

[75] P.S. Garcia, F.D.B. de Sousa, de Lima, S.A. Cruz, C.H. Scuracchio, Devulcanization of ground tire rubber: physical and chemical changes after different microwave exposure times, Express Polym. Lett. 9 (11) (2015) 1015–1026. https://doi.org/10.3144/expresspolymlett.2015.91.

[76] K. Aoudia, S. Azem, N. Aït Hocine, M. Gratton, V. Pettarin, S. Seghar, Recycling of waste tire rubber: microwave devulcanization and incorporation in a thermoset resin, Waste Manag. 60 (2017) 471–481. https://doi.org/10.1016/j.wasman.2016.10.051.

[77] M. Molanorouzi, S.O. Mohaved, Reclaiming waste tire rubber by an irradiation technique, Polym. Degrad. Stabil. 128 (2016) 115–125. https://doi.org/10.1016/j.polymdegradstab.2016.03.009.

[78] F.D.B. de Sousa, C.H. Scuracchio, G.H. Hu, S. Hoppe, Devulcanization of waste tire rubber by microwaves, Polym. Degrad. Stabil. 138 (2017) 169–181. https://doi.org/10.1016/j.polymdegradstab.2017.03.008.

[79] S. Seghar, L. Asaro, M. Rolland-Monnet, N. Aït Hocine, Thermo-mechanical devulcanization and recycling of rubber industry waste, Resour. Conserv. Recycl. 144 (2019) 180–186. https://doi.org/10.1016/j.resconrec.2019.01.047.

[80] S. Ghorai, S. Bhunia, M. Roy, D. De, Mechanochemical devulcanization of natural rubber vulcanizate by dual function disulfide chemicals, Polym. Degrad. Stabil. 129 (2016) 34–46. https://doi.org/10.1016/j.polymdegradstab.2016.03.024.

[81] Y. Suzuki, M. Owaki, M. Mouri, N. Sato, H. Honda, K. Nakashima, Recycling technology using a twin screw reactive extruder for vulcanized EPDM rubber, Toyota Tech. Rev. 48 (1998) 53.

[82] Mouri, N. Sato, H. Okamoto, Matsushita, H. Honda, K. Nakashima, et al., NG 99/05/278 continuous devulcanisation by shear flow stage reaction control technology for rubber recycling. Part 3. Study of the devulcanisation process for EPDM, Int. Polym. Sci. Technol. 27 (2000).

[83] S. Li, J. Lamminmäki, K. Hanhi, Effect of ground rubber powder and devulcanizates on the properties of natural rubber compounds, J. Appl. Polym. Sci. 97 (1) (2005) 208–217. https://doi.org/10.1002/app.21748.

[84] Z.Z. Xiu, Z.S. Ling, K.J. Kuk, Evaluation of mechanical, morphological and thermal properties of waste rubber tire powder/LLDPE blends, E-polymers 8 (1) (2008). https://doi.org/10.1515/epoly.2008.8.1.687.

[85] H. Si, T. Chen, Y. Zhang, Effects of high shear stress on the devulcanization of ground tire rubber in a twin-screw extruder, J. Appl. Polym. Sci. 128 (4) (2013) 2307–2318. https://doi.org/10.1002/app.38170.

[86] X. Wang, C. Shi, L. Zhang, Y. Zhang, Effects of shear stress and subcritical water on devulcanization of styrene-butadiene rubber based ground tire rubber in a twin-screw extruder, J. Appl. Polym. Sci. 130 (3) (2013) 1845–1854. https://doi.org/10.1002/app.39253.

[87] G. Tao, Q. He, Y. Xia, G. Jia, H. Yang, W. Ma, The effect of devulcanization level on mechanical properties of reclaimed rubber by thermal-mechanical shearing devulcanization, J. Appl. Polym. Sci. 129 (5) (2013) 2598–2605. https://doi.org/10.1002/app.38976.

[88] K. Formela, M. Cysewska, J. Haponiuk, The influence of screw configuration and screw speed of co-rotating twin screw extruder on the properties of products obtained by thermomechanical reclaiming of ground tire rubber, Polimery/Polymers 59 (2) (2014) 170–177. https://doi.org/10.14314/polimery.2014.170.

[89] J. Shi, H. Zou, L. Ding, X. Li, K. Jiang, T. Chen, X. Zhang, L. Zhang, D. Ren, Continuous production of liquid reclaimed rubber from ground tire rubber and its application as reactive polymeric plasticizer, Polym. Degrad. Stabil. 99 (1) (2014) 166–175. https://doi.org/10.1016/j.polymdegradstab.2013.11.010.

[90] K. Jin Kuk, J.W. Park, Biological and chemical desulfurization of crumb rubber for the rubber compounding, J. Appl. Polym. Sci. 72 (12) (1999) 1543–1549. https://doi.org/10.1002/(SICI)1097-4628(19990620)72:12<1543::AID-APP6>3.0.CO;2–3.

[91] C. Yao, S. Zhao, Y. Wang, B. Wang, M. Wei, M. Hu, Microbial desulfurization of waste latex rubber with Alicyclobacillus sp, Polym. Degrad. Stabil. 98 (9) (2013) 1724–1730. https://doi.org/10.1016/j.polymdegradstab.2013.06.002.

[92] R. Romine, L.J. Snowden-Swan, Method for the Addition of Vulcanized Waste Rubber to Virgin Rubber Products, 1997.

[93] K. Bredberg, B. Erik Andersson, E. Landfors, O. Holst, Microbial detoxification of waste rubber material by wood-rotting fungi, Bioresour. Technol. 83 (3) (2002) 221–224. https://doi.org/10.1016/S0960-8524(01)00218-8.

[94] Y. Li, S. Zhao, Y. Wang, Microbial desulfurization of ground tire rubber by *Thiobacillus ferrooxidans*, Polym. Degrad. Stabil. 96 (9) (2011) 1662–1668. https://doi.org/10.1016/j.polymdegradstab.2011.06.011.

[95] M. Löffler, Microbial surface desulfurization of scrap rubber crumb-a contribution towards material recycling of scrap rubber, Kautsch. Gummi Kunstst. 48 (1995).

[96] S. Sato, Y. Ohashi, M. Kojima, T. Watanabe, Y. Honda, T. Watanabe, Degradation of sulfide linkages between isoprenes by lipid peroxidation catalyzed by manganese peroxidase, Chemosphere 77 (6) (2009) 798–804. https://doi.org/10.1016/j.chemosphere.2009.08.014.

[97] X. Cui, S. Zhao, B. Wang, Microbial desulfurization for ground tire rubber by mixed consortium-Sphingomonas sp. and Gordonia sp, Polym. Degrad. Stabil. 128 (2016) 165–171. https://doi.org/10.1016/j.polymdegradstab.2016.03.011.

[98] A. Tsuchii, Y. Tokiwa, Two-step cultivation method for microbial disintegration of tire rubber particles, J. Polym. Environ. 13 (1) (2005) 75–80. https://doi.org/10.1007/s10924-004-1231-6.

[99] A. Ali Shah, F. Hasan, Z. Shah, N. Kanwal, S. Zeb, Biodegradation of natural and synthetic rubbers: a review, Int. Biodeterior. Biodegrad. 83 (2013) 145–157. https://doi.org/10.1016/j.ibiod.2013.05.004.

[100] Y. Li, S. Zhao, L. Zhang, Y. Wang, W. Yu, The effect of different Fe^{2+} concentrations in culture media on the recycling of ground tyre rubber by Acidithiobacillus ferrooxidans YT-1, Ann. Microbiol. 63 (1) (2013) 315–321. https://doi.org/10.1007/s13213-012-0476-x.

[101] D. Raghavan, R. Guay, A.E. Torma, A study of biodegradation of polyethylene and biodesulfurization of rubber, Appl. Biochem. Biotechnol. (1990) 387–396. https://doi.org/10.1007/bf02920262.

[102] I. Mangili, M. Oliveri, M. Anzano, E. Collina, D. Pitea, M. Lasagni, Full factorial experimental design to study the devulcanization of ground tire rubber in supercritical carbon dioxide, J. Supercrit. Fluids 92 (2014) 249–256. https://doi.org/10.1016/j.supflu.2014.06.001.

[103] R. Piñero-Hernanz, C. Dodds, J. Hyde, J. García-Serna, M. Poliakoff, E. Lester, M.J. Cocero, S. Kingman, S. Pickering, K.H. Wong, Chemical recycling of carbon fibre reinforced composites in near critical and supercritical water, Compos. A Appl. Sci. Manuf. 39 (3) (2008) 454–461. https://doi.org/10.1016/j.compositesa.2008.01.001.

[104] Q. Zhang, C. Tzoganakis, Devulcanization of recycled tire rubber using supercritical carbon dioxide, in: Annual Technical Conference - ANTEC, Conference Proceedings, vol. 3, 2004, pp. 3509–3513.

[105] M. Kojima, M. Tosaka, E. Funami, K. Nitta, M. Ohshima, S. Kohjiya, Phase behavior of crosslinked polyisoprene rubber and supercritical carbon dioxide, J. Supercrit. Fluids 35 (3) (2005) 175–181. https://doi.org/10.1016/j.supflu.2005.02.004.

[106] T. Luo, A.I. Isayev, Rubber/plastic blends based on devulcanized ground tire rubber, J. Elastomers Plastics 30 (2) (1998) 133–160. https://doi.org/10.1177/009524439803000204.

[107] A.M. Joseph, K.N. Madhusoodanan, R. Alex, B. George, Incorporation of devulcanised rubber in fresh rubber compounds: impact of filler correction on vulcanisate properties, Prog. Rubber Plast. Recycl. Technol. 33 (4) (2017) 281–302. https://doi.org/10.1177/147776061703300405.

[108] V. Lapkovskis, V. Mironovs, D. Goljandin, Suitability of devulcanized crumb rubber for oil spills remediation, Energy Procedia 147 (2018) 351–357. https://doi.org/10.1016/j.egypro.2018.07.103.

[109] M. Allouch, M. Kamoun, J. Mars, M. Wali, F. Dammak, Experimental investigation on the mechanical behavior of recycled rubber reinforced polymer composites filled with aluminum powder, Construct. Build. Mater. 259 (2020). https://doi.org/10.1016/j.conbuildmat.2020.119845.

[110] V.K. Gupta, M.R. Ganjali, A. Nayak, B. Bhushan, S. Agarwal, Enhanced heavy metals removal and recovery by mesoporous adsorbent prepared from waste rubber tire, Chem. Eng. J. 197 (2012) 330–342. https://doi.org/10.1016/j.cej.2012.04.104.

[111] G. Zhang, H. Feng, K. Liang, Z. Wang, X. Li, X. Zhou, B. Guo, L. Zhang, Design of next-generation cross-linking structure for elastomers toward green process and a real recycling loop, Sci. Bull. 65 (11) (2020) 889–898. https://doi.org/10.1016/j.scib.2020.03.008.

[112] F. Karabork, A. Akdemir, Friction and wear behavior of styrene butadiene rubber-based composites reinforced with microwave-devulcanized ground tire rubber, J. Appl. Polym. Sci. 132 (33) (2015). https://doi.org/10.1002/app.42419.

[113] A.R. Tripathy, D.E. Williams, R.J. Farris, Rubber plasticizers from degraded/devulcanized scrap rubber: a method of recycling waste rubber, Polym. Eng. Sci. 44 (7) (2004) 1338–1350. https://doi.org/10.1002/pen.20129.

[114] A. Gugliemotti, C. Lucignano, F. Quadrini, Production of rubber parts by tyre recycling without using virgin materials, Plast., Rubber Compos. 41 (1) (2012) 40–46. https://doi.org/10.1179/1743289811Y.0000000010.

[115] C.M.B. Lehmann, M. Rostam-Abadi, M.J. Rood, J. Sun, Reprocessing and reuse of waste tire rubber to solve air-quality related problems, Energy Fuels 12 (6) (1998) 1095–1099. https://doi.org/10.1021/ef9801120.

[116] Y. Sasson, U. Stoin, M. Kopylov, V. Goldshtein, Devulcanized Rubber, Method for Its Preparation and Its Use as an Absorbent, 2018.

[117] G.S. Miguel, G.D. Fowler, C.J. Sollars, Pyrolysis of tire rubber: porosity and adsorption characteristics of the pyrolytic chars, Ind. Eng. Chem. Res. 37 (6) (1998) 2430–2435. https://doi.org/10.1021/ie970728x.

[118] M. Gandoman, M. Kokabi, Sound barrier properties of sustainable waste rubber/geopolymer concretes, Iran. Polym. J. 24 (2) (2015) 105–112. https://doi.org/10.1007/s13726-014-0304-1.

[119] D.G. Jamison, Method for Using Scrap Rubber; Scrap Synthetic and Textile Material to Create Particle Board Products with Desirable Thermal and Acoustical Insulation Values, 1995.

[120] L.C. Jia, Y.K. Li, D.X. Yan, Flexible and efficient electromagnetic interference shielding materials from ground tire rubber, Carbon 121 (2017) 267–273. https://doi.org/10.1016/j.carbon.2017.05.100.

[121] L. Simon-Stőger, C. Varga, E. Greczula, B. Nagy, A journey into recycling of waste elastomers via a novel type of compatibilizing additives, Express Polym. Lett. 13 (5) (2019) 443–455. https://doi.org/10.3144/expresspolymlett.2019.37.

[122] Y. Tanaka, T. Watanabe, T. Okita, M. Matsushita, H. Okamoto, K. Fukumori, et al., The Technology to Produce Thermoplastic Elastomer Based on Waste Rubber (SAE Technical Paper), 2003.

[123] V.K. Gupta, A. Nayak, S. Agarwal, I. Tyagi, Potential of activated carbon from waste rubber tire for the adsorption of phenolics: effect of pre-treatment conditions, J. Colloid Interface Sci. 417 (2014) 420–430. https://doi.org/10.1016/j.jcis.2013.11.067.

[124] X. Zhang, Z. Lu, D. Tian, H. Li, C. Lu, Mechanochemical devulcanization of ground tire rubber and its application in acoustic absorbent polyurethane foamed composites, J. Appl. Polym. Sci. 127 (5) (2013) 4006–4014. https://doi.org/10.1002/app.37721.

[125] W. Bastian, S. Andreas, A. Vincent, K. Joachim, H. Christopher, S. Dirk, Recycling and reprocessing of thermoplastic polyurethane materials towards nonwoven processing, Polymers 1917 (2020). https://doi.org/10.3390/polym12091917.

Further reading

[1] D. Losanno, A. Calabrese, I.E. Madera-Sierra, M. Spizzuoco, J. Marulanda, P. Thomson, G. Serino, Recycled versus natural-rubber fiber-reinforced bearings for base isolation: review of the experimental findings, J. Earthq. Eng. (2020) 1–20. https://doi.org/10.1080/13632469.2020.1748764.

[2] A.I. Isayev, T. Liang, T.M. Lewis, Effect of particle size on ultrasonic devulcanization of tire rubber in twin-screw extruder, Rubber Chem. Technol. 87 (1) (2014) 86–102. https://doi.org/10.5254/RCT.13.87926.

[3] Y. Li, S. Zhao, Y. Wang, Improvement of the properties of natural rubber/ground tire rubber composites through biological desulfurization of GTR, J. Polym. Res. 19 (5) (2012). https://doi.org/10.1007/s10965-012-9864-y.

[4] J. Oh, J. Suh, J. Kim, Floor noise isolation system of the residential buildings using waste rubbers, Appl. Chem. Eng. 28 (4) (2017) 427–431. https://doi.org/10.14478/ace.2017.1032.

[5] S. Seghar, N.A. Hocine, V. Mittal, S. Azem, F. Al-Zohbi, B. Schmaltz, N. Poirot, Devulcanization of styrene butadiene rubber by microwave energy: effect of the presence of ionic liquid, Express Polym. Lett. (2015) 1076–1086. https://doi.org/10.3144/expresspolymlett.2015.97.

[6] J. Wang, S. Chen, T. Lin, J. Ke, T. Chen, X. Wu, C. Lin, A catalyst-free and recycle-reinforcing elastomer vitrimer with exchangeable links, RSC Adv. 10 (64) (2020) 39271–39276. https://doi.org/10.1039/d0ra07728c.

[7] Chem. Biol. 1 (1) (2015) 1–9.

15

Applications of elastomer blends and composites

Sudheer Kumar[1], Sukhila Krishnan[2], Smita Mohanty[1]

[1]SCHOOL FOR ADVANCED RESEARCH IN PETROCHEMICALS (SARP), LABORATORY FOR ADVANCED RESEARCH IN POLYMERIC MATERIALS (LARPM), CENTRAL INSTITUTE OF PETROCHEMICALS ENGINEERING AND TECHNOLOGY (CIPET), BHUBANESWAR, ODISHA, INDIA; [2]SAHRDAYA COLLEGE OF ENGINEERING AND TECHNOLOGY, DEPARTMENT OF APPLIED SCIENCE AND HUMANITIES, KODAKARA, KERALA, INDIA

15.1 Introduction

Elastomers are versatile materials prevalent to researchers, to be used in various industrial applications, such as shape memory, flame resistance, sensing, and self-healing, astronautics devices, automobiles, medical implants, and soft robots, etc. as depicted in Fig. 15.1 owing to its tremendous performances, solvent resistance, high elastic properties, outstanding abrasion resistance, toughness, and good stability as well as radiation resistance [52] [11,13,23,41,69,90]. The incorporation of solid fillers into elastomeric materials mostly leads to improvement in its various properties as well as reduces the overall cost.

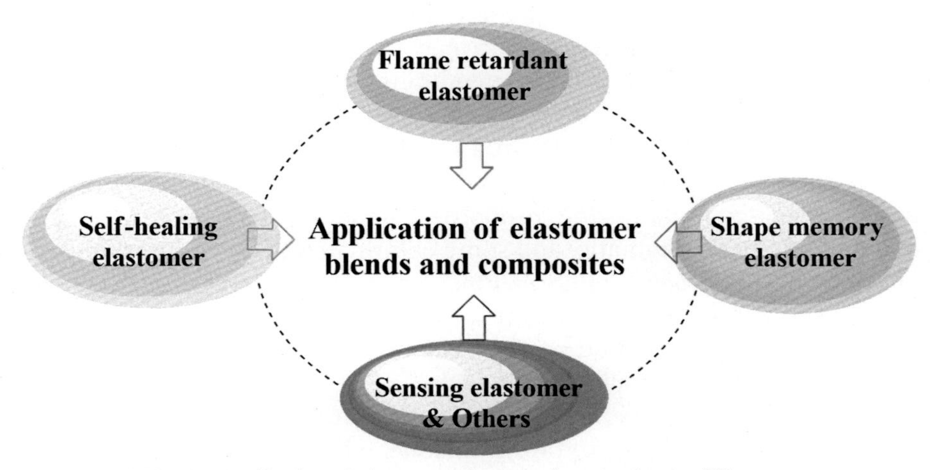

FIGURE 15.1 Application of elastomer blends and composites in different areas.

Elastomer Blends and Composites. https://doi.org/10.1016/B978-0-323-85832-8.00003-1

Polyurethanes (PUs) are generally obtained from the exothermic reaction of polyol and polyisocyanate and other components, e.g., long-chain polyol diisocyanate, and chain extender used to control their thermoplastics properties [1,35,65,79]. Further, PU-based elastomers were synthesized with different characteristics by changing the proportions of soft and hard segments. They are a particular set of polymeric materials that are different approaches than general other polymer materials and are also included in several applications, such as coating, paints, elastomer, elastic fiber, insulator, foam, etc. PU is a versatile material and is suitable to substitute for other deficient constituents, leading to its additional uses. Among all elastomers, PU is one of the most significant kinds and is extensively incorporated into the various engineering products and bestows the highest impeccable properties [5,93].

On the other hand, silicone elastomers have often been reported as high-performance polymers with a wide range of potential in different areas such as aerospace, lithography, hydrophobic coatings, electronics, and biomedicine, attributed to their flexible, tough, and biocompatible nature. Higher than their glass transition temperatures, silicone polymers have an elastic response that rises abruptly on cross-linking with an elastomer, because of the network's covalent nature. Although silicone elastomers are with good electrical insulating properties, and so, basically a weak conductor, their application as highly flexibile and electrically conducting elastomers cannot be excluded, attributed to the easy processability in the presence of conductive fillers.

Further, EPDM comes under a saturated synthetic elastomeric polymer whose behavior in the presence of light, oxygen, heat, and especially the ozone atmosphere is prominent because of its stable and saturated polymer backbone structure. It possesses excellent electrical resistance, good resistivity toward polar solvents, such as acids, alkalis, phosphate, esters, many ketones, and alcohols and is nonpolar by nature. Various fillers such as carbon black, silica, clays, etc., are incorporated into the EPDM matrix to attain the required mechanical, thermal, and electrical properties. They are found to exhibit many applications, such as automotive, electrical insulation, steam hoses, seals, radiators, weather stripping, and roofing membranes.

However, fluoroelastomers are also mostly utilized to withstand harsh temperatures and strong chemicals, which makes them a crucial polymer in industries. Due to the presence of strong C—F bonds, the polymer backbone furnishes good thermal and chemical resistivity, good antiaging properties, superior resistance toward greases and mineral oils, aliphatic, aromatic, and some chlorinated hydrocarbons, fuels, greases and silicone oils. But, fluoroelastomers exhibit weak resistance to bases. They are found either in the form of a copolymer, a terpolymer, or a tetrapolymer. Chlorosulfonated polyethylene compounds give good ozone, oxidation, sunlight, and weather resistivity, along with better resistivity for alkalis and acids.

The current chapter emphasizes providing elaborate information on various elastomer blends and composites derived from silicone, ethylene-propylene-diene (EPDM), fluorocarbon, and chlorosulfonated polyethylene rubber and their characteristic properties in various applications. This book chapter aims at imparting elastomeric blends and composites varying applications in the aspects of its various synthesis methods and its incorporated additives.

15.2 Polyurethane-based elastomer blends and composites

The elastomer blend and composites were synthesized by PUs incorporation with various engineered products. Many studies revealed Pus' unique and excellent performance by blending them with various novel polymeric materials or additives via various methods. PUs with varying ratio of hard and soft segments also alter their properties immensely.

15.2.1 Polyurethane-based flame-retardant elastomer

Due to its ease of synthesis and versatile nature, PUs have been utilized in various applications, by incorporating functional groups in the polymer matrix, causing an enhancement in many of their properties such as flame retardancy. Therefore, the flame retarder with better safety in some applications has become more and more prominent [8,15,16]. Improved mechanical and thermal properties, along with lower cost TPU composites by the addition of mica and aluminum trihydrate, were reported. The incorporation of aluminum trihydrate leads to a slight improvement in the fire resistance of the composites, and with 70 and 80 phr of aluminum trihydrate, its fire retardancy is prominent [6].

Aluminum-12 silicon's conductive coatings on PU elastomer through a flame spraying method have also been studied. The study indicates that during thermal spraying on elastomers, along with velocity and the temperature dispersal inside the matrix, impacting particles temperature is a vital factor that can alter the properties of the deposited coating [9].

Xu et al. [81] prepared graphene (MoO_3-GNS) hybrids with molybdenum trioxide (MoO_3) by hydrothermal method, whereas they prepared graphene (Cu_2O-GNS) hybrids with cuprous oxide (Cu_2O) via one-pot coprecipitation method. After that, the polyurethane elastomer (PUE) matrix is mixed with both these hybrids respectively. Due to the collaborated phenomena among graphene sheets, barrier performance, and MoO_3 or Cu_2O, catalytic charring activity, PUE composites' flame retardancy and smoke reducing behavior show much improvement in the presence of MoO_3-GNS or Cu_2O-GNS hybrids as depicted in Fig. 15.2. Simultaneously, the outer matrix area of graphene with MoO_3 or Cu_2O coatings prevents the burning within the polymer matrix and averts the fire-related accidents of elastomers either by Friedel—Crafts alkylation mechanism or via reductive coupling reaction.

Yang et al. [84] by melt polycondensation synthesized a flame retardant based on Schiff-base polyphosphate ester (SPE) as depicted in Fig. 15.3 and incorporated the thermoplastic polyurethane elastomer (TPU) with flame-retarding properties. Results confirmed that 5 wt% SPE demonstrated better flame retardancy and dripping behavior compared to the virgin TPU and had passed the UL-94-0 rating, along with 29% limited oxygen index (LOI). Further, the cone calorimeter test also demonstrated that after the addition of 5 wt% SPE, TPU/SPE correspondingly exhibits decreased average mass loss, heat release rate, as well as maximum heat release emission from 61.7% to 41.2% in contrast to virgin TPU. Thus, for TPU, the SPE is a good polyphosphonate ester flame retardant.

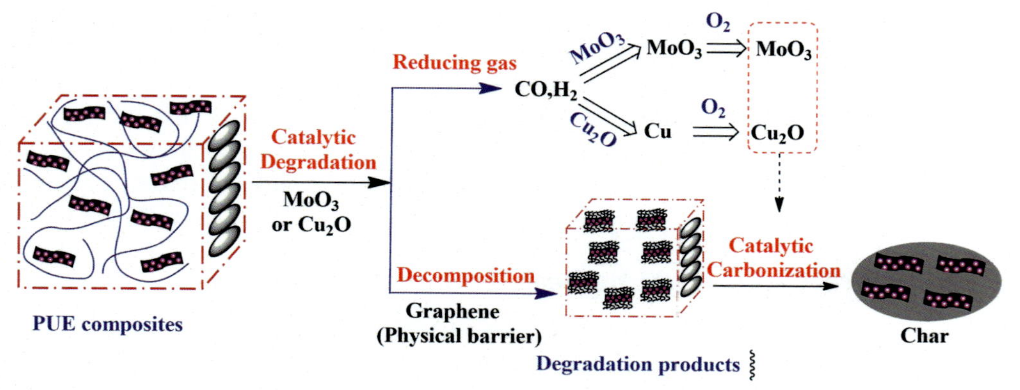

FIGURE 15.2 Schematic representation of flame-retardant behavior and collaborated result of MoO_3-GNS and Cu_2O-GNS hybrids.

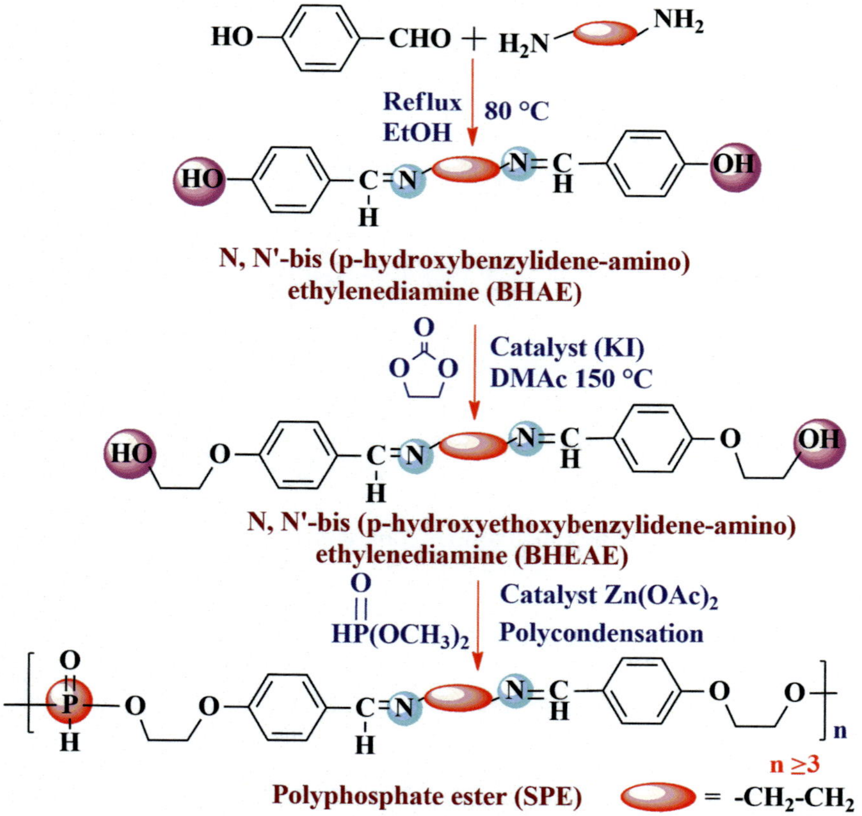

HO–C₆H₄–CHO + H₂N–⬭–NH₂

Reflux EtOH 80 °C

HO–C₆H₄–CH=N–⬭–N=CH–C₆H₄–OH

N, N'-bis (p-hydroxybenzylidene-amino) ethylenediamine (BHAE)

Catalyst (KI) DMAc 150 °C

N, N'-bis (p-hydroxyethoxybenzylidene-amino) ethylenediamine (BHEAE)

HP(O)(OCH₃)₂ Catalyst Zn(OAc)₂ Polycondensation

n ≥3

Polyphosphate ester (SPE) ⬭ = -CH₂-CH₂

FIGURE 15.3 Schematic representation of polyphosphate ester flame retardant (SPE).

Huang et al. [28] reported low-sulfur expandable graphite (EG) by a two-step intercalation process in the presence of H_2O_2, which is a green oxidant. Additionally, 30 wt% EG in the prepared EG/EVM composite exhibits better flame retardancy with 30.4% of LOI. Through in situ radical polymerizations, EG is grafted with poly(vinyl acetate) (PVAc) to enhance the adhesion among EG and EVM. The EG-g-PVAc/EVM composite illustrates good flame-resistant properties, because of enhanced interfacial interaction with polymeric material with good control of fly ash and more applications toward eco-friendly polymeric fire resistance in the coming years. Miao et al. [58] introduced PUE with improved fire retardancy and mechanical strength by interpenetrating polymer networks (IPNs) within PU and porous organic polymer consisting of phosphorus and nitrogen elements (PNPOP) as represented in Fig. 15.4. Through in situ polymerization, IPN structured PNPOP/PU composites are synthesized with PU monomer as polyester diol by inserting inside the pores of PNPOP by the diffusion method.

The PNPOP/PU composites exhibit good thermal properties, increased mechanical performance, and stiffness in contrast to the PU sample. With the rise in PNPOP amounts, there is a significant increase in the fire protection of PNPOP/PU composites. Due to the formation of IPNs, the compatibility between PNPOP and PU improved greatly, resulting in the easy preparation of PU composites, with flame retardant condensed phase as described in Fig. 15.5. In general, the preparation of an IPN, with the polymer matrix of PNPOP-based flame-retardant additives, gives a type of motivation as well as a method for the preparation of flame-resistant polymeric materials of good performance.

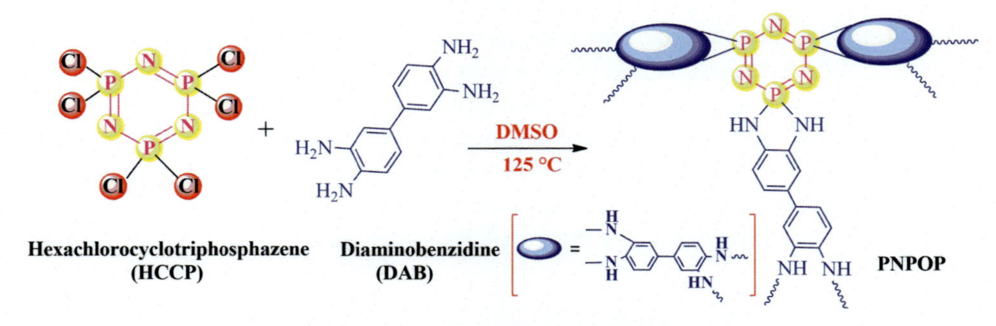

FIGURE 15.4 Schematic representation of PNPOP flame retardant.

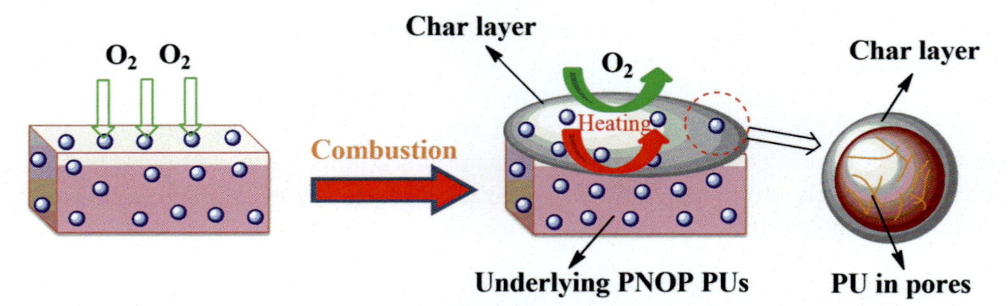

FIGURE 15.5 Schematic representation of PNPOP in PNPOP/PU composites flame-retardant mechanism.

15.2.2 Polyurethane-based self-healing elastomer

Self-healing polymers have attained a high attraction due to PUs' high-performance applications in different areas. Therefore, the synthesis of elastomers with self-healing and good mechanical behavior is highly recommended. Moreover, during usage, the flammability of elastomers makes them dangerous. On damage activation of polymer network by a chitosan or chitosan derivatives based self-healing polyurethane materials were studied by Urban et al. [19–21,85] afterward, the growth of self-healing PU elastomers was progressed.

PU networks are quite effective in regaining mechanical properties during mechanical damage by adding the catalyst zinc acetate ($Zn(OAc)_2$) to methyl α-D-glucopyranoside (MGP). Amine functionalities due to mechanical effects cause covalent rebonding to form urea bonding owing to the regaining of its mechanical behavior. Because of their ability to modify covalent linkages, the nucleophilic amines make this system particularly attractive, hence making them attractive in various systems [85].

Heo et al. [26] reported a viscous resin by the addition of cross-linked products between 9-anthracenemethanol and N-(2- hydroxyethyl)-maleimide (HEM) with hexamethylene diisocyanate (HDI) via common thermosetting polymerization methods, resulting in initial data on Diel's Alder (DA) cross-links within anthracene and maleimide with bulk polymers [44,89]. Self-healing composites processing for good thermal structural applications, the liquid resin is very important, as vacuum-assisted resin transfer molding (VARTM) was used. The results reveal that the polymer and its composite have mechanical properties similar to aerospace-grade epoxies.

Chen et al. [35] synthesized and investigated the self-healing PU elastomer based on van der Waals and hydrogen bonds. The results reveal these noncovalent bonds' high dependence on temperature. CO_2 is a key ingredient for the preparation of polypropylene carbonate, which is a great contribution toward carbon dioxide recycling. A similar type of constituent can be substituted for the isocyanate and chain extender. The preparation method is quite easy and is quite helpful for industrial application in various types of PUs and has become a key resource for different applications of self-healing materials.

Transparent self-healing PU elastomer, which is an imine modified, was synthesized with HDI, polytetramethylene ether glycol (PTMEG), and chain extender imine-diol. At room temperature, good mechanical behavior was exhibited, attributed to the symmetrical structure of imine-diol and C=N double bonds rigidity, whereas at elevated temperatures, the exchange mechanism of the imine bond controls self-healing behavior. By regulating the mixing proportion of HMDI and PTMEG initially, the hard segment materials, as well as the dimension of the hard segments, were altered, and their variation in self-healing and mechanical behavior was studied. Elastomer shows a phase separation approach with a specially designed hard phase structure, which plays a vital role in fabricating strong self-healing materials [27].

In another study, PUE exhibiting better tensile strength (37.11 ± 1.89 MPa) along with a 91.8% self-healing efficiency, consisting of thermo-reversible cross-linking linkages as well as many hydrogen linkages, has also been reported. To synthesize self-healing, recyclable, and flame-retardant polyurethane, the furan-terminated phosphorus as monomer and tri(2-furyl) phosphoramide (TFP) was prepared and subsequently added with a maleimide-terminated linear segmented PU main chain. The prepared elastomer exhibits improved self-healing properties and mechanical behavior with uses in the area of fire-prevention coatings as well as construction constituents [86].

By digital light processing 3D printing, a novel PUE with better self-healing properties was prepared. Initially, PU acrylate with disulfide bonds is prepared and finally mixed by reactive diluent and photoinitiators to obtain a photopolymer resin. The photopolymer resin can be used in DLP 3D printing and many complex structured 3D objects with good printing accuracy because of its high fluidity and good curing rate. The PUE shows 3.39 ± 0.09 MPa tensile strength and 400.38 ± 14.26%, elongation at break, and 95% healing properties can be reached after healing for 12 h at 80°C, and many times it can be healed. Due to its easy processing and good properties, the PUE shows huge DLP 3D printing applications ranging from flexible electronics, soft robotics, and sensors [42].

With hydroxyl-terminated polybutadiene, cross-linking agent of three functional amino alcohols and chain extending and healing agents of disulfide containing compounds, self-healable PU, and polyurethane-ureas coatings have been prepared. From the results, it was found that with chain extender 2-hydroxyethyl disulfide (0.8 equivalents) and cross-linking agent triethanolamine (0.15 equivalents) in the polymer matrix gives a 41% regaining capability. Similarly, thiol-ene side reactions in the PU matrix improve the tensile strength by which the cross-linking density enhances, whereas it reduces the healing performance, attributed to the removal of the metathesis reaction source. Healing ability studies demonstrate that with triethanolamine and 2-hydroxyethyl disulfide, the PUE sample exhibits good healing ability with any type of scrapes on the PU surface [88].

PU/polydopamine composite with dynamic DA bonds was also reported to shape memory and self-healing [87]. The PU composite shows ultrafast near-infrared light responsiveness (NIR) by adding polydopamine, having a significant photothermal effect. At 1 wt% of polydopamine concentration, the tensile strength (TS) of 22 MPa and a percentage elongation of 1385% were reported. A healing efficiency of about 96% can be attained while irradiating the damaged polymer material with a 0.75 W/cm^2 NIR power output to approximately 60 s. Though, the healing efficiency was reduced to 76% by 1 h heating at a temperature of 110°C and then 24 h heating at a temperature of 80°C. Moreover, a conductive self-healing composite was also reported containing CNTs and cross-linked PU with DA bonds [63]. The composite mechanical behavior improved with 25.4 MPa TS and 1113% elongation at break. Electricity and NIR composites are by the incorporation of CNTs attributed to the electrothermal and photothermal effects, leading to crack identifying and self-healing behavior. Electricity treatment for 180 s exhibits a healing performance of 98%. Whereas the healing efficiency of 83% is obtained after three damage healing cycles. Kim et al. [31] prepared a set of hard as well as quick self-healing transparent PUs, with a chain extender of bis(4-hydroxyphenyl) disulfide. The obtained self-healing elastomer reveals 6.8 MPa tensile stress and 900% percentage of elongation. After keeping it together, the torn PUE can be healed and afterward for about 2 h, it was kept at 27°C, obtaining the recovery of 86% break stress and a full revival of elongation at break. A novel self-healing PUE through phase-locked disulfide bonds with colorless, good transparency, and highest tensile stress of 25 MPa, and an elongation at break of about 1600%, was investigated by Lai et al. [37].

15.2.3 Polyurethane-based shape memory elastomer

Shape memory polymers (SMPs) are a stimulating group of stimuli-responsive smart materials, which exhibit reactive and altering mechanical behavior, mostly through varying various external stimuli conditions. Kumar et al. [33] synthesized SMP foam based on PU for better pressure rearrangement, which reveals regular alteration in dynamic pressure redistribution property at 24°C mostly matching with the variations in body contact pressure for dynamic pressure release. The foam system demonstrates good responsivity during plantar pressure variation playing an important role in pressure ulcer curing.

Zheng et al. [91] prepared shape memory materials consisting of alternate layers of thermoplastic polyurethane (TPU) as well as polycaprolactone (PCL) by layer-multiplying extrusion method. With similar compositions, the multilayer-assembled system, in contrast to commercial polymer blends, shows better shape fixing and regaining proportions, which can be enhanced more through increasing the layer numbers. Results reveal that poor-hardness TPU or substituting virgin PCL layers by TPU/PCL blend with cocontinuous morphology is important for attaining excellent shape memory properties and to boost the recovery method owing to the enhancement within the interfaces.

Zhou et al. [92] successfully prepared the stereo complexed and homochiral PUE with controllable multishape memory effects (multi-SMEs) via cross-linking the triblock prepolymers containing the poly(L-lactic acid) (PLLA) and poly(D-lactic acid) (PDLA) enantiomeric segments. The homochiral PU is almost amorphous, but the stereo complexed PU develops into extremely crystalline owing to the stereo complexation of enantiomeric domains. Besides, the glass and melting transitions of PLLA (or PDLA) segments in PUs are combined to develop the thermally induced triple and quadruple SMEs. Tuning the enantiomeric segmental proportions permits control of the crystallinity, mechanical and thermal behavior, and multi-SMEs of PUs. It's an effective and easy method to fabricate functional multi-SMPs which exhibit various potential applications.

Kurahashi et al. [34] prepared the better shape memory performance blend through blending elastic PU with crystalline Poly(oxyethylene) (POE). In the blend, as the POE amount increases, the shape fixing proportion lower than crystallization temperature

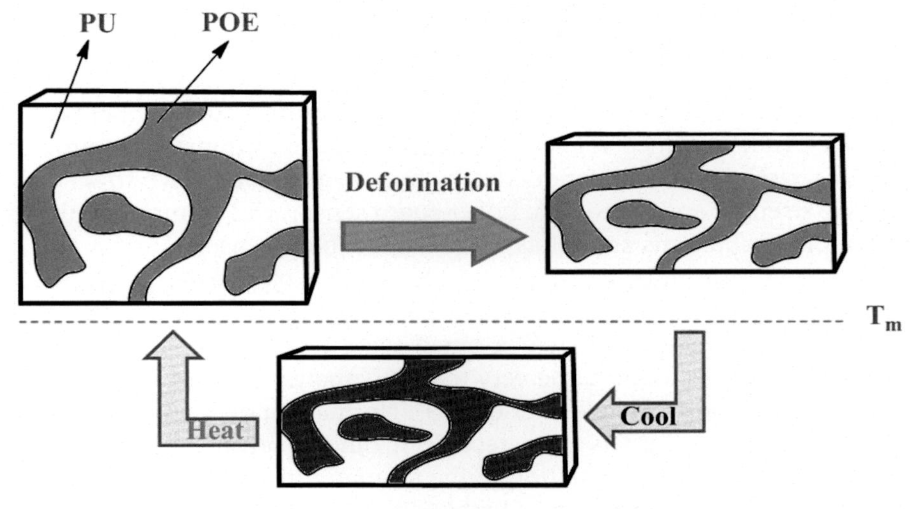

FIGURE 15.6 Shape memory mechanism of PU/POE polymer blends.

shows enhancement of about 100%, although shape recovery higher than melting temperature happens very gradually, attributed to the reduced cross-linking of the PU matrix as POE content increases. PU/POE polymer blends exhibit excellent tunable shape memory behavior (demonstrated in Fig. 15.6) and exhibit significant uses in biomedical constituents.

15.2.4 Polyurethane-based sensing elastomer

Nowadays, huge scientific and industrial attention is drawn toward stretchable electronics, such as wearable electronic sensors for humans, transparent flexible displays, and feedback sensors in soft robots [14,22,45–47,68,76]. Through depositing electronic conductors, e.g., carbon nanotubes (CNTs), graphene, and silver, on the elastic materials (such as polyethylene terephthalate) (PET) and poly(dimethylsiloxane) (PDMS), flexible sensors are prepared that are mostly applied via manual assembly of cast segments [18,48,54,74]. But, its applications have been limited, owing to the huge difficulty in the fabrication of complex sensing systems by traditional processes. So, novel production technologies are taken into consideration to fabricate elastomer substrates for stretchable electronic sensors. For the flourishing soft robots, elastomers that are mechanically strong and extremely stretchable are very crucial and which can endure high pressures and more deformations to deliver heavy works [67,78]. Simultaneously, good transparency for polymer materials is needed for various usage of transparent touch panels and sensors. [32,68]]

Peng et al. [60,61] had reported three types of PU acrylate oligomers by adding an acrylate monomer (isobornyl acrylate) for synthesizing photocurable resins having less viscosity for a digital light processing 3D printer in the absence of custom equipment. The resin with poly(tetrahydrofuran) segments (PPTMGA-40) shows good mechanical performance and shape retainability with a TS of 15.7 MPa and elongation at a break of 414.3%. Without fracture and with 80% strain, it can resist 100 compression cycles, exhibiting 89.4% transmittance at 550 nm, exhibiting its good transparent properties. In addition to this, stretchable and conductive hydrogels, a piezoresistive strain sensor has a 6 MPa tensile strength and a wearable finger guard sensor, which improves the usage of wearable sensors for humans as well as for robot feedback sensors.

Composites such as armors, solid rocket motors, and explosive designs show the huge significance of localized temperature sensing. For studying the safety of these materials, impact and shock loading resulted in a small-scale temperature rise, which is very sensitive. However, it was found to be very tough to analyze experimentally, because of its low size as well as quick time evolution [73]. Normally, polymer-bonded explosives that consist of explosive crystals immersed in a binding agent, e.g., urethane-cross-linked polybutadiene, in-situ calculation of the temperature without altering the micro- and mesoscale mechanical behaviors is highly demanding.

Mason et al. [53] synthesized a cross-linked PUE for local temperature analysis by a thermochromic molecular sensor. Into the polymer blend, in a very small quantity, an altered donor—acceptor Stenhouse adduct (DASA) known as thermochrome is spread uniformly. For sensing localized temperature rise, resulting in an impact or fast heating, DASA shows appropriate kinetics. In materials such as explosive and propellant composites, the thermochrome maintains the maximum temperature in the polymer, which resulted in a postmortem mapping of micron-scale temperature localization.

15.3 Silicone-based elastomer blends and composites

Silicone elastomers exhibit enormous applications, such as in astronaut devices, automobiles, medical implants, and soft robots, attributed to their chemically inert nature, resistance to harsh environments, and also their easy processability. Silicone polymers and silicone networks are mostly amorphous and so, in the elastomer, there is no wear due to microcrystalline domains. Integral stability, as well as the reliability of suitably designed silicone elastomers, is one of its important properties. Mostly silicone elastomers are prepared either via platinum-catalyzed hydrosilylation, tin catalyzed condensation or by peroxide-initiated radical reactions as shown in Fig. 15.7. The self-healing properties of silicone materials make the materials lifelong and diminish fuel waste [36]. Few limitations are reported in self-healing silicone materials, such as poor tensile performance and self-healing capacity. So, it's very crucial to prepare good strength and self-healing, efficient self-healing silicone materials.

In high-performance polymers, polydimethylsiloxane (PDMS) exhibits various properties such as excellent thermal and oxidation performance, poor surface energy, easy fabrication, varying hardness, and better biocompatibility. Wang et al. [75] prepared an imine-bond-based transparent healable PDMS (HPDMS) elastomer from amino-

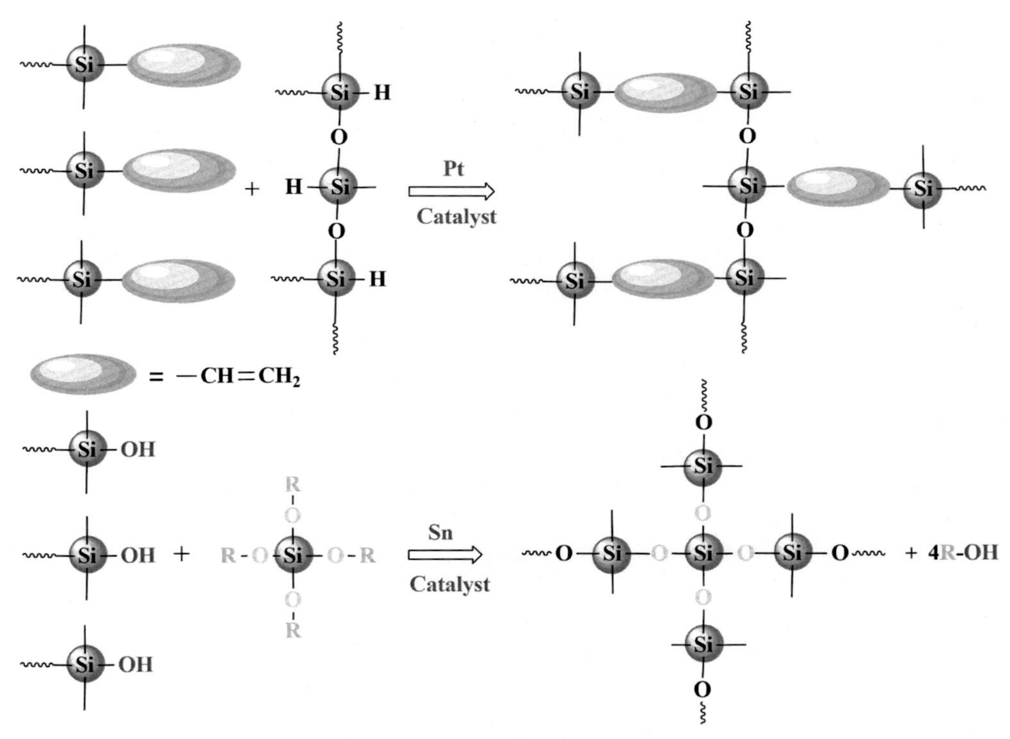

FIGURE 15.7 Schematic representation of silicon elastomer cross-linking.

modified polydimethylsiloxane (A-PDMS) and 1,4-diformylbenzene (DFB). The elastomer shows more than 80% of optical transmittance. Without the presence of any interventions such as heat or light, the reversible imine bond causes the contact of the fracture surfaces as well as fast and repeatable self-healing. Besides, with a 0.5 wt% DFB and HPDMS-50, the elastomer resulted in recyclable and reprocessable samples with 25.58 kPa TS and a poor elongation at a break of 23.65%. These elastomers are mostly hard but brittle in comparison to the HPDMS-35 samples. As the amount of DFB increases, the tensile strength is enhanced, whereas the elongation at break decreased owing to the presence of numerous imine bonds by the reaction with amino groups of PDMS. The mechanical strength improves by imine bonds behaving as physical cross-linker and a reinforcing element, in contrast to tough PU-based elastomers [17,29,37]. At room temperature when the elastomers were healed for 5 min, they regained the 42.6% strength and 47% elongation. However, when elastomers are kept for repairing to 1 h, they attain a TS of 38.68 kPa, which is very near to the original elastomers value, denoting an enhanced self-healing property of the developed elastomer as well as that the destructed elastomers can regain within 1 h after contact. Self-healing, mechanically strong, and transparent PDMS elastomers have many uses for artificial skins, protective coatings, and wearable devices.

Liu et al. [49] synthesized a novel cross-linked silicone elastomer, which is better self-healable, stretchable, and thermally stable using aminopropyl terminated polydimethylsiloxane (AP-PDMS), thioctic acid (TA), and 2,6-pyridine dialdehyde (Py). Dynamic disulfide bonds are contributed by the addition of TA, whereas metal–ligand active areas are because of the insertion of Py. Three different forms of chemical bonding were reported in silicone elastomers: the disulfide bonds, hydrogen bonds, and metal–ligand bonds having excellent anticorrosion and adhesion properties. These silicone elastomers exhibit many contributions toward soft robotics, wearable electronics, and stretchable circuits. Li et al. [43] prepared an elastomer via amino-ene Michael addition reaction in the presence of the basic catalyst tetramethylammonium hydroxide to get a dynamically cross-linked network. Elastomers were reported with increased mechanical behavior with about 1.08 MPa tensile strength and 206% elongation at break, indicating their good mechanical properties. With cross-linked networks and 10%–15% nanosilica, the mechanical performance was improved.

Lei et al. [38] reported an innovative self-healing, recyclable silicone elastomer by a two-step process. AP-PDMS is developed initially through copolymerization of dimethylsiloxane cyclic tetramer and 3-aminopropyl(diethoxy)methylsilane and then, AP-PDMS, pendant amino groups react with 1,4-phthalaldehyde (PTA) forming an imine-based cross-linked silicone elastomer (PTAAPDMS) as depicted in Fig. 15.8. PTA acts as a cross-linking source binding two imine bonds with benzene ring to improve a

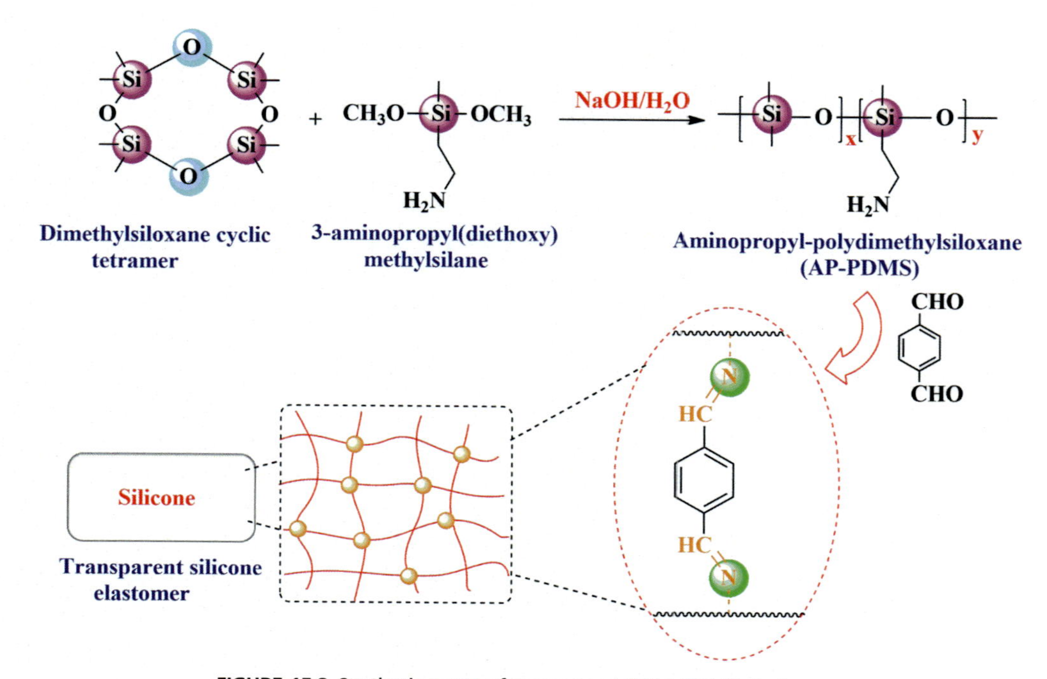

FIGURE 15.8 Synthetic route of transparent PTAA-PDMS elastomer.

conjugation effect. These silicon elastomers exhibit good self-healing properties with 95% healing efficiency at room temperature. In addition, they illustrate excellent optical transmittance of 91% and selective fluorescence quenching to Fe^{3+} ion with good potential in the area of the optical field.

Mazurek et al. [56] synthesized a double-phase glycerol–silicone hybrid elastomer, which, based on formulation, consists of a bicontinuous or closed-cell foam structure. Both of which are found to be biocompatible and harmless for various usages in the biomedical industries [50,59,71,80]. In presence of high shear forces, the glycerol–silicone composites are prepared to effectively nonhomogeneous mixtures of glycerol and silicone prepolymer [55].The silicone portion of the composites with a discrete glycerol phase morphology is spread as the glycerol section expands through the water absorption method. So, the silicone arrangement within the nearby droplets was converted into narrow, easing quick passage inside the material, thereby causing rapid drain of substances from the material. The elastomers with more glycerol amounts exhibit a zero-order release property with special drug delivery technology. In addition to biocompatibility, cost-effectiveness, and quick implementation, this technology exhibits releasing property, which is dependent on biobased materials found appropriate for the wound dressing application in which smart functionalized materials are in much demand.

Yan et al. [83] prepared a thermally reversible hyperbranched polysiloxane (HPSis) consisting of furan rings and double bonds by Diels–Alder (DA) in addition to enhancing the mechanical property of the elastomer. In the absence of HPSis, the mechanical performance of polysiloxane was about 0.20 MPa; although, it may be enhanced to 0.87 MPa in the presence of HPSis. As HPSis consists of multifuran rings needed for thermal reactions that are reversible, the addition of HPSis doesn't show an adverse impact on its self-healing properties. Within a stipulated time, the crevices on the surface of polysiloxane can be self-healed, attaining 85% self-healing efficiency. In addition, polysiloxane matrix is mixed with conductive filler carbon black (CB) to prepare flexible conductive composites PDMS/carbon black, which can be used in wearable flexible sensors. The mechanical property of the polysiloxane elastomer improves as the amount of γ-methacryloxypropyltrimethoxysilane increases. The elastomer exhibits good solvent resistance to almost all solvents, and the healed samples regain mechanical properties and the self-healing performance depends on the amount of HPSis and C=C bonds. Through the incorporation of hyperbranched and covalent bonds into the polysiloxane elastomer, the mechanical behavior of the polysiloxane elastomer is enhanced to 0.87 MPa.

15.4 Ethylene-propylene-diene monomer (EPDM)-based elastomer

EPDM copolymer is a form of nonpolar polyolefin rubber, with extensive application in industries such as waterproofing rolls, wires, sealing strips, cables, and automotive due to its good heat and ozone resistivity, excellent weather resistivity, and electrical insulation, etc. [40] As EPDM has weak tensile strength and fracture toughness and to rectify its limitations, EPDM is reinforced with fillers [7,12]. Nowadays, most added rubber filler is carbon black. But energy consumption is high for its production and is mostly dependent on costly petrochemical raw materials [62]. So, it has drawn great attention to finding alternative cost-efficient green materials for the traditional rubber filler. In silicone elastomer, the Si–O bond has a low number of alkyl groups and bond length is higher in comparison to the C=C bond, because of which it has poor molecular force in polymeric chain network. Consequently, silicone exhibits weak mechanical performance although with good thermal stability, weathering resistivity, poor surface tension, and excellent dielectric performance. So, silicone finds its application in the area of thermal shielding material, high-performance gaskets as well as weather-resistant coating [64,82]. The combination of these two polymers, EPDM and silicone rubber, is an alternative solution to prepare a novel material with properties of both. A gamma radiation-induced methacrylic acid (MAA)-grafted EPDM elastomer was mixed with (g-EPDM) in silicone rubber (SiR)/EPDM elastomer blends by the process of the electron beam (EB) cross-linking (Fig. 15.9). The gel content is intended to raise amount of EPDM, the presence of g-EPDM, and radiation dose. The incorporation of g-EPDM and EB cross-linking resulted in an increase in shrinkability, tensile, and electrical performance. Blends with more EPDM content exhibit a great degree of cross-linking (gel content) [10]. The improvement of tensile strength and Young's modulus indicates the improved interfacial strength resulting in a large stress transfer due to grafted EPDM. So, toughness and ductility show an improvement by the addition of g-EPDM in SiR/EPDM blends [25]. On EB irradiation, all blends' tensile strength as well as modulus increases, due to the development of intraphase and interphase 3D cross-linked domains of silicone and EPDM [70].Whereas elongation at break declines, owing to the cross-linking, which hinders the chain mobility reducing its energy dissipation [24].The virgin EPDM exhibits a hardness of 65. In the blend ratio, as the amount of SiR rises, the hardness reduces. However, the hardness remains unaltered by the addition of g-EPDM. Besides, the EB cross-linking boosts the dielectric property, volume, and surface resistivity of all SiR/EPDM blends. The 3D cross-linking network reduces the development of the charge carrier under applied electric field and the material shows insulating behavior.

Leone et al. [39] fabricated the EPDM via terpolymerization of ethylene with propylene and 5-ethylidene-2-norbornene (ENB), in which ethylene is added nonstop to the reaction mixture, whereas propylene and ENB are supplied in the initial step only. The prepared EPDMs were a macromolecular mixture, each of which has a uniform commoner distribution as well as a nonuniform structure with excellent elongation at break, strain, shape healing properties, and can be again melted for processing without altering its any properties, even during recycling and reuse. Recently, Sumimoto Chemical Co. scientists found EPDMs with excellent weather resistivity, which can be prepared in the presence of modified vanadium catalysts through narrowing the Chemical Composition Distribution [77].The terpolymers exhibit poor modulus and tensile strength, but with

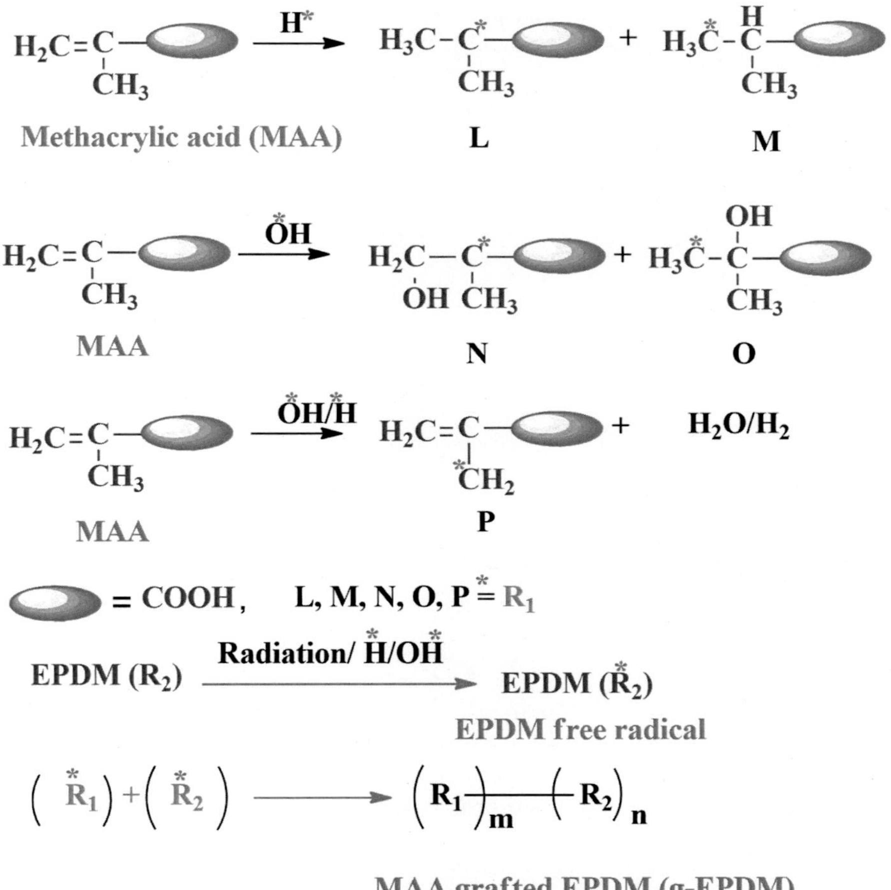

FIGURE 15.9 Representation of MAA grafted EPDM mechanism.

the enhanced percentage of elongation by reducing the amount of ENB and simultaneously increasing the percentage of ethylene.

Mei et al. [57] used a modified compatibilizer 3-amino-1,2,4-triazole (ATA) to introduce dynamic hydrogen bonds within the lignin/EPDM composite material to improve the interfacial linkages between EPDM and lignin as depicted in Fig. 15.10. The prepared biorenewable-based high-performance elastomer composites act as reinforcing agents. Polyethylene elastomer grafted maleic anhydride (POE-MA) was employed as the reactive compatibilizer while ATA was added to form a bond with POE-MA, because POE-MA is miscible in EPDM. The schematic mechanism of POE-MA and ATA as well as that among lignin and polymer matrix H-bonding interactions by amide triazolecarboxylic acid groups generated from the reaction of POE-MA and ATA is shown in Fig. 15.10.

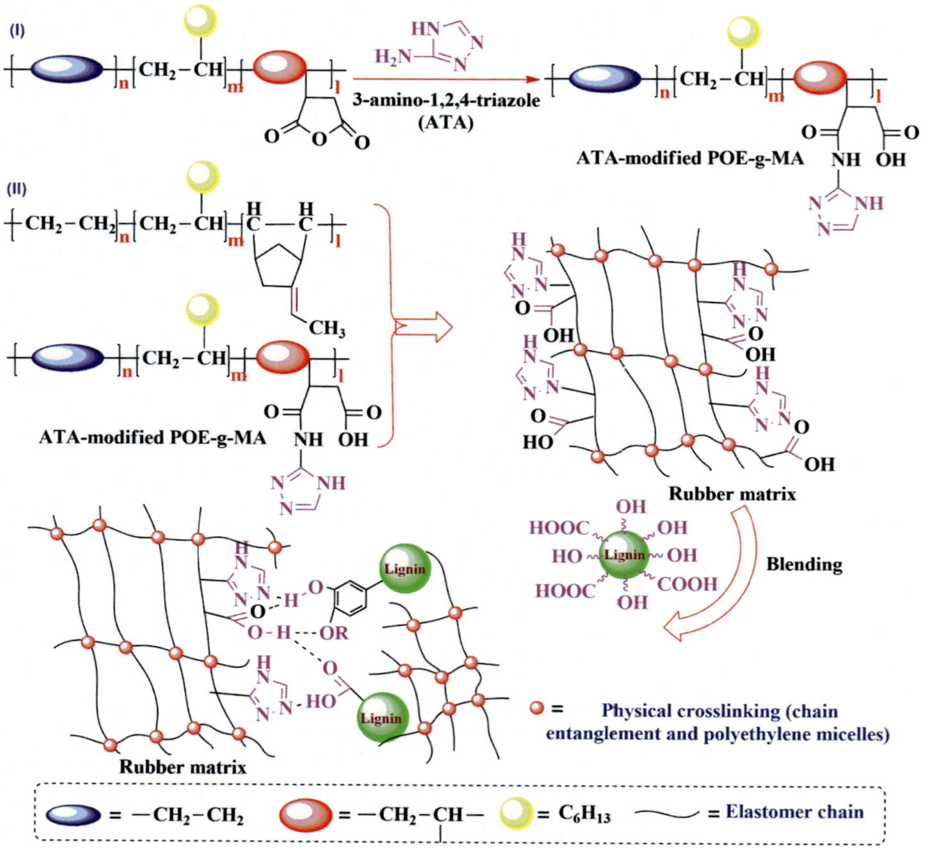

FIGURE 15.10 Synthesis mechanism of lignin and EPDM elastomer.

In the presence of POE-MA, weak particle aggregation and removal of interface exfoliation are attained, indicating improved interfacial compatibility. This was attributed to the formation of H-bonding by maleic anhydride groups in POE-MA with lignin. More content of ATA in addition to the composite resulted in an extra increment in the stress at 300% strain of 4.3 MPa and the tensile toughness attaining 45.3 MJ/m^3. Due to the sacrificial H-bonds within the lignin and POE matrix, tensile strength, as well as toughness of lignin/EPDM composites, was improved. Further, the addition of POE-MA and ATA increases the storage modulus again, because of the strong cross-linking network due to H-bonding interactions. However, as ATA content reached 7 phr, the modulus, as a result of the stress at 300%, was reduced, indicating that the additional H-bonds have an opposite impact on lignin and EPDM compatibility. At 20 phr lignin, the elongation at break of 911% was attained, which was near to the virgin EPDM. So, by tuning the amount of POE-MA, ATA, and lignin, the mechanical performance of lignin/EPDM composites can be tailored, with maximum property achieved at 20 phr content of POE-MA, 5 phr of ATA, and 20–40 phr of lignin in 100 phr EPDM. Lignin's availability is abundant as its the main by-product from the paper pulping industry and is cost-effective in comparison to carbon black with good antiaging and UV resistivity. Hence, these lignin-based high-performance elastomer composites are an alternative for commercial reinforcing filler such as carbon black.

Karaagac et al. [30] employed natural rubber, traditional EPDM, and a new grade EPDM to prepare linear molecular structure as a substitute for EPDM elastomers by temperature scanning stress relaxation (TSSR) method. These elastomers exhibit enhanced tensile strength as well as isothermal stress relaxation performance. Minor traces of process oil incorporation also increased their service temperatures owing to excellent processing settings. An annealing study indicated that strain-induced crystallization plays a crucial role in regulating service conditions of EPDM. In addition, TSSR method was a quick and influential method to calculate cross-link density and to find out service temperatures of natural rubber and EPDM elastomers.

Peng et al. [60,61] investigated the behavior and cross-linked networks of silane cross-linked polypropylene (PP)/EPDM elastomer blends. The blend consistent with the gel point shows highly enhanced mechanical performance. In the presence of EPDM, the tensile strength reduces abruptly along with the improvement in the elongation at break of simple PP/EPDM blends. Due to the presence of a weak degree of cross-linking, it shows a negligible effect on tensile strength and elongation at the break of silane cross-linked PP/EPDM blend at 10% EPDM. Whereas for 20% EPDM, the silane cross-linked PP/EPDM blends exhibit strong and stable cross-linking linkages with increment in tensile strength and reaching near to neat PP, while elongation at break boosts abruptly and attains 25 times more in contrast to neat PP. As the amount of EPDM increases, the thermal stability of silane cross-linked PP/EPDM blends improves as compared to PP and further increases slowly. In addition to the cross-link between PP and EPDM, they can also be linked together by PP—Si—O—Si—EPDM cross-linking bonds, which boosts their interfacial interaction as well as adhesion. During the tensile tests, PP—Si—O—Si—EPDM cross-linking networks behave like a bridge that can transmit stress between PP and EPDM matrixes and prevent cracking. Cross-linking networks causes neck formation and attainment of the strain hardening. Thus, the tensile strength, as well as elongation at the break of silane cross-linked PP/EPDM blends, exhibits excellent improvement.

15.5 Other elastomers

15.5.1 Fluorocarbon elastomer

Fluorocarbon (FKM) is a family of elastomer materials described under ISO 1629 and ASTM D1418 standards [2]. The monomer unit of the fluorocarbon elastomer is vinylidene fluoride. On the other hand, these types of elastomers are synthesized via free radical and emulsion polymerization with a high-pressure combination of two dissimilar monomer units for illustrated vinylidene fluoride (VDF), tetrafluoroethylene (TEF), and ethylene monomers. These monomers exhibited bulky side groups such as perfluoromethylvinylether (PMVE), hexafluoropropylene (HFP), and propylene in addition to the synthesized amorphous polymer. After the insertion of the bulky material, it bestows excellent low-temperature flexibility contrasted to copolymer and terpolymer Fig. 15.11 demonstrates the four main categories of monomers employed to synthesize the fluoroelastomers.

On the other hand, fluoroelastomers are more costly compared to the traditional elastomers due to their higher characteristics such as tremendous heat and chemical resistance. The classification of the FKMs is based on fluorine content, cross-linking mechanism as well as chemical constitution. The widely used FKM elastomer exhibited 65%–68% fluorine content and plays a vital role in defining its performance. Mostly, the higher content of fluorine provides higher chemical resistance. Frequently, terpolymer demonstrates more fluorine amounts in contrast to copolymers. Conversely, fluorocarbon polymer chains are comparatively stronger than hydrocarbons and reveal lower flexibility as well as slow relaxation and recovery from the strain.

However, their properties were significant owing to the existence of a very large energy bond (442 kJ/mol) such as a C–F bond and weak chemical reactivity contrasted to other chemical bonds. Therefore, FKMs show exclusive resistance to oxygen, chemicals,

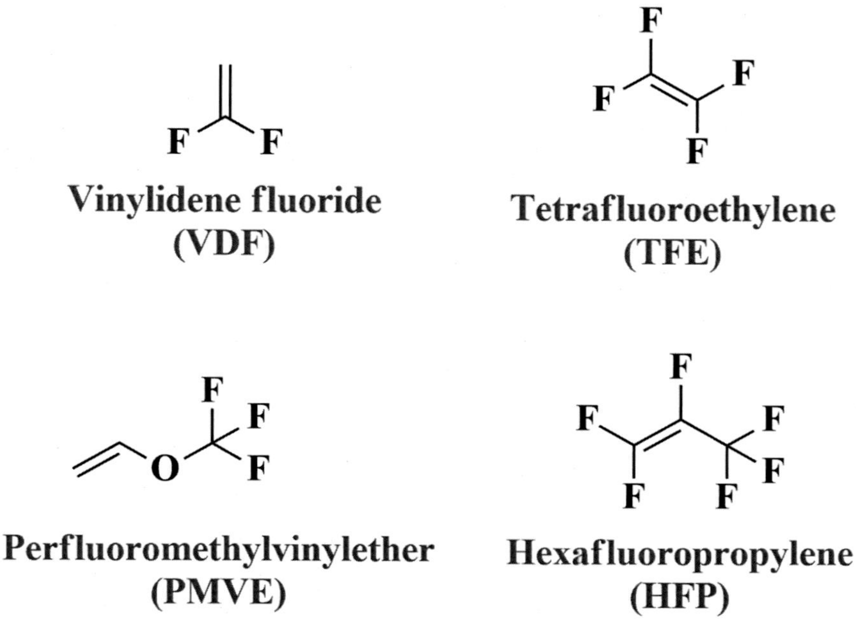

Vinylidene fluoride (VDF)

Tetrafluoroethylene (TFE)

Perfluoromethylvinylether (PMVE)

Hexafluoropropylene (HFP)

FIGURE 15.11 Fluorocarbon elastomer synthesis monomer.

and ozone at high temperatures. These elastomers are polar and have outstanding resistance to oils, fuels, aliphatic and aromatic, chlorinated hydrocarbons along with concentrated acids. However, FKMs are not resistant to ketones or esters.

These types of elastomers are mostly employed in the automobile industry for numerous applications, such as engines, fuel system components, particularly seals, hoses, O-rings, and gaskets. This type of elastomer enhanced the demand for the high-temperature elastomer for alcohol-containing fuels and aggressive lubricants. Further, FKMs are also used in various sectors: petroleum, petrochemical, industrial pollution control, industrial pneumatic and hydraulic etc. Thus, the fluoroelastomer is more costly compared to other speciality elastomers. Nevertheless, they provide a longer service life when used in chemical and heat resistance applications [51].

15.5.2 Chlorosulfonated polyethylene rubber elastomer

Cholorosulfonated polyethylene (CSM) is a significant elastomer synthesized from the chlorination simultaneously with chlorosulfonation of polyethylene providing the help of cross-linking of the elastomer with sulfur vulcanization behavior [66]. The CSM was already commercialized by DuPont with the registered trade name of Hypalon with different grades of synthetic rubber [51]. These polymer elastomers were synthesized with completely saturated backbone and pendant functional groups with various routes to pact with the vulcanization the monomer structure depicted in Fig. 15.12 Conversely, cross-linked CSM exhibited high stability toward the declining effect of oxygen, ozone, weather, heat, and chemicals. Additionally, it also exhibited extreme resistance against discoloration on contact with light and is widely employed in light-colored vulcanizates as well as blended with polymer and endows the higher mechanical properties, abrasion resistivity, and weather resistivity of the developed blend [52,72]. SCM elastomer is used in various applications such as coatings, industrial hoses, insulation, automotive electrical wiring, oil and chemical resistance, and inflatable applications, etc. (*Fuel Bladder Material: TPU* versus *Hypalon, EREZ Tech. Text,* [4]. Https://Ereztherm. Com/Fuel-Bladder-Material-Tpu-Hypalon/. *Accessed August 2, 2021, n.d.; Hypalon* versus, *TPU Coating for RIBs - What Builds a Better Boat? EREZ Tech. Text,* [3]).

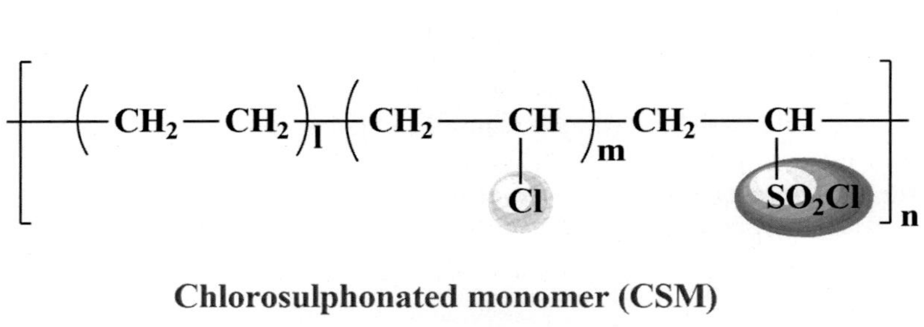

$$\left[\left(CH_2-CH_2 \right)_l \left(CH_2-\underset{Cl}{CH} \right)_m CH_2-\underset{SO_2Cl}{CH} \right]_n$$

Chlorosulphonated monomer (CSM)

FIGURE 15.12 Monomer of CSM elastomer.

15.6 Conclusions

Elastomer materials are extensively used in various applications such as shape memory, flame retardant, sensing, and self-healing. All the elastomer blends and composites are versatile materials that demonstrate exclusive properties after blending and enhancing flame retardancy, self-healing efficiency, tremendous shape memory, and sensor properties. In this chapter, various elastomer blends and composites, tunable properties, and applications based on their different synthesis methods and their incorporated additives have been described. However, all elastomer blends and composites reveal the huge potential to explore new industrial applications.

References

[1] B.A. Dombrow, Polyurethanes, 1965.

[2] Standard Practice for Rubber and Rubber Latices – Nomenclature, 2017, https://doi.org/10.1520/D1418-17. ASTM, D1418.

[3] Hypalon vs, Tpu Coating for Ribs – What Builds a Better Boat? Jul 2, 2020, Welding, 2020. EREZ Tech. Text.

[4] Fuel Bladder Material: TPU vs Hypalon, EREZ Tech. Text, 2019, n.d. https://ereztherm.com/fuel-bladder-material-tpu-hypalon/.(Accessed 2 August 2021.

[5] J.O. Akindoyo, M.D.H. Beg, S. Ghazali, M.R. Islam, N. Jeyaratnam, A.R. Yuvaraj, Polyurethane types, synthesis and applications-a review, RSC Adv. 6 (115) (2016) 114453–114482, https://doi.org/10.1039/c6ra14525f.

[6] U. Almeida Pinto, L.L.Y. Visconte, J. Gallo, R.C.R. Nunes, Flame retardancy in thermoplastic polyurethane elastomers (TPU) with mica and aluminum trihydrate (ATH), Polym. Degrad. Stabil. 69 (3) (2000) 257–260, https://doi.org/10.1016/S0141-3910(00)00047-1.

[7] S. Araby, Q. Meng, L. Zhang, I. Zaman, P. Majewski, J. Ma, Nanotechnology 26 (2015).

[8] R.M. Aseeva, G.E. Zaikov, Flame retardancy of cellular polymeric materials, Int. J. Polym. Mater. Polym. Biomater. 31 (1–4) (1996) 237–256, https://doi.org/10.1080/00914039608029378.

[9] H. Ashrafizadeh, A. McDonald, P. Mertiny, Deposition of electrically conductive coatings on castable polyurethane elastomers by the flame spraying process, J. Therm. Spray Technol. 25 (3) (2016) 419–430, https://doi.org/10.1007/s11666-015-0376-2.

[10] L. Bazli, A. Khavandi, M.A. Boutorabi, M. Karrabi, Correlation between viscoelastic behavior and morphology of nanocomposites based on SR/EPDM blends compatibilized by maleic anhydride, Polymer 113 (2017) 156–166.

[11] S. Benli, Ü. Yilmazer, F. Pekel, S. Özkar, Effect of fillers on thermal and mechanical properties of polyurethane elastomer, J. Appl. Polym. Sci. 68 (7) (1998) 1057–1065. https://doi.org/10.1002/(SICI)1097-4628(19980516)68:7<1057::AID-APP3>3.3.CO;2-E.

[12] N. Bitinis, M. Hernandez, R. Verdejo, J.M. Kenny, M.A. Lopez-Manchado, Adv. Mater. 23 (2011).

[13] S. Bourbigot, T. Turf, S. Bellayer, S. Duquesne, Polyhedral oligomeric silsesquioxane as flame retardant for thermoplastic polyurethane, Polym. Degrad. Stabil. 94 (8) (2009) 1230–1237. https://doi.org/10.1016/j.polymdegradstab.2009.04.016.

[14] D. Chen, Q. Pei, Electronic muscles and skins: a review of soft sensors and actuators, Chem. Rev. 117 (17) (2017) 11239–11268. https://doi.org/10.1021/acs.chemrev.7b00019.

[15] D. Dieterich, E. Grigat, W. Hahn, H. Hespe, H.G. Schmelzer, Polyurethane Handbook, 1985, pp. 19–22.

[16] H.J. Fabris, J.G. Sommer, Flammability of elastomeric materials, Rubber Chem. Technol. 50 (3) (1977) 523–569. https://doi.org/10.5254/1.3535157.

[17] C.-J. Fan, Z.-C. Huang, B. Li, W.-X. Xiao, E. Zheng, K.-K. Yang, Y.-Z. Wang, A robust self-healing polyurethane elastomer: from H-bonds and stacking interactions to well-defined microphase morphology, Sci. China Mater. 62 (8) (2019) 1188–1198. https://doi.org/10.1007/s40843-019-9422-7.

[18] Y.F. Fu, Y.Q. Li, Y.F. Liu, P. Huang, N. Hu, S.Y. Fu, High-performance structural flexible strain sensors based on graphene-coated glass fabric/silicone composite, ACS Appl. Mater. Interfaces 10 (41) (2018) 35503–35509. https://doi.org/10.1021/acsami.8b09424.

[19] B. Ghosh, M.W. Urban, Self-repairing oxetane-substituted chitosan polyurethane networks, Science 323 (5920) (2009) 1458–1460. https://doi.org/10.1126/science.1167391.

[20] B. Ghosh, K.V. Chellappan, M.W. Urban, Self-healing inside a scratch of oxetane-substituted chitosan-polyurethane (OXE-CHI-PUR) networks, J. Mater. Chem. 21 (38) (2011) 14473–14486. https://doi.org/10.1039/c1jm12321a.

[21] B. Ghosh, K.V. Chellappan, M.W. Urban, UV-initiated self-healing of oxolane-chitosan-polyurethane (OXO-CHI-PUR) networks, J. Mater. Chem. 22 (31) (2012) 16104–16113. https://doi.org/10.1039/c2jm31126g.

[22] G. Gu, H. Xu, S. Peng, L. Li, S. Chen, T. Lu, X. Guo, Integrated soft ionotronic skin with stretchable and transparent hydrogel-elastomer ionic sensors for hand-motion monitoring, Soft Robot. 6 (3) (2019) 368–376. https://doi.org/10.1089/soro.2018.0116.

[23] P.A. Gunatillake, G.F. Meijs, S.J. Mccarthy, R. Adhikari, Poly(dimethylsiloxane)/Poly(hexamethylene oxide) mixed macrodiol based polyurethane elastomers. I. Synthesis and properties, J. Appl. Polym. Sci. 76 (14) (2000) 2026–2040. https://doi.org/10.1002/(SICI)1097-4628(20000628)76:14<2026::AID-APP5>3.0.CO;2-X.

[24] S. Gupta, R. Chowdhury, J.K. Mishra, C.K. Das, P.K. Patra, A.K. Tripathy, W. Millins, M.S. Banerjee, Mater. Lett. 46 (2000).

[25] A. Hassan, Wahit, Polym. Test. 22 (2003).

[26] Y. Heo, M.H. Malakooti, H.A. Sodano, Self-healing polymers and composites for extreme environments, J. Mater. Chem. 4 (44) (2016) 17403–17411. https://doi.org/10.1039/c6ta06213j.

[27] J. Hu, R. Mo, X. Sheng, X. Zhang, A self-healing polyurethane elastomer with excellent mechanical properties based on phase-locked dynamic imine bonds, Polym. Chem. 11 (14) (2020) 2585–2594. https://doi.org/10.1039/d0py00151a.

[28] J. Huang, Q. Tang, W. Liao, G. Wang, W. Wei, C. Li, Green preparation of expandable graphite and its application in flame-resistance polymer elastomer, Ind. Eng. Chem. Res. 56 (18) (2017) 5253–5261. https://doi.org/10.1021/acs.iecr.6b04860.

[29] J.H. Kang, D. Son, G.J.N. Wang, Y.X. Liu, J. Lopez, Y. Kim, J.Y. Oh, T. Katsumata, J.W. Mun, Y. Lee, L. H. Jin, J.B.H. Tok, Z.N. Bao, Tough and water insensitive self-healing elastomer for robust electronic skin, Adv. Mater. 30 (13) (2018).

[30] B. Karaağaç, S.C. Cengiz, T. Bayram, M. Şen, Identification of temperature scanning stress relaxation behaviors of new grade ethylene propylene diene elastomers, Adv. Polym. Technol. 37 (8) (2018) 3027–3037. https://doi.org/10.1002/adv.21973.

[31] S.M. Kim, H. Jeon, S.H. Shin, S.A. Park, J. Jegal, S.Y. Hwang, D.X. Oh, J. Park, Superior toughness and fast self-healing at room temperature engineered by transparent elastomers, Adv. Mater. 30 (1) (2018). https://doi.org/10.1002/adma.201705145.

[32] C.C. Kim, H.H. Lee, K.H. Oh, J.Y. Sun, Highly stretchable, transparent ionic touch panel, Science 353 (6300) (2016) 682–687. https://doi.org/10.1126/science.aaf8810.

[33] B. Kumar, N. Noor, S. Thakur, N. Pan, H. Narayana, S.C. Yan, F. Wang, P. Shah, Shape memory polyurethane-based smart polymer substrates for physiologically responsive, dynamic pressure (re) distribution, ACS Omega 4 (13) (2019) 15348–15358. https://doi.org/10.1021/acsomega.9b01167.

[34] E. Kurahashi, H. Sugimoto, E. Nakanishi, K. Nagata, K. Inomata, Shape memory properties of polyurethane/poly(oxyethylene) blends, Soft Matter 8 (2) (2012) 496–503. https://doi.org/10.1039/c1sm06585h.

[35] J. Chen, F. Li, Y. Luo, Y. Shi, X. Ma, M. Zhang, D.W. Boukhvalov, Z. Luo, A self-healing elastomer based on an intrinsic non-covalent cross-linking mechanism, J. Mater. Chem. 7 (25) (2019) 15207–15214. https://doi.org/10.1039/c9ta03775f.

[36] J.C. Lai, J.F. Mei, X.Y. Jia, C.H. Li, X.Z. You, Z. Bao, A stiff and healable polymer based on dynamic-covalent boroxine bonds, Adv. Mater. 28 (37) (2016) 8277–8282. https://doi.org/10.1002/adma.201602332.

[37] Y. Lai, X. Kuang, P. Zhu, M. Huang, X. Dong, D. Wang, Colorless, transparent, robust, and fast scratch-self-healing elastomers via a phase-locked dynamic bonds design, Adv. Mater. 30 (38) (2018) 1802556. https://doi.org/10.1002/adma.201802556.

[38] X. Lei, Y. Huang, S. Liang, X. Zhao, L. Liu, Preparation of highly transparent, room-temperature self-healing and recyclable silicon elastomers based on dynamic imine bond and their ion responsive properties, Mater. Lett. 268 (2020) 127598. https://doi.org/10.1016/j.matlet.2020.127598.

[39] G. Leone, G. Zanchin, R. Di Girolamo, F. De Stefano, C. Lorber, C. De Rosa, G. Ricci, F. Bertini, Semibatch terpolymerization of ethylene, propylene, and 5-Ethylidene-2-norbornene: heterogeneous high-ethylene EPDM thermoplastic elastomers, Macromolecules 53 (14) (2020) 5881–5894. https://doi.org/10.1021/acs.macromol.0c01123.

[40] A.-M. Lepadatu, S. Asaftei, N. Vennemann, J. Appl. Polym. Sci. 132 (2015).

[41] H. Li, J.T. Sun, C. Wang, S. Liu, D. Yuan, X. Zhou, J. Tan, L. Stubbs, C. He, High modulus, strength, and toughness polyurethane elastomer based on unmodified lignin, ACS Sustain. Chem. Eng. 5 (9) (2017) 7942–7949. https://doi.org/10.1021/acssuschemeng.7b01481.

[42] X. Li, R. Yu, Y. He, Y. Zhang, X. Yang, X. Zhao, W. Huang, Self-healing polyurethane elastomers based on a disulfide bond by digital light processing 3D printing, ACS Macro Lett. 8 (11) (2019) 1511–1516. https://doi.org/10.1021/acsmacrolett.9b00766.

[43] X. Li, R. Yu, T. Zhao, Y. Zhang, X. Yang, X. Zhao, W. Huang, A self-healing polysiloxane elastomer based on siloxane equilibration synthesized through amino-ene Michael addition reaction, Eur. Polym. J. 108 (2018) 399–405. https://doi.org/10.1016/j.eurpolymj.2018.09.021.

[44] Y.L. Liu, T.W. Chuo, Self-healing polymers based on thermally reversible Diels-Alder chemistry, Polym. Chem. 4 (7) (2013) 2194–2205. https://doi.org/10.1039/c2py20957h.

[45] H. Liu, H. Qing, Z. Li, Y.L. Han, M. Lin, H. Yang, A. Li, T.J. Lu, F. Li, F. Xu, Paper: a promising material for human-friendly functional wearable electronics, Mater. Sci. Eng. R Rep. (2017) 1–22. https://doi.org/10.1016/j.mser.2017.01.001.

[46] H. Liu, M. Li, S. Liu, P. Jia, X. Guo, S. Feng, T.J. Lu, H. Yang, F. Li, F. Xu, Spatially modulated stiffness on hydrogels for soft and stretchable integrated electronics, Mater. Horiz. 7 (1) (2020a) 203–213. https://doi.org/10.1039/c9mh01211g.

[47] H. Liu, M. Li, C. Ouyang, T.J. Lu, F. Li, F. Xu, Biofriendly, stretchable, and reusable hydrogel electronics as wearable force sensors, Small 14 (36) (2018) 1801711. https://doi.org/10.1002/smll.201801711.

[48] Z.F. Liu, S. Fang, F.A. Moura, J.N. Ding, N. Jiang, J. Di, M. Zhang, X. Lepró, D.S. Galvão, C.S. Haines, N.Y. Yuan, S.G. Yin, D.W. Lee, R. Wang, H.Y. Wang, W. Lv, C. Dong, R.C. Zhang, M.J. Chen, et al., Hierarchically buckled sheath-core fibers for superelastic electronics, sensors, and muscles, Science 349 (6246) (2015) 400–404. https://doi.org/10.1126/science.aaa7952.

[49] Y. Liu, J. Yuan, K. Zhang, K. Guo, L. Yuan, Y. Wu, C. Gao, A novel type of self-healing silicone elastomers with reversible cross-linked network based on the disulfide, hydrogen and metal-ligand bonds, Prog. Org. Coating 144 (2020b). https://doi.org/10.1016/j.porgcoat.2020.105661.

[50] R.K. Malcolm, S.D. McCullagh, A.D. Woolfson, S.P. Gorman, D.S. Jones, J. Cuddy, Controlled release of a model antibacterial drug from a novel self-lubricating silicone biomaterial, J. Contr. Release 97 (2) (2004) 313–320. https://doi.org/10.1016/j.jconrel.2004.03.029.

[51] N. Mandlekar, M. Joshi, B.S. Butola, A review on specialty elastomers based potential inflatable structures and applications, Adv. Industr. Eng. Polym. Res. (2021). https://doi.org/10.1016/j.aiepr.2021.05.004. In press.

[52] G. Marković, M. Marinović-Cincović, V. Jovanović, S. Samaržija-Jovanović, J. Budinski-Simendić, NR/CSM/biogenic silica rubber blend composites, Compos. B Eng. 55 (2013) 368–373. https://doi.org/10.1016/j.compositesb.2013.06.045.

[53] B.P. Mason, M. Whittaker, J. Hemmer, S. Arora, A. Harper, S. Alnemrat, A. McEachen, S. Helmy, J. Read De Alaniz, J.P. Hooper, A temperature-mapping molecular sensor for polyurethane-based elastomers, Appl. Phys. Lett. 108 (4) (2016). https://doi.org/10.1063/1.4940750.

[54] N. Matsuhisa, D. Inoue, P. Zalar, H. Jin, Y. Matsuba, A. Itoh, T. Yokota, D. Hashizume, T. Someya, Printable elastic conductors by in situ formation of silver nanoparticles from silver flakes, Nat. Mater. 16 (8) (2017) 834–840. https://doi.org/10.1038/nmat4904.

[55] P. Mazurek, S. Hvilsted, A.L. Skov, Green silicone elastomer obtained from a counterintuitively stable mixture of glycerol and PDMS, Polymer 87 (2016) 1–7. https://doi.org/10.1016/j.polymer.2016.01.070.

[56] P. Mazurek, M.A. Brook, A.L. Skov, Glycerol-silicone elastomers as active matrices with controllable release profiles, Langmuir 34 (38) (2018) 11559–11566. https://doi.org/10.1021/acs.langmuir.8b02039.

[57] J. Mei, W. Liu, J. Huang, X. Qiu, Lignin-reinforced ethylene-propylene-diene copolymer elastomer via hydrogen bonding interactions, Macromol. Mater. Eng. 304 (4) (2019). https://doi.org/10.1002/mame.201800689.

[58] J. Miao, Y. Fang, Y. Guo, Y. Zhu, A. Hu, G. Wang, Interpenetrating polymer networks of porous organic polymers and polyurethanes for flame resistance and high mechanical properties, ACS Appl. Polym. Mater. (2019) 2692–2702. https://doi.org/10.1021/acsapm.9b00633.

[59] P.S. Murphy, G.R.D. Evans, Advances in wound healing: a review of current wound healing products, Plast. Surg. Int. (2012) 1–8. https://doi.org/10.1155/2012/190436.

[60] S. Peng, Y. Li, L. Wu, J. Zhong, Z. Weng, L. Zheng, Z. Yang, J.T. Miao, 3D printing mechanically robust and transparent polyurethane elastomers for stretchable electronic sensors, ACS Appl. Mater. Interfaces 12 (5) (2020a) 6479–6488. https://doi.org/10.1021/acsami.9b20631.

[61] H. Peng, M. Lu, H. Wang, Z. Zhang, F. Lv, M. Niu, W. Wang, Comprehensively improved mechanical properties of silane crosslinked polypropylene/ethylene propylene diene monomer elastomer blends, Polym. Eng. Sci. 60 (5) (2020b) 1054–1065. https://doi.org/10.1002/pen.25361.

[62] S. Praveen, P.K. Chattopadhyay, P. Albert, V.G. Dalvi, B.C. Chakraborty, S. Chattopadhyay, Effect of nanoclay on the mechanical and damping properties of aramid short fibre-filled styrene butadiene rubber composites, Polym. Int. 2 (59) (2009) 187–197. https://doi.org/10.1002/pi.2706.

[63] W. Pu, D. Fu, Z. Wang, X. Gan, X. Lu, L. Yang, H. Xia, Realizing crack diagnosing and self-healing by electricity with a dynamic crosslinked flexible polyurethane composite, Adv. Sci. 5 (5) (2018) 1800101. https://doi.org/10.1002/advs.201800101.

[64] V. Rajini, R. Deepalaxmi, Property enhancement of SiR-EPDM blend using electron beam irradiation, J. Electr. Eng. Technol. 9 (3) (2014) 984–990.

[65] D. Ratna, J. Karger-Kocsis, Recent advances in shape memory polymers and composites: a review, J. Mater. Sci. 43 (1) (2008) 254–269. https://doi.org/10.1007/s10853-007-2176-7.

[66] B. Rodgers, Rubber compounding: chemistry and applications, in: Rubber Compounding: Chemistry and Applications, second ed., CRC Press, 2015, pp. 1–600. https://doi.org/10.1201/b18931.

[67] D. Rus, M.T. Tolley, Design, fabrication and control of soft robots, Nature 521 (7553) (2015) 467–475. https://doi.org/10.1038/nature14543.

[68] M.S. Sarwar, Y. Dobashi, C. Preston, J.K.M. Wyss, S. Mirabbasi, J. David, W. Madden, Bend, stretch, and touch: locating a finger on an actively deformed transparent sensor array, Sci. Adv. 3 (3) (2017). https://doi.org/10.1126/sciadv.1602200.

[69] X. Shan, P. Zhang, L. Song, Y. Hu, S. Lo, Compound of nickel phosphate with Ni(OH)(PO4)2- layers and synergistic application with intumescent flame retardants in thermoplastic polyurethane elastomer, Ind. Eng. Chem. Res. 50 (12) (2011) 7201–7209. https://doi.org/10.1021/ie2001555.

[70] K. Sharma, Chowdhury, Ray, S. Jha, A. Samanta, A.K. Mahanwar, P. Sarma, K. S, J. Appl. Polym. Sci. 134 (2017).

[71] C.L. Silva, J.C. Pereira, A. Ramalho, A.A.C.C. Pais, J.J.S. Sousa, Films based on chitosan polyelectrolyte complexes for skin drug delivery: development and characterization, J. Membr. Sci. 320 (1–2) (2008) 268–279. https://doi.org/10.1016/j.memsci.2008.04.011.

[72] M. Tariq, S. Nisar, A. Shah, S. Akbar, M.A. Khan, S.Z. Khan, Effect of hybrid reinforcement on the performance of filament wound hollow shaft, Compos. Struct. 184 (2018) 378–387. https://doi.org/10.1016/j.compstruct.2017.09.098.

[73] C.M. Tarver, S.K. Chidester, A.L. Nichols, Critical conditions for impact- and shock-induced hot spots in solid explosives, J. Phys. Chem. 100 (14) (1996) 5794–5799. https://doi.org/10.1021/jp953123s.

[74] Y. Wan, Z. Qiu, Y. Hong, Y. Wang, J. Zhang, Q. Liu, Z. Wu, C.F. Guo, A highly sensitive flexible capacitive tactile sensor with sparse and high-aspect-ratio microstructures, Adv. Electr. Mater. 4 (4) (2018) 1700586. https://doi.org/10.1002/aelm.201700586.

[75] P. Wang, L. Yang, B. Dai, Z. Yang, S. Guo, G. Gao, L. Xu, M. Sun, K. Yao, J. Zhu, A self-healing transparent polydimethylsiloxane elastomer based on imine bonds, Eur. Polym. J. 123 (2020). https://doi.org/10.1016/j.eurpolymj.2019.109382.

[76] S. Wang, J.Y. Oh, J. Xu, H. Tran, Z. Bao, Skin-inspired electronics: an emerging paradigm, Acc. Chem. Res. 51 (5) (2018) 1033–1045. https://doi.org/10.1021/acs.accounts.8b00015.

[77] K. Watanabe, S. Nakano, Development of EPDM with Excellent Cold Resistance, 2018. R&D Report Sumitomo Chemical Asia Pte Ltd. SUMITOMO KAKUGO.

[78] G.M. Whitesides, Soft robotics, Angew. Chem. Int. Ed. 57 (16) (2018) 4258–4273. https://doi.org/10.1002/anie.201800907.

[79] G. Woods, ICI Polyurethane New York (NY), 1990.

[80] A.D. Woolfson, R.K. Malcolm, R.J. Gallagher, Design of a silicone reservoir intravaginal ring for the delivery of oxybutynin, J. Contr. Release 91 (3) (2003) 465–476. https://doi.org/10.1016/S0168-3659(03)00277-3.

[81] W. Xu, L. Liu, B. Zhang, Y. Hu, B. Xu, Effect of molybdenum trioxide-loaded graphene and cuprous oxide-loaded graphene on flame retardancy and smoke suppression of polyurethane elastomer, Ind. Eng. Chem. Res. 55 (17) (2016) 4930–4941. https://doi.org/10.1021/acs.iecr.6b00383.

[82] Q. Xu, M. Pang, L. Zhu, Y. Zhang, S. Feng, Mechanical properties of silicone rubber composed of diverse vinyl content silicone gums blending, Mater. Des. 9 (31) (2010) 4083–4087. https://doi.org/10.1016/j.matdes.2010.04.052.

[83] Q. Yan, L. Zhao, Q. Cheng, T. Zhang, B. Jiang, Y. Song, Y. Huang, Self-healing polysiloxane elastomer based on integration of covalent and reversible networks, Ind. Eng. Chem. Res. 58 (47) (2019) 21504–21512. https://doi.org/10.1021/acs.iecr.9b04355.

[84] A.H. Yang, C. Deng, H. Chen, Y.X. Wei, Y.Z. Wang, A novel Schiff-base polyphosphate ester: highly-efficient flame retardant for polyurethane elastomer, Polym. Degrad. Stabil. 144 (2017) 70–82. https://doi.org/10.1016/j.polymdegradstab.2017.08.007.

[85] Y. Yang, M.W. Urban, Self-healing of glucose-modified polyurethane networks facilitated by damage-induced primary amines, Polym. Chem. 8 (1) (2017) 303–309. https://doi.org/10.1039/c6py01221c.

[86] S. Yang, S. Wang, X. Du, Z. Du, X. Cheng, H. Wang, Mechanically robust self-healing and recyclable flame-retarded polyurethane elastomer based on thermoreversible crosslinking network and multiple hydrogen bonds, Chem. Eng. J. 391 (2020) 123544. https://doi.org/10.1016/j.cej.2019.123544.

[87] L. Yang, X. Lu, Z. Wang, H. Xia, Diels-Alder dynamic crosslinked polyurethane/polydopamine composites with NIR triggered self-healing function, Polym. Chem. 9 (16) (2018) 2166–2172. https://doi.org/10.1039/c8py00162f.

[88] M. Yarmohammadi, M. Shahidzadeh, B. Ramezanzadeh, Designing an elastomeric polyurethane coating with enhanced mechanical and self-healing properties: the influence of disulfide chain extender, Prog. Org. Coating 121 (2018) 45–52. https://doi.org/10.1016/j.porgcoat.2018.04.009.

[89] N. Yoshie, S. Saito, N. Oya, A thermally-stable self-mending polymer networked by Diels-Alder cycloaddition, Polymer 52 (26) (2011) 6074–6079. https://doi.org/10.1016/j.polymer.2011.11.007.

[90] H. Zhang, W. Li, X. Yang, L. Lu, X. Wang, X. Sun, Y. Zhang, Development of polyurethane elastomer composite materials by addition of milled fiberglass with coupling agent, Mater. Lett. 61 (6) (2007) 1358–1362. https://doi.org/10.1016/j.matlet.2006.07.031.

[91] Y. Zheng, R. Dong, J. Shen, S. Guo, Tunable shape memory performances via multilayer assembly of thermoplastic polyurethane and polycaprolactone, ACS Appl. Mater. Interfaces 8 (2) (2016) 1371–1380. https://doi.org/10.1021/acsami.5b10246.

[92] J. Zhou, H. Cao, R. Chang, G. Shan, Y. Bao, P. Pan, Stereocomplexed and homochiral polyurethane elastomers with tunable crystallizability and multishape memory effects, ACS Macro Lett. 7 (2) (2018) 233–238. https://doi.org/10.1021/acsmacrolett.7b00995.

[93] K.M. Zia, M. Zuber, M.J. Saif, M. Jawaid, K. Mahmood, M. Shahid, M.N. Anjum, M.N. Ahmad, Chitin based polyurethanes using hydroxyl terminated polybutadiene, part III: surface characteristics, Int. J. Biol. Macromol. 62 (2013) 670–676. https://doi.org/10.1016/j.ijbiomac.2013.10.001.

16

Properties of elastomer—biological phenolic resin composites

Kushairi Mohd Salleh[1], Marhaini Mostapha[1,2], Kam Sheng Lau[1], Sarani Zakaria[1]

[1]BIORESOURCE & BIOREFINERY RESEARCH GROUP, DEPARTMENT OF APPLIED PHYSICS, FACULTY OF SCIENCE AND TECHNOLOGY, UNIVERSITI KEBANGSAAN MALAYSIA, BANGI, SELANGOR, MALAYSIA; [2]HICOE-CENTRE FOR BIOFUELS AND BIOCHEMICAL RESEARCH (CBBR), INSTITUTE OF SELF-SUSTAINABLE BUILDING (ISB), UNIVERSITI TEKNOLOGI PETRONAS, SERI ISKANDAR, PERAK, MALAYSIA

16.1 Introduction

Elastomer is a high-molar-mass polymeric material that deformed at room temperature and quickly reverts to nearly original size when the load causing the deformation is removed (ISO 1382:1996). Meanwhile, IUPAC defines elastomer as a polymer that displays rubber-like elasticity [1]. They usually exemplify a loosely cross-linked polymer that wholly suppresses irreversible flow with low intermolecular forces exhibiting amorphous molecular arrangements at a glass transition temperature (T_g) below room temperature. The delineated definition is permitting them to behave flexibly at high degrees of elongation. Relatively rubber-like materials have a long polymeric chain with a high degree of flexibility and mobility, joined into a network structure [2]. Though being soft and flexible, they can withstand and maintain their original form when subjected to external stresses. If the material presents severe deformation after the load is removed; thus, it cannot be deemed an elastomeric material. Compared to other polymers, elastomer has a low Young's modulus and very high elongation at break. The elastomer can typically be thermoset or thermoplastic, further classified into three distinct groups: diene, non-diene, and thermoplastic elastomers. Their bonding nature characterizes the distinction between diene and non-diene. For instance, diene is a polymerized monomer of two sequential double bonds where the typical example for this class is polyisoprene, polychloroprene, and polybutadiene. Meanwhile, non-diene structurally exhibits no double bonds where the cross-linking process requires non-vulcanization of polymer condensation, free radical polymerization, or copolymerization process. A typical example for non-diene elastomers is polyisobutylene, polysiloxanes, and polyurethane. Thermoplastic elastomers or often referred to as TPEs are a

Elastomer Blends and Composites. https://doi.org/10.1016/B978-0-323-85832-8.00005-5

copolymer of physical mixtures of plastics and rubber that intermediately exemplify thermoplastic and elastomeric behavior. For example, blends of polyolefins (plastic) with diene rubber (i.e., butadiene rubber) are TPEs with comparable tensile property and thermal stability than that of thermoset rubber materials, with the exception of the network; hence, TPEs are thermally reversible [3]. Customarily, TPEs are relatively easier to be used for manufacturing than thermoset elastomers (TSE), such as melt extrusion and injection molding, due to the principal differences in their cross-linking method. Nonetheless, the fundamental nature of both classes is similar—elastomeric behavior. Furthermore, it is important to note, based on the definitions outlined by the IUPAC, any polymer that exhibits elastomeric behavior is considered an elastomer [1]. Generally, elastomers are used as inert material to task-specific complex forms as energy-absorbing, damping that can uphold their structural integrity and functionality from harsh conditions; they can exhibit extreme hardness and strength with appropriate flexibility under high temperatures and pressures. In addition, elastomers are also known to exemplify high abrasion resistance, impermeable, high toughness values, and excellent adhesion properties. Even though elastomer has demonstrated immense potential in a wide array of functionalities and applications, room for advancements is still vivid. Thus, elastomer blended with biological phenolic resin (BPR) is established.

By definition, phenol is a compound having one or more hydroxyl groups attached to a benzene or other arene ring [4]. Petroleum-based phenol is perhaps widely used to produce phenolic resin composite, but its high and fluctuating price and depletion of these fossil-based resources reveal the instability of its future. Thus, non-petroleum-based sources such as plant biomass are therefore of great interest. The use of biomass for energy has caught researchers' attention because it is readily available, renewable, and naturally balancing the CO_2 concentration in the atmosphere. During the biomass conversion to bio-oil, a colossal amount of carbon released is neutralized by the plants' naturally occurring photosynthesis cycle. Thus, biomass's wonder and stability showed promising status as a reliable energy source, principally composing mainly cellulose, hemicellulose, and lignin [5,6]. They are available in huge quantities without competing and affecting the food chain. The conversion process of biomass to bio-oil is by thermochemical conversions of pyrolysis (i.e., phase separation, ultrasonic irradiation solid-state separation, and enzymatic modification) or liquefaction process. Pyrolysis is the chemical decomposition method of organic materials at high temperatures with an absence of oxygen. Meanwhile, liquefaction is a liquid fuel generation process of solid biopolymeric structure thermochemically converted to mainly liquid components by exceeding its melting point, often called a liquefaction point in a pressurized water environment to a phase transition from vapor to liquid. Bio-oil is a complex mixture of organic compounds such as anhydro-sugars, alcohol, ketones, aldehydes, carboxylic acids, and phenols [7]. Among these precious chemicals, phenolic compounds can be used as a precursor to produce a phenolic resin. Phenolic is an aromatic compound that forms a strong hydrogen bond, typically mildly acidic, easily dissolves in sodium hydroxide and polar organic solvents but insoluble in sodium carbonate and lesser soluble in aliphatic hydrocarbons. Generally, phenols appear colorless solid-like-prism, and its color changes when it is in contact with air or if it contains iron or copper [8].

Biological phenols or biophenols are sourced from plants, distinguished by their hydroxyl group in the aromatic ring, classified as polyphenols (tannins and flavonoids) and simple phenolics (phenolic acids and coumarins). Tannins are water-soluble polyphenols of yellowish or brownish bitter-tasting organic substances that occur in many wood species that interact directly with cellulose with a molecular weight between 500 and 3000 Da valuable for plants to combat fungi and insect [9]. Proanthocyanidins are another name for condensed tannins, are the most versatile and reactive polyphenol with a freer hydroxyl group and molecular weight smaller than lignin [10]. Meanwhile, flavonoids are the largest group of naturally occurring phenolic compounds with two benzene rings separated by propane unit. Flavones and flavonols are the most widely distributed of all phenolics [11]. Phenolic acids are secondary metabolic distinguished by a carbon group attached to a benzene ring (C6–C1), are typically identified by gallic, syringic acids, p-hydroxybenzoic, protocatechuic, and vanillic [10]. Phenolic acids are also the most widely distributed nonflavonoid phenolic compounds in the tree, conjugated-soluble, and insoluble-bound forms. Lastly, coumarins are a colorless crystalline solid that architecturally fused by a benzene ring used in the α-pyrone ring with two adjacent hydrogen atoms replaced by a lactone-like chain categorized as lactose [12].

In the thermochemical conversion of biomass to bio-oil, phenols are degradation products of lignin. Lignin is a rich source of phenolic compounds, a foremost compound in woody biomass, and the most profuse aromatic biopolymer on earth [13]. Lignin is an amorphous three-dimensional phenyl-propanol of three phenyl-propanols: p-coumaryl alcohol (p-hydroxyl-phenyl propanol), coniferyl-alcohol (guaiacyl-propanol), and sinapyl-alcohol (syringyl-propanol). During the fast pyrolysis or liquefaction process, the lignin phenyl-propanol depolymerized into phenolic chemicals. The extracted biophenols can then mix with water or formaldehyde to form BPR. In producing BPR with water, mixtures of 90% phenol and 10% water are preferably used. Above 65.3°C, phenol can be mixed at any ratio with water developing two phases of phenol/water and water/phenol [14]. Nevertheless, phenol is largely used in the production of phenol-formaldehyde (PF) resins. Generally, BPR is defined as permanently bonded molecules of large networks between phenol and formaldehyde compounds via the polycondensation process. The resin properties can be varied depending on the balance between phenol and formaldehyde and whether the catalyst is an acid or a base. However, the BPR can be solidified without the presence of catalysts using only high temperatures at high pressure [10]. Generally, BPR can be divided into two; resoles (single stage phenolics) and novolac (two-stage phenolics). Habitually, BPR has high thermal stability (resistant to high temperature), chemicals and biological resistance, moisture, and flame resistance. Therefore, they widely used as an impregnating resin forming an excellent composite with coveted properties. Physically, they are opaque forming dark yellow tones to medium yellow tones. Nevertheless, the fracture, low toughness, and low elongation make them brittle limiting their submission to provide in wide array of applications. Thus, elastomer blended with BPR is a practical step to overcome the deficiency shown on both polymers in forming composites of an enhanced properties.

The "composite material" implies materials comprising of two or more starting materials with distinct physical, thermal, and/or chemical properties. A composite material's properties vary entirely from its constituent components [15]. However, the complete differences are pursued intermediately by the constituent components. For instance, polymeric BPR and elastomer unite the reinforcement materials (i.e., fibers) and foster the improvement and boost of stress transfer between the fibers uniformly. Practically, composites formation involves many complex physicals, chemical, and physicochemical changes, consequently affecting the matrix of composite materials and their ensuing properties. Elastomer–BPR composites blends composed of two or more components without permanent linkages of a covalent bond between them with clear distinctive boundary surfaces and interfaces. The polymer blend is defined as intimate mixtures between two or more polymers with an absence of covalent bonds. Thus, elastomer–BPR composites are a multicomponent polymeric system that forms heterogeneous mixtures of macro- and microphase separation, offering a further robust performance than its precursors. The matrix system can be loaded with other materials such as fillers to exploit composites' key role. In other words, elastomer–BPR composites are an intricate multicomponent system whose properties hinge on the type of elastomer and BPR used. Owing to exceptional properties shown by the elastomer–BPR composites, they have been subjected to many high-end applications (i.e., aerospace, high-speed train, medicine, car industry, military equipment, etc.). Contemplating this, elastomer–BPR composites represent a gratifying standpoint of efficacy by which to address properties offered by composites. Hence, in this chapter, the resulted properties of elastomer–BPR blends and their composites are reviewed comprehensively.

16.2 Biological phenolic resin

Bio-based composites are reinforced polymeric materials in which one of the matrixes and reinforcement components or both are from bio-based origins. Meanwhile, BPR is a phenol that extracts or originated from biological particularly plants substitute for petroleum-based phenol. The biophenolic composite industry has recently drawn significant attention for diverse applications, from household articles to automobiles. This is due to their low cost, biodegradability, lightweight, availability, and environmental concerns over synthetic and nonrenewable materials derived from limited resources such as fossil fuel. Regardless of the current popularity, wide prominence, and expansiveness of fossil-based PF resins, there has been a significant drive toward partial or completely supplanting the raw materials of PF in its processing synthetically by using BPR. There are three fundamental motivations for such a substitution: moving toward more sustainable biological usage of green raw materials, demand for improving health and security during processing by the manufacturer and ending consumer, and upgrading the performance and quality of the BPR composite products [16]. These biorefineries strategies are essential to encourage the readily available biomass waste material, which is the best alternative for phenol production, and it is renewable. The conversion of this unattainable waste to valuable products simultaneously becomes one of the solutions for biomass waste management [17,18].

16.2.1 Phenolic compounds from biomass-based

Phenolic compounds are abundant in nature because various sources include the environment, domestic, industrial, and often associated with toxic materials that may cause harm to humans and animals, either short- or long-term exposure. Thus, biological phenols become an alternative to overcome this toxicity concern for any phenol-based materials or composite application. Plant biomass has drawn great attention as phenolic resources since phenol is widely distributed in the plant, including the polyphenols (lignin, tannins, phenolic acids, and stilbenes), bi-phenols, and simple phenolics (thymol, hydroquinone, resorcinol, and coumarins). This phenolic compound can be found within the leaves cell wall, the roots, and stem of the plant and can be either soluble (low molecular weight such as phenol, flavonoids, and tannins) or insoluble (with high-molecular-weight tannins and phenolic acids).

Plant biomass is a potential candidate for the green synthesis of renewable BPR. Lignin is one of the most abundant (15%–25%) natural polymer by-products in lignocellulose biomass, potentially used as phenol substitutes due to the availability of its phenolic structure. Plant phenols have been used as plasticizers in the cement and concrete industry, activated carbon products, and carbon fiber [19]. Tannins are a natural water-soluble phenolic resource available in lignocellulose biomass, fruits, woods, and vegetable oil in a moderate amount (0.2%–25%). Tannins were divided into three main groups: hydrolyzable tannins, condensed tannins, and phlorotannins applicable as biomaterial, nutraceuticals, biopesticide, and antioxidant, fabrics dye, and adhesive for the composite industry. However, the constituent of phenolic sources in biomass-based may differ in plants and depends on its lignin and tannins content. Lignin and tannin present in various plant biomass include palm oil, coconut shell, sisal, palm kernels as the biological raw materials that are highly potential as a natural substitute to synthetic PF. Table 16.1 is the list of BPR plant biomass resources and their characteristics.

Pyrolysis and liquefaction are two important techniques/processes for the conversion of biomass lignin and tannins into BPR. The conversion involved a critical process in manufacturing carbon or related composites using PF resin at high temperature and pressure, known as pyrolysis. The resin matrix was converted into amorphous carbon. This pyrolysis process mainly involved three stages, including additional cross-link resulting from a condensation reaction of phenol functional group, followed by scissored cross-link as temperature increase formed gas of methane, hydrogen, and carbon monoxide, and finally, the release of hydrogen atoms as gases from the aromatic structure of phenol [28]. Meanwhile, in the absence of air biomass, biomass pyrolysis was heated at 400–500°C, thus generating vapor and condensed into a liquid. A previous

Table 16.1 Biological phenolic resin resources and their contribution/improvement on polymers' material characteristics.

Plants biomass resources	Substitute	Type of BPR	Material characteristics	References
Wheat straw	Lignin	Resole	Comparable pH, viscosity, and solid content specification as PF resin	[20]
Pine	Lignin	Resole	Lower resin curing temperature	[21]
Corn stover	Lignin	Resole	Comparable mechanical properties as PF resin	[22]
Kraft pine	Lignin	Novolac	Improved flexural modulus	[23]
Chesnut	Tannin	Novolac	Gelation time reduce	[24]
Larch	Tannin	Resole	Improved friability and thermal conductivity	[25]
Chestnut	Tannin	Resole	High compressive strength	[26]
Valonea	Tannin	Resole	Faster curing, good bonding strength, low formaldehyde emission	[27]

study has reported the fractional pyrolysis of cellulosic biomass using zeolite catalyst to decompose lignin into phenols; simultaneously depolymerization of lignocellulosic biomass yields gaseous products [29]. However, bio-based biomass pyrolysis into phenol is still challenging since agglomeration tends to form as pyrolysis reaction that occurs at high temperature, thus reducing its end yield. More research is needed to explore pyrolysis of lignocellulose biomass over conventional pyrolysis of pure lignin, which can generate a variety of methoxy-substituted aromatic compounds [13]. Once pyrolysis was completed, lignin phenyl-propanol was then depolymerized into phenolic chemical (biophenols) followed by mixing with water or formaldehyde to form BPR by polycondensation process resulting in a permanent bond of PF.

In medicine, biophenolic compounds are commonly being used as antiseptics (i.e., 4-hexylresorcinol), disinfectants, vitamins, and neurotransmitters. In considerably safe dose, biophenols also one of the active ingredients in mouthwash and throat lozenges. Biophenols are well known for their powerful antioxidant power due to their hydrogen donating ability to terminate the reactive oxygen species (ROS), thus stopping the new generation of harmful free radicals. Biophenols are also a useful precursor in the synthesis of pharmaceuticals products, flavoring agents, food preservatives, polymers, resins, and adhesives. Meanwhile, in the making of composite, biophenols are useful when they turn into bioresins or well known as BPR to work as an adhesive, providing moldability, structural stability, enhancing biological resistance, and promoting overall functionality. Depending on the techniques to produce the BPR, it could be thermoplastics or thermoset, fostering similar purposes with dissimilar characteristics. Therefore, the manufacturability of biophenolic compounds is limitless that solely confines to human creativity.

16.2.2 Thermoplastic versus thermoset biological phenolic resin

BPR can be classified into two: thermoplastic and thermosetting phenol resin. These classifications are majorly influenced by several factors: type of catalyst used (acid/base), pH medium, phenol to formaldehyde ratio (molar), and the raw material and chemical structure. The principle of synthesis of thermoplastic phenolic resin occurs under acid reaction conditions (pH 0.5–1.5) with the presence of an acid catalyst such as hydrochloric acid, sulfuric acid, oxalic acid, and benzenesulfonic acid with 1:(0.7–0.9) PF (<1) molar ratio, commonly known as novolac BPR. A much faster reaction between phenol-hydroxymethyl and the hydrogen atom on the phenol ring occurs under acidic conditions. At >90°C, methylene and dimethylene amino bridges were cross-linked by hexamine (hardener for novolac). The thermoplastic phenolic resin contains only methylene chains and phenolic hydroxyl groups linked by methylene and/or ether group resulting in much poor solubility and higher molecular weight of 500–900 [30].

Thermoset phenolic resin is synthesized in the presence of the base catalyst, phenol, formaldehyde, and water known as resole BPR (formaldehyde to phenol ratio >1). The principle for thermoset phenolic resin synthesis is via hydroxylation reaction and polycondensation reaction. During hydroxymethylation, phenols reacted with a base catalyst (NaOH) and heated, followed by the formation of phenoxy anion that acts as a nucleophilic reagent to attack the carbonyl group of formaldehyde, thus forming phenol-hydroxymethyl. The addition reaction product is phenol-monomethylol and phenol-polymethylol [30]. The polycondensation reaction then occurred among the phenol-hydroxymethyl with its additional product phenol-monomethylol and phenol-polymethylol. Two condensation reactions occurred during heating in base conditions, resulting in an increase of molecular weight, linked by the ether chain and methine chain. Since polycondensation is lower, polymers contain more hydrophilic groups (hydroxymethyl group) and appear as water-soluble and with low viscosity. However, as the condensation occurs between the hydrophilic groups, they simultaneously produce a highly viscous, less water-soluble, or immiscible, and become a viable solid material.

Generally, a phenolic thermoset composite/material must be able to provide high mechanical performance. Thus, its modification is crucial to overcome and improve brittleness, strength, thermal stability, and biocompatibility, which depend on the preparation of phenolic resin temperature and its curing process. It is important to note that even though thermoset phenolic resin is irreversibly hardened by curing, it also shows substantial elastomeric characteristics, making them an elastomer. Thermoset phenolic resin with elastomeric properties is also known as TSEs. Meanwhile, thermoplastic phenolic resin with elastomeric property is called TPEs. Extrusion and injection molding are two manufacturing methods for TPEs composite [31]. TPEs' elasticity makes it suitable and available for 3D printing, which becomes economically advantageous for large-scale production. TPEs can also be processed by blow molding, melt calendaring, thermoforming, and heat welding. Both thermoplastic (novolac) and thermoset resins (resole) with elastomeric properties are beneficial in composites industries, where they not only enhancing the overall characteristics but also guaranteeing security.

16.2.3 Elastomeric properties of thermoset and thermoplastic BPR

Elastomer polymer chains are held together by relatively weak intermolecular bonds that allow the polymers to easily stretch when under macroscopic stresses, known as elasticity properties of rubber-like solids as displayed in Fig. 16.1. The elasticity of these materials depends on the ability of long chains to reconfigure themselves to distribute applied stress and return to their original configuration via covalent cross-linkages, known as TPEs. As a result of this, extreme flexibility of elastomers can reversibly extend from 5% to 700%, depending on the specific material [32]. For nonelastic/less elasticity materials, the absence/lower amount of the cross-linkages within the material hence uneasily reconfigured chains from the applied stress resulted in permanent deformation, a nonreversible structure known as TSEs.

The elasticity of polymers formed was greatly influenced by its glass transition temperature (Tg). A material/composite formed at a lower temperature will have many glassy properties of the crystalline phase, becoming less elastic due to the reduction of its mobile chains. Meanwhile, many elastic properties of materials produced when it manipulated at higher than its glass transition temperature, thus increasing its mobile chains and cross-links resulted in many elastic polymers as in Fig. 16.2. Elastomers are usually thermosets (TSEs) requiring vulcanization but may also be thermoplastic (TPEs).

Thermoplastic Elastomer

A

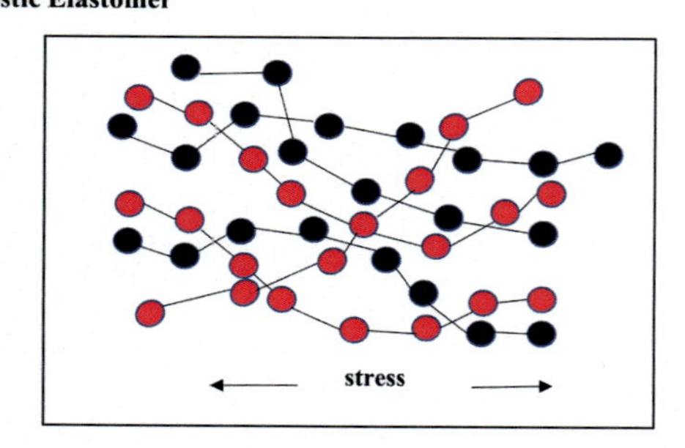

Thermoset Elastomer

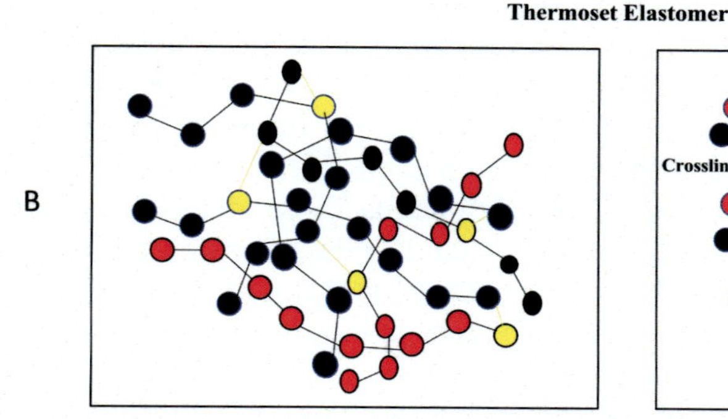

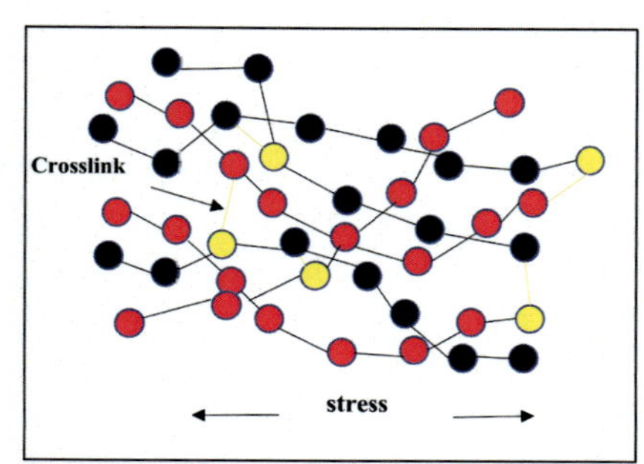

FIGURE 16.1 The polymer chain and its cross-link (intermolecular bonds) of thermoplastic elastomer (A), and thermoset elastomer (B) upon applied stress.

The long polymer chains cross-link occurs while curing upon vulcanization process. The elastomers' molecular structure can be imagined as a "spaghetti and meatball" structure, with the meatballs signifying cross-links as seen in Fig. 16.2.

Thermoset elastomeric properties of BPR reinforced with glass fiber (BPEC) successfully fabricated using BPR resin synthesized from oil palm empty fruit bunches (EFB) fibers [33]. A higher impact strength at 47.71 kJ/m^2 and 65.18 MPa flexural strain was obtained from BPEC. The author confirmed an elastomeric property to the produced composite due to epoxidized natural rubber (ENR) blended into the system by improving the BPR and glass fiber compatibility. It is described that an elastomeric behaving gel from wood bark via auto condensation of flavonoid tannins in water with ethylene glycol as plasticizer (thermoplastic novolac phenolic resin) [34]. The obtained results demonstrated the produced elastomer–BPR composite substantially improved elastomeric behaviors, from hard and brittle to ultrasoft (deformable) with increasing internal plasticizer (EG). Interestingly, the BPR-based gel produce having comparable characteristics with the commercial rubber spring in its stress–strain (20–25 MPa), elastic moduli (68–141 MPa), and substantial decrease of compressive strength (10%–15%) as the plasticizer content increases from 20 to 47 wt.%, which corresponded to softer materials. Its reversibility compression–release curves were completed after 30 cycles of up to 10% strain and demonstrated the positive behavior of elastomer–BPR composite as rubber springs materials.

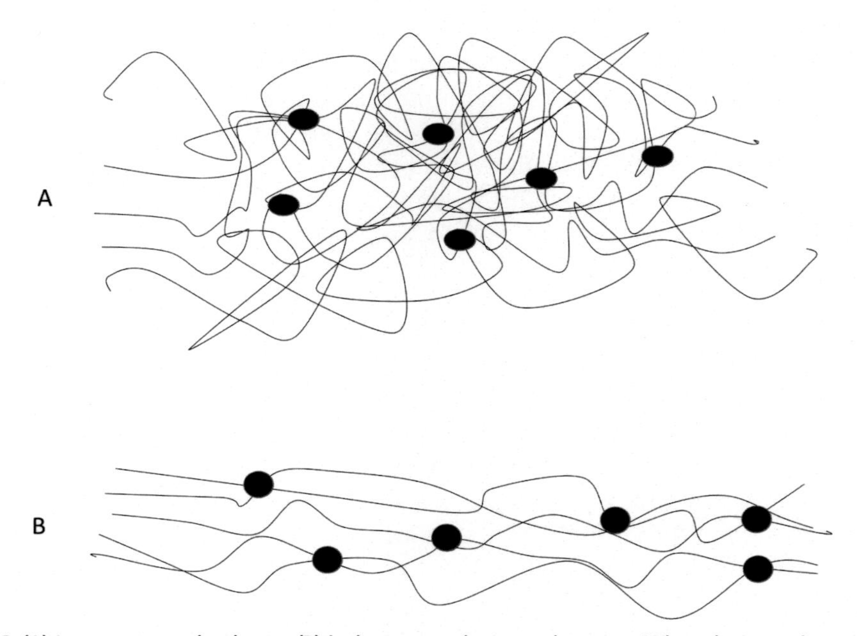

FIGURE 16.2 (A) Is an unstressed polymer; (B) is the same polymer under stress. When the stress is removed, it will return to the A configuration. (The dots represent cross-links).

16.3 Properties of blended composite

BPR had its limitation in mechanical properties due to its brittleness. Therefore, elastomeric materials can be added as a toughening agent to improve its resistance toward impact [33]. BPR due to amorphous properties will not affect the soft rubbery matrix of the composites throughout the processing, testing, and end use. Besides, the functional groups in the BPR are useful in processing and blending with other elastomeric resin that can undergo cross-linking and vulcanization processes due to its high carbon and hydroxyl content, thus enhancing the mechanical properties, rheological and viscoelastic properties of the resins [35,36]. Elastomer-BPR is a new type of material that also can provide flexibility, UV resistance, thermal resistance, chemical structural robustness, and moisture vapor barrier properties, which have a wide application such as flexible printed, electronic substrates, surface coating, etc. [37–39].

16.3.1 Rheological characteristics

Generally, rheological properties for polymer resin are used to determine its flow behavior for processing and manufacturing ease. Further understanding of these properties can better understand the resin's flow behavior during high volume complex manufacturing techniques [40]. By measuring the changes in complex viscosity (η^*), storage modulus (G'), and loss modulus (G'') at a different applied shear rate, stress, and angular frequency, it can provide information during the curing reaction process such as cross-linking, and vulcanization process occurred in the polymeric matrix. The curing kinetics is often used to fit with the kinetic model that can explain the curing mechanism of the polymeric resin during polymer blending and interaction of the resin molecules in the polymeric matrix [41]. The gelling point information can be extracted from rheology analysis, which shows the transition of liquid-like behavior to solid-like resin; the melt rheology for measuring the die swell index using capillary at various shear rates provides information such as flowability of the polymeric resin and filler properties that can understand the dispersion quality related to surface property, shape, size, and particle–matrix interaction [42]. Besides, the melt flow behavior is essential for resin materials to predict the optimum temperature for extrusion processability obtained during these rheology measurements [38].

The mixing characteristics of terpene phenolic resin functionalization into natural rubber (NR) by measuring the mixer torque during the reaction were studied [43]. When the terpene phenolic resin was firstly added into the NR mixture, the mixer torque increased. After that, it suddenly decreased due to the internal plasticization of NR molecules. When the initiator was added, the torque value substantially increased due to the functionalization of BPR into the rubber matrix, finally achieving a maximum torque value indicating complete functionalization. Furthermore, the curing characteristics of the BPR functionalized NR/ENR blends with the addition of highly dispersed silica (HDS) were investigated. The addition of HDS significantly increased the cure torque of the blends with a higher specific area of HDS particles. These phenomena are due to the presence of the phenolic hydroxyl group and epoxy group present in BPR functionalized NR and ENR, allowing vulcanization reaction between HDS particles. The BPR functionalized NR resin shows a shear-thinning behavior decrease in die swell properties results in substantial cure characteristics favorable in extrusion processes that can enhance the physico-mechanical properties of these materials [43,44]. Meanwhile, synthesis of biophenol monosaccharide resin/styrene-butadiene rubber mixture to study its processability by comparing with commercial resins was done [38]. The resins were tested from lower angular frequency toward higher value, the complex viscosity decreased, and storage modulus increased. These phenomena are due to the increase of free hydrodynamic volume and softer flowable volume fraction of the mixture with the addition of the resins. Besides, these biophenol monosaccharide resins experience a decrease in loss factor when the angular frequency increases due to the hardening effect at a higher frequency. Furthermore, the die swell test indicated that biophenol monosaccharide resins act as plasticizers that improve the flow and improve rubber's flow property. The biophenol monosaccharide resin also exhibits a lower die swell index compared with commercial resins of PF, indicating good processability of this resin.

Rheological properties of the elastomer–BPR composite are useful to understand the curing mechanism and role of the components added into the mixing phase of the composite materials. Besides, BPR resin that acts as plasticizer also affects the flow property and is the composite material that can predict the optimum processability parameters, which is crucial for the long-term production process of these elastomer–BPR composite materials.

16.3.2 Physical attributes

Surface morphology and roughness of BPR composites can be used to study the resin matrix interactions between polymeric matrix and filler. The results can correlate with the physiomechanical properties of the composite materials. On the BPR matrix, the surface morphology depends on the polymerization process and the solubility of the BPR produced [45]. When it forms composites with other polymeric matrix and filler, the surface morphology shows agglomeration and particle-like dispersion systems due to the immiscibility caused by surface energies and chemical composition on the surface of these components [38,46].

The surface hydrophilicity and hydrophobicity of the BPR composites are dependent on the interaction between the components in the resin matrix. The formation of BPR consumes the hydroxyl group on the phenol during the condensation polymerization process that causes the increase in hydrophobicity of the BPR composite surface [47]. For biophenol molecules that are rich in phenolic hydroxyl will form a more polar surface on the resin matrix, allowing more hydrogen bonding interaction to occur, thus improving its surface hydrophilicity [46]. When the biophenol molecules were functionalized with a nonpolar polymer matrix, the hydroxyl group in the biophenol molecules will be consumed and hindered, reducing their surface hydrophilicity. However, surface hydrophilicity can be improved by adding hydrophilic fillers that improve resin dispersion in the composites and allow hydrogen bonding interactions to occur [44].

Water absorption properties of these elastomer–BPR composites at different ENR composition were studied [48]. The results showed a significant reduction of water absorption of the composites up to 35% ENR composition. This is due to the compatibility and dispersive properties of ENR that can fill up the microvoids in the phenolic resin, which can reduce the amount of absorbed water trapped in the composites. Besides, the water-resistance properties of biophenol substituted phenol-formaldehyde resin exhibit lower performance than nonsubstituted resin. Furthermore, a substantial improvement is demonstrated in the resin's water resistance properties by solvent treatment with benzene and hexane [49].

An elastomer–BPR composite successfully prepared from BPR synthesized using liquefied oil palm EFB fibers incorporate with NR with different types of ENR [48]. The results demonstrated the addition of ENR provides an elastomeric property to overcome the brittleness of the thermoset BPR. An increase in composites strength and impact strength increased from 72.40 kJ/m^2 to 128.65 kJ/m^2 with the increasing amount of ENR-25 and ENR-50, respectively. These findings proved that the epoxides group in the ENR backbone could provide compatibility by allowing the rubber–BPR interaction. The elastomer–BPR composites' flexural properties were found to be decreased when rubber was introduced indicating that fabricated composites possessed elastomeric properties.

Revisions on the recent progress in hybrid biocomposite showed that currently slowly shifted from traditional biocomposite systems, including thermoplastic polymers (synthetic) reinforced with natural fibers or vice versa [50]. This hybrid biocomposite is important to overcome the shortcomings of natural fibers or matrices. This approach leads to improved mechanical and physical properties of biocomposite, including strength properties, water, and burning resistance. Fabrication on biocomposite with PF as a matrix reinforces with areca fine (AF)/calotropis gigantea (CG) via hand lay-up process with mechanical properties of 17.5 wt.% CG and 17.5 wt.% AF fibers had maximum tensile, flexural, and impact properties [51]. A study on hybrid of PF resin reinforces with areca/sisal or areca/glass or areca/roselle via hand lay-up process cross-link divinylbenzene was attempted [52]. The produced biocomposite presented the highest tensile strength and tensile modulus; meanwhile, flexural strength and flexural modulus increased by hybridization of sisal, roselle, and glass fibers with areca. Biocomposite hybrid of polyurethane foam reinforced with natural fiber of roselle fibers (RF) with spherical silica (silica-A) and amorphous silica (silica-B) via liquid molding resulted in improved of FM with increasing wt.% of silica-A and silica-B; meanwhile, the tensile strength increased with the increase of silica-B and RF [53].

16.3.3 Mechanical performances

Physical properties of the materials when external forces are applied are known as mechanical properties of materials. In general, it discusses a materials' behavior by interpreting its tensile strength, elongation, Young's modulus, impact strength, and flexural strength [47]. Tensile strength and modulus of the elastomeric materials can be tested using a universal tensile machine; whereby its sample dimension/geometry and test method are based on standards such as ISO/DIS37-1994, EN ISO 527:1996, ASTM D412, ASTM 3379-75, etc. [41]. Moreover, these elastomeric resins can be used in impact and bending tests based on standards such as ASTM D256 and DIN EN ISO 178, respectively [48,54]. Furthermore, dynamic mechanical analysis (DMA) experiments are another method to determine the tensile strength and modulus of the materials that explain the materials' viscoelastic behavior [55].

The effect of mechanical properties on styrene-butadiene rubber resin with the incorporation of different content of biophenol to carbohydrate monomer ratio resins was studied [38]. The tensile strength and tensile modulus of the resins improve with the higher phenol to carbohydrate monomer ratio due to the better flowability of the resin molecules that enhanced molecular diffusion, which causes improved contact between bonding surfaces. Besides, the tack strength of the resins increased with a higher carbohydrate to phenol ratio due to the lower glass transition temperature of the resins that allow chain movements of the rubber compound. Meanwhile, preparation of TPEs by using enzymatic hydrolysis lignin and maleic anhydride grafted polyolefin elastomer to study its mechanical properties of these elastomers [56]. The tensile strength of the elastomers improves with increasing lignin content. This phenomenon is due to the oxygenated functional groups in lignin and can have hydrogen bonding interactions with the maleic anhydride groups in polyolefin. Besides, the addition of zinc species in the elastomers can allow ligand coordination in the oxygenated functional groups in lignin, thus improving the strength and toughness of the polymer matrix. Moreover, stress relaxation analysis concluded that zinc species coordination in the elastomer shows a slower relaxation process implying the strong restriction of the coordination bonds in the polymer chains.

The elastic modulus and compression strength can be influenced by the addition of plasticizers during the curing process of the tannin-based BPR. It was reported that the decrement of the elastic modulus and compress strength was managed by adding various ethylene glycol content, which acts as an internal plasticizer in the resin, which also enables it to withstand shape deformation [34]. It can recover back to its initial height after compression at 40% strain, showing a better performance than rubber springs. Meanwhile, the usage of terpene phenolic resin in silica functionalized NR improves the tensile strength, modulus, elongation at break, and hardness. This is due to its rich in hydroxyl groups promote better silica dispersion that able to reduce filler–filler interaction in the composites; thus, efficiently reinforcing the composite matrix [57]. Besides, some studies also show that the improvement of BPR composites' compression

Table 16.2 Comparison of various elastomer–BPR composites on their mechanical properties based on the previous works.

Materials	Tensile strength (MPa)	Young's modulus (MPa)	Impact strength (kJ/m^2)	Flexural strength (MPa)	References
Glass fiber reinforced biophenolic composite/ epoxidized natural rubber	28.86	–	35.32	6.25	[33]
Biophenolic resin/epoxidized natural rubber/ glass fiber prepregs	–	–	128.65	24.64	[48]
Phenol furfural resin/styrene butadiene rubber	0.35	0.32	–	–	[38]
Lignin-polyol resin/polyurethane composites	81.6	1399	–	–	[46]
Maleic anhydride grafted polyolefin elastomer/ lignin composite	21.4	14.6	–	–	[56]
Bio-phenol-hydroxymethylfurfural resin/ hexamethylenetetramine/glass fiber composite	89	–	–	–	[61]
Terpene phenolic resin/functionalized natural rubber/epoxidized natural rubber	21	–	–	–	[43]

strength can be enhanced by physically adding fillers and optimizing the solvent ratio for resin during the mixing process [58–60].

Elongation properties of elastomer BPR composites are mainly contributed by the influence of the BPR domain due to its surface properties and chemical interactions between the polymer matrix and filler composition. The presence of the BPR matrix in the elastomeric composites allows a plasticizing effect that dilutes the composites' polymeric system and improves the flowability of the matrix that provides better mixing processes [38]. These composites will form more ductile materials, which enable higher elongation at break to occur [46]. The presence of BPR in elastomeric composite materials tunes the mechanical properties according to the desired application. The improvement in mechanical strength is highly influenced by the composition and chemical interaction between the BPR and elastomer materials matrix. These composite materials can foresee as new bio-based materials that reduce the usage of man-made synthetic raw materials without sacrificing the physiomechanical properties that can maintain the sustainability of the elastomeric materials industry. Table 16.2 shows comparisons of different polymer blending and their mechanical performances.

16.3.4 Thermal properties

Elastomer–BPR composite dynamic properties such as thermal resistance, nonflammable, good electrical insulator, chemical resistance, and dimensional stability were interest advantages in various applications/industries. The elastomer–BPR composite's thermal stability and degradation are essential characteristics for its utilization at high temperature-related exposure. For elastomer–BPR composites of kenaf/phenolic and pineapple leaf fiber (PALF)/phenolic, generally, at an early stage of decomposition in the thermogravimetric analysis (TGA) ranging between 278.57 and 293.14°C portrays prominent weight loss corresponding to the evaporation of water or volatile matter from the fiber itself. Meanwhile, hybrid composite obtained lower weight loss (%) ranging from 20.84 (PALF/kenaf/phenolic) to 24.91 (1:1 kenaf/phenolic), reflecting thermal stability of the elastomer–BPR composite. Amaral-Labat and coworkers reported that an outstanding fire retardancy and elastomeric behaving gel from wood bark phenols blends of tannin/phenols/ethylene glycol able to endure in a fire-heated at 1000°C for 5 min without any flame propagation [34]. This elastomer–BPR composite tends to decompose into glassy carbon. Its poor flammability was due to high aromatic that cross-linked on the elastomer–BPR composite at higher Tg, approximately more than 1000°C. In another study reported by Roslan and coworkers, thermosets resole-phenolic resin (phenol/formaldehyde/oil palm EFB fibers) via liquefaction TGA curve was first to decompose at 150°C, followed by 150–350°C and final decomposition at 350–600°C [62]. TGA thermal decomposition shows an increment of char yield (59.9%–65.1%) as the PF ratio increases from 1.8 to 2.2, reflecting an improvement in thermal stability. The functional groups from this elastomer–BPR composite of oil palm EFB fibers are identical to commercial PF resin, ranging from 1.70 to 2.53 MPa with a maximum shear strength of 2.93 MPa, fulfilling the specified requirement of commercial PF in JIS K-6852. However, this elastomer–BPR composite's physical and thermal characteristic is mainly influenced by the PF ratio and viscosity of PF resin. Meanwhile, reduction of flame propagation rate from 36 to 28 mm and weight loss from 4.69 to 6.98 gms were observed on the elastomer–BPR composite of woven coconut tea leaf sheath (CLS) reinforces with PF under alkaline treatment of 5% NaOH–resole BPR. The trends shown were due to good adhesion and strong internal bonding between fiber and matrix in CLS resulted from a lower limiting oxygen index (33.96%–34.65%), subsequently require higher oxygen level and needed longer time for ignition (16–27s) [63]. Tao and coresearchers reported an elastomer–BPR composite with good fire-retardant characteristics is a blending of pine flake/PF composite filled with different molarity calcium carbonate (0.5–1.0 M) [64]. The higher accumulation of char at the end thermal decomposition analysis corresponds to higher flame retardancy. However, the produced elastomer–BPR composite showed a slight reduction of internal bonding occurs with substantial improvement of about 182.9% of its modulus elasticity and 63.5% of its modulus rupture.

The elastomer–BPR composite's ability to resist heat action at a given temperature instantaneously maintains its structure, strength, and elasticity, referred to as a high thermal stable material. In common synthetic elastomer material, the thermal conductivity at 25°C was 0.14–0.15 W/mK (NR), silicone rubber reinforces with glass fiber 0.35 W/mK, and polyurethane rubber 0.29 W/mK. Meanwhile, BPR itself has a very low thermal conductivity with 0.28 W/mK (cured phenolic resin). As BPR reinforces with several other materials, especially elastomer, the thermal conductivity marginally increases in forming into a composite. For instance, the elastomer–BPR composite of vapor grew carbon fiber VGCF/phenolic with 35.2 MPa flexural strength and 2.21 GPa flexural modulus has a thermal conductivity of 0.54–0.62 W/mK in line with the increment of fiber content from 0 to 45 wt.% [65]. However, the usage of shorter fiber for composite making is suggestible as it will have a better heat insulator.

In the aerospace industry, AIRBUS has recommended a combination of phenolic resin/honeycomb cores/woven flax fabric (from flax fiber) composite as their compartment. In the making of this composite, the flax fiber was pretreated with numerous chemicals, namely phosphates, tetrakis (hydroxymethyl), phosphonium salts (THPX), boron, nitrogen compounds, tin, and stannic oxide to enhance its flame-retardant treatment and smoke suppressants. The produced elastomer–BPR composite panel did not ignite at all, having the lowest fire growth rate of 0.1 Kw/s, heat release rate of 6.87 Kw/m^2, peak heat release rate of 158 s, and total heat release of 1.7 MJ/m^2 [66]. The characteristics shown by the elastomer–BPR composite have proved to be gratifying as a promising material for aerospace application. This successively marked the potentials of the elastomer-BPR composite for many high-end applications and uses.

16.4 Conclusion

The elastomer–BPR composites promised an exceptional bountiful property benefitting desired applications amid the involved methods are simple and straightforward. The compatibility of elastomer and BPR in forming composites is acknowledged. From blending to its composites, elastomer–BPR composites have shown different properties than their precursor. The distinction is more pronounced when physical, chemical, and physicochemical processes are involved during its formation. One thing is certain, the produced composite, irrespective of the materials' techniques and types, has proven to have an outstanding mechanical strength, dimensional stability, thermal, flame, friction, chemical, and biological resistance. Up to an extent, the applications are immense to an inconceivable use where they are outstandingly working as a component in spaceships as they exhibit a very good thermal, flame, and friction resistance upon atmospheric entry. The composites' predetermined behavior is conferred by the elastomer and BPR properties, empowering them to be emitted by many researchers.

16.5 Future trend

The elastomer–BPR composites' future embarks a similar tune with the worlds' adversity—the scarcity of renewable and sustainable technology. The sheer pressure is driven by the community demanding lower production costs with adequate properties, successively neglecting the importance of green technology as green technology is not always illuminating cost-effectiveness. Refusal of the value offered by green technology can only be tamed by intersecting them with cost efficiency. Green technology employed in producing elastomer–BPR composites molded the processing techniques whose attention is centered on utilizing plant biowaste with zero-waste by-products. Regardless of the enhancements made on the produced product, the trait will be presumably similar. It will be swiveling with the insignificant portrayal of improvements and revolving with certain limitations. This is due to the lingering convoluted apprehension on renewability, sustainability, and green technology will continually be there—unintendedly creating boundaries of improvements. However, sometimes, diminutive improvements are good enough only if the proposed technology plunges into each benefitting category we have outlined off. Nevertheless, the constraints will drive researchers to find a better alternative and a more straightforward route or direction to convey future needs. A robust and bold intervention is required to mark elastomer–BPR composite, which is still a realistic composite.

Acknowledgments

We want to acknowledge the Ministry of Higher Education (MOHE) of Malaysia for research grant LRGS/1/2019/UKM-UKM/5/1 and Universiti Kebangsaan Malaysia (UKM) for MI-2020-006.

References

[1] J. Alemán, A.V. Chadwick, J. He, M. Hess, K. Horie, R.G. Jones, P. Kratochvíl, I. Meisel, I. Mita, G. Moad, S. Penczek, R.F.T. Stepto, Definitions of terms relating to the structure and processing of sols, gels, networks, and inorganic-organic hybrid materials (IUPAC recommendations 2007), Pure Appl. Chem. 79 (10) (2007) 1801–1829, https://doi.org/10.1351/pac200779101801.

[2] J.G. Drobny, Handbook of Thermoplastics Elastomers, second ed., 2007.

[3] R.J. Spontak, N.P. Patel, Thermoplastic elastomers: fundamentals and applications, Curr. Opin. Colloid Interface Sci. 5 (5–6) (2000) 333–340, https://doi.org/10.1016/s1359-0294(00)00070-4.

[4] G.P. Moss, P.A.S. Smith, D. Tavernier, Glossary of class names of organic compounds and reactive intermediates based on structure (IUPAC recommendations 1995), Pure Appl. Chem. 67 (8–9) (1995) 1307–1375, https://doi.org/10.1351/pac199567081307.

[5] K.M. Salleh, S. Zakaria, S. Gan, K.W. Baharin, N.A. Ibrahim, R. Zamzamin, Interconnected macropores cryogel with nano-thin crosslinked network regenerated cellulose, Int. J. Biol. Macromol. 148 (2020) 11–19, https://doi.org/10.1016/j.ijbiomac.2019.12.240.

[6] K.M. Salleh, S. Zakaria, M.S. Sajab, S. Gan, H. Kaco, Superabsorbent hydrogel from oil palm empty fruit bunch cellulose and sodium carboxymethylcellulose, Int. J. Biol. Macromol. 131 (2019) 50–59, https://doi.org/10.1016/j.ijbiomac.2019.03.028.

[7] G.G. Choi, S.J. Oh, S.J. Lee, J.S. Kim, Production of bio-based phenolic resin and activated carbon from bio-oil and biochar derived from fast pyrolysis of palm kernel shells, Bioresour. Technol. 178 (2015) 99–107, https://doi.org/10.1016/j.biortech.2014.08.053.

[8] M. Asim, N. Saba, M. Jawaid, M. Nasir, M. Pervaiz, O. Alothman, A review on phenolic resin and its composites, Curr. Anal. Chem. 13 (2017) 1–13, https://doi.org/10.2174/1573411013666171003154410.

[9] K. Khanbabaee, T. van Ree, Tannins: classification and definitions, Nat. Prod. Rep. 18 (6) (2001) 641–649.

[10] M. Jawaid, M. Asim, Phenolic polymers based composite materials, in: Phenolic Polymers Based Composite Materials, 2021. https://doi.org/10.1007/978-981-15-8932-4.

[11] P.B. Kaufman, L.J. Cseke, C.S. Warber, A.S. James, H.L. Brielmann, Natural Products from Plants, CRC Press, London, 1999.

[12] T. Bor, S.O. Aljaloud, R. Gyawali, S.A. Ibrahim, Antimicrobials from herbs, spices, and plants, in: Fruits, Vegetables, and Herbs: Bioactive Foods in Health Promotion, Elsevier Inc, 2016, pp. 551–578. https://doi.org/10.1016/B978-0-12-802972-5.00026-3.

[13] A.E. Vithanage, E. Chowdhury, L.D. Alejo, P.C. Pomeroy, W.J. DeSisto, B.G. Frederick, W.M. Gramlich, Renewably sourced phenolic resins from lignin bio-oil, J. Appl. Polym. Sci. 134 (19) (2017) 1–10. https://doi.org/10.1002/app.44827.

[14] A. Knop, L.A. Pilato, Phenolic Resins: Chemistry, Applications and Performance-Future Directions, 1987, p. 100.

[15] N. Dishovsky, M. Mihaylov, Elastomer-based composite materials: mechanical, dynamic, and microwave properties and engineering applications, J. Chem. Inf. Model. 53 (9) (2019) 1689–1699.

[16] P.R. Sarika, P. Nancarrow, A. Khansaheb, T. Ibrahim, Bio-based alternatives to phenol and formaldehyde for the production of resins, Polymers 12 (10) (2020) 2237. https://doi.org/10.3390/polym12102237.

[17] M. Mostapha, N.A. Jahar, K.A. Azizan, S. Zakaria, W.M. Aizat, S.N.S. Jaafar, Proteomic analysis of stored core oil palm trunk (COPT) sap identifying proteins related to stress, disease resistance and differential gene/protein expression, Sains Malays. 47 (6) (2018) 1259–1268. https://doi.org/10.17576/jsm-2018-4706-22.

[18] K.M. Salleh, S. Zakaria, M.S. Sajab, S. Gan, C.H. Chia, S.N.S. Jaafar, U.A. Amran, Chemically crosslinked hydrogel and its driving force towards superabsorbent behaviour, Int. J. Biol. Macromol. 118 (2018) 1422–1430. https://doi.org/10.1016/j.ijbiomac.2018.06.159.

[19] T. Varila, H. Romar, T. Luukkonen, T. Hilli, U. Lassi, Characterization of lignin enforced tannin/furanic foams, Heliyon 6 (1) (2020) e03228. https://doi.org/10.1016/j.heliyon.2020.e03228.

[20] N. Tachon, B. Benjelloun-Mlayah, M. Delmas, Organosolv wheat straw lignin as a phenol substitute for green phenolic resins, Bioresources 11 (3) (2016) 5797–5815.

[21] M. Wang, M. Leitch, C. Xu, Synthesis of phenol-formaldehyde resol resins using organosolv pine lignins, Eur. Polym. J. 45 (12) (2009) 3380–3388.

[22] S. Kalami, M. Arefmanesh, E. Master, M. Nejad, Replacing 100% of phenol in phenolic adhesive formulations with lignin, J. Appl. Polym. Sci. 134 (30) (2017) 45124.

[23] A. Tejado, G. Kortaberria, C. Pena, M. Blanco, J. Labidi, J.M. Echeberria, I. Mondragon, Lignins for phenol replacement in novolac-type phenolic formulations. II. Flexural and compressive mechanical properties, J. Appl. Polym. Sci. 107 (2008) 159–165.

[24] C. Peña, M.D. Martin, A. Tejado, J. Labidi, J.M. Echeverria, I. Mondragon, Curing of phenolic resins modified with chestnut tannin extract, J. Appl. Polym. Sci. 101 (3) (2006) 2034–2039.

[25] J. Li, A. Zhang, S. Zhang, Q. Gao, W. Zhang, J. Li, Larch tannin-based rigid phenolic foam with high compressive strength, low friability, and low thermal conductivity reinforced by cork powder, Compos. B Eng. 156 (2019) 368–377. https://doi.org/10.1016/j.compositesb.2018.09.005.

[26] S. Jahanshaei, T. Tabarsa, J. Asghari, Eco-friendly tannin-phenol formaldehyde resin for producing wood composites, Pigment Resin Technol. 41 (2012) (2012) 296–301.

[27] C. Li, W. Wang, Y. Mu, J. Zhang, S. Zhang, J. Li, W. Zhang, Structural properties and copolycondensation mechanism of valonea tannin-modified phenol-formaldehyde resin, J. Polym. Environ. 26 (2018) 1297–1309.

[28] H. Jiang, J. Wang, S. Wu, Z. Yuan, Z. Hu, R. Wu, Q. Liu, The pyrolysis mechanism of phenol formaldehyde resin, Polym. Degrad. Stabil. 97 (2012) 1527–1533. https://doi.org/10.1016/j.polymdegradstab.2012.04.016.

[29] S. Meesuk, J.P. Cao, K. Sato, Y. Ogawa, T. Takarada, Study of catalytic hydropyrolysis of rice husk under nickel-loaded brown coal char, Energy Fuels 25 (2011) 5438–5443. https://doi.org/10.1021/ef201266b.

[30] Y. Xu, L. Guo, H. Zhang, H. Zhai, H. Ren, Research status, industrial application demand and prospects of phenolic resin, RSC Adv. 9 (50) (2019) 28924–28935. https://doi.org/10.1039/c9ra06487g.

[31] R.A. Shanks, I. Kong, General Purpose Elastomers: Structure, Chemistry, Physics and Performance, Springer-Verlag, 2013, pp. 11–45. https://doi.org/10.1007/978-3-642-20925-3_2.

[32] P.C. Zhao, W. Li, W. Huang, C.-H. Li, A self-healing polymer with fast elastic recovery upon stretching, Molecules 25 (3) (2020) 597. https://doi.org/10.3390/molecules25030597.

[33] Z. Zakaria, S. Zakaria, R. Roslan, C.H. Chia, S.N.S. Jaafar, U.A. Amran, S. Gan, Physico-mechanical properties of glass fibre reinforced biophenolic elastomer composite, Sains Malays. 47 (10) (2018) 2573–2580. https://doi.org/10.17576/jsm-2018-4710-34.

[34] G. Amaral-Labat, L.I. Grishechko, G.F.B. Lenz e Silva, B.N. Kuznetsov, V. Fierro, A. Pizzi, A. Celzard, Rubber-like materials derived from biosourced phenolic resins, J. Phys. Conf. 879 (2017) (2017) 12013. https://doi.org/10.1088/1742-6596/879/1/012013.

[35] D. Barana, S.D. Ali, A. Salanti, M. Orlandi, L. Castellani, T. Hanel, L. Zoia, Influence of lignin features on thermal stability and mechanical properties of natural rubber compounds, ACS Sustain. Chem. Eng. 4 (10) (2016) 5258–5267. https://doi.org/10.1021/acssuschemeng.6b00774.

[36] L. Musilová, A. Mráček, A. Kovalcik, P. Smolka, A. Minařík, P. Humpolíček, R. Vícha, P. Ponížil, Hyaluronan hydrogels modified by glycinated Kraft lignin: morphology, swelling, viscoelastic properties and biocompatibility, Carbohydr. Polym. 181 (2018) 394–403. https://doi.org/10.1016/j.carbpol.2017.10.048.

[37] H. Kaneko, R. Ishii, A. Suzuki, T. Nakamura, T. Ebina, T.T. Nge, T. Yamada, Flexible clay glycol lignin nanocomposite film with heat durability and high moisture-barrier property, Appl. Clay Sci. 132 (133) (2016) 425–429. https://doi.org/10.1016/j.clay.2016.07.009.

[38] R. Koley, R. Kasilingam, S. Sahoo, S. Chattopadhyay, A.K. Bhowmick, Synthesis and characterization of phenol furfural resin from Moringa Oleifera gum and biophenol and its application in styrene butadiene rubber, Ind. Eng. Chem. Res. 58 (40) (2019) 18519–18532. https://doi.org/10.1021/acs.iecr.9b03684.

[39] K. Ono, O. Tanaike, R. Ishii, T. Nakamura, K. Shikinaka, T. Ebina, T.T. Nge, T. Yamada, Solvent-free fabrication of an elastomeric epoxy resin using glycol lignin from Japanese cedar, ACS Omega 4 (17) (2019) 17251–17256. https://doi.org/10.1021/acsomega.9b01884.

[40] A. Chaloupka, T. Pflock, R. Horny, N. Rudolph, S.R. Horn, Dielectric and rheological study of the molecular dynamics during the cure of an epoxy resin, J. Polym. Sci. B Polym. Phys. 56 (12) (2018) 907–913. https://doi.org/10.1002/polb.24604.

[41] W. Yang, B. Liang, W. Tan, X. He, J. Lv, H. Xiao, K. Zeng, J. Hu, G. Yang, Rheological study on the cure kinetics of dicyanimidazole resin, Thermochim. Acta 694 (2020). https://doi.org/10.1016/j.tca.2020.178801.

[42] J. Liao, N. Brosse, A. Pizzi, S. Hoppe, X. Xi, X. Zhou, Polypropylene blend with polyphenols through dynamic vulcanization: mechanical, rheological, crystalline, thermal, and UV protective property, Polymers 11 (7) (2019). https://doi.org/10.3390/polym11071108.

[43] P. Manoharan, K. Naskar, Plant based poly-functional hindered phenol derivative as an alternative for silane in cis-polyisoprene and its functionalized derivative systems containing nano-sized particulates, J. Polym. Environ. 26 (4) (2018) 1355–1370. https://doi.org/10.1007/s10924-017-1036-z.

[44] P. Manoharan, K. Naskar, Biologically sustainable rubber resin and rubber-filler promoter: a precursor study, Polym. Adv. Technol. 28 (12) (2017) 1642–1653. https://doi.org/10.1002/pat.4034.

[45] B. Subramaniam, J.R. Silverman, A.M. Danby, Facile prepolymer formation with ozone-pretreated grass lignin by in situ grafting of endogenous aromatics, ACS Sustain. Chem. Eng. 8 (46) (2020) 17001–17007. https://doi.org/10.1021/acssuschemeng.0c03811.

[46] Y. Cao, Z. Liu, B. Zheng, R. Ou, Q. Fan, L. Li, C. Guo, T. Liu, Q. Wang, Synthesis of lignin-based polyols via thiol-ene chemistry for high-performance polyurethane anticorrosive coating, Compos. B Eng. 200 (2020). https://doi.org/10.1016/j.compositesb.2020.108295.

[47] W. Zhang, Y. Ma, C. Wang, S. Li, M. Zhang, F. Chu, Preparation and properties of lignin-phenol-formaldehyde resins based on different biorefinery residues of agricultural biomass, Ind. Crop. Prod. 43 (1) (2013) 326–333. https://doi.org/10.1016/j.indcrop.2012.07.037.

[48] U.A. Amran, S. Zakaria, C.H. Chia, S.N.S. Jaafar, R. Roslan, Mechanical properties and water absorption of glass fibre reinforced bio-phenolic elastomer (BPE) composite, Ind. Crop. Prod. 72 (2015) 54–59. https://doi.org/10.1016/j.indcrop.2015.01.054.

[49] S. Zabelkin, A. Valeeva, A. Sabirzyanova, A. Grachev, V. Bashkirov, Neutrals influence on the water resistance coefficient of phenol-formaldehyde resin modified by wood pyrolysis liquid products, Biomass Convers. Biorefin. (2020). https://doi.org/10.1007/s13399-020-01025-0.

[50] M. Bahrami, J. Abenojar, M.Á. Martínez, Recent progress in hybrid biocomposites: mechanical properties, water absorption, and flame retardancy, Materials 13 (22) (2020) 5145. https://doi.org/10.3390/ma13225145.

[51] S. Venkatarajan, B. V.Bhuvaneswari, A. Athijayamani, S. Sekar, Effect of addition of areca fine fibers on the mechanical properties of Calotropis Gigantea fiber/phenol formaldehyde biocomposites, Vacuum 166 (2019) 6–10. https://doi.org/10.1016/j.vacuum.2019.04.022.

[52] A. Athijayamani, M. Chrispin Das, S. Sekar, K. Ramanathan, Mechanical properties of phenol formaldehyde hybrid composites reinforced with natural cellulose fibers, Bioresources 12 (1) (2017) 1960–1967. https://doi.org/10.15376/biores.12.1.1960-1967.

[53] A. Soundhar, M. Rajesh, K. Jayakrishna, M.T.H. Sultan, A.U.M. Shah, Investigation on mechanical properties of polyurethane hybrid nanocomposite foams reinforced with roselle fibers and silica nanoparticles, Nanocomposites 5 (1) (2019) 1–12. https://doi.org/10.1080/20550324.2018.1562614.

[54] J. Dörrstein, R. Scholz, D. Schwarz, D. Schieder, V. Sieber, F. Walther, C. Zollfrank, Effects of high-lignin-loading on thermal, mechanical, and morphological properties of bioplastic composites, Compos. Struct. 189 (2018) 349–356. https://doi.org/10.1016/j.compstruct.2017.12.003.

[55] H. Li, G. Sivasankarapillai, A.G. McDonald, Lignin valorization by forming toughened thermally stimulated shape memory copolymeric elastomers: evaluation of different fractionated industrial lignins, J. Appl. Polym. Sci. 132 (5) (2015). https://doi.org/10.1002/app.41389.

[56] J. Huang, W. Liu, X. Qiu, High performance thermoplastic elastomers with biomass lignin as plastic phase, ACS Sustain. Chem. Eng. 7 (7) (2019) 6550–6560. https://doi.org/10.1021/acssuschemeng.8b04936.

[57] P. Manoharan, N. Chandra Das, K. Naskar, On-demand tuned hazard free elastomeric composites: a green approach, Biopolymers 107 (7) (2017). https://doi.org/10.1002/bip.23019.

[58] M. Culebras, A. Beaucamp, Y. Wang, M.M. Clauss, E. Frank, M.N. Collins, Biobased structurally compatible polymer blends based on lignin and thermoplastic elastomer polyurethane as carbon fiber precursors, ACS Sustain. Chem. Eng. 6 (7) (2018) 8816–8825. https://doi.org/10.1021/acssuschemeng.8b01170.

[59] H. Li, J.T. Sun, C. Wang, S. Liu, D. Yuan, X. Zhou, J. Tan, L. Stubbs, C. He, High modulus, strength, and toughness polyurethane elastomer based on unmodified lignin, ACS Sustain. Chem. Eng. 5 (9) (2017) 7942–7949. https://doi.org/10.1021/acssuschemeng.7b01481.

[60] J. Zhang, Y. Chen, P. Sewell, M.A. Brook, Utilization of softwood lignin as both crosslinker and reinforcing agent in silicone elastomers, Green Chem. 17 (3) (2015) 1811–1819. https://doi.org/10.1039/c4gc02409e.

[61] Y. Zhang, Z. Yuan, N. Mahmood, S. Huang, C.C. Xu, Sustainable bio-phenol-hydroxymethylfurfural resins using phenolated de-polymerized hydrolysis lignin and their application in bio-composites, Ind. Crop. Prod. 79 (2016) 84–90.

[62] R. Roslan, S. Zakaria, C.H. Chia, R. Boehm, M.P. Laborie, Physico-mechanical properties of resol phenolic adhesives derived from liquefaction of oil palm empty fruit bunch fibres, Ind. Crop. Prod. 62 (2014) 119–124. https://doi.org/10.1016/j.indcrop.2014.08.024.

[63] K.N. Bharath, S. Basavarajappa, Flammability characteristics of chemical treated woven natural fabric reinforced phenol formaldehyde composites, Procedia Mater. Sci. 5 (2014) 1880–1886. https://doi.org/10.1016/j.mspro.2014.07.507.

[64] Y. Tao, P. Li, L. Cai, S.Q. Shi, Flammability and mechanical properties of composites fabricated with $CaCO_3$-filled pine flakes and phenol formaldehyde resin, Compos. B Eng. 167 (2019) 1–6. https://doi.org/10.1016/j.compositesb.2018.12.005.

[65] R.D. Patton, C.U. Pittman, L. Wang, J.R. Hill, A. Day, Ablation, mechanical and thermal conductivity properties of vapor grown carbon fiber/phenolic matrix composites, Compos. A Appl. Sci. Manuf. 33 (2) (2002) 243–251. https://doi.org/10.1016/S1359-835X(01)00092-6.

[66] R.D. Anandjiwala, M.J. John, P. Wambua, S. Chapple, T. Klems, M. Doecker, M. Goulain, L. Erasmus, Bio-based structural composite materials for aerospace applications, in: SAIAS Symposium, September, 1–6, 2008. https://www.researchgate.net/publication/30510924.

17

Advances in stimuli-responsive and functional thermoplastic elastomers

Jiaqi Yan[1], Richard J. Spontak[1,2]

[1]DEPARTMENT OF CHEMICAL & BIOMOLECULAR ENGINEERING, NORTH CAROLINA STATE UNIVERSITY, RALEIGH, NC, UNITED STATES; [2]DEPARTMENT OF MATERIALS SCIENCE & ENGINEERING, NORTH CAROLINA STATE UNIVERSITY, RALEIGH, NC, UNITED STATES

17.1 Overview of thermoplastic elastomers and their applications

Polymers are organic macromolecules that consist of a large number of repeating units and are frequently classified by their thermomechanical properties [1], as indicated in Fig. 17.1. While thermoplastics are generally solid (glassy or semicrystalline) at application temperature, they are processable as polymeric liquids (melts) at temperatures above their glass transition temperature (T_g) or melting temperature (T_m), respectively. As examples of polymers that are not melt processable after they are formed, both

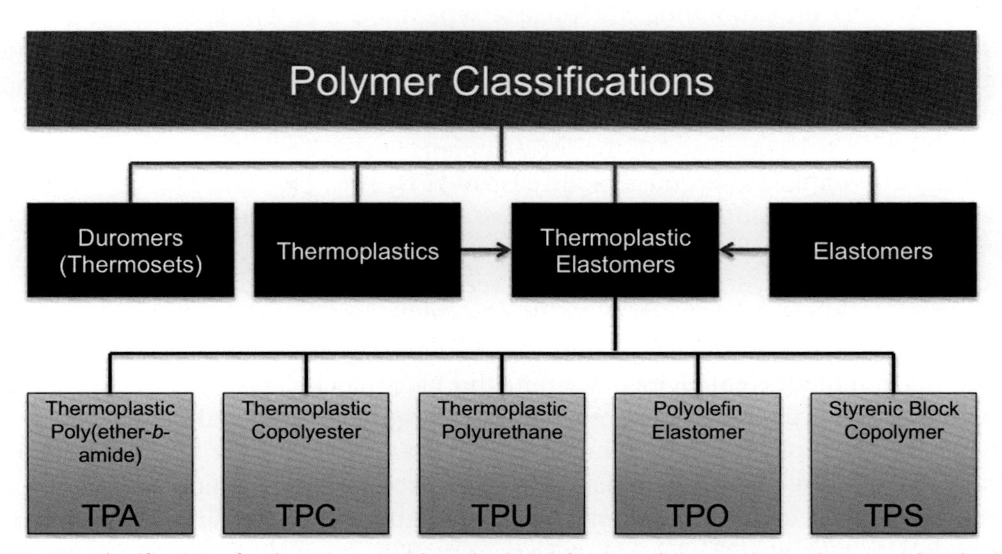

FIGURE 17.1 Classification of polymeric materials and subclassification of thermoplastic elastomers (TPEs), along with abbreviated designations.

Elastomer Blends and Composites. https://doi.org/10.1016/B978-0-323-85832-8.00006-7

elastomers and duromers (also referred to as thermosets) consist of networks generated by irreversible chemical reactions, and they display markedly different properties: elastomers are low-modulus, elastic polymers, whereas duromers are high-modulus, brittle materials. Thermoplastic elastomers (TPEs) constitute an increasingly important class of hybrid polymers that combine the elasticity of chemically cross-linked elastomers with the melt processability of thermoplastics. To achieve this dual functionality, most TPEs are block (or "blocky") copolymers that are synthesized with hard (glassy or semicrystalline) and soft (elastomeric) chemical units [2–7]. Because of this molecular architecture and the range of moieties that can be integrated, TPEs afford significant advantages over conventional elastomers derived primarily from polydienes and polysiloxanes due to their tunable nanoscale structures and corresponding bulk properties, in addition to low-waste processing, low energy consumption, and excellent quality control [8]. Major commercial uses of TPEs include automotive components [9], personal electronics [10], consumer goods [11], and biomedical devices [12,13]. The current global demand for TPEs is rapidly expanding and diversifying, and is forecasted to reach a production level [14] of 7.4 million metric tons by 2022 and a market size [15] of 27.8 billion USD by 2024. The discovery of elasticity in plasticized polyvinyl chloride (PVC), commonly referred to as "plastisol," by Semon [16] in 1926, is credited for opening the door to TPEs. The subsequent development of polyurethane, polyester, polyamide, polyolefin, and polystyrene thermoplastics yields the hard segments that distinguish the families of TPEs listed in Fig. 17.1: thermoplastic polyurethane (TPU), copolyester (TPC), poly(ether-*b*-amide) (TPA), polyolefin elastomer (TPO), and styrenic block copolymer (TPS).

The combination of chemically incompatible hard and soft segments in TPE molecules is essential for promoting *micro*phase separation (i.e., phase separation that is restricted to parts of polymer molecules instead of entire molecules) under application conditions [17–20]. Upon melt processing and ensuing microphase separation, the hard segments segregate to form phase domains that serve as physical cross-links to stabilize a molecular network composed of the soft segments in conceptually similar fashion as a chemically cross-linked elastomer. In the case of TPU, TPC, TPA, and TPO polymers, the hard segments are *crystallizable* and, because of the polymerization routes employed, vary in both length and number along the copolymer chains. Synthesized by step-growth polycondensation, polyurethane is widely recognized as a highly versatile polymer [21,22]. The segmented TPU variant is the first polymeric material to combine thermoplastic and elastomeric characteristics as a randomly coupled (in contrast to perfectly alternating) multiblock copolymer wherein the hard blocks crystallize while the soft blocks remain amorphous. The morphology of microphase-separated TPU is illustrated in Fig. 17.2 and couples interphase mixing with intra/intermolecular hydrogen bonding in both microphases. This material possesses excellent mechanical properties for abrasion, puncture, and tear resistance, in combination with substantial elasticity and good hydrolytic stability [23–25]. For these reasons, TPU is widely used as protective and abrasion-resistant coatings. Similarly, TPCs also consist of high melting segments that

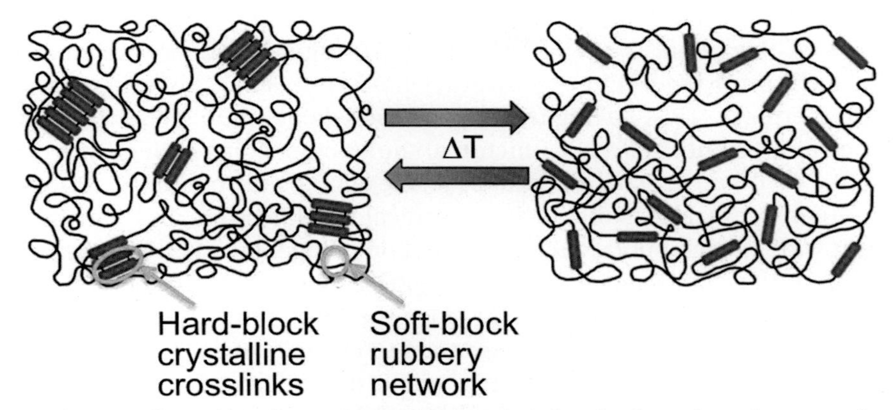

FIGURE 17.2 Schematic illustration of a semicrystalline TPE depicting the formation of an amorphous rubbery network stabilized by crystalline physical cross-links as the temperature is lowered from the molten state (where the hard and soft blocks form a homogeneous solution).

are capable of crystallization and soft segments with a low T_g. In this case, the hard segments often derive from short-chain esters (e.g., tetramethylene terephthalate), and the soft segments are either polyester glycols or aliphatic polyethers [26,27]. One TPC example, Hytrel, is widely popular because of its durability in mechanical gears [28], medical devices [29], railway technology [30], and cable insulation [31]. The TPAs are high-performance block copolymers consisting of hard and soft segments joined by amide linkages. Unlike other TPEs, these polymers possess excellent heat and solvent resistance due to strong hydrogen bonding, and they are frequently employed in medical devices requiring good flexibility, breathability, and sterilization [32–34]. Lastly, TPOs possess physical properties that can be continuously maintained at elevated temperatures that are typically below those of the other crystallizable TPEs. This TPE class is frequently applied as roofing materials since they do not degrade upon exposure to UV radiation [35].

The alternative to a crystallizable hard segment is a *vitrifiable* (i.e., glass forming) hard segment, and TPEs formulated according to this synthetic strategy rely primarily on atactic polystyrene (S), although acrylic TPEs developed from poly(methyl methacrylate) (PMMA) have also been developed for this purpose [36]. Unlike the other TPEs discussed above, styrenic TPEs are synthesized by chain-growth polymerization mechanisms (free radical or ionic), in which case their molecular architectures can be precisely controlled, their molecular weights can be relatively high, and their dispersities can be very low (<1.2) [37]. In fact, commercial TPEs synthesized as triblock copolymers with styrenic endblocks and a polydiene/olefin midblock by living anionic polymerization (with dispersities <1.1) constitute the most ubiquitous family of TPEs worldwide and account for nearly 33% of the commercial TPE market globally [38]. Rather than forming semicrystalline domains to serve as physical cross-links, this class of TPEs stabilizes a subambient-T_g molecular network with glassy S microdomains having a low

T_g (≈100°C) relative to the melting points of the crystallizable TPEs. While commercially relevant TPEs of this genre typically contain an saturated (olefinic) midblock such as poly(ethylene-*alt*-propylene) (EP) or poly(ethylene-*co*-butylene) (EB) to ensure chemical and reaction inertness, TPSs having unsaturated polydiene midblocks afford numerous opportunities to introduce different functional moieties, such as maleic anhydride [39], polyester [40], or liquid crystalline grafts [41,42], along the midblock. Moreover, the styrenic endblocks are also suitable for chemical modification and can be used to incorporate a wide range of polar moieties based on halogenated [43] or charged species [44]. By incorporating such species into a microphase-separated styrenic TPE (which can exhibit highly ordered nanostructures [45] since the hard/soft sequence characteristics of these copolymers are well defined in contrast to crystallizable TPEs), these materials can exhibit tunable amphiphilicity, wherein some microdomains remain hydrophobic while others become hydrophilic [46,47]. Hydrophilic channels capable of permeating polar liquids (including water) in these TPEs open the door to new stimuli-responsive and functional technologies [48–50], as described in later sections.

17.2 Introduction to model block copolymers as TPEs

Block copolymers with precisely controlled molecular characteristics represent a technological breakthrough as TPEs in the rubber industry. Block copolymers are macromolecules that consist of at least two long, contiguous sequences ("blocks") of chemically dissimilar repeat units [2,7,51,52]. Due to thermodynamic incompatibility between the blocks, these materials can microphase-separate and spontaneously self-assemble into a wide variety of soft nanostructures [17,19,53,54]. Current polymer synthetic methods provide growing opportunities to generate a broad portfolio of block copolymer families, ranging from perfectly alternating linear multiblock copolymers [55–58] to star [59,60], cyclic [61], branched [62], tapered [63–66], and bottlebrush block copolymers [67,68]. Several excellent sources addressing morphological development in various block copolymers (including those with more than two constituent chemical species) are available and focus on relevant topics such as the effects of molecular architecture, composition, and weight [19,69–74]. Increases in molecular complexity greatly expand the property sets that can be realized through systematic variation in molecular design, which can be guided by predictive theory and computer simulation. The self-consistent field phase diagram of molecularly symmetric ABA triblock copolymers (with identical A endblocks) predicted by Matsen [19], for example, is presented in Fig. 17.3 and reveals several important features. First, the thermodynamic incompatibility is expressed as the coupled parameter χN, where χ is the Flory–Huggins interaction parameter (which commonly scales as $1/T$, where T denotes absolute temperature), and N represents the number of statistical units along the copolymer backbone. Values of χ for several chemical species relative to S are ranked in Fig. 17.4. Copolymer composition, given by the number (volume) fraction of one block type, is

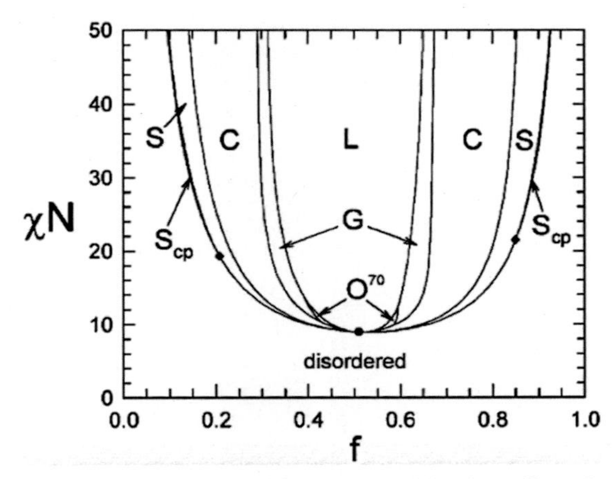

FIGURE 17.3 Phase diagram of ABA triblock copolymers predicted by the self-consistent field theory of Matsen [19]. Included here are the close-packed spherical (S_{cp}), body-centered spherical (S), hexagonally packed cylindrical (C), gyroid (G), *Fddd* (O^{70}), and lamellar (L) morphologies. *Reproduced with permission from M.W. Matsen, Effect of architecture on the phase behavior of AB-type block copolymer melts, Macromolecules 45 (2012) 2161–2165. Copyright 2012 American Chemical Society.*

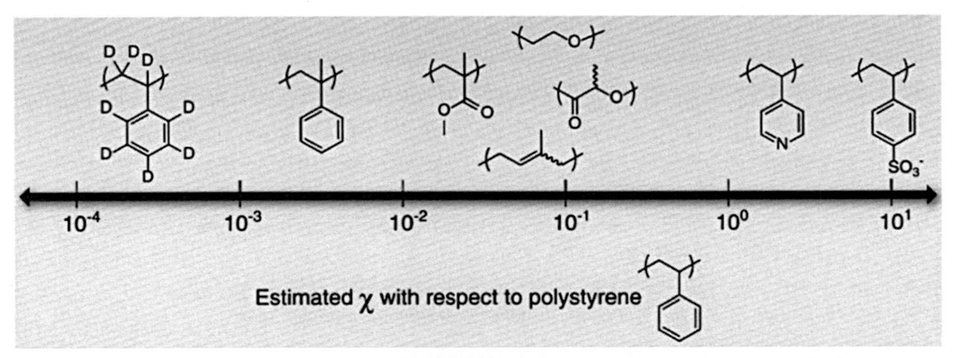

FIGURE 17.4 Estimated values of χ between polystyrene and other species commonly found in styrenic block copolymers. *Reproduced with permission from W. Wang, W. Lu, A. Goodwin, H. Wang, P. Yin, N.G. Kang, K. Hong, J.W. Mays, Recent advances in thermoplastic elastomers from living polymerizations: macromolecular architectures and supramolecular chemistry, Prog. Polym. Sci. 95 (2019) 1–31. Copyright 2019 Elsevier.*

designated as f in Fig. 17.3. Within the ordered state, copolymer molecules can self-assemble into classical nanostructures: A(B) spheres typically positioned on a body-centered-cubic (bcc) or face-centered-cubic (fcc) lattice in a continuous B(A) matrix, A(B) hexagonally packed cylinders arranged in a B(A) matrix and coalternating lamellae [2,7,72]. Alternatively, more spatially complex morphologies include the bicontinuous gyroid [75–77] (displayed in 3D for an ABA triblock copolymer [78,79] in Fig. 17.5), double-diamond [80,81], O^{70} [82,83], and the Frank–Kasper σ phase [84]. In the event that more than two species are incorporated into a block copolymer, numerous

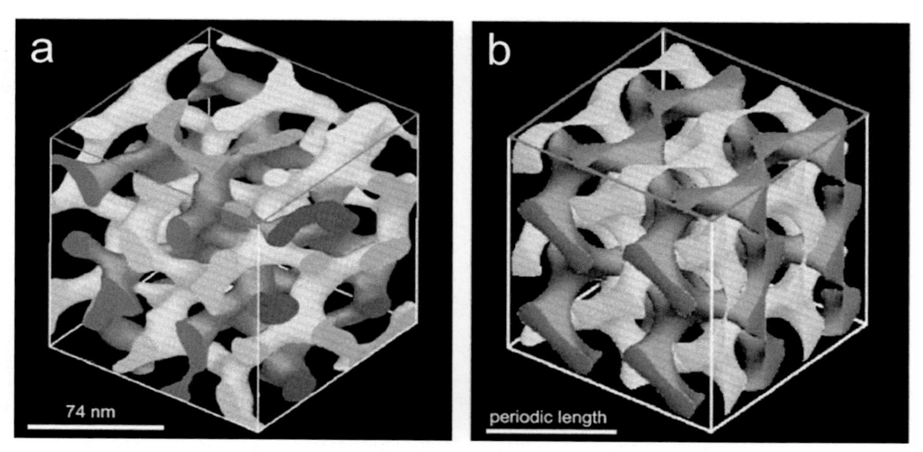

FIGURE 17.5 3D renditions of the gyroid morphology in a microphase-ordered ABA triblock copolymer: (A) experimental reconstruction from transmission electron microtomography (TEMT) and (B) calculated Schoen's gyroid surface corrected for composition. In both cases, the isoprenic matrix is transparent so that the bicontinuous styrenic channels are visible. *Reproduced with permission from H. Jinnai, R.J. Spontak, Transmission electron microtomography in polymer research, Polymer 50 (2009) 1067–1087. Copyright 2009 Elsevier.*

additional hybrid and unique morphologies have been predicted, simulated, and observed [69,71,72]. Unless otherwise specified, however, only bicomponent linear block copolymers will be initially considered here.

Controlling block copolymer phase behavior and properties is routinely achieved by varying chemical species and temperature (through χ), molecular weight (through N), and block composition (through f). Several additional avenues by which to alter block copolymer attributes are now examined. The first focuses on block purity [85]. As previously mentioned, bicomponent block copolymers consist of two chemically dissimilar species, A and B, in which case $\chi(T)$ is fixed for each polymer pair since χ is a pairwise-specific parameter. One way to alter χ without changing the two species is to generate blocks that are random copolymers, rather than pure sequences, of A and B [86–88]. This special case, referred to as block random copolymers (BRCs), introduces a new experimentally tunable variable, namely composition contrast (Δ), since each block can now contain both chemical species but at different compositions. Application of the copolymer equation for a bicomponent copolymer yields [85] $\Delta = |w_{A,1} - w_{A,2}|$, where $w_{A,i}$ corresponds to the mass fraction of A in the ith block ($i = 1$ or 2), so that an effective interaction parameter for the BRC (χ_{eff}) can be directly related to χ between A and B species: $\chi_{eff} = \chi\Delta^2$. As evinced by the transmission electron microscopy (TEM) images provided for two diblock BRCs composed of S and polyisoprene (I) in Fig. 17.6, increasing Δ from 0.25 to 0.50 at constant temperature (χ), molecular weight ($\sim$N), and block fraction (f) is accompanied by the formation of a lamellar morphology. Corresponding results from dynamic rheology confirm the existence of only one, albeit broad, T_g (from the single maximum in tan δ) when $\Delta = 0.25$ but two distinct T_g values

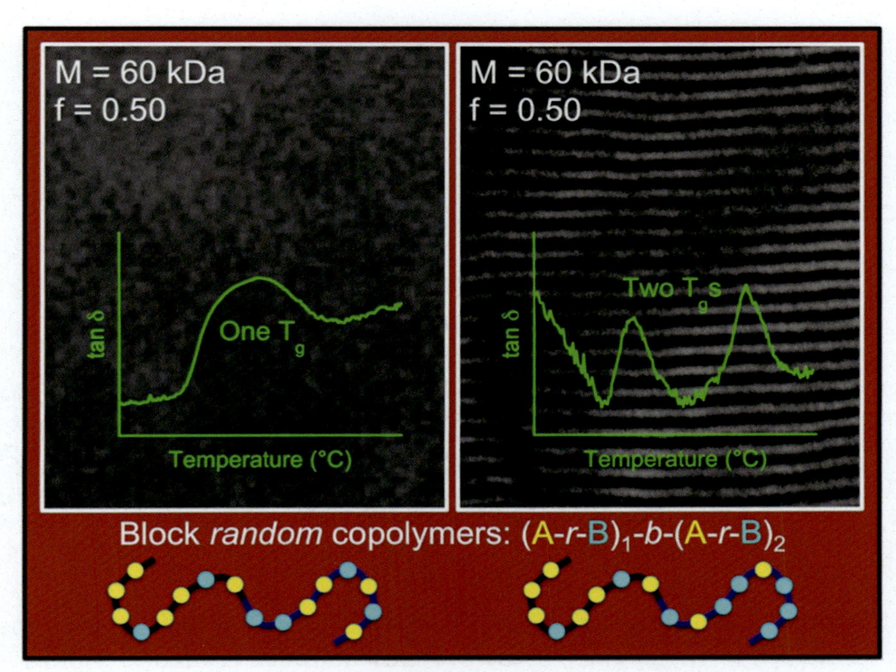

FIGURE 17.6 TEM images of two block random copolymers (BRCs) with identical chemical species, molecular weights, and block fractions under isothermal conditions, but different block purities, as schematically depicted. The copolymer on the left possesses a single T_g according to rheology, which is indicative of a homogeneous material, whereas the one on the right possesses 2 T_gs expected from a microphase-separated copolymer (in agreement with the TEM images). *Adapted from A.R. Ashraf, J.J. Ryan, M.M. Satkowski, B. Lee, S.D. Smith, R.J. Spontak, Bicomponent block copolymers derived from one or more random copolymers as an alternative route to controllable phase behavior, Macromol. Rapid Commun. 38 (2017) 1700207. Copyright 2017 Wiley.*

due to the microphase-separated blocks when $\Delta = 0.50$. Values of the order–disorder transition temperature (T_{ODT}) discerned for several low-dispersity BRCs (varying in N and f) from optical and rheological measurements provide values of $(\chi N)_{ODT}$, which can be corrected for composition via Δ to yield $\chi_{eff}N$, as displayed in Fig. 17.7. In Fig. 17.7A, $(\chi_{eff}N)_{ODT}$ is presented as a function of Δ at nearly constant f (=0.54) to demonstrate that its average is 10.5, which corresponds well to the mean-field value of $(\chi N)_{ODT}$ for diblock copolymers at $f = 0.50$ in the absence of critical fluctuations commonly associated [89,90] with block copolymers at their ODT. The results provided in Fig. 17.7B indicate that the ODT predicted for pure diblock copolymers can be approximated as functions of only f and Δ with diblock BRCs at constant temperature (χ) and molecular weight ($\sim$N). It should be noted that the midblock of TPS triblock copolymers can likewise be synthesized as a random copolymer (with added S) to controllably increase the T_g of the midblock [85].

Another synthesis-based route by which to alter the phase behavior and corresponding properties of TPE block copolymers requires changes to molecular architecture

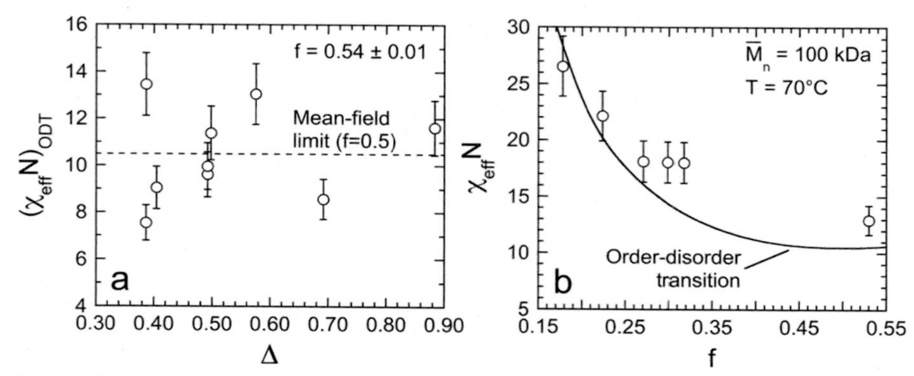

FIGURE 17.7 In (A), $(\chi_{eff}N)_{ODT}$ presented as a function of the composition contrast (Δ) for several different BRCs possessing comparable block fractions ($f \approx 0.5$). The average of the data is 10.5, which agrees favorably with the value of $(\chi N)_{ODT}$ in the mean-field limit for a comparable diblock copolymer with pure blocks (*dashed line*). In (B), the dependence of $\chi_{eff}N$ on f for copolymers possessing identical molecular weights under isothermal conditions. The theoretical ODT for diblock copolymers is included for reference (*solid line*). *Adapted from A.R. Ashraf, J.J. Ryan, M.M. Satkowski, B. Lee, S.D. Smith, R.J. Spontak, Bicomponent block copolymers derived from one or more random copolymers as an alternative route to controllable phase behavior, Macromol. Rapid Commun. 38 (2017) 1700207. Copyright 2017 Wiley.*

through the incorporation of additional blocks. The first case considered here provides a segue from diblock to triblock copolymers as an additional endblock is progressively grown from a parent AB diblock copolymer to generate asymmetric A_1BA_2 triblock copolymers wherein the endblocks possess the same chemistry but differ in length (i.e., $N_{A1} \neq N_{A2}$) [18,91]. In addition to exploring the utility of this approach to adjust phase behavior and properties, the systematic transition from nonnetworking diblock copolymers, a model material archetype capable of identifying the key elements of molecular self-assembly, to triblock copolymers, another model material archetype associated with physical network formation, can elucidate the conditions under which a triblock copolymer is first able to form an equilibrium network. To facilitate the discussion here, an asymmetry parameter (τ), defined as $N_{A1}/(N_{A1} + N_{A2})$, is introduced to differentiate the degree of asymmetry in this molecular progression: $\tau = 1$ is the maximum permissible value and corresponds to a diblock copolymer, whereas $\tau = {}^1/_2$ identifies a molecularly symmetric triblock copolymer (with $N_{A1} = N_{A2}$). It is possible, however, for $\tau < {}^1/_2$ when $N_{A2} > N_{A1}$. Experimental values of T_{ODT} are presented as a function of the molecular weight of the new endblock (M_{A2}) for two different copolymer systems in Fig. 17.8A and immediately reveal that T_{ODT} initially decreases at low M_{A2} before increasing as expected. When M_{A2} is recast in terms of τ and temperature is related to χ for SI block copolymers, the data are in excellent quantitative agreement with the mean-field theoretical predictions of Mayes and Olvera de la Cruz [92] (*cf.* Fig. 17.8B). Similar results acquired from small-angle X-ray scattering (SAXS) and self-consistent field analysis of ordered morphologies, as well as dynamic melt rheology [93], confirm that the diblock→triblock transition is not a simple interpolatable process, especially when M_{A2} is relatively small (and τ lies between 1 and ${}^1/_2$).

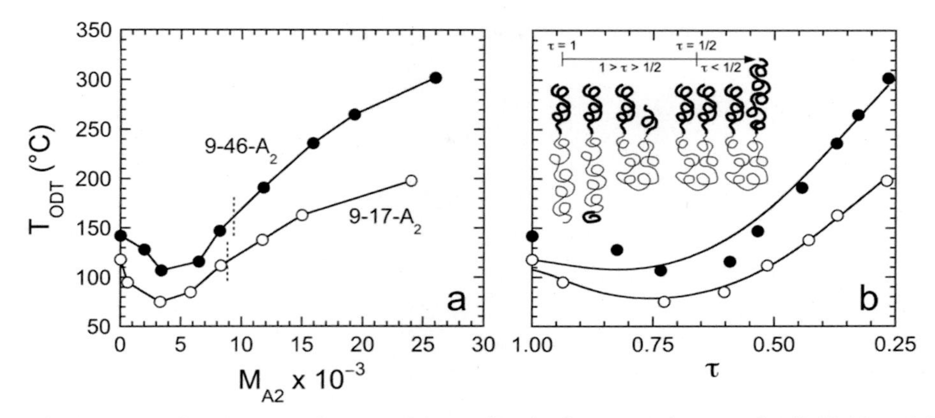

FIGURE 17.8 The order–disorder transition temperature (T_{ODT}) of two copolymer series (9-17-A2 and 9-46-A2) measured by dynamic rheology as functions of (A) M_{A2} and (B) τ in molecularly asymmetric triblock copolymers progressively grown from two parent diblock copolymers. The *solid lines* in (A) serve to connect the data, whereas those in (B) correspond to mean-field predictions. The diagram included in (B) illustrates several different copolymer architectures differing in τ. *Adapted from M.W. Hamersky, S.D. Smith, A.O. Gozen, R.J. Spontak, Phase behavior of triblock copolymers varying in molecular asymmetry, Phys. Rev. Lett. 95 (2005) 168306. Copyright 2005 American Institute of Physics.*

An improved understanding of this molecular transition can be realized through the use of computer simulations and their unique ability to quantify the different types of accessible chain conformations [94–96]. Fig. 17.9A is an illustration of the conformations that a microphase-separated ABA triblock copolymer can adopt: bridges (B), loops (L), and dangles (D). Depending on the magnitude of χN, some chains can also remain unsegregated (un). Bridges are associated with midblock network formation since they connect neighboring (glassy) microdomains serving as physical cross-links (knotted loops are not considered here). Included in Fig. 17.9B and C are the fractions of these conformations (F_B, F_L, F_D, and F_{un}) extracted from dissipative particle dynamics (DPD) simulations as functions of (i) χN at constant f and (ii) f at constant χN, respectively. Two important observations are apparent from these results: (1) as χN increases and the copolymer eventually enters the strong-segregation regime ($\chi N > 100$), F_B increases to a plateau and becomes independent of χN; and (2) as the block fraction (f) increases and the copolymer morphologies transform from spheres to cylinders and finally to lamellae, F_B monotonically decreases since the interfacial area/volume likewise decreases. These findings establish that simulations can be used to monitor the molecular conformations of network-forming block copolymers in moderate-to-strong segregation regimes, and that this methodology can be applied to molecularly asymmetric A_1BA_2 copolymers [97]. When $\tau = 1.00$, $F_D = 1.0$, whereas $F_B = F_L = 0.0$, in Fig. 17.10A. A reduction in τ is initially accompanied by increases in F_B and F_L and a systematic decrease in F_D, according to DPD simulations. At τ_N, F_B becomes independent of τ, indicating the formation of a thermodynamically equilibrated network. The value of F_B in this limit agrees well with Monte Carlo (MC) simulations and self-consistent field calculations (the latter of which

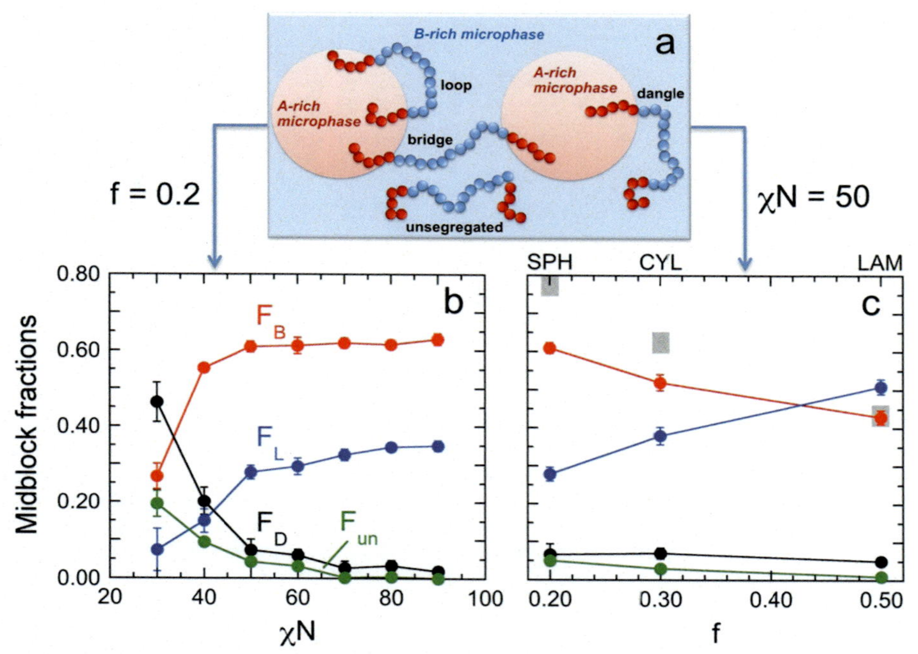

FIGURE 17.9 In (A), an illustration of a microphase-separated ABA triblock copolymer identifying the microphases and matrix present, along with the four possible midblock conformations: bridges (B), loops (L), dangles (D), and unsegregated (un). Corresponding midblock fraction (F) values extracted from DPD simulations are provided as functions of (B) χN at constant f and (C) f at constant χN (color-coded, labeled). The *solid lines* serve to connect the data, and the *gray boxes* in (B) correspond to predictions from self-consistent field theory (which does not account for dangles or unsegregated chains). *Adapted from S.S. Tallury, R.J. Spontak, M.A. Pasquinelli, Dissipative particle dynamics of triblock copolymer melts: a midblock conformational study at moderate segregation, J. Chem. Phys. 141 (2014) 244911. Copyright 2014 American Institute of Physics.*

do not, however, account for the presence of dangles). The increase in F_L beyond τ_N reflects the continuing decrease in F_D. The value of τ^* (=0.79) highlighted in Fig. 17.10A identifies the molecular asymmetry corresponding to pronounced minima in independent experimental measurements of the ODT temperature [18], the microdomain period, and the melt modulus [93], as well as a maximum in the invariant from SAXS [91].

Taken together, the results provided in Fig. 17.10A indicate that short A_2 endblocks remain mixed in the B matrix and only occasionally serve as bridges between A_1 microdomains, thereby yielding flocs that become increasingly prevalent as both M_{A2} and F_B increase (and τ decreases). At sufficiently high M_{A2} ($<M_{A1}$), these flocs ultimately merge to form the contiguous, load-bearing network expected of a TPE. Molecular asymmetry in a triblock copolymer can also be used to drive morphological transitions at constant copolymer composition, expressed as either β ($=N_B/N_{A1}$) or f_B (the B block fraction) in Fig. 17.10B. At very high χN and very low M_{A2} levels, a somewhat unexpected phenomenon is also observed: dual morphologies coexisting in a single bicomponent

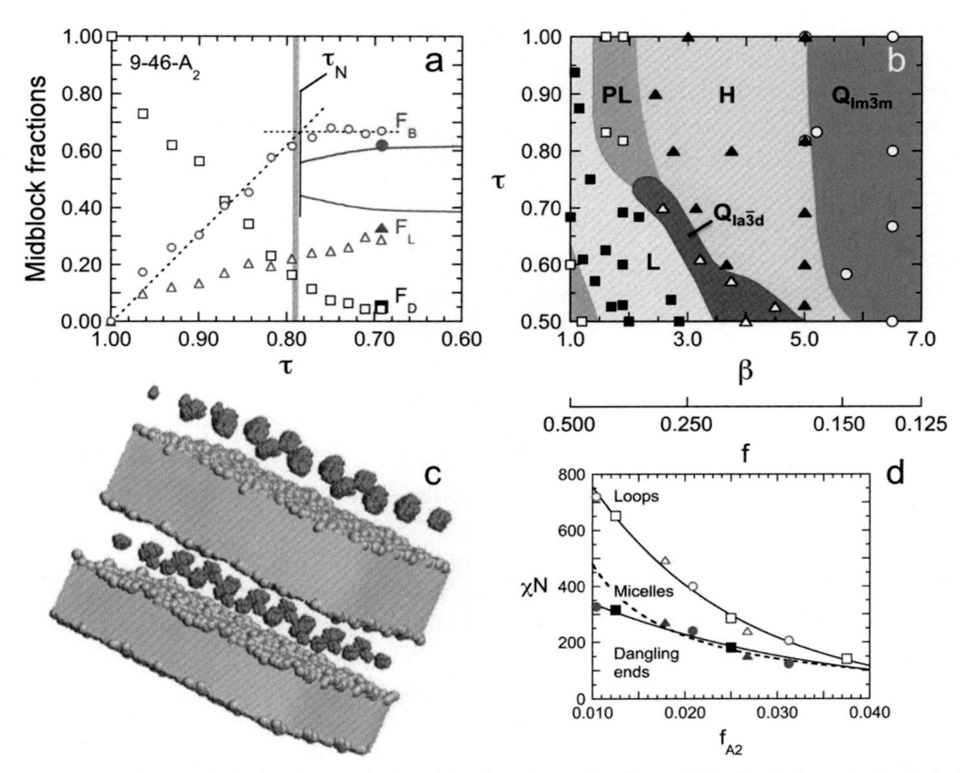

FIGURE 17.10 In (A), midblock fractions presented as functions of τ from DPD simulations (*color-coded open symbols*, labeled), whereas those from MC simulations are designated by *filled symbols* and those from self-consistent field theory are provided as *solid lines*. The value of τ at which F_B identifies the formation of a network is given by τ_N, and the value of τ discerned from the most pronounced changes in phase behavior, nanostructural dimensions, and melt rheology (τ*) is displayed as the gray region. In (B), molecular asymmetry is used in conjunction with β (or f) to control morphological development in molecularly asymmetric triblock copolymers. In the superstrong-segregation regime, coexisting lamellae and micelles can develop in A_1BA_2 triblock copolymers with short A_2 blocks (C), and the phase diagram (from MC simulations) showing the range over which such interstitial micelles form is included in (D). The *solid lines* in (D) serve as guides for the eye, and the dashed curve corresponds to a theory [98] for free micelles. *Adapted from S.S. Tallury, K.P. Mineart, S. Woloszczuk, D.N. Williams, R.B. Thompson, M.A. Pasquinelli, M. Banaszak, R.J. Spontak, Molecular-level insights into asymmetric triblock copolymers: network and phase development, J. Chem. Phys. 141 (2014) 121103. S. Woloszczuk, K.P. Mineart, R.J. Spontak, M. Banaszak, Dual modes of self-assembly in super strongly segregated bicomponent triblock copolymer melts, Phys. Rev. E 91 (2015) 010601. Copyright 2014 and 2015 American Institute of Physics.*

block copolymer [99]. Although coexisting morphologies have been reported in neat (i.e., nonmixed) bicomponent block copolymers, they typically arise due to either nonequilibrium processing or close proximity to an order–order transition (OOT). They also develop in tricomponent block terpolymers [69,71,72]. Here, coexistence involves the independent self-assembly of the A_1 and A_2 endblocks under high χN conditions in the superstrong-segregation regime. The outcome, displayed in Fig. 17.10C, consists of alternating A_1 lamellae separated by a single array of hexagonally arranged A_2 micelles that exist within the (transparent) B lamellae. The boundary conditions between which these micelles form are included in Fig. 17.10D and indicate that they become less stable as the fraction of A_2 endblocks (f_{A2}) increases since longer endblocks will ultimately start to interact with the A_1 microdomains by adopting bridge or loop conformations. Interestingly, the initial formation of these molecularly confined micelles (relative to dangling ends) can be accurately modeled as free (i.e., nonconfined) micelles according to the theoretical framework proposed by Semenov and coworkers [98]. Although this predicted coexistence discerned from MC simulations has not yet been experimentally validated, it suggests that TPEs based on the design pictured in Fig. 17.10C could exhibit greater extensibility. Through the implementation of cutting-edge synthetic protocols, this strategy of adding blocks to copolymer molecules to control the phase behavior and property development of TPSs has successfully spawned new materials such as perfectly alternating linear multiblock copolymers possessing either constant block weight (and variable molecular weight) [55] or constant molecular weight (and variable block weight) [56], as well as perfectly alternating tapered multiblock copolymers arranged in linear [57,58] and nonlinear [66] architectures. Another intriguing variation of triblock copolymers introduces bottlebrush blocks [74], as depicted in Fig. 17.11.

All of the previous modifications to block copolymers discussed here involve the synthesis of designer molecules, which might not always be economically viable. An alternative approach often chosen to vary block copolymer phase behavior and morphology/property development involves postsynthesis chemical functionalization [100–102]. While this topic will be discussed in detail later, one particular reaction is commonly utilized. As mentioned earlier, typical TPSs include triblock copolymers in which the midblock is either a polydiene or its saturated analog. For this reason, hydrogenation is considered here since it is routinely used to generate saturated midblocks from unsaturated ones [103]. Hydrogenation is a catalytic reaction that can be performed with a homogeneous (unsupported) or heterogeneous (supported) catalyst solution [104]. The results discussed below employ nickel(II) bis(2-ethylhexanoate) dissolved in cyclohexane to yield a homogeneous solution. Representative proton nuclear magnetic resonance (^{1}H NMR) spectra of S–I copolymers at different degree of hydrogenation (DOH) levels are supplied in Fig. 17.12A to demonstrate the gradual and quantifiable

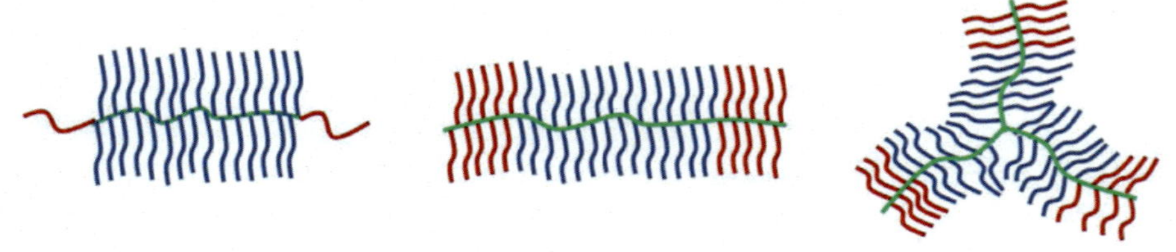

FIGURE 17.11 Schematic diagrams of three TPE molecular architectures containing bottlebrush (bb) blocks: (left) A-B_{bb}-A, (middle) A_{bb}-B_{bb}-A_{bb}, and (right) $(A_{bb}$-$B_{bb})_3$ star. *Reproduced with permission from W. Wang, W. Lu, A. Goodwin, H. Wang, P. Yin, N.G. Kang, K. Hong, J.W. Mays, Recent advances in thermoplastic elastomers from living polymerizations: macromolecular architectures and supramolecular chemistry, Prog. Polym. Sci. 95 (2019) 1–31. Copyright 2019 Elsevier.*

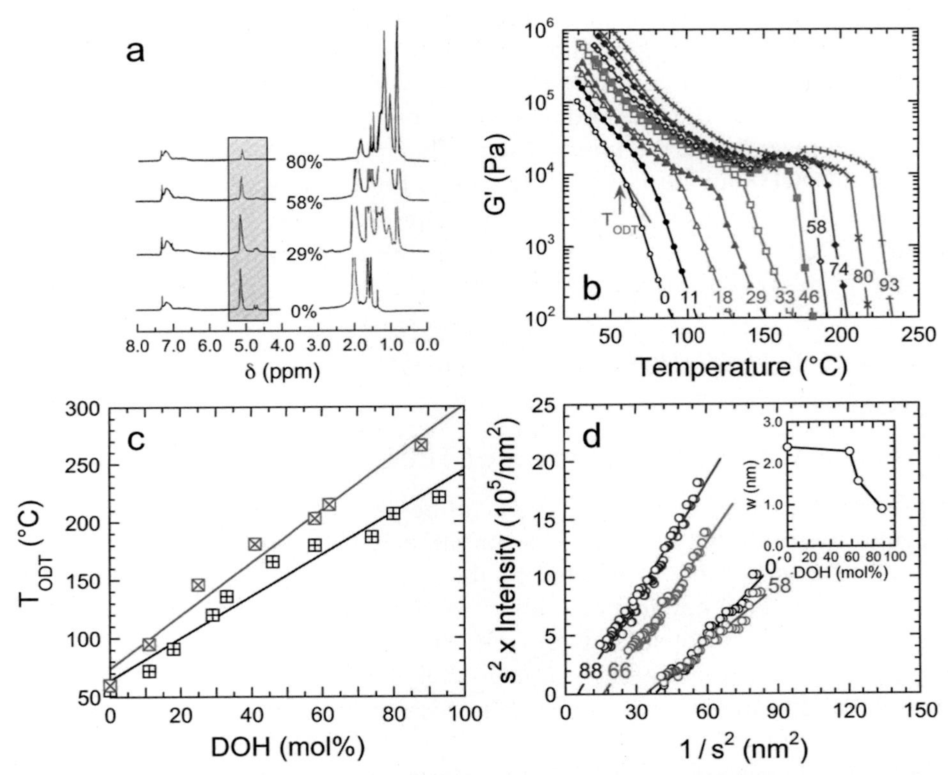

FIGURE 17.12 In (A), ^{1}H NMR spectra acquired from a BRC at different DOH levels (labeled), as quantified by the disappearance of unsaturated bonds (*shaded area*). In (B), the dynamic storage modulus (G′) is displayed as a function of temperature for a BRC varying in DOH (color-coded, labeled). Values of T_{ODT} extracted from such rheological data for two different BRCs (color-differentiated) are supplied in (C) as a function of DOH and reveal a linear dependence (*solid lines*). In (D) SAXS data collected from four BRCs differing in their DOH are provided in the form suggested by Vonk [105] to extract the interphase thickness (w), values of which are given as a function of DOH in the inset of (D). *Adapted from A.R. Ashraf, J.J. Ryan, M.M. Satkowski, S.D. Smith, R.J. Spontak, Effect of systematic hydrogenation on the phase behavior and nanostructural dimensions of block copolymers, ACS Appl. Mater. Interfaces 10 (2018) 3186–3190. Copyright 2018 American Chemical Society.*

disappearance of unsaturated bonds. Although hydrogenated SIS and poly(styrene-*b*-butadiene-*b*-styrene) (SBS) triblock copolymers are the most commercially relevant TPEs since they yield SEPS and SEBS copolymers, respectively, we monitor here the hydrogenation of S–I BRCs since their properties can be precisely adjusted [106] to provide a more complete picture of hydrogenation (otherwise, hydrogenating SIS and SBS copolymers could yield relatively little insightful information before degradation). For instance, the unaltered BRC possesses a relatively low T_{ODT}, as ascertained by dynamic rheology in Fig. 17.12B. Progressive hydrogenation to a DOH level of 93 mol% clearly signifies that the T_{ODT} is systematically increased to over 200°C (due to increased incompatibility) and that this dependence, measured by both dynamic rheology and

SAXS, is unexpectedly linear for two different BRCs in Fig. 17.12C. By casting associated SAXS profiles into the form proposed by Vonk [105] in Fig. 17.12D, it is possible to extract information regarding the interphase thickness (w) that exists between chemically dissimilar microdomains. The dependence of w on DOH is included in the inset of Fig. 17.12D and reveals that the interphase is relatively broad (>3 nm) for the unaltered BRC (at 0 mol% DOH) but abruptly narrows when the DOH exceeds 58 mol% (eventually approaching ~ 1 nm). Since $w \sim \chi^{-1/2}$ in the limit of strong-segregation behavior according to Helfand and Tagami [107], this dependence corroborates the results from T_{ODT}(DOH), verifying that increasing hydrogenation boosts copolymer incompatibility.

17.3 Physical modification of nonpolar TPEs and their applications

17.3.1 Fabrication and properties of TPEGs

Recent progress has established that TPEs can be physically modified to achieve properties beyond their inherent mechanical limitations [108–111]. For instance, the lowest modulus achievable in TPSs is dictated by the plateau modulus of the middle rubbery block. In some emerging applications, however, a lower modulus might be desirable, in which case a sound physical strategy must be developed to achieve a tunable reduction in modulus while still maintaining control over morphological development. In this section, we examine one approach to attain this objective through the physical incorporation of a low-volatility aliphatic oil, which serves as a midblock-selective solvent, to yield thermoplastic elastomer gels (TPEGs). These soft, stretchy and self-healing materials are of tremendous general interest due to their unique abilities to, for example, mimic mammalian skeletal muscle [112], show little evidence of mechanical hysteresis [113], become responsive to electrical stimulation [114–118], and dampen compression and vibrations [119]. They can be used as anthropomorphic surrogates [120,121] to test ballistics [122,123] or crash/explosion survival [124] or replace cadavers in surgical training [125], and they can be used for controlled drug delivery [126] and microfluidics [127]. As shock-absorbing media, they are suitable for underground housing for fragile fiber optics in the telecommunications industry [128,129]. Moreover, as electro-responsive media, they can be used for soft robotics [130,131], haptic devices [132], and even energy-harvesting media. Lastly, this general materials design can be readily extended to yield pressure-sensitive adhesives [133,134] with tunable properties, high-precision shape-memory materials [135–137], and unparalleled polymeric systems that exhibit time–composition rheological equivalence under isothermal conditions [138,139]. While the TPEGs of most interest here consist of a TPS that is selectively swollen with a primarily aliphatic oil (e.g., mineral oil, MO), the phase diagram [140] displayed in Fig. 17.13, generated from SAXS analysis, confirms that the morphologies observed in neat block copolymer systems are retained in ternary physical blends composed of a nonpolar SEPS copolymer with a midblock-selective MO and an

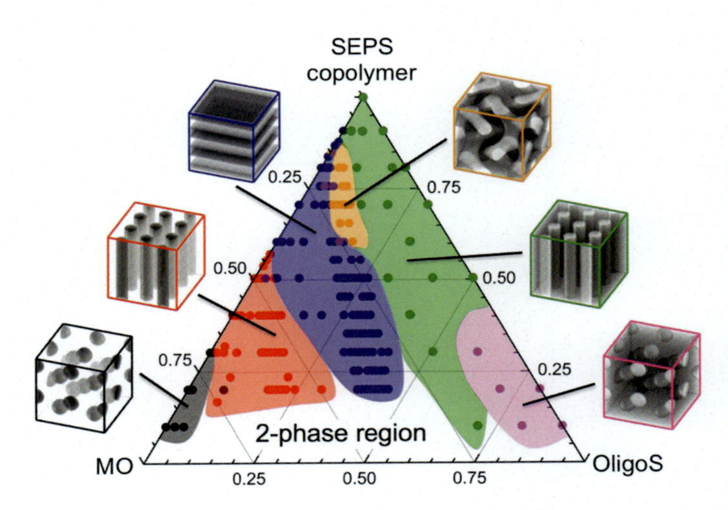

FIGURE 17.13 Ternary phase diagram generated from synchrotron SAXS for a TPE (SEPS copolymer) physically modified with a midblock-selective mineral oil (MO) and an endblock-selective oligostyrene (OligoS). The colored regions correspond to the color-coded morphologies. *Adapted from A.S. Krishnan, S.D. Smith, R.J. Spontak, Ternary phase behavior of a triblock copolymer in the presence of an endblock-selective homopolymer and a midblock-selective oil, Macromolecules 45 (2012) 6056–6067. Copyright 2012 American Chemical Society.*

endblock-selective oligostyrene. This level of morphological consistency substantiates our contention that mixing paradigms developed for miscible block copolymer blends and amphiphilic block copolymers apply to these TPEGs.

Most styrenic TPEs of commercial relevance commonly possess 20–33 wt% S at surprisingly few molecular weights. This shortage of available, well-defined TPSs severely limits experimental inquiry, in which case computer simulations again demonstrate their obvious value by exploring key parameters in phase space [141–143]. The MC simulations in Fig. 17.14A–D, for instance, focus on the effect of midblock solvation for ABA triblock copolymers differing in f (the A fraction). While (perforated) lamellae, cylinders, rods, and spherical micelles constitute most of the morphologies identified, a unique nanostructure, referred to as the truncated octahedron (OCT, depicted in Fig. 17.14D), is predicted to be stable in TPEGs when the copolymer is 80% endblock (which is more representative of a rubber-reinforced plastic than a true TPE). The corresponding bridge fractions evaluated at different isotherms are included in Fig. 17.15 and corroborate that discrete micellar morphologies tend to exhibit the highest level of midblock bridging. Dynamic rheological analysis of TPEGs at relatively low ABA triblock copolymer loading levels (φ) in Fig. 17.16A reveals that network behavior (as evidenced by the dynamic storage modulus, G′, being independent of oscillatory frequency, ω, and greater than the dynamic loss modulus, G″) is observed down to 2 wt% copolymer. In addition, the low-ω G′ measured for TPEGs composed of SEBS copolymers possessing similar molecular compositions but different molecular weights scales as φ^k, where $k \approx 2$, suggesting that entanglements dominate the rheological contribution to G′, in favorable agreement with previous independent studies [144,145], as well as the theoretical studies of de Gennes [146] proposed for entangled homopolymers in solution. Associated simulations of the midblock bridging fraction for TPEGs containing ABA

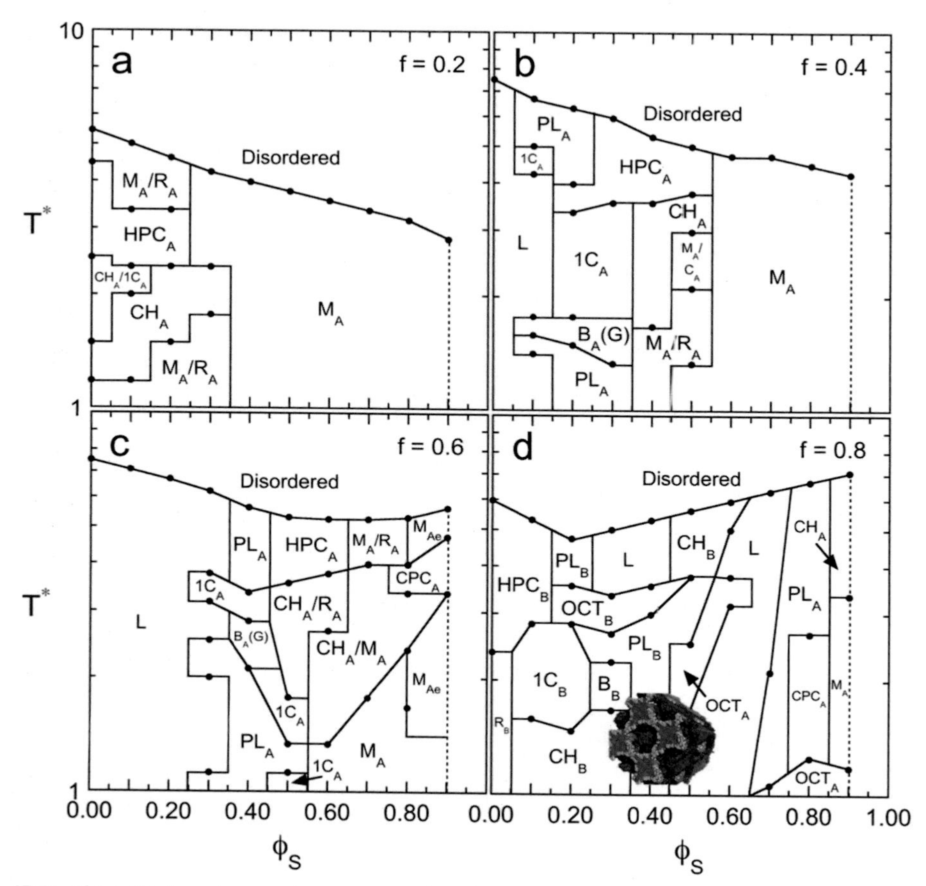

FIGURE 17.14 Dimensionless temperature–composition (T*-φ_S, where φ_S denotes the fraction of midblock-selective solvent) phase diagrams from MC computer simulations for triblock copolymers varying in molecular composition (f): (A) 0.2, (B) 0.4, (C) 0.6, and (D) 0.8. In addition to lamellae (L), perforated lamellae (PL), bicontinuous (B), cylinders and rods (C and R), and spherical micelles (M), a new morphology denoted by OCT (for octahedron) is identified and depicted. *Adapted from S. Woloszczuk, M.O. Tuhin, S.R. Gade, M.A. Pasquinelli, M. Banaszak, R.J. Spontak, Complex phase behavior and network characteristics of midblock-solvated triblock copolymers as physically cross-linked soft materials, ACS Appl. Mater. Interfaces 9 (2017) 39940–39944. Copyright 2017 American Chemical Society.*

triblock copolymers varying in chain length at constant composition (*cf.* Fig. 17.16B) indicate that the copolymer composition (φ_{eq}) required to achieve equilibrated molecular networks at $F_{B,net}$ varies inversely with copolymer chain length, and the scaling relationship extracted from the inset in Fig. 17.16B is given by $\varphi_{eq} \sim N^{-1.4}$.

A series of DPD simulations generated [143] from linear multiblock copolymers of the form $(ABA)_n$, where n varies from 1 to 3 to include triblock (n = 1), pentablock (n = 2), and heptablock (n = 3), at constant chain length is presented at different midblock-solvation levels in Fig. 17.17A. Due to the starting composition of these copolymers

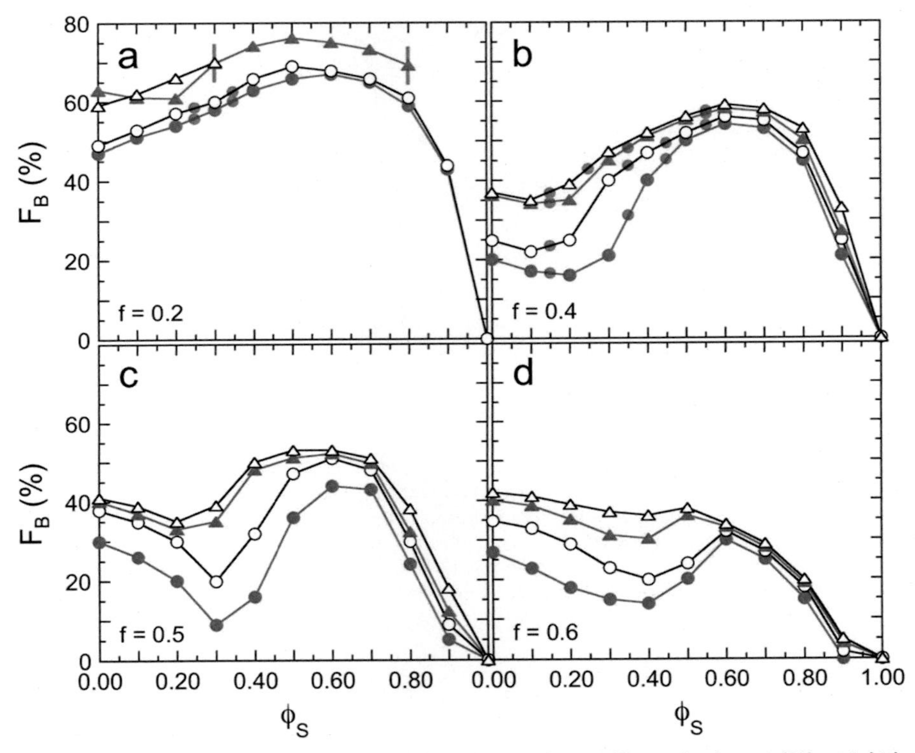

FIGURE 17.15 Midblock bridging fraction as a function of φ_S at four different isotherms (T*) – 1 (●), 2 (○), 3 (▲), and 4 (Δ) (see Fig. 17.14) – for four triblock copolymers varying in molecular composition (f): (A) 0.2, (B) 0.4, (C) 0.5, and (D) 0.6. The *solid lines* serve to connect the data, and the *vertical red lines* indicate the onset of the ODT. The *red dots* identify order–order transitions from the corresponding phase diagrams displayed. *Reproduced with permission from S. Woloszczuk, M.O. Tuhin, S.R. Gade, M.A. Pasquinelli, M. Banaszak, R.J. Spontak, Complex phase behavior and network characteristics of midblock-solvated triblock copolymers as physically cross-linked soft materials, ACS Appl. Mater. Interfaces 9 (2017) 39940–39944. Copyright 2017 American Chemical Society.*

(20 wt% hard block), they all exhibit spherical micelles. The ensemble of molecular conformations possible in these three TPEG series is depicted in Fig. 17.17B, and each can be uniquely differentiated to yield a midblock conformation index (MCI), which is expressed by four numbers: [number of connected micelles, number of bridges, number of loops, number of dangles]. Careful examination of the chain trajectories in computer simulations yields the most probable copolymer conformation in each TPEG. These are provided in schematic form and as functions of φ in Fig. 17.18 and signify that the pentablock and heptablock molecules prefer not to be fully extended (since each exhibits a loop). Also included in this figure is the total fraction of bridged molecules in each TPEG series. Similarly detailed topological analyses of networks have been reported [147–149] for several different polymer systems. Rheological investigation of TPEGs fabricated from matched pentablock and heptablock copolymers implies that $G' \sim \varphi$, in

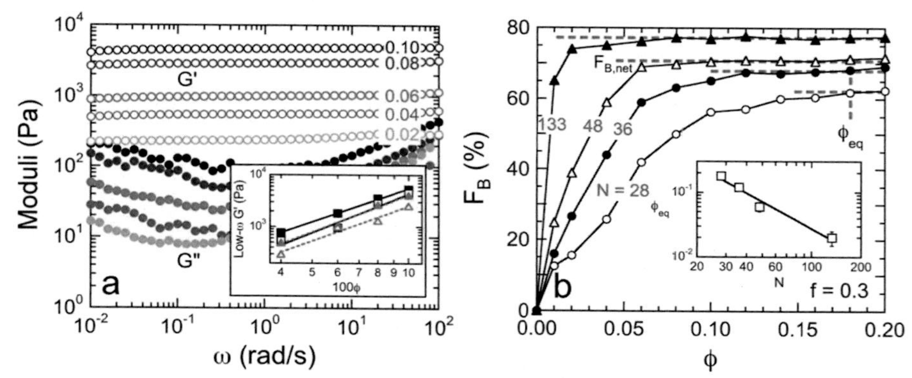

FIGURE 17.16 In (A) frequency spectra of G′ (*open symbols*) and G″ (*filled symbols*) at different copolymer fractions (color-coded, labeled). The inset in (A) displays G′ at low ω as a function of φ for four SEBS copolymers possessing similar composition (f ≈ 0.3) but different molecular weights. The *lines* in the inset correspond to power-law regressions to the data. In (B), the midblock bridging fraction extracted from DPD simulations is presented as a function of φ for copolymers differing in chain length (N, labeled) at f = 0.3. The value of F_B signaling network formation is identified ($F_{B,net}$), as is the composition at which it occurs (φ_{eq}), and the *solid lines* serve to connect the data. The dependence of φ_{eq} on N in included in the inset in (B), and the *solid line* is a power-law regression. *Adapted from M.O. Tuhin, S. Woloszczuk, K.P. Mineart, M.A. Pasquinelli, J.D. Sadler, S.D. Smith, M. Banaszak, R.J. Spontak, Communication: molecular-level description of constrained chain topologies in multiblock copolymer gel networks, J. Chem. Phys. 148 (2018) 231101. Copyright 2018 American Institute of Physics.*

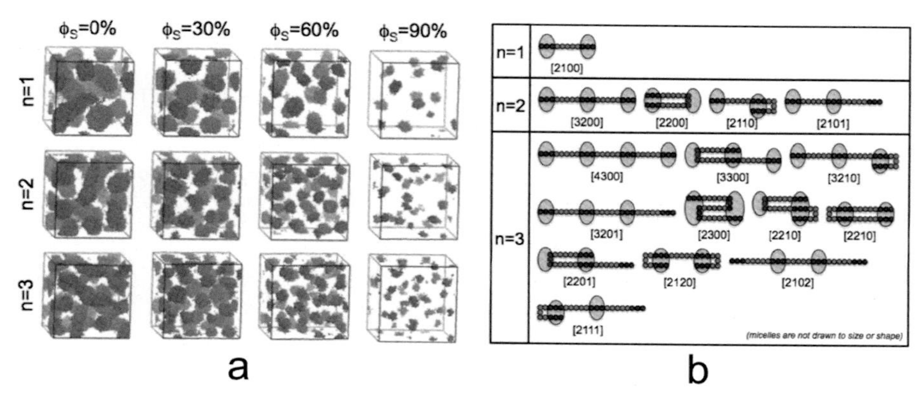

FIGURE 17.17 In (A), DPD simulations of TPEGs composed of $(AB)_nA$ multiblock copolymers (n varies from 1 to 3) at different solvent compositions (labeled). The midblock conformations displayed in (B) yield the midblock conformation index (MCI) values (included) developed to describe the topologies of TPEG physical networks. A description of the MCI is provided in the text. *Adapted from M.O. Tuhin, J.J. Ryan, J.D. Sadler, Z. Han, B. Lee, S.D. Smith, M.A. Pasquinelli, R.J. Spontak, Microphase-separated morphologies and molecular network topologies in multiblock copolymer gels, Macromolecules 51 (2018) 5173–5181. Copyright 2018 American Chemical Society.*

which case it follows that physical cross-links dominate G′ (in marked contrast to the triblock-based TPEG governed by entanglements). Complementary quasistatic uniaxial tensile tests of these TPEGs can be analyzed in the context of the Slip-Tube Network (STN) model proposed by Rubinstein and Panyukov [150] to distinguish the

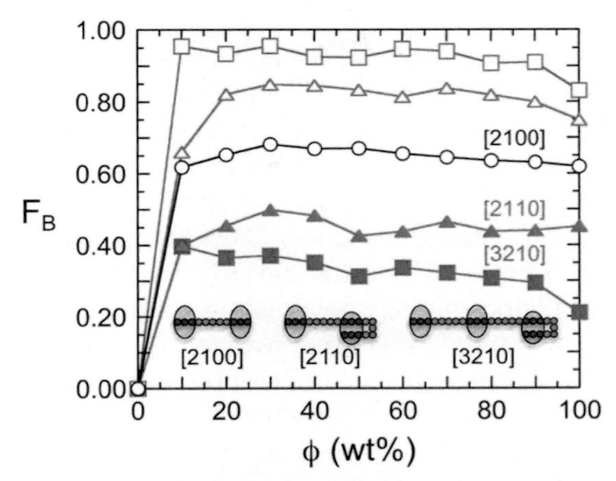

FIGURE 17.18 Midblock bridging fractions determined from DPD simulations (see Fig. 17.18) as a function of copolymer composition (φ) for TPEGs composed of triblock (*circles*), pentablock (*diamond*), and heptablock (*square*) copolymers of constant composition and molecular weight. Except for the case of triblock-based TPEGs, these results identify the most probable conformation (*filled symbols*, depicted with its MCI) and total values (*open symbols*). The *solid lines* serve to connect the data. *Reproduced with permission from M.O. Tuhin, S. Woloszczuk, K.P. Mineart, M.A. Pasquinelli, J.D. Sadler, S.D. Smith, M. Banaszak, R.J. Spontak, Communication: molecular-level description of constrained chain topologies in multiblock copolymer gel networks, J. Chem. Phys. 148 (2018) 231101. Copyright 2018 American Institute of Physics.*

contributions of elastic cross-links and chain entanglements to the shear modulus (G_c and G_e, respectively) in polymer networks. Results obtained [151] by fitting tensile data at ambient temperature are presented in Fig. 17.19 and verify that the physical cross-links of these TPEGs are important at low copolymer levels but entanglements

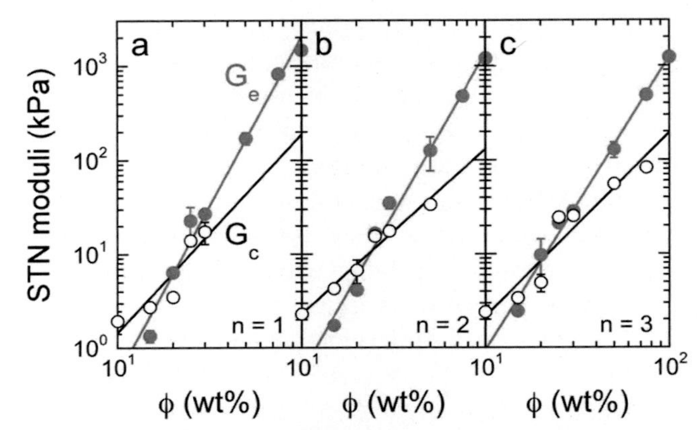

FIGURE 17.19 Values of G_c and G_e (color-coded, labeled) obtained from regressions of the Slip-Tube Network model [150] to quasistatic uniaxial tensile test data and presented as a function of φ for TPEGs composed of triblock (n = 1), pentablock (n = 2), and heptablock (n = 3) copolymers of constant composition and molecular weight. The *solid lines* are power-law fits to the data. *Reproduced with permission from J. Yan, M.O. Tuhin, J.D. Sadler, S.D. Smith, M.A. Pasquinelli, R.J. Spontak, Network topology and stability of homologous multiblock copolymer physical gels, J. Chem. Phys. 153 (2020) 124904. Copyright 2020 American Institute of Physics.*

dictate the response of all TPEGs at high φ, with the crossover consistently between ~20 and 25 wt% copolymer. While similar behavior is observed from dynamic rheology for the TPEGs composed of pentablock and heptablock copolymers, the apparent difference in the case of the triblock-based TPEGs is attributed to the extent to which the samples are deformed (the rheological tests were performed on a stress-controlled rheometer with a constant stress amplitude of 1 Pa in the linear viscoelastic regime, whereas the tensile data were examined up to 200% strain). Detailed mechanical analyses such as these provide a starting point to develop a more fundamental and in-depth understanding of how network topology governs the mechanical properties of TPEGs.

One final aspect of styrenic TPEGs that warrant mention is that their dynamic properties can exhibit isothermal time–composition superpositioning (tCS) [138,139] (also referred to as rheological equivalence), in contrast to conventional time–temperature superpositioning (tTS) [1]. In this case, a cycloaliphatic resin (CR) that behaves as a tackifying resin with a T_g above ambient temperature is added to MO to yield ternary, rather than binary, TPEGs. While TPEGs formulated with MO consistently exhibit network behavior at copolymer loading levels above the critical gel concentration according to dynamic rheology, those prepared with CR display evidence of viscoelastic behavior in which G'' exceeds G' up to a crossover point in frequency spectra. To ensure that these midblock-selective cosolvents are miscible at all compositions, T_g values of MO/CR mixtures have been measured over the entire composition range and subsequently analyzed according to the Couchman equation to confirm infinite miscibility [152]. Frequency spectra acquired from ternary TPEGs containing a commercial SEBS copolymer physically swollen with MO/CR mixtures are included in Fig. 17.20A and reveal that a gel network is retained up to 60 wt% CR in the cosolvent. At and above 80 wt % CR in Fig. 17.20B, the TPEGs exhibit viscoelastic behavior. Shifting the frequency spectra in Fig. 17.20A and B yields the master G' and G'' curves as functions of adjusted frequency over 10 decades (up to 12 decades have been recorded) in Fig. 17.20C. While the master curve for G' appears relatively seamless, the one for G'' displays variation in the low-frequency limit. This characteristic is attributed to the onset of a second relaxation mechanism. Rheological equivalence derives from a single relaxation mechanism, typically due to chain entanglements. The second mechanism evident in Fig. 17.20C reflects endblock hopping when a copolymer endblock pulls out from one glassy micelle and reenters another one [153]. For this reason, this manifestation becomes amplified (and superpositioning fails) with copolymers possessing short endblocks, but disappears altogether with copolymers having long endblocks. The shift factor obtained from tTS (a_T) is commonly related to temperature through an Arrhenius expression or the Williams–Landel–Ferry (WLF) equation, whereas the one from tCS (a_C) is found to scale with the zero-shear cosolvent viscosity, and the corresponding scaling exponents vary linearly with copolymer loading level [139].

While most studies of TPEGs have focused on those produced from TPSs, Shull and coworkers [154–157] have pioneered TPEGs derived from acrylic triblock copolymers (these TPEGs are designated here as ATPEGs to reflect their acrylic content). Those

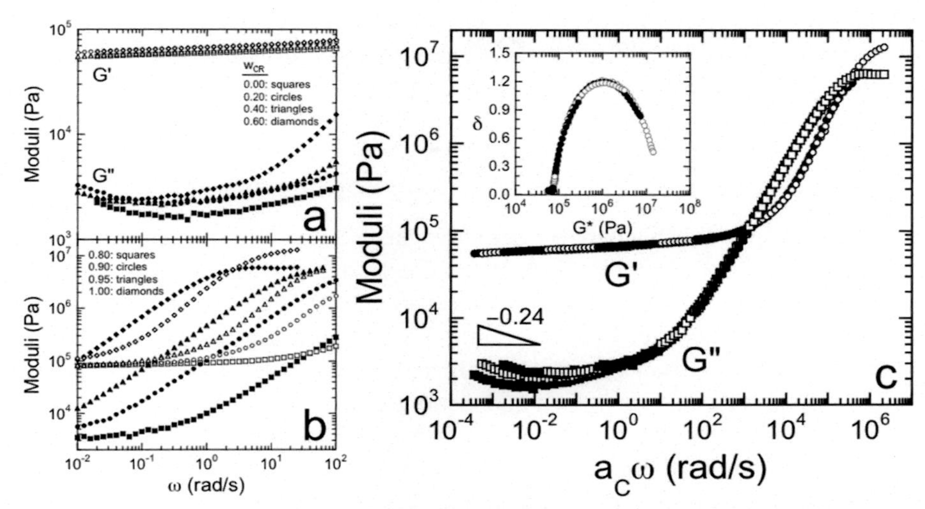

FIGURE 17.20 In (A) and (B), frequency spectra of G′ (*open symbols*) and G″ (*filled symbols*) for TPEGs consisting of triblock copolymers with a midblock-selective cosolvent composed of MO and a cycloaliphatic resin (CR, the fraction of which is provided in the legend of each panel). Time–composition superpositioning yields the *master curves* displayed in (C), as well as the van Gurp–Palmen plot of phase angle (δ) versus complex modulus (G*), which assesses the quality of superpositioning by how well the data superimpose. Variations evident in superpositioned G″ at low frequency imply the existence of a second relaxation mechanism (due to endblock hopping [153]). *Adapted from A.S. Krishnan, S. Seifert, B. Lee, S.A. Khan, R.J. Spontak, Cosolvent-regulated time–composition rheological equivalence in block copolymer solutions, Soft Matter 6 (2010) 4331–4334. Copyright 2010 Royal Society of Chemistry.*

efforts have focused on various mechanical attributes, as well as important thermal transitions. An important aspect of ATPEGs that immediately distinguishes them from styrenic TPEGs is that they can incorporate polar solvents [116] (e.g., dioctyl phthalate, DOP), which can be highly beneficial in stimuli-responsive applications (discussed below). Alternatively, TPEGs can also be generated from TPOs. For example, addition of MO to polyolefin TPEs composed of randomly coupled high-density polyethylene hard blocks and poly(ethylene-*co*-α-octene) soft blocks yields olefinic TPEGs (OTPEGs) that can exhibit altogether different property sets due to the ability of this TPE to crystallize [117,158]. Since this copolymer only contains polyethylene, it is suitable for various biomedical applications [159]. As before, the MO serves to swell the soft blocks, while the hard blocks stabilize the network composed of swollen soft blocks. Thermal calorimetry of OTPEGs differing in copolymer concentration confirms that the degree of copolymer crystallinity, which can be controllably varied by adjusting the hard/soft segment ratio, only undergoes a noticeable reduction from its neat copolymer limit at low copolymer loading levels (<10 wt%). This observation indicates that the presence of MO has little influence on copolymer crystallization. Moreover, unlike styrenic TPEGs that exhibit low mechanical hysteresis (with little irrecoverable strain) during cyclic tensile deformation, OTPEGs initially suffer from substantial irrecoverable strain (that can exceed 100%) [158], accompanied by a large hysteretic energy, due to plastic deformation.

Incorporation of relatively little MO (~20 wt%) into these TPOs, however, promotes an unexpected property development. While styrenic TPEGs are generally hyperelastic and routinely surpass over 1500% strain, OTPEGs can achieve strain levels exceeding 4000% before failure.

17.3.2 Stimuli-responsive and electrically conductive TPEGs

Electroactive polymers (EAPs) can be broadly classified as either electronic or ionic, depending on the mechanism by which they electroactuate (other EAP types are discussed elsewhere [160]). In the case of electronic EAPs, stimulation is achieved by applying an external electric field so that the mechanism relies on electrostatics [120,161–163]. Materials that are commonly categorized as electronic EAPs include ferroelectric polymers, such as poly(vinylidene fluoride) (PVDF) and its copolymers [164,165], which respond to an electric field because of a field-induced phase transition that results in a change in lattice dimensions, as well as dielectric elastomers (D-EAPs) [166–168], the operating mechanism of which will be detailed below. Ionic EAPs, on the other hand, require the presence of a polar liquid (e.g., water) that permits the migration of solvated ionic species in the presence of a potential, with the result that an electro-osmotic gradient promotes bending actuation. Two examples of ionic EAPs include single-wall carbon nanotubes [169] and ionic polymer–metal composites (IPMCs) [170–175], the latter of which will be discussed in a later section of this chapter. Generally speaking, these types of stimuli-responsive materials have garnered tremendous attention since they can provide rapid and significant actuation strains with relatively low power consumption, and they possess low mass densities relative to, for instance, shape-memory alloys or electroactive ceramics. In a broad overview of materials considered in this field, the performance metrics presented in Fig. 17.21

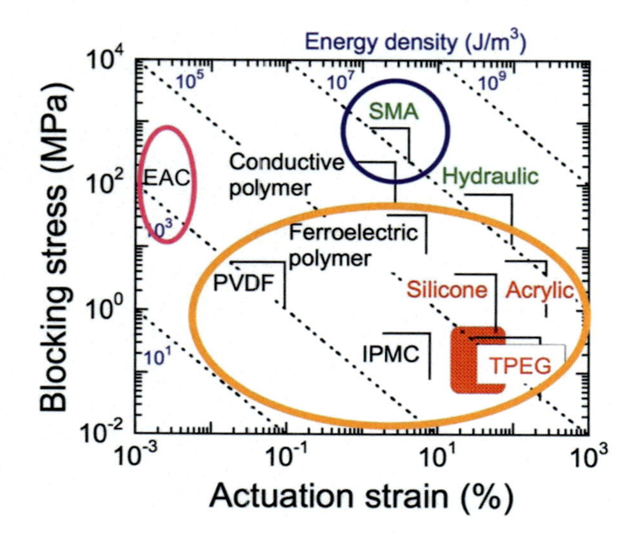

FIGURE 17.21 Blocking stress as a function of electroactuation strain for several different actuator classifications (color-coded), including electronic and ionic EAPs (large gold oval). Of particular interest here are IPMCs and D-EAPs (silicone, acrylic, and TPEG elastomers). The *red-shaded* region corresponds to mammalian skeletal muscle.

immediately confirm that traditional actuators, such as hydraulics (as well as pneumatics and DC motors), afford high actuation strains capable of high blocking stresses (to push/pull) and therefore possess very high energy density levels (approaching GJ/m^3). Their shortcoming, however, is their high mass density, which translates into heavy devices. In light of current efforts intended to minimize energy use so as to mitigate environmental pollution and conserve natural resources, weight becomes a critical consideration. For this reason, polymeric actuators such as the EAPs included in Fig. 17.21 show promise as lightweight replacements for their heavier counterparts. Several of the electronic and ionic EAPs described above are included in this figure (silicone and acrylic elastomers join the TPEGs as examples of D-EAPs), and the D-EAPs reside in close proximity to the highlighted region corresponding to mammalian skeletal muscle. Since this section focuses on TPEGs, we now address the operational mechanism of D-EAPs and the performance of TPEGs as stimuli-responsive materials.

Compared to many other electroactive materials, D-EAPs are lightweight, flexible, tough, shape-processable, scalable, available, and inexpensive [121,176,177]. The mechanism by which these elastomers actuate can be divided into two contributions, which depend sensitively on the D-EAP configuration. In typical electroactuation tests, conformal electrodes (e.g., silver nanoparticles embedded in silicone grease or carbon nanotubes/fibers) that do not impede the ability of the D-EAP to electroactuate are applied to the top and bottom surfaces of a D-EAP film to yield an aligned circular active area with conductive leads. Once an electric field is applied and the electrodes become oppositely charged, the electrodes attract each other, thereby introducing a compressive Maxwell stress over the active area, as schematically illustrated in Fig. 17.22. Since electroactuation of D-EAPs is considered to be an isochoric process, squeezing the film in the z-direction is accompanied by lateral expansion. In addition, charge buildup along each electrode likewise promotes lateral expansion. At sufficiently high electric fields in this test configuration, the result of electroactuation is that the D-EAP thins in the normal direction and expands laterally. For an ideal D-EAP, the magnitude of the Maxwell stress (σ_M) is given [160,161] by $\varepsilon_o \varepsilon E^2$, where ε_o is the dielectric permittivity of vacuum (8.85×10^{-12} F/m), ε is the relative dielectric permittivity (dielectric constant) of the D-EAP, and E is the magnitude of the electric field. While this relationship suggests that electroactuation could be greatly enhanced by using polar D-EAPs with high dielectric constants, this is not necessarily the case, since the modulus of the D-EAP likewise plays a critical, but implicit, role. For materials that are capable of giant electroactuation strains, constitutive equations are used to determine the corresponding thickness strain (s_z). Two other important performance metrics associated with ideal D-EAPs include the energy density, given by $\sigma_M \ln(1 + s_z)$, and the electromechanical coupling efficiency (a measure of how much electrical energy is converted to mechanical energy), obtained from $-2s_z - s_z^2$ [178].

Low-dielectric TPEGs derived from a wide variety of styrenic TPEs selectively swollen with MO have been examined as D-EAPs, and their highly tunable modulus provides a tremendously facile means by which to control electroactuation performance, ranging

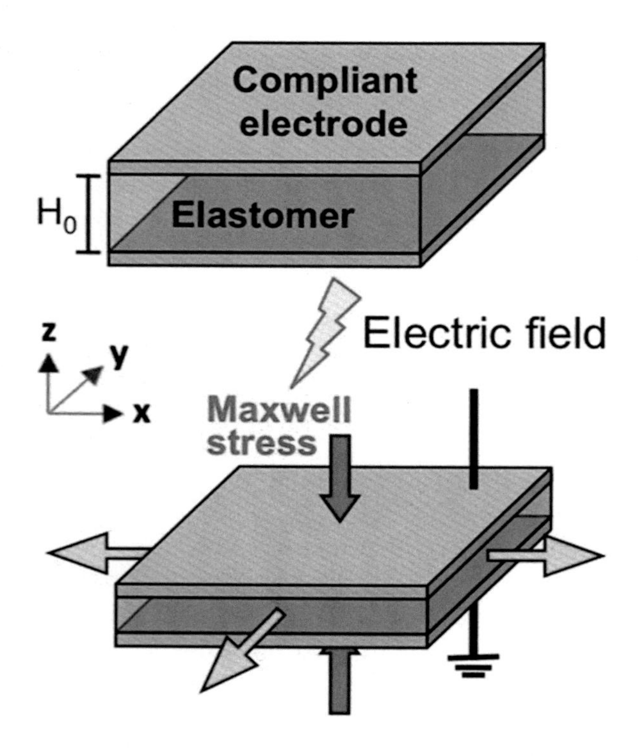

FIGURE 17.22 Schematic illustration of the mechanism by which D-EAPs electroactuate in the presence of an applied electric field. An elastomer of initial thickness (H_0) is coated with conformal electrodes that laterally expand with the D-EAP as it is compressed due to the onset of a normal Maxwell stress along the z-axis. *Reproduced with permission from D.P. Armstrong, R.J. Spontak, Designing dielectric elastomers over multiple length scales for 21st century soft materials technologies, Rubber Chem. Technol. 90 (2017) 207–224. Copyright 2017 American Chemical Society.*

from high actuation strain at low electric fields to low actuation strains at high electric fields [114]. To put some of these metrics into perspective, Fig. 17.23 displays the maximum actuation strain as a function of electric field for several important D-EAP classes [179] (acrylic and silicone elastomers, as well as TPEGs) and clearly establishes that styrenic TPEGs possess the most highly tailorable electroresponsiveness. In addition, TPEGs composed of different TPSs with MO can, depending on copolymer molecular weight and composition, achieve energy densities of up to $\sim 2\ MJ/m^3$ and electromechanical coupling efficiencies exceeding 90%, while exhibiting relatively low cycling hysteresis and short response times ($\sim$20 ms) [180]. From data such as those shown in Fig. 17.23, electroactuation stress–strain curves can be readily generated and used to extract electromechanical moduli [130]. Although the magnitude of these moduli depends on the extent of mechanical prestrain required (to thin specimen films so that lower voltages can be safely used to achieve necessarily high electric fields), they exhibit scaling behavior with respect to composition that is comparable to that previously observed for TPEGs subjected to low-strain dynamic rheology (*cf.* Fig. 17.16A), implying that the electromechanical properties of TPEGs directly relate to their mechanical analogs. Even without prestrain, these TPEGs exhibit higher actuation strain than the gold standard, an acrylic elastomer that is manufactured as an adhesive. If the low-viscosity MO is substituted with the viscoelastic CR mentioned earlier, the actuation and

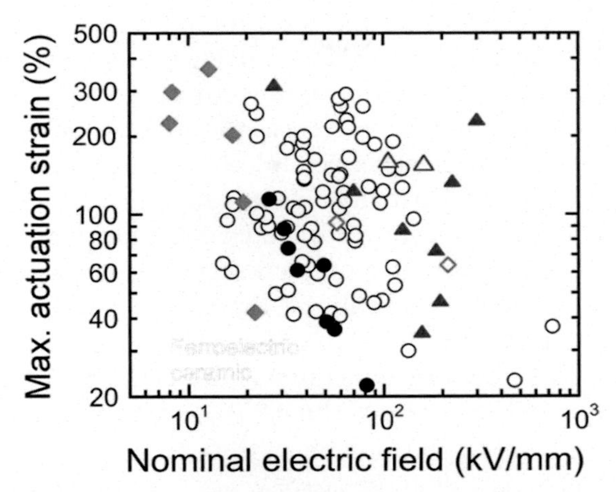

FIGURE 17.23 Compilation of maximum actuation strain values as a function of nominal electric field for styrenic TPEGs (*open circles*) and acrylic TPEGs (*filled circles*). Included for comparison are data from unmodified acrylic elastomers (*open triangles*) and chemically modified acrylic elastomers (*filled triangles*), as well as linear silicone elastomers (*open diamonds*) and bottlebrush silicone elastomers (*filled diamonds*). *Reproduced with permission from D.P. Armstrong, R.J. Spontak, Designing dielectric elastomers over multiple length scales for 21st century soft materials technologies, Rubber Chem. Technol. 90 (2017) 207–224. Copyright 2017 American Chemical Society.*

relaxation response times are considerably longer due to the higher solvent viscosity, but the energy density can increase beyond 8 MJ/m^3. Other TPEGs provided in Fig. 17.23 include ATPEGs swollen with DOP. One of their signature characteristics is that, because of the high dielectric constant of DOP, these materials can attain electroactuation strains and efficiencies beyond 100% and 75%, respectively, in the presence of relatively low electric fields (<30 kV/mm) without any mechanical prestrain [116]. This feature makes ATPEGs particularly well suited for dynamic applications such as active displays. While crystallizable OTPEGs are not included in Fig. 17.23, they afford a unique opportunity to induce highly anisotropic actuation upon crystal alignment due to mechanical conditioning [117].

While styrenic TPEGs can serve as stimuli-responsive D-EAPs possessing highly tunable electroactuation properties, they can also be of considerable value as flexible and extendable conductive soft materials for use as stretchable wires/antennae [181,182]. For this purpose, fully enclosed channels can be thermally molded into a TPEG film in the same fashion as reported [127] for microfluidics, and a liquid metal such as eutectic gallium indium (EGaIn) can be injected into the channels to yield a conductive medium that exhibits both low mechanical and electrical hysteresis upon strain cycling. In comparison to similar constructs fabricated from other elastomers [183], the metal-containing TPEG is capable of achieving much larger strains (at least 600%, which was limited by the measurement and not by the material). A more conventional means to transform a TPEG into a conductive material is through physical addition of conductive nanoparticles, such as carbon nanomaterials [184]. The scanning electron microscopy (SEM) images displayed in Fig. 17.24 demonstrate how vapor-grown carbon nanofiber (CNF) is distributed at different loading levels throughout a TPEG. As expected, the tensile modulus increases, whereas the resistivity decreases, with increasing CNF content. Cyclic electromechanical measurements indicate that these materials exhibit negative piezoresistivity, in which the resistivity systematically decreases (and the conductivity increases) with increasing strain until the maximum strain is reached and the strain reverses to complete each cycle. [The electrical signal is, however, very noisy, reflecting the likelihood that the MO coats the CNF.] Negative piezoresistivity is commonly encountered in many conductive polymer nanocomposites and indicates that additional conductive contacts are created as the polymer matrix is deformed. If a nonsolvated TPS replaces the TPEG, strain-reversible piezoresistivity is

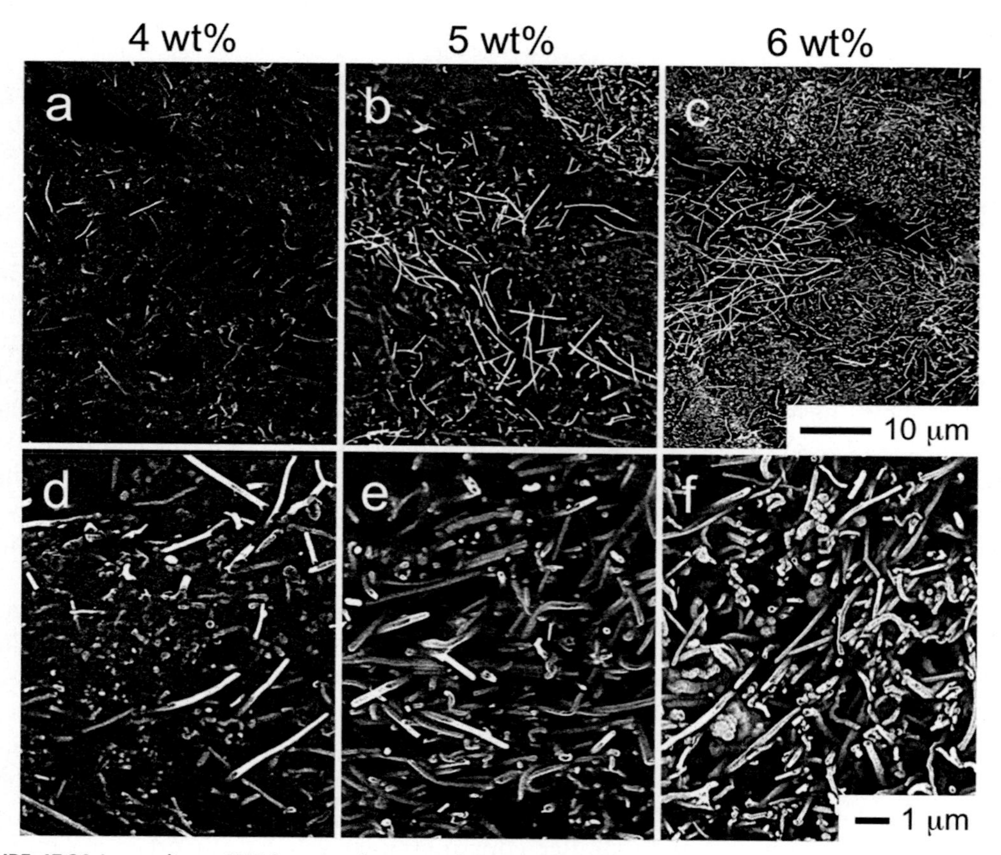

FIGURE 17.24 Low-voltage SEM images of uncoated TPEGs (*dark matrix*) containing vapor-grown carbon nanofiber (CNF) at different loading levels (labeled). Included here are two different magnifications in (A–C) and (D–F). *Reproduced with permission from A. Turgut, M.O. Tuhin, O. Toprakci, M.A. Pasquinelli, R.J. Spontak, H.A.K. Toprakci, Thermoplastic elastomer systems containing carbon nanofibers as soft piezoresistive sensors, ACS Omega 3 (2018) 12648–12657. Copyright 2018 American Chemical Society.*

observed from low-noise cyclic electromechanical measurements that are highly reproducible at low and high stretch rates. In this case, negative piezoresistivity is evident as new conductive contacts develop but positive piezoresistivity ensues as the anisotropic CNF separates (and contacts lessen) before the cycle half-time when the applied deformation is reversed, as indicated in Fig. 17.25 for two maximum strain levels. Such strain-reversible piezoresistivity is uncommon but has been previously reported [185] in other elastomers.

These applications of responsive TPEGs require electricity for the materials to respond cyclically. In this section, the stimulus needed to trigger a single material response is heat. One class of such stimuli-responsive polymers is said to possess shape memory, since they can be deformed from one strain state, fixed in a different strain state, and then stimulated to return to a prior strain state. This mechanism requires two material properties: net points (from a network) that permit the material to return from one strain state to another, and switch points that activate upon exposure to an external stimulus. While different forms of stimulation (e.g., light [186] or pH [187]) could be used for this purpose, the most common stimulus is related to heat, directly [188,189] or through the use of an intermediate species such as nanoparticles that emit heat locally when stimulated by light at their absorption wavelength [190,191]. To extend the concept of TPEGs to shape-memory polymers, they can be swollen with a crystallizable midblock-selective solvent, such as a long-chain hydrocarbon (HCn) with a T_m that depends on the number of carbon units per molecule (n) [136]. In this case, each TPS/HCn blend will behave as a typical, hyperelastic TPEG at $T > T_m$, but as a solid at $T < T_m$, as illustrated in Fig. 17.26. A complete time-dependent strain cycle over the course of slightly less than 5 min is presented in Fig. 17.27A and verifies that the strain fixity and strain recovery levels are both very high

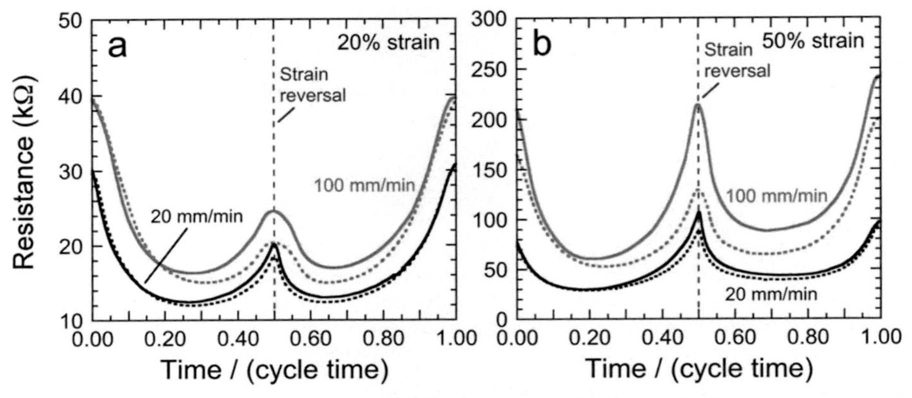

FIGURE 17.25 Single-cycle results from cyclic electromechanical measurements of a styrenic TPE embedded with CNF at strain levels of (A) 20% and (B) 50%. These test results correspond to two different crosshead speeds (in mm/min)—20 (black) and 100 (red)—for the first and final cycles (*solid and dotted lines*, respectively). *Adapted from A. Turgut, M.O. Tuhin, O. Toprakci, M.A. Pasquinelli, R.J. Spontak, H.A.K. Toprakci, Thermoplastic elastomer systems containing carbon nanofibers as soft piezoresistive sensors, ACS Omega 3 (2018) 12648–12657. Copyright 2018 American Chemical Society.*

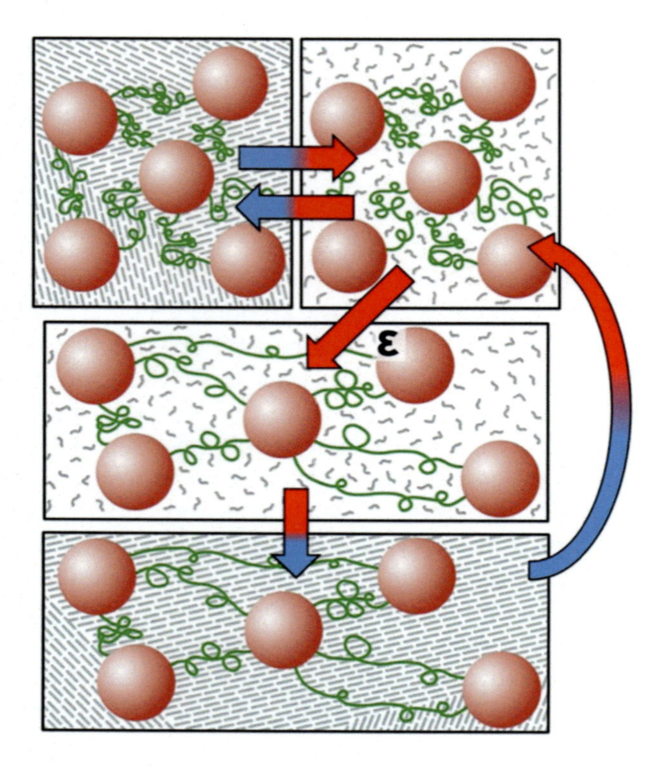

FIGURE 17.26 Schematic illustration of the shape-memory mechanism involving TPEGs. Materials are produced with a crystallizable midblock-selective species (e.g., a long-chain hydrocarbon in the case of styrenic TPEs) above its T_m to ensure TPEG formation. Upon cooling below T_m, the material reversibly solidifies. At temperatures above T_m, the resulting TPEG can be deformed to a new strain state and fixed in that state upon cooling below T_m. Subsequent heating above T_m causes the strained material to return to a previous strain state. *Reproduced with permission from K.P. Mineart, S.S. Tallury, T. Li, B. Lee, R.J. Spontak, Phase-change thermoplastic elastomer blends for tunable shape memory by physical design, Ind. Eng. Chem. Res. 55 (2016) 12590–12597. Copyright 2016 American Chemical Society.*

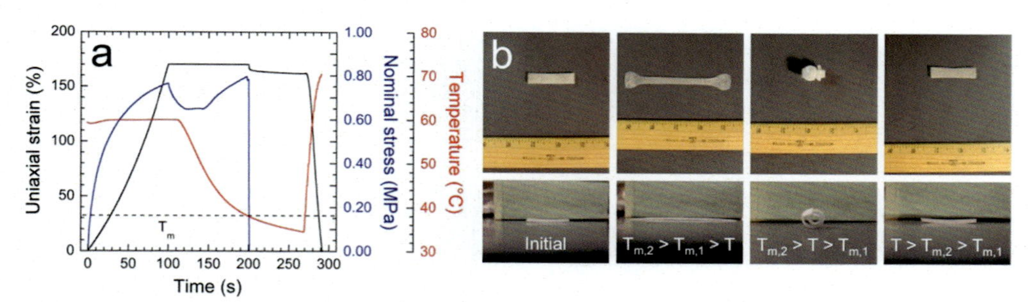

FIGURE 17.27 In (A), a time-resolved shape-memory cycle for a commercial SEBS copolymer containing a midblock-selective hydrocarbon with a known T_m. Initially, the temperature (T) is increased so that $T > T_m$, after which the material is strained to a target level and the stress is monitored. The material is then cooled, and the stress is removed once $T < T_m$. The corresponding change in strain yields the strain fixity. The material is subsequently heated to above T_m and the strain level drops, which provides the strain recovery. In (B), examples of a bilayered laminate composed of two welded shape-memory materials with different T_m values and the intermediate structures achievable when T lies between these two values upon reheating. *Adapted from K.P. Mineart, S.S. Tallury, T. Li, B. Lee, R.J. Spontak, Phase-change thermoplastic elastomer blends for tunable shape memory by physical design, Ind. Eng. Chem. Res. 55 (2016) 12590–12597. Copyright 2016 American Chemical Society.*

(>95%). Thermal calorimetry likewise confirms that T_m remains virtually unaffected when mixed with the TPS so that precise trigger points can be ensured on the basis of the HCn selected. Since the TPEG is thermally weldable above the highest T_m in a given system, laminates, or sandwich structures, can be produced in which each layer possesses a different HCn and, thus, a different T_m. In this fashion, intermediate strain states and structures can be accessed by heating to an intermediate temperature (*cf.* Fig. 17.27B). Incorporation of a liquid metal into these crystallizable TPEGs yields shape-memory wires or antennae, and addition of a CR promotes a significant and controllable reduction in recovery kinetics. These modified TPEGs differ greatly from conventional shape-memory polymers in that the trigger temperature can be accurately varied through the choice of integrated HCn and not by the chemistry of designer macromolecules that do not permit tunable trigger temperatures.

17.4 Chemical modification of nonpolar TPEs and their applications

Block copolymers are often used as compatibilizing agents to reduce interfacial tension and the size of dispersed phase domains in phase-separated polymer blends [192–194]. By localizing along the interface, they promote adhesion through chain entanglements and thus improve mechanical properties. Unfortunately, as evinced in Fig. 17.28, straightforward use of styrenic TPEs do not always promote compatibilization if the timescale for diffusion to the polymer/polymer interface is insufficiently short, in which case the TPE molecules self-assemble instead [195]. Since few block copolymers are available for specific polymer blends, customizable compatibilizers have become increasingly important. One successful example incorporates maleic anhydride into TPEs so as to introduce polarity to a nonpolar block copolymer [196]. In this spirit, the strategy discussed here relies on the postfunctionalization of premade TPEs possessing well-defined molecular attributes. In the case of TPSs, both of the styrenic endblocks and unsaturated polydiene midblocks are amenable to modification. First, we consider chemically altering the midblock of a partially hydrogenated SIS copolymer (partial hydrogenation helps to prevent undesirable chemical crosslinking) [197]. One promising chemical approach to do this relies on thiol–ene click chemistry, which affords mild reaction conditions, relatively high yields, and rapid reaction rates [198]. By applying thiol–ene click chemistry, a library of polar moieties can be grafted onto nonpolar TPEs to enhance their compatibility with various polymers and polymer blends [199,200]. One illustrative exemplar uses the thiol–ene click chemistry depicted in Fig. 17.29A to generate esterified SEPS, which is then suitable for toughening poly(lactic acid) (PLA). The inherent brittleness and poor mechanical properties of PLA prevent this sustainable bioplastic from being used to replace conventional plastics derived from fossil fuel sources in numerous commodity applications. The esterified SEPS constitutes a facile and low-cost route by which to rubber-toughen PLA and improve its mechanical

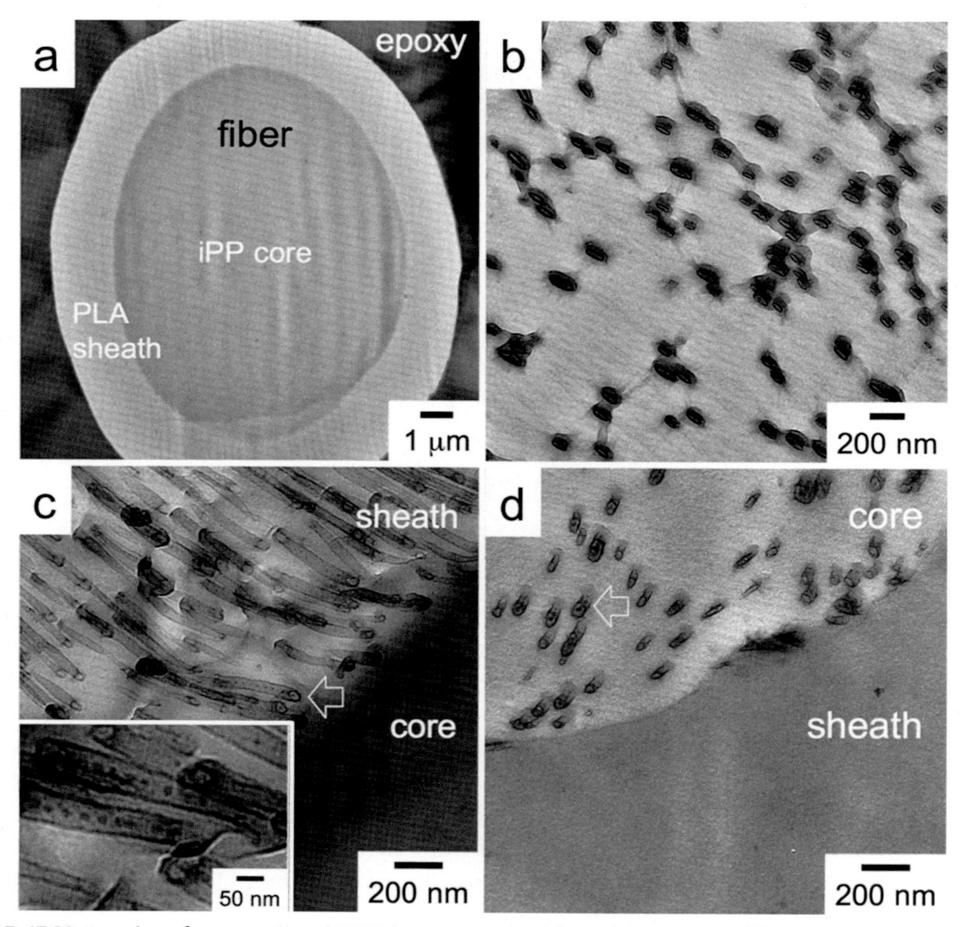

FIGURE 17.28 A series of cross-sectional TEM images acquired from bicomponent fibers composed of isotactic polypropylene (iPP) and poly(lactic acid) (PLA). The entire fiber is visible in (A), and a reference displaying an SEBS copolymer in PLA is included in (B). The unexpected morphologies of the copolymer in (C) PLA-sheath and (D) PLA-core fibers reveal the effect of nonequilibrium processing on copolymer self-assembly. The "peas-in-a-pod" (micelles within nanotubes) arrangement is evident in the inset of (C). *Reproduced with permission from S.A. Arvidson, K.E. Roskov, J.J. Patel, R.J. Spontak, S.A. Khan, R.E. Gorga, Modification of melt-spun isotactic polypropylene and poly(lactic acid) bicomponent filaments with a premade block copolymer, Macromolecules 45 (2012) 913–925. Copyright 2012 American Chemical Society.*

properties [197]. Representative ^{1}H NMR spectra confirming ester substitution on the midblock are provided in Fig. 17.29B before reaction and Fig. 17.29C after reaction, and several mechanical performance metrics are included in Fig. 17.29D and E. According to these results, a Goldilocks composition range exists from 1 to 5 wt% copolymer wherein the tensile strength does not vary noticeably, but the elongation at break and fracture toughness increase by about 2300% and 1500%, respectively, at 1 wt%. These results are not only competitive with more exotic means of toughening PLA, but the design paradigm is sufficiently general and can also be extended to other polyesters as well.

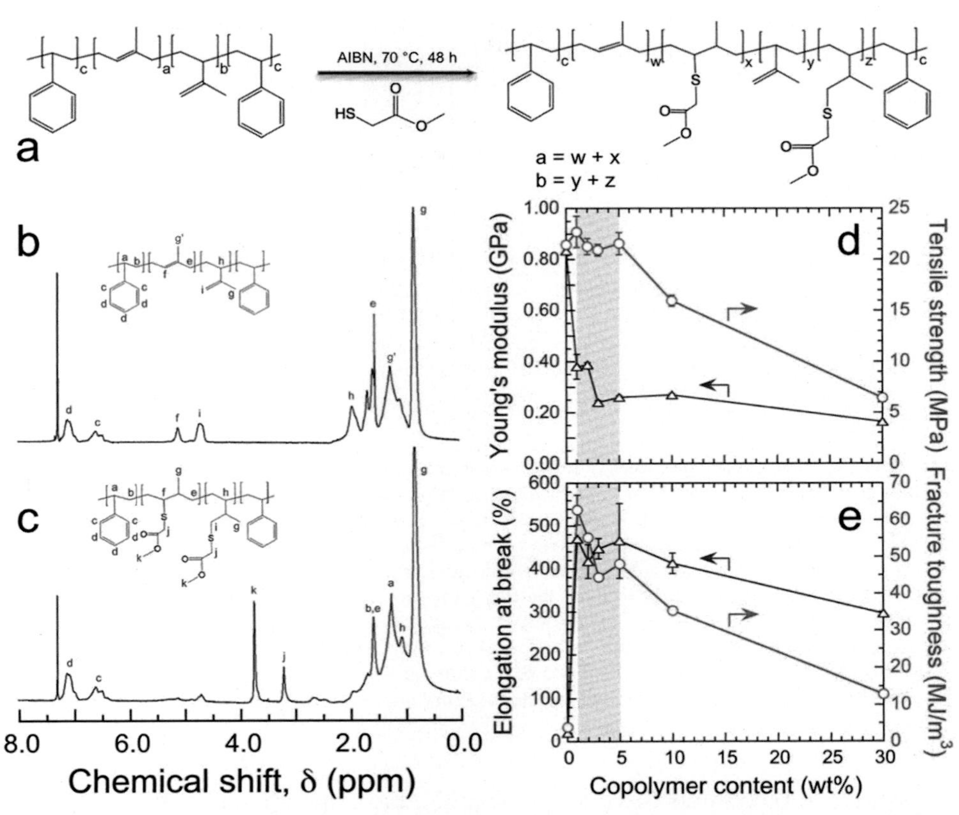

FIGURE 17.29 In (A), the thiol–ene click reaction used to esterify the midblock of a TPE SIS copolymer (the actual material used is partially hydrogenated to prevent undesirable side reactions). The ^{1}H NMR spectra presented in (B) and (C) confirm the chemical conversion. Resultant mechanical properties of PLA modified with the esterified copolymer are included in (D) and (E) and confirm that properties associated with fracture resistance are greatly improved. *Adapted from J. Yan, R.J. Spontak, Toughening poly(lactic acid) with thermoplastic elastomers modified by thiol–ene click chemistry, ACS Sustain. Chem. Eng. 7 (2019) 10830–10839. Copyright 2019 American Chemical Society.*

Similarly diverse methods designed to functionalize the styrenic endblocks of TPSs are available, and one of the most versatile routes involves chloromethylation [201]. Although the presence of a halogen on a TPS is advantageous for functional membranes and ion exchange resins [202], this modification is amenable to further chemical reaction to introduce, for example, phosphonium or ammonium cations, which could interact strongly with polyamides. The 90 degrees peel strength results provided in Fig. 17.30A indicate that these chemical alterations to an SEBS copolymer greatly benefit the compatibilization of polar nylon-6 and nonpolar linear low-density polyethylene (LLDPE) [203]. Selective sulfonation of styrenic TPEs, however, constitutes one of the most mature routes to alter the functionality of inherently nonpolar macromolecules by making them amphiphilic [204,205]. While most sulfonation efforts (including the

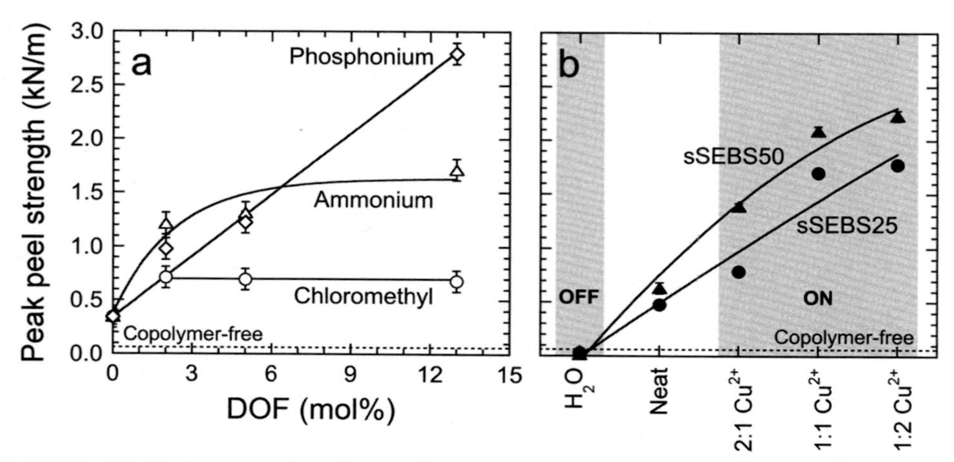

FIGURE 17.30 In (A), 90 degrees peak peel strength results from laminates of polar nylon-6 and nonpolar LLDPE separated by a thin film of SEBS copolymer in which the endblocks are chemically functionalized (labeled) and homopolymer adhesion is clearly improved. Similar results measured from the same laminates separated by an endblock-sulfonated SEBS copolymer (sSEBSn, where n denotes the DOS) are displayed in (B). While the neat sSEBS copolymer moderately improves interfacial strength, two other possibilities are possible: "easy-off" (shaded blue), referring to the addition of water to swell the copolymer and reduce interfacial adhesion by decompatibilization, and "easy-on" (shaded gold), referring to the addition of Cu^{+} ions to the copolymer to improve adhesion by compatibilization. *Adapted from R.J. Spontak, J.J. Ryan, Polymer blend compatibilization by the addition of block copolymers, in: S. Thomas, A.R. Ajitha (Eds.), Compatibilization of Polymer Blends: Micro and Nano Scale Phase Morphologies, Interphase Characterization and Ultimate Properties; Elsevier, Amsterdam, (2020) 57–102. Copyright 2020 Elsevier.*

seminal studies of Weiss and coworkers [206,207]) have specifically targeted the styrenic endblocks of TPSs, targeted sulfonation of unsaturated midblocks has also been successfully achieved (without chemical cross-linking) through the use of a 1,4-sulfur trioxide-1,4-dioxane complex [208], and the resulting material is capable of substantial swelling in water without dissolving since the properties of the nonpolar styrenic endblocks are not compromised (as they would be in the case of endblock-sulfonated materials). The major drawback of endblock-sulfonated TPSs is that, in the presence of a polar liquid, the endblocks responsible for network stabilization either become plasticized or, depending on the degree of sulfonation (DOS), dissolve so that the network altogether fails. A relatively recent solution to this shortcoming requires the introduction of a multivalent cation that can complex with and physically cross-link the sulfonated styrenic endblocks [209]. This strategy permits styrenic TPEGs to remain stable at temperatures far above the T_g of polystyrene. An alternative solution focuses on a TPS composed of a midblock-sulfonated pentablock polymer: poly[*tert*-butylstyrene-*b*-(ethylene-*alt*-propylene)-*b*-(styrene-*co*-styrenesulfonate)-*b*-(ethylene-*alt*-propylene)-*b*-*tert*-butylstyrene] (TESET). Since the *tert*-butylstyrene endblocks cannot be sulfonated, they remain intact in the presence of a polar liquid and maintain the integrity of the swollen molecular network. The EP intermediate blocks serve a vital role in that they prevent this material from becoming brittle, as is the case of an analogous triblock copolymer [210]: poly(*tert*-butylstyrene-*b*-(styrene-*co*-styrenesulfonate)-*b*-*tert*-butylstyrene) (TST).

While sulfonation can promote the development of unexpected, but nonetheless interesting, properties in TPSs, here we first consider the use of sulfonated copolymers for compatibilization and decompatibilization, as discerned from peel tests similar to those mentioned above. We refer to this class of compatibilizers in Fig. 17.30B as "easy-on/easy-off" since this designation accurately describes the uniquely contrary ability of these materials to either promote or reduce interfacial adhesion between chemically dissimilar polymers. Endblock sulfonation of an SEBS triblock copolymer, for example, introduces charged species that can interact with a polar polymer (such as nylon-6), whereas the olefinic midblock would prefer to interact with a nonpolar polymer (such as LLDPE). The result, as with the cationic TPSs generated from endblock-chloromethylated SEBS, is a considerable net reduction in interfacial tension and, hence, improved compatibilization between the two incompatible homopolymers (the 90 degrees peak peel strength between nylon-6 and LLDPE without a compatibilizer is very low, 0.06 kN/m). A practical challenge that arises with sulfonated TPSs is, however, that they are sensitive to thermal degradation at elevated temperatures. One way to overcome this drawback and concurrently improve compatibilization is through metal ion complexation [209]. Neutralization of a sulfonated SEBS copolymer with Cu^{2+} cations yields a processable material that increases the nylon-6/LLDPE peak peel strength in Fig. 17.30B to 2.5 kN/m, which is comparable to the phosphonium modification in Fig. 17.30A. In marked contrast, the inherent hydrophilicity of nonneutralized sulfonic acid groups is likewise suitable for a vastly different purpose. By introducing water to trilayered laminates containing a thin middle layer of sulfonated SEBS, the laminates can be readily separated (as the peak peel strength drops to 0.01 kN/m). In this case, selective swelling of the sulfonated SEBS copolymer serves to increase interfacial tension and reduce interfacial adhesion, thereby facilitating separation of the homopolymers. Such decompatibilization is of commercial interest in the separation of bicomponent fibers to yield hollow or nanoscale fibers, depending on the cross-sectional fiber geometry.

17.5 Morphological development and applications of charged TPEs

Here, we examine the morphological characteristics and emerging applications of ion-containing TPSs, specifically those modified with sulfonic acid groups, to identify relevant thermodynamic considerations and opportunities for technological breakthroughs. For this purpose, we focus on the TESET TPS due to its ability to imbibe polar liquids because of the sulfonated midblock and behave as a tough physical hydrogel due to retention of the intact (nonsulfonated) endblocks. While nonsulfonated TPSs can be easily melt-processed without loss of properties, the TESET materials must be cast from solvent, in which case the choice of solvent is critically important. Since the polar and nonpolar blocks are highly incompatible (*cf.* Fig. 17.4), identifying a common solvent is challenging, but tetrahydrofuran (THF) satisfies the requirements. Alternatively, a polar/nonpolar cosolvent of, for instance, toluene and isopropyl alcohol (referred to here as TIPA) can also be used to achieve dissolution, but care must be exercised with regard to solvent templating. Mineart et al. [211] have used a combination of SAXS and small-angle neutron scattering (SANS) to explore the size and composition dependence of TESET micelles in TIPA varying in composition and report that the micellar size increases with increasing toluene content due to nonpolar coronal swelling. Concurrently, the extent to which the isopropyl alcohol partitions between the bulk solvent and the polar core also increases so that resultant films retain this micellar morphology. In marked contrast, THF-cast films exhibit a mixed morphology composed of lamellae and hexagonally packed cylinders [212]. Electron microscopy images of these distinctively different morphologies are displayed for comparison in Fig. 17.31. Results from a DPD computer simulation [45] of the TESET TPS are included in this figure and indicate that the anticipated equilibrium morphology is lamellar, which suggests that the experimental morphologies are nonequilibrated, although the THF-cast films possess lamellae, as discerned from cross-sectional curvature analysis of the corresponding 3D reconstructions. For nonsulfonated TPEs, equilibration is commonly achieved by thermal annealing, but this approach cannot be used here.

Solvent-vapor annealing (SVA) is frequently used to controllably alter the morphologies of block copolymer thin films [213,214] (often measuring <50 nm thick) and occasionally thick films [215,216]. In the same fashion as thermal annealing conducted above the upper T_g of a TPS to enhance molecular mobility, SVA relies on solvent swelling and subsequent plasticization to achieve the same objective. While several solvents are suitable candidates for SVA, the vapor of THF has proven to be the most effective for TESET films cast from either THF or TIPA. In fact, a time sequence of SAXS profiles acquired [45] from a TIPA-cast TESET bulk film is provided in Fig. 17.32A and confirms that the initially micellar morphology rapidly and completely transforms into lamellae after an exposure time of just a few minutes. Longer exposure times serve to refine the lamellar morphology, which is confirmed by the TEM image shown in Fig. 17.32B, and simultaneously increase the grain size. An unexpected and added benefit of SVA is that the orientation of the lamellae also improves with increasing exposure time, as evidenced by the 2D SAXS pattern included in Fig. 17.32C. Since equilibration of charged block copolymers is a well-known challenge, the use of SVA to overcome this difficulty and yield near-equilibrium morphologies in bulk polymer films represents a notable breakthrough that is not limited to the present sulfonated TPS. Two other points warrant mention at this juncture. First, the SVA-induced transformation process is much faster and complete for TIPA-cast films than for THF-cast films since the former are kinetically trapped much further from equilibrium. Second, according to SAXS measurements of SVA-equilibrated films, an increase in the DOS from 26 to 52 mol % yields a systematic increase in the extent of microdomain swelling (as evidenced by a

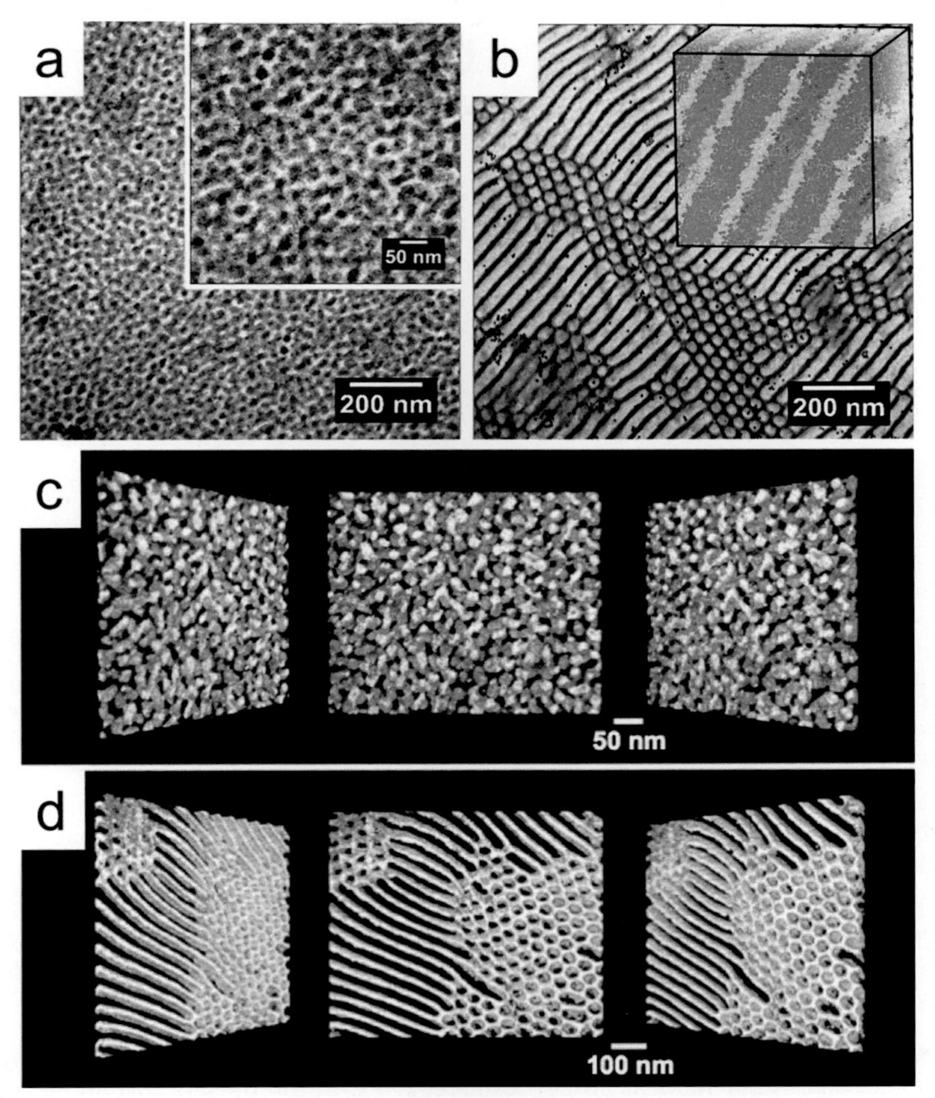

FIGURE 17.31 TEM (A,B) and TEMT (C,D) images collected from the TESET charged pentablock polymer cast from the TIPA cosolvent (A,C) and THF (B,D). The contrast between the TEM and TEMT images is reversed. Included for comparison in (B) are results from a DPD simulation of the morphology under equilibrium conditions. *Adapted from K.P. Mineart, X. Jiang, H. Jinnai, A. Takahara, R.J. Spontak, Morphological investigation of midblock-sulfonated block ionomers prepared from solvents differing in polarity, Macromol. Rapid Commun. 36 (2015) 36, 432–438. K.P. Mineart, B. Lee, R.J. Spontak, A solvent-vapor approach toward the control of block ionomer morphologies, Macromolecules 49 (2016) 3126–3137. Copyright 2015 Wiley and 2016 American Chemical Society.*

corresponding increase in the lamellar period from 38 to 45 nm) [45]. A distinctively different and irreversible transformation that occurs in TIPA-cast TESET films is induced by exposure to liquid water. Upon swelling, the morphology drastically evolves [217]

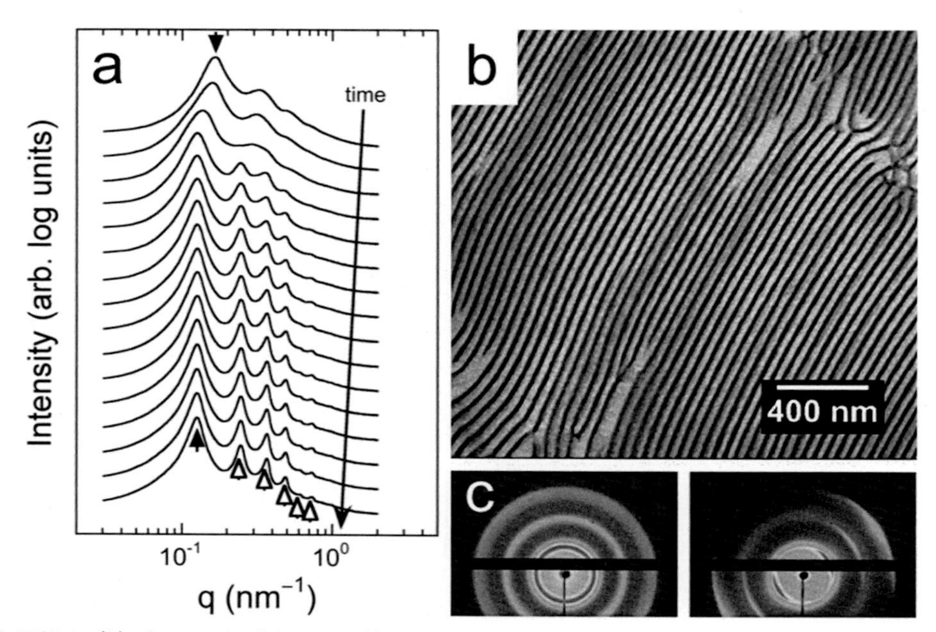

FIGURE 17.32 In (A), time-resolved SAXS profiles in 2 min intervals obtained from the TESET material cast from TIPA and subsequently subjected to SVA in THF at ambient temperature. The peak positions (identified by *arrows*) indicate the transformation to a highly ordered lamellar morphology, which is confirmed by the TEM image displayed in (B). The 2D SAXS patterns in (C) also reveal that the newly formed lamellae can be highly oriented. *Adapted from K.P. Mineart, B. Lee, R.J. Spontak, A solvent-vapor approach toward the control of block ionomer morphologies, Macromolecules 49 (2016) 3126–3137. Copyright 2016 American Chemical Society.*

from discrete micelles to highly interconnected, but irregular, channels that promote superabsorbency at elevated temperature [218].

Now that the factors governing morphological development in the TESET material have been elucidated, we turn our attention to the broad spectrum of applications suitable for this sulfonated TPS. Several applications are classified below according to three contemporary global concerns:

(a) *Energy*. As mentioned earlier, midblock-swollen TPEGs without sulfonation exhibit excellent electromechanical properties as D-EAPs. In the presence of an electrolyte solution, the TESET material can similarly function as an IPMC, a type of ionic EAP [173]. In this scenario, a film is plated with electrodes and solvated Li^+ ions migrate to the oppositely charged electrode upon application of a potential, resulting in solvent-rich and solvent-lean regions of the film and, consequently, a bending motion. Alternatively, they can be fabricated into organic photovoltaic devices that either mimic the performance of natural leaves [219] or operate on the principle of dye-sensitized solar cells (with added TiO_2) in the presence of photosensitive Ru-based dye molecules [220]. Representative photocurrent density–voltage curves are displayed for each design in Fig. 17.33 and reveal that relatively

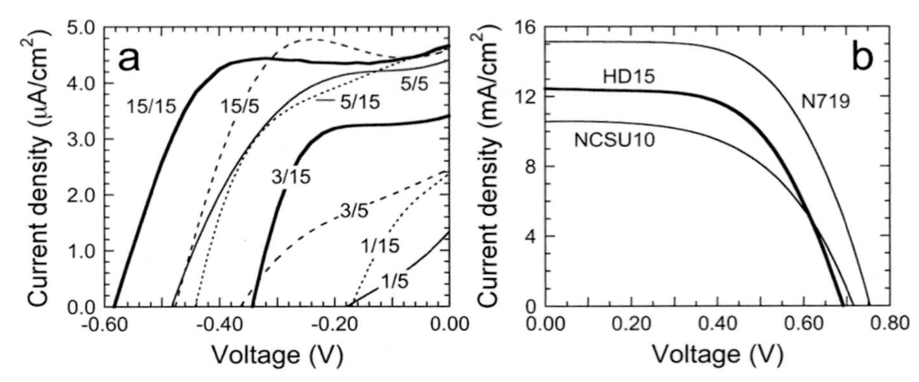

FIGURE 17.33 Photocurrent densities of photosensitive dye-containing TESET materials presented as a function of voltage for two solar cell designs: (A) leaf surrogates, employing only a cationic Ru dye and a sulfonated anionic dye; and (B) dye-sensitized cells, each containing one of three different Ru dyes in the presence of TiO_2. Dye ratios are varied in (A), whereas the HD15 dye in (B) is highlighted (*thick solid line*) because it is hydrophobic, unlike the other two dyes. Note the difference in photocurrent density scale. *Adapted from H.A. Al-Mohsin, K.P. Mineart, R.J. Spontak, Highly flexible aqueous photovoltaic elastomer gels derived from sulfonated block ionomers, Adv. Energy Mater. 5 (2015) 1401941. H.A. Al-Mohsin, K.P. Mineart, D.P. Armstrong, A. El-Shafei, R.J. Spontak, Quasi-solid-state dye-sensitized solar cells containing a charged thermoplastic elastomeric gel electrolyte and hydrophilic/phobic photosensitizers, Sol. RRL 2 (2018) 1700145. Copyright 2015 and 2018 Wiley.*

high fill factors can be achieved in both, but the short-circuit photocurrent density level and net efficiency differ significantly. The highest fill factor, short-circuit current density, and efficiency values reported [220] for nonoptimized dye-sensitized solar cells based on the TESET TPS are 0.67, 15.1 mA/cm^2, and 7.0%, respectively.

(b) *Environment.* One of the most critical concerns facing the world today pertains to global climate change due to the emission of greenhouse gases, particularly CO_2, from various industrial sources [221]. While ethoxylated TPAs have been successfully employed [222,223] to remove CO_2 from mixed gases (since polyethers have an inherently high chemical affinity for CO_2), the TESET TPS, as an anionic macromolecule, operates on a different principle to selectively permeate CO_2. Recall that this material swells considerably, but remains intact, in the presence of water and that water possesses a high CO_2 solubility. Relative to another amphiphilic polyelectrolyte of commercial interest for use in fuel cells (Nafion), the diffusivity and solubility of water are higher in the TESET material, especially at high humidity levels [224]. Since permeation equals the product of diffusivity and solubility in the solution-diffusion regime, this combination results in reasonably high CO_2 permeation and CO_2/N_2 selectivity, where ideal selectivity denotes the ratio of the two penetrant species being compared. Through hydrothermal treatment and the irreversible morphological transformation mentioned earlier [217], the permeation of CO_2 can be increased substantially from less than 100 Barrer to nearly 500 Barrer at high humidity levels. Incorporation of a hydrophilic/CO_2-philic ionic liquid such as 1-butyl-3-methylimidazolium tetrafluoroborate ([Bmim][BF_4]) into TESET membranes provides a facile avenue to improve CO_2/N_2 selectivity up to ~130 [225]. In addition to selectively permeating CO_2, TESET membranes are remarkably effective at permeating a basic gas such as NH_3 [224]. At high humidity levels, the permeability of NH_3 exceeds 5000 Barrer and the NH_3/N_2 selectivity approaches 1900, making the TESET membrane suitable for "methane sweetening" of biogas.

(c) *Healthcare*. The global healthcare system is presently facing one of its greatest challenges, the COVID-19 pandemic [226], which has paralyzed much of the world since early 2020 and resulted in over 5.4 million lost lives as of the time of this writing [227]. While efforts to apply antibacterial methods to thwart the spread of the SARS-CoV-2 virus (prior to the production of safe vaccines) have been met with sporadic success, a new healthcare paradigm has emerged and focuses on infection prevention through the use of self-disinfecting materials. [50,228]' Although the SARS-CoV-2 virus is spread primarily by the droplets and aerosols emanating from a person's nose and/or mouth upon coughing or sneezing, independent studies have established [229,230] that the virus can survive for extended periods of times (days or more) on common surfaces, where it can be transmitted due to direct contact. Indeed, this is the route by which many infective microbes proliferate, in which case the development of self-disinfecting coatings can help to prevent the spread of disease, especially in healthcare facilities and highly populated locales. A plethora of different materials have been developed for this purpose, and many of them have been successful at combating specific microbes [231–233]. The fight against pathogens must, however, address a wide range of bacteria (including those that are or are becoming antibiotic-resistant [234]), (non) enveloped viruses and fungi, but the inactivation mechanisms employed by current antimicrobial materials simply cannot meet this requirement, a fact that becomes suddenly alarming when pathogens undergo mutation and successful vaccinations fail to protect [235]. As an anionic macromolecule, the TESET material possesses a truly unique antimicrobial mechanism: in the presence of moisture, protons from the sulfonic acid groups migrate to the surface and create a very hostile, low-pH (<1.0) environment that comprehensively and nonspecifically inactivates a broad spectrum of microbes, such as Gram-positive/negative bacteria (including the commonly fatal *C. difficile*), antibiotic-resistant bacteria (including methicillin-resistant *S. aureus*, also known as the "superbug" MRSA), highly contagious viruses (including SARS-CoV-2 and influenza-A), and dangerous fungi (including *A. niger*, which is responsible for black mold) [50,228]. The remarkable aspect of this self-disinfecting material is that the exposure needed to achieve minimum detection (often relating to 99.9999% inactivation) is 5 min or less. The survivability of various microbes illustrating this exceptional level of performance is presented in Fig. 17.34. This discovery by Peddinti et al. [50] completely dismisses the contention that anionic macromolecules cannot be antimicrobial (in comparison to their cationic counterparts) and introduces a revolutionary design paradigm that brings a TPE to centerstage in the escalating war against pathogens. This approach has been successfully translated [228] to selectively sulfonated

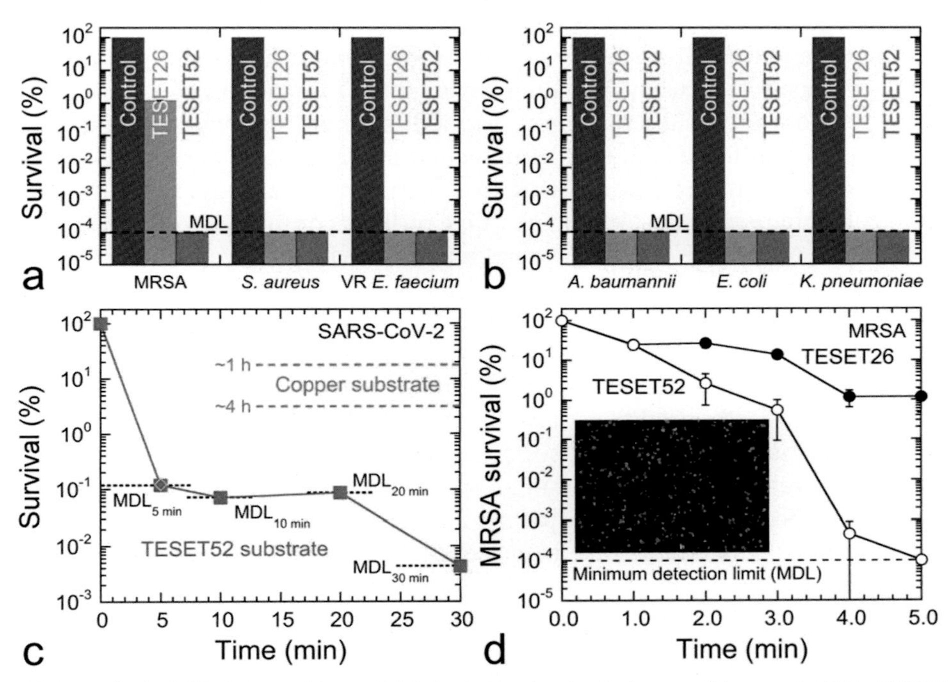

FIGURE 17.34 Survival of (A) Gram-positive and (B) Gram-negative bacteria comprising most of the *ESKAPE* family of bacteria [228] primarily responsible for nosocomial infections after exposure to two different TESET surfaces (TESETn, where n denotes the DOS, labeled) for 5 min. Survival kinetics are also provided in (C) and (D) for SARS-CoV-2 and methicillin-resistant *S. aureus* (MRSA), respectively. Included in (C) are values measured for the virus on a Cu substrate [229], and a life-dead assay is provided for MRSA on TESET52 after 5 min in (D). *Adapted from B.S.T. Peddinti, F. Scholle, M.G. Vargas, S.D. Smith, R.A. Ghiladi, R.J. Spontak, Inherently self-sterilizing charged multiblock polymers that kill drug-resistant microbes in minutes, Mater. Horiz. 6 (2019) 2056–2062. B.S.T. Peddinti, S.N. Downs, J. Yan, S.D. Smith, R.A. Ghiladi, V. Mhetar, R. Tocchetto, A. Griffiths, F. Scholle, R.J. Spontak, Rapid and repetitive inactivation of SARS-CoV-2 and human coronavirus on self-disinfecting anionic polymers, Adv. Sci. 8 (2021) 2003503. Copyright 2019 Royal Society of Chemistry and 2021 Wiley.*

bicomponent TPSs, as well as to Nafion. If the sulfonic acid moieties become neutralized due to exposure to free cations during application, the TESET material can be fully rejuvenated upon exposure to a weak aqueous acid. Moreover, since the pH drop responsible for killing microbes does not target specific functionalities on microbes in the same fashion as medications or metals, future microbial resistance is also highly unlikely.

17.6 Concluding remarks

Since their inception, new TPEs have been emerging, and many of them afford excellent properties for a wide range of mature applications, as well as new opportunities for the development of currently needed technologies, such as gas-separation membranes, microfluidic and haptic devices, soft robotics, solar cells, and self-disinfecting surfaces.

This class of materials can effectively serve as building blocks that are simultaneously capable of spontaneous self-assembly and network formation, with little, if any, process waste and no need for curing. At their core, the TPEs considered in this chapter essentially rely on physical cross-links formed by hard (glassy or semicrystalline) microdomains to stabilize a soft, elastic molecular network upon microphase separation of various copolymer molecules differing in chemistry and architecture, in addition to other molecular characteristics. While the range of TPEs continues to expand, physical or chemical modification of existing copolymers provides novel routes to stimuli-responsive and functional macromolecules that benefit from possessing both a nano-structure and a network. A key feature of TPEs responsible for making this class of materials particularly popular in both numerous and diverse applications is their facile melt processability, which can be retained in many of the applications discussed herein if the TPEs are physically modified through the use of additives. In the event that melt processability is compromised because of chemical functionalization, the advantage of the new material relative to its processing requirements must be assessed. In the case of creating highly effective and broad-spectrum antimicrobial materials from sulfonated TPSs, for instance, the need for solvent-based processing represents a small price to pay to improve global healthcare. Similarly, the production of membranes capable of high CO_2 permeability and selectivity via the same process route is clearly worthwhile if climate change can be mitigated. While applications such as these are unequivocally important, the advances in polymer physics and chemistry responsible for such new materials and their properties likewise offer guidance and yield new strategies to design and develop materials that will overcome future challenges. Block copolymers, and TPEs in particular, introduce abundant opportunities for the exploration of next-generation (multi)functional materials that will become increasingly important in the 21st century.

Acknowledgments

This work has been supported by the NC State University Nonwovens Institute and Kraton Corporation, and we thank all of our collaborators for their valuable assistance and insights that have successfully forged TPEs as a new generation of stimuli-responsive and functional soft materials.

References

[1] R.J. Young, P.A. Lovell, Introduction to Polymers, third ed., CRC Press, Boca Raton, FL, 2011.

[2] I.W. Hamley, The Physics of Block Copolymers, Oxford University Press, New York, 1998.

[3] R.J. Spontak, N.P. Patel, Thermoplastic elastomers: fundamentals and applications, Curr. Opin. Colloid Interface Sci. 5 (2000) 333–340.

[4] G. Holden, H.R. Kricheldorf, R.P. Quirk, Thermoplastic Elastomers, third ed., Hanser, Munich, 2004.

[5] A. Hotta, E. Cochran, J. Ruokolainen, V. Khanna, G. Fredrickson, E. Kramer, Y.-W. Shin, F. Shimizu, A. Cherian, P. Hustad, J. Rose, G. Coates, Semicrystalline thermoplastic elastomeric polyolefins:

advances through catalyst development and macromolecular design, Proc. Natl. Acad. Sci. U.S.A. 103 (2006) 15327–15332.

[6] Forum on block copolymers for nanotechnology applications, in: P. Müller-Buschbaum (Ed.), ACS Appl. Mater. Interfaces 9 (2017) 31213–32412.

[7] C.M. Bates, F.S. Bates, 50th anniversary perspective: block polymers—pure potential, Macromolecules 50 (2017) 3–22.

[8] S.I. Rosenbloom, D.T. Gentekos, M.N. Silberstein, B.P. Fors, Tailor-made thermoplastic elastomers: customisable materials via modulation of molecular weight distributions, Chem. Sci. 11 (2020) 1361–1367.

[9] J. Markarian, Processing and recycling advantages drive growth in thermoplastic elastomers, Plast. Adhes. Compound. 6 (2004) 22–25.

[10] C.-C. Kim, H.-H. Lee, K.H. Oh, J.-Y. Sun, Highly stretchable, transparent ionic touch panel, Science 353 (2016) 682–687.

[11] C. Creton, G.J. Hu, F. Deplace, L. Morgret, K.R. Shull, Large-strain mechanical behavior of model block copolymer adhesives, Macromolecules 42 (2009) 7605–7615.

[12] J.E. Puskas, Y. Chen, Biomedical application of commercial polymers and novel polyisobutylene-based thermoplastic elastomers for soft tissue replacement, Biomacromolecules 5 (2004) 1141–1154.

[13] M. El Fray, P. Prowans, J.E. Puskas, V. Alstadt, Biocompatibility and fatigue properties of polystyrene-polyisobutylene-polystyrene, an emerging thermoplastic elastomeric biomaterial, Biomacromolecules 7 (2006) 844–850.

[14] Cathy DZBH, Global thermoplastic elastomers (TPEs) market 2020 with top countries data, BEC Materials (2020). www.becmaterials.com/news/601Global_ Thermoplastic_Elastomers_Market_ 2020_With_Top_Countries_Data.

[15] Thermoplastic elastomer market to reach USD 27.8 billion by 2024, Forencis Research, September 27, 2019. www.forencisresearch.com/press-release/thermoplastic-elastomer-market/.

[16] W.L. Semon, G.A. Stahl, History of vinyl chloride polymers, J. Macromol. Sci. A: Chem. 15 (1981) 1263–1278.

[17] L. Leibler, Theory of microphase separation in block copolymers, Macromolecules 13 (1980) 1602–1617.

[18] M.W. Hamersky, S.D. Smith, A.O. Gozen, R.J. Spontak, Phase behavior of triblock copolymers varying in molecular asymmetry, Phys. Rev. Lett. 95 (2005) 168306.

[19] M.W. Matsen, Effect of architecture on the phase behavior of AB-type block copolymer melts, Macromolecules 45 (2012) 2161–2165.

[20] W. Jiang, Y. Qiang, W. Li, F. Qiu, A.-C. Shi, Effects of chain topology on the self-assembly of AB-type block copolymers, Macromolecules 51 (2018) 1529–1538.

[21] J.O. Akindoyo, M.D.H. Beg, S. Ghazali, M.R. Islam, N. Jeyaratnam, A.R. Yuvaraj, Polyurethane types, synthesis and applications – a review, RSC Adv. 6 (2016) 114453–114482.

[22] J. Datta, P. Kasprzyk, Thermoplastic polyurethanes derived from petrochemical or renewable resources: a comprehensive review, Polym. Eng. Sci. 58 (2018) E14–E35.

[23] L. Bartolomé, J. Aurrekoetxea, M.A. Urchegui, W. Tato, The influences of deformation state and experimental conditions on inelastic behaviour of an extruded thermoplastic polyurethane elastomer, Mater. Des. 49 (2013) 974–980.

[24] M. Charlon, B. Heinrich, Y. Matter, E. Couzigné, B. Donnio, L. Avérous, Synthesis, structure and properties of fully biobased thermoplastic polyurethanes, obtained from a diisocyanate based on modified dimer fatty acids, and different renewable diols, Eur. Polym. J. 61 (2014) 197–205.

[25] K.L. Kull, R.W. Bass, G. Craft, T. Julien, E. Marangon, C. Marrouat, J.P. Harmon, Synthesis and characterization of an ultra-soft poly(carbonate urethane), Eur. Polym. J. 71 (2015) 510–522.

[26] R.J. Cella, Morphology of segmented polyester thermoplastic elastomers, J. Polym. Sci., Polym. Symp. 42 (1973) 727–740.

[27] S.M. Grayson, J.M.J. Fréchet, Synthesis and surface functionalization of aliphatic polyether dendrons, J. Am. Chem. Soc. 122 (2000) 10335–10344.

[28] T. Nishi, T.K. Kwei, Improvement of the impact strength of a blend of poly(vinyl chloride) with copolyester thermoplastic elastomer by heat treatment, J. Appl. Polym. Sci. 20 (1976) 1331–1337.

[29] L. Vogt, F. Ruther, S. Salehi, A.R. Boccaccini, Poly(glycerol sebacate) in biomedical applications—a review of the recent literature, Adv. Health. Mater. (2021) 2002026.

[30] I.A. Carrascal, A. Pérez, J.A. Casado, S. Diego, J.A. Polanco, D. Ferreño, J.J. Martín, Experimental study of metal cushion pads for high speed railways, Construct. Build. Mater. 182 (2018) 273–283.

[31] S. Seyedin, P. Zhang, M. Naebe, S. Qin, J. Chen, X. Wang, M. Razal, Textile strain sensors: a review of the fabrication technologies, performance evaluation and applications, Mater. Horiz. 6 (2019) 219–249.

[32] N.J. Sijbrandi, A.J. Kimenai, E.P.C. Mes, R. Broos, G. Bar, M. Rosenthal, Y. Odarchenko, D.A. Ivanov, P.J. Dijkstra, J. Feijen, Synthesis, morphology, and properties of segmented poly(ether amide)s with uniform oxalamide-based hard segments, Macromolecules 45 (2012) 3948–3961.

[33] S. Armstrong, B. Freeman, A. Hiltner, E. Baer, Gas permeability of melt-processed poly(ether-*block*-amide) copolymers and the effects of orientation, Polymer 53 (2012) 1383–1392.

[34] K.A. Murray, J.E. Kennedy, B. McEvoy, O. Vrain, D. Ryan, R. Cowman, C.L. Higginbotham, Effects of gamma ray and electron beam irradiation on the mechanical, thermal, structural and physicochemical properties of poly(ether-*block*-amide) thermoplastic elastomers, J. Mech. Behav. Biomed. Mater. 17 (2013) 252–268.

[35] A.K. Bhowmick, J.R. White, Thermal, UV- and sunlight aging of thermoplastic elastomeric natural rubber-polyethylene blends, J. Mater. Sci. 37 (2002) 5141–5151.

[36] J. Feldthusen, B. Ivan, A.H.E. Müller, Synthesis of linear and star-shaped block copolymers of isobutylene and methacrylates by combination of living cationic and anionic polymerizations, Macromolecules 31 (1998) 578–585.

[37] S.L. Aggarwal, Structure and properties of block polymers and multiphase polymer systems: an overview of present status and future potential, Polymer 17 (1976) 938–956.

[38] TPE market approaches maturity, Plast. Rubber Wkly. (October 18, 2016).

[39] A.N. Wilkinson, M.L. Clemens, V.M. Harding, The effects of SEBS-*g*-maleic anhydride reaction on the morphology and properties of polypropylene/PA6/SEBS ternary blends, Polymer 45 (2004) 5239–5249.

[40] J.K. Oh, Polylactide (PLA)-Based amphiphilic block copolymers : synthesis, self-assembly, and biomedical applications, Soft Matter 7 (2011) 5096–5108.

[41] H. Fischer, S. Poser, Liquid crystalline block and graft copolymers, Acta Polym. 47 (1996) 413–428.

[42] P. Figueiredo, S. Geppert, R. Brandsch, G. Bar, R. Thomann, R.J. Spontak, W. Gronski, R. Samlenski, P. Müller-Buschbaum, Ordering of cylindrical microdomains in thin films of hybrid isotropic/liquid crystalline triblock copolymers, Macromolecules 34 (2001) 171–180.

[43] C.S. Marvel, G.E. Inskeep, R. Deanin, A. Juve, C. Schroeder, M. Goff, Copolymers of butadiene with halogenated styrenes, Ind. Eng. Chem. 39 (1947) 1486–1490.

[44] X. Wang, M. Goswami, R. Kumar, B.G. Sumpter, J. Mays, Morphologies of block copolymers composed of charged and neutral blocks, Soft Matter 8 (2012) 3036.

[45] K.P. Mineart, B. Lee, R.J. Spontak, A solvent-vapor approach toward the control of block ionomer morphologies, Macromolecules 49 (2016) 3126–3137.

[46] D. Park, C.J. Weinman, J.A. Finlay, B.R. Fletcher, M.Y. Paik, H.S. Sundaram, M.D. Dimitriou, K.E. Sohn, M.E. Callow, J.A. Callow, D.L. Handlin, C.L. Willis, D.A. Fischer, E.J. Kramer, C.K. Ober, Amphiphilic surface active triblock copolymers with mixed hydrophobic and hydrophilic side chains for tuned marine fouling-release properties, Langmuir 26 (2010) 9772–9781.

[47] M.L. Adams, A. Lavasanifar, G.S. Kwon, Amphiphilic block copolymers for drug delivery, J. Pharmacol. Sci. 92 (2003) 1343–1355.

[48] H.A. Al-Mohsin, K.P. Mineart, D.P. Armstrong, R.J. Spontak, Tuning the performance of aqueous photovoltaic elastomer gels by solvent polarity and nanostructure development, J. Polym. Sci. B Polym. Phys. 55 (2017) 85–95.

[49] Z. Dai, J. Deng, H. Aboukeila, J. Yan, L. Ansaloni, K.P. Mineart, M. Giacinti Baschetti, R.J. Spontak, L. Deng, Highly CO_2-permeable membranes derived from a midblock-sulfonated multiblock polymer after submersion in water, NPG Asia Mater. 11 (2019) 1–7.

[50] B.S.T. Peddinti, S.N. Downs, J. Yan, S.D. Smith, R.A. Ghiladi, V. Mhetar, R. Tocchetto, A. Griffiths, F. Scholle, R.J. Spontak, Rapid and repetitive inactivation of SARS-CoV-2 and human coronavirus on self-disinfecting anionic polymers, Adv. Sci. 8 (2021) 2003503.

[51] I.W. Hamley (Ed.), Developments in Block Copolymer Science and Technology, Wiley, Chichester, 2004.

[52] T. Smart, H. Lomas, M. Massignani, M.V. Flores-Merino, L.R. Perez, G. Battaglia, Block copolymer nanostructures, Nano Today 3 (2008) 38–46.

[53] F. Bates, G. Fredrickson, Block copolymer thermodynamics: theory and experiment, Annu. Rev. Phys. Chem. 41 (1990) 525–557.

[54] M.W. Matsen, M. Schick, Self-assembly of block copolymers, Curr. Opin. Colloid Interface Sci. 1 (1996) 329–336.

[55] S.D. Smith, R.J. Spontak, M.M. Satkowski, A. Ashraf, A.K. Heape, J.S. Lin, Microphase-separated poly(styrene-*b*-isoprene)$_n$ multiblock copolymers with constant block lengths, Polymer 35 (1994) 4527–4536.

[56] R.J. Spontak, S.D. Smith, Perfectly-alternating linear $(AB)_n$ multiblock copolymers: effect of molecular design on morphology and properties, J. Polym. Sci. B Polym. Phys. 39 (2001) 947–955.

[57] M. Steube, T. Johann, E. Galanos, M. Appold, C. Ruttiger, M. Mezger, M. Gallei, A.H.E. Müller, G. Floudas, H. Frey, Isoprene/styrene tapered multiblock copolymers with up to ten blocks: synthesis, phase behavior, order, and mechanical properties, Macromolecules 51 (2018) 10246–10258.

[58] C. Wahlen, J. Blankenburg, P. von Tiedemann, J. Ewald, P. Sajkiewicz, A.H.E. Müller, G. Floudas, H. Frey, Tapered multiblock copolymers based on farnesene and styrene: impact of biobased polydiene architectures on material properties, Macromolecules 53 (2020) 10397–10408.

[59] G.M. Grason, R.D. Kamien, Interfaces in diblocks: a study of miktoarm star copolymers, Macromolecules 37 (2004) 7371–7380.

[60] N.A. Lynd, F.T. Oyerokun, D.L. O'Donoghue, D.L. Handlin, G.H. Fredrickson, Design of soft and strong thermoplastic elastomers based on nonlinear block copolymer architectures using self-consistent-field theory, Macromolecules 43 (2010) 3479–3486.

[61] J.E. Poelma, K. Ono, D. Miyajima, T. Aida, K. Satoh, C.J. Hawker, Cyclic block copolymers for controlling feature sizes in block copolymer lithography, ACS Nano 6 (2012) 10845–10854.

[62] G.M. Grason, R.D. Kamien, Self-consistent field theory of multiply branched block copolymer melts, Phys. Rev. E 71 (2005) 051801.

[63] S.J. Diamanti, V. Khanna, A. Hotta, R.C. Coffin, D. Yamakawa, E.J. Kramer, G.H. Fredrickson, G.C. Bazan, Tapered block copolymers containing ethylene and a functionalized comonomer, Macromolecules 39 (2006) 3270–3274.

[64] N. Singh, M.S. Tureau, T.H. Epps III, Manipulating ordering transitions in interfacially modified block copolymers, Soft Matter 5 (2009) 4757–4762.

[65] V. Sethuraman, V. Ganesan, Segmental dynamics in lamellar phases of tapered copolymers, Soft Matter 12 (2016) 7818–7823.

[66] P. von Tiedemann, J. Yan, R.D. Barent, R.J. Spontak, G. Floudas, H. Frey, R.A. Register, Tapered multiblock star copolymers: synthesis, selective hydrogenation, and properties, Macromolecules 53 (2020) 4422–4434.

[67] R. Fenyves, M. Schmutz, I.J. Horner, F.V. Bright, J. Rzayev, Aqueous self-assembly of giant bottlebrush block copolymer surfactants as shape-tunable building blocks, J. Am. Chem. Soc. 136 (2014) 7762–7770.

[68] J.L. Self, C.S. Sample, A.E. Levi, K. Li, R. Xie, J.R. de Alaniz, C.M. Bates, Dynamic bottlebrush polymer networks: self-healing in super-soft materials, J. Am. Chem. Soc. 142 (2020) 7567–7573.

[69] R. Stadler, C. Auschra, J. Beckmann, U. Krappe, I. Voight-Martin, L. Leibler, Morphology and thermodynamics of symmetric poly(A-*block*-B-*block*-C) triblock copolymers, Macromolecules 28 (1995) 3080–3097.

[70] C. Lee, S.P. Gido, M. Pitsikalis, J.W. Mays, N.B. Tan, S.F. Trevino, N. Hadjichristidis, Asymmetric single graft block copolymers: effect of molecular architecture on morphology, Macromolecules 30 (1997) 3732–3738.

[71] V. Abetz, P.F.W. Simon, Phase behaviour and morphologies of block copolymers, in: V. Abetz (Ed.), Block Copolymers I, Springer, Berlin, 2005, pp. 125–212.

[72] F.S. Bates, G.H. Fredrickson, Block copolymers—designer soft materials, Phys. Today 52 (2008) 32.

[73] J. Lu, F.S. Bates, T.P. Lodge, Remarkable effect of molecular architecture on chain exchange in triblock copolymer micelles, Macromolecules 48 (2015) 2667–2676.

[74] W. Wang, W. Lu, A. Goodwin, H. Wang, P. Yin, N.-G. Kang, K. Hong, J.W. Mays, Recent advances in thermoplastic elastomers from living polymerizations: macromolecular architectures and supramolecular chemistry, Prog. Polym. Sci. 95 (2019) 1–31.

[75] D.A. Hajduk, P.E. Harper, S.M. Gruner, C.C. Honeker, G. Kim, E.L. Thomas, L.J. Fetters, The gyroid: a new equilibrium morphology in weakly segregated diblock copolymers, Macromolecules 27 (1994) 4063–4075.

[76] J.H. Laurer, D.A. Hajduk, J.C. Fung, J.W. Sedat, S.D. Smith, S.M. Gruner, D.A. Agard, R.J. Spontak, Microstructural analysis of a cubic bicontinuous morphology in a neat SIS triblock copolymer, Macromolecules 30 (1997) 3938–3941.

[77] A.J. Meuler, M.A. Hillmyer, F.S. Bates, Ordered network mesostructures in block polymer materials, Macromolecules 42 (2009) 7221–7250.

[78] H. Jinnai, Y. Nishikawa, R.J. Spontak, S.D. Smith, D.A. Agard, T. Hashimoto, Direct measurement of interfacial curvature distributions in a bicontinuous block copolymer morphology, Phys. Rev. Lett. 84 (2000) 518–521.

[79] H. Jinnai, R.J. Spontak, Transmission electron microtomography in polymer research, Polymer 50 (2009) 1067–1087.

[80] C.Y. Chu, X. Jiang, H. Jinnai, R.Y. Pei, W.F. Lin, J.C. Tsai, H.L. Chen, Real-space evidence of the equilibrium ordered bicontinuous double diamond structure of a diblock copolymer, Soft Matter 11 (2015) 1871–1876.

[81] C.-H. Lin, T. Higuchi, H.-L. Chen, J.-C. Tsai, H. Jinnai, T. Hashimoto, Stabilizing the ordered bicontinuous double diamond structure of diblock copolymer by configurational regularity, Macromolecules 51 (2018) 4049–4058.

[82] C.A. Tyler, D.C. Morse, Orthorhombic *Fddd* network in triblock and diblock copolymer melts, Phys. Rev. Lett. 94 (2005) 208302.

[83] M. Takenaka, T. Wakada, S. Akasaka, S. Nishitsuji, K. Saijo, H. Shimizu, M.I. Kim, H. Hasegawa, Orthorhombic *Fddd* network in diblock copolymer melts, Macromolecules 40 (2007) 4399–4402.

[84] S. Lee, M.J. Bluemle, F.S. Bates, Discovery of a frank-kasper σ phase in sphere-forming block copolymer melts, Science 330 (2010) 349–353.

[85] A.R. Ashraf, J.J. Ryan, M.M. Satkowski, B. Lee, S.D. Smith, R.J. Spontak, Bicomponent block copolymers derived from one or more random copolymers as an alternative route to controllable phase behavior, Macromol. Rapid Commun. 38 (2017) 1700207.

[86] J.H. Laurer, A. Ashraf, S.D. Smith, R.J. Spontak, Complex phase behavior of a disordered "random" diblock copolymer in the presence of a parent homopolymer, Langmuir 13 (1997) 2250–2258.

[87] U. Staudinger, B.K. Satapathy, M. Thunga, R. Weidisch, A. Janke, K. Knoll, Enhancement of mechanical properties of triblock copolymers by random copolymer middle blocks, Eur. Polym. J. 43 (2007) 2750–2758.

[88] B.S. Beckingham, R.A. Register, Regular mixing thermodynamics of hydrogenated styrene–isoprene block–random copolymers, Macromolecules 46 (2013) 3084–3091.

[89] G.H. Fredrickson, E. Helfand, Fluctuation effects in the theory of microphase separation in block copolymers, J. Chem. Phys. 87 (1987) 697–705.

[90] P. Medapuram, J. Glaser, D.C. Morse, Universal phenomenology of symmetric diblock copolymers near the order–disorder transition, Macromolecules 48 (2015) 819–839.

[91] A.O. Gozen, M.K. Gaines, M.W. Hamersky, P. Maniadis, K.Ø. Rasmussen, S.D. Smith, R.J. Spontak, Controlling the phase behavior of block copolymers via sequential block growth, Polymer 51 (2010) 5304–5308.

[92] A.M. Mayes, M. Olvera de la Cruz, Microphase separation in multiblock copolymer melts, J. Chem. Phys. 91 (1989) 7228–7235.

[93] S.D. Smith, M.W. Hamersky, M.K. Bowman, K.Ø. Rasmussen, R.J. Spontak, Molecularly asymmetric triblock copolymers as a single-molecule route to ordered bidisperse polymer brushes, Langmuir 22 (2006) 6465–6468.

[94] M. Nguyen-Misra, W.L. Mattice, Micellization and gelation of symmetric triblock copolymers with insoluble end blocks, Macromolecules 28 (1995) 1444–1457.

[95] T.L. Chantawansri, T.W. Sirk, Y.R. Sliozberg, Entangled triblock copolymer gel: morphological and mechanical properties, J. Chem. Phys. 138 (2013) 024908.

[96] S.S. Tallury, R.J. Spontak, M.A. Pasquinelli, Dissipative particle dynamics of triblock copolymer melts: a midblock conformational study at moderate segregation, J. Chem. Phys. 141 (2014) 244911.

[97] S.S. Tallury, K.P. Mineart, S. Woloszczuk, D.N. Williams, R.B. Thompson, M.A. Pasquinelli, M. Banaszak, R.J. Spontak, Molecular-level insights into asymmetric triblock copolymers: network and phase development, J. Chem. Phys. 141 (2014) 121103.

[98] P.G. Khalatur, A.R. Khokhlov, I.A. Nyrkova, A.N. Semenov, Aggregation processes in self-associating polymer systems: computer simulation study of micelles in the superstrong segregation regime, Macromol. Theory Simul. 5 (1996) 713–747.

[99] S. Woloszczuk, K.P. Mineart, R.J. Spontak, M. Banaszak, Dual modes of self-assembly in super strongly segregated bicomponent triblock copolymer melts, Phys. Rev. E 91 (2015) 010601.

[100] T. Wright, A.S. Jones, H.J. Harwood, Enhancement of the high-temperature properties of an SEBS thermoplastic elastomer by chemical modification, J. Appl. Polym. Sci. 86 (2002) 1203–1210.

[101] H. Nandivada, X. Jiang, J. Lahann, Click chemistry: versatility and control in the hands of materials scientists, Adv. Mater. 19 (2007) 2197–2208.

[102] J. Romulus, J.T. Henssler, M. Weck, Postpolymerization modification of block copolymers, Macromolecules 47 (2014) 5437–5449.

[103] Y. Araki, D. Shimizu, Y. Hori, K. Nakatani, H. Saito, Mechanical properties and microphase structure of hydrogenated S-SB-S triblock copolymers, Polym. J. 45 (2013) 1140–1145.

[104] R. Velichkova, V. Toncheva, C. Antonov, V. Alexandrov, S. Pavlova, L. Dubrovina, E. Gladkova, Styrene–Isoprene block copolymers. II. Hydrogenation and solution properties, J. Appl. Polym. Sci. 42 (1991) 3083–3090.

[105] C.G. Vonk, Investigation of non-ideal two-phase polymer structures by small-angle X-ray scattering, J. Appl. Crystallogr. 6 (1973) 81–86.

[106] A.R. Ashraf, J.J. Ryan, M.M. Satkowski, S.D. Smith, R.J. Spontak, Effect of systematic hydrogenation on the phase behavior and nanostructural dimensions of block copolymers, ACS Appl. Mater. Interfaces 10 (2018) 3186–3190.

[107] E. Helfand, Y. Tagami, Theory of the interface between immiscible polymers, J. Chem. Phys. 57 (1972) 1812–1813.

[108] A.S. Krishnan, J.H. van Zanten, S. Seifert, B. Lee, R.J. Spontak, Selectively solvated triblock copolymer networks under biaxial strain, Appl. Phys. Lett. 99 (2011) 101908.

[109] Y. Shi, H. Ha, A. Al-Sudani, C.J. Ellison, G. Yu, Thermoplastic elastomer-enabled smart electrolyte for thermoresponsive self-protection of electrochemical energy storage devices, Adv. Mater. 28 (2016) 7921–7928.

[110] M.-S. Park, H.-S. Woo, J.-M. Heo, J.-M. Kim, R. Thangavel, Y.-S. Lee, D.-W. Kim, Thermoplastic polyurethane elastomer-based gel polymer electrolytes for sodium-metal cells with enhanced cycling performance, ChemSusChem 12 (2019) 4645–4654.

[111] L.A. Rankin, B. Lee, K.P. Mineart, Effect of network connectivity on the mechanical and transport properties of block copolymer gels, J. Polym. Sci. 59 (2021) 34–42.

[112] C. Larson, B. Peele, S. Li, S. Robinson, M. Totaro, L. Beccai, B. Mazzolai, R. Shepherd, Highly stretchable electroluminescent skin for optical signaling and tactile sensing, Science 351 (2016) 1071–1074.

[113] R. Shankar, A.K. Krishnan, T.K. Ghosh, R.J. Spontak, Triblock copolymer organogels as high-performance dielectric elastomers, Macromolecules 41 (2008) 6100–6109.

[114] R. Shankar, T.K. Ghosh, R.J. Spontak, Electroactive nanostructured polymers as tunable actuators, Adv. Mater. 19 (2007) 2218–2223.

[115] B. Kim, Y.D. Park, K. Min, J.H. Lee, S.S. Hwang, S.M. Hong, B.H. Kim, S.O. Kim, C.M. Koo, Electric actuation of nanostructured thermoplastic elastomer gels with ultralarge electrostriction coefficients, Adv. Funct. Mater. 21 (2011) 3242–3249.

[116] P.H. Vargantwar, A.E. Özçam, T.K. Ghosh, R.J. Spontak, Prestrain-free dielectric elastomers based on acrylic thermoplastic elastomer gels: a morphological and (Electro)Mechanical property study, Adv. Funct. Mater. 22 (2012) 2100–2113.

[117] D.P. Armstrong, R.J. Spontak, Crystallization-directed anisotropic electroactuation in selectively solvated olefinic thermoplastic elastomers: a thermal and (Electro)Mechanical property study, Adv. Funct. Mater. 28 (2018) 1803467.

[118] B.T. White, T.E. Long, Advances in polymeric materials for electromechanical devices, Macromol. Rapid Commun. 40 (2019) 1800521.

[119] D.T. Casem, A.K. Dwivedi, R.A. Mrozek, J.L. Lenhart, Compression response of a thermoplastic elastomer gel tissue surrogate over a range of strain-rates, Int. J. Solid Struct. 51 (2014) 2037–2046.

[120] R. Shankar, T.K. Ghosh, R.J. Spontak, Dielectric elastomers as next-generation polymeric actuators, Soft Matter 3 (2007) 1116–1129.

[121] Z.I. Kalcioglu, M. Qu, K.E. Strawhecker, T. Shazly, E. Edelman, M.R. VanLandingham, J.F. Smith, K. J. Van Vliet, Dynamic impact indentation of hydrated biological tissues and tissue surrogate gels, Phil. Mag. 91 (2011) 1339–1355.

[122] A.I. Uzar, M. Dakak, T. Ozer, G. Ogunc, T. Yigit, C. Kayahan, K. Oner, D. Sen, A new ballistic simulant “transparent gel candle” (experimental study), Ulus. Travma. Acil. Cerrahi. Derg. 9 (2003) 104–106.

[123] R.A. Mrozek, B. Leighliter, C.S. Gold, I.R. Beringer, J.H. Yu, M.R. VanLandingham, P. Moy, M.H. Foster, J.L. Lenhart, The relationship between mechanical properties and ballistic penetration depth in a viscoelastic gel, J. Mech. Behav. Biomed. Mater. 44 (2015) 109–120.

[124] S. Mishra, R.M.B. Prado, T.E. Lacy, S. Kundu, Investigation of failure behavior of a thermoplastic elastomer gel, Soft Matter 14 (2018) 7958–7969.

[125] J.W. East, B. Knutson, E.V. East Jr., Spinal Injection Trainer and Methods Therefor, U.S. Patent #9,275,556 B1, March 1, 2016.

[126] K.P. Mineart, W.W. Walker, J. Mogollon-Santiana, I.A. Coates, C. Hong, B. Lee, Nanocarrier-loaded block copolymer dual domain organogels, Polymer 214 (2021) 123246.

[127] A.P. Sudarsan, J. Wang, V.M. Ugaz, Thermoplastic elastomer gels: an advanced substrate for microfluidic chemical analysis systems, Anal. Chem. 77 (2005) 5167–5173.

[128] A.E. Covington, K. Dawes, T.A. Hunter, Barrier Material for Gel Sealant-cable Jacket Interface, Eur. Pat. #EP0355108A1, February 28, 1990.

[129] R. Bening, R. Ding, H. Yang, C. Maris, Block Copolymers for Gel Compositions, U.S. Patent #10,336,884 B2, July 2, 2019.

[130] A.S. Krishnan, P.H. Vargantwar, T.K. Ghosh, R.J. Spontak, Electroactuation of solvated triblock copolymer dielectric elastomers: decoupling the roles of mechanical prestrain and specimen thickness, J. Polym. Sci. B Polym. Phys. 49 (2011) 1569–1582.

[131] D. Chen, Q. Pei, Electronic muscles and skins: a review of soft sensors and actuators, Chem. Rev. 117 (2017) 11239–11268.

[132] S. Li, H. Bai, R.F. Shepherd, H. Zhao, Bio-inspired design and additive manufacturing of soft materials, machines, robots, and haptic interfaces, Angew. Chem. Int. Ed. 58 (2019) 11182–11204.

[133] J. Courtois, I. Baroudi, N. Nouvel, E. Degrandi, S. Pensec, G. Ducouret, C. Chaneac, L. Bouteiller, C. Creton, Supramolecular soft adhesive materials, Adv. Funct. Mater. 20 (2010) 1803–1811.

[134] R.A. Mrozek, M.C. Berg, C.S. Gold, B. Leighliter, J.T. Morton, J.L. Lenhart, Highly compliant shape memory polymer gels for tunable damping and reversible adhesion, Smart Mater. Struct. 25 (2016) 025004.

[135] Q. Zhang, S. Song, J. Feng, P. Wu, A new strategy to prepare polymer composites with versatile shape memory properties, J. Mater. Chem. 22 (2012) 24776–24782.

[136] K.P. Mineart, S.S. Tallury, T. Li, B. Lee, R.J. Spontak, Phase-change thermoplastic elastomer blends for tunable shape memory by physical design, Ind. Eng. Chem. Res. 55 (2016) 12590–12597.

[137] S.Y. Chen, Q.L. Zhang, J.C. Feng, 3D printing of tunable shape memory polymer blends, J. Mater. Chem. C 5 (2017) 8361–8365.

[138] A.S. Krishnan, S. Seifert, B. Lee, S.A. Khan, R.J. Spontak, Cosolvent-regulated time–composition rheological equivalence in block copolymer solutions, Soft Matter 6 (2010) 4331–4334.

[139] A.S. Krishnan, R.J. Spontak, Factors affecting time–composition equivalence in ternary block copolymer/cosolvent systems, Soft Matter 8 (2012) 1334–1343.

[140] A.S. Krishnan, S.D. Smith, R.J. Spontak, Ternary phase behavior of a triblock copolymer in the presence of an endblock-selective homopolymer and a midblock-selective oil, Macromolecules 45 (2012) 6056–6067.

[141] S. Woloszczuk, M.O. Tuhin, S.R. Gade, M.A. Pasquinelli, M. Banaszak, R.J. Spontak, Complex phase behavior and network characteristics of midblock-solvated triblock copolymers as physically cross-linked soft materials, ACS Appl. Mater. Interfaces 9 (2017) 39940–39944.

[142] M.O. Tuhin, S. Woloszczuk, K.P. Mineart, M.A. Pasquinelli, J.D. Sadler, S.D. Smith, M. Banaszak, R. J. Spontak, Communication: molecular-level description of constrained chain topologies in multiblock copolymer gel networks, J. Chem. Phys. 148 (2018) 231101.

[143] M.O. Tuhin, J.J. Ryan, J.D. Sadler, Z. Han, B. Lee, S.D. Smith, M.A. Pasquinelli, R.J. Spontak, Microphase-separated morphologies and molecular network topologies in multiblock copolymer gels, Macromolecules 51 (2018) 5173–5181.

[144] J.H. Laurer, J.F. Mulling, S.A. Khan, R.J. Spontak, R. Bukovnik, Thermoplastic elastomer gels. I. Effects of composition and processing on morphology and gel behavior, J. Polym. Sci. B Polym. Phys. 36 (1998) 2379–2391.

[145] D.A. Vega, J.M. Sebastian, Y.-L. Loo, R.A. Register, Phase behavior and viscoelastic properties of entangled block copolymer gels, J. Polym. Sci. B Polym. Phys. 39 (2001) 2183–2197.

[146] P.G. de Gennes, Scaling Concepts in Polymer Physics, Cornell University Press, New York, 1979.

[147] Y. Mori, L.S. Lim, F.S. Bates, Consequences of molecular bridging in lamellae-forming triblock/pentablock copolymer blends, Macromolecules 36 (2003) 9879–9888.

[148] M. Zhong, R. Wang, K. Kawamoto, B.D. Olsen, J.A. Johnson, Quantifying the impact of molecular defects on polymer network elasticity, Science 353 (2016) 1264–1268.

[149] A.K. Pervaje, J.C. Tilly, A.T. Detwiler, R.J. Spontak, S.A. Khan, E.E. Santiso, Molecular simulations of thermoset polymers implementing theoretical kinetics with top-down coarse-grained models, Macromolecules 53 (2020) 2310–2322.

[150] M. Rubinstein, S. Panyukov, Elasticity of polymer networks, Macromolecules 35 (2002) 6670–6686.

[151] J. Yan, M.O. Tuhin, J.D. Sadler, S.D. Smith, M.A. Pasquinelli, R.J. Spontak, Network topology and stability of homologous multiblock copolymer physical gels, J. Chem. Phys. 153 (2020) 124904.

[152] W. Brostow, R. Chiu, I.M. Kalogeras, A. Vassilikou-Dova, Prediction of glass transition temperatures: binary blends and copolymers, Mater. Lett. 62 (2008) 3152–3155.

[153] A.S. Krishnan, R.J. Spontak, Deviation from time-composition equivalence in polymer solutions with selective cosolvents, AIP Adv. 1 (2011) 042159.

[154] C.M. Flanigan, A.J. Crosby, K.R. Shull, Structural development and adhesion of acrylic ABA triblock copolymer gels, Macromolecules 32 (1999) 7251–7262.

[155] M.E. Seitz, W.R. Burghardt, K.T. Faber, K.R. Shull, Self-assembly and stress relaxation in acrylic triblock copolymer gels, Macromolecules 40 (2007) 1218–1226.

[156] M.E. Seitz, D. Martina, T. Baumberger, V.R. Krishnan, C.-Y. Hui, K.R. Shull, Fracture and large strain behavior of self-assembled triblock copolymer gels, Soft Matter 5 (2009) 447–456.

[157] M.E. Seitz, K.R. Shull, K.T. Faber, Acrylic triblock copolymer design for thermoreversible gelcasting of ceramics: rheological and green body properties, J. Am. Ceram. Soc. 92 (2009) 1519–1525.

[158] D.P. Armstrong, K.P. Mineart, B. Lee, R.J. Spontak, Olefinic thermoplastic elastomer gels: combining polymer crystallization and microphase separation in a selective solvent, ACS Macro Lett. 5 (2016) 1273–1277.

[159] F. Carpi, E. Smela (Eds.), Biomedical Applications of Electroactive Polymer Actuators, Wiley, New York, 2009.

[160] F. Carpi (Ed.), Electromechanically Active Polymers: A Concise Reference, Springer, Switzerland, 2016.

[161] R. Pelrine, R. Kornbluh, Q. Pei, J. Joseph, High-speed electrically actuated elastomers with strain greater than 100%, Science 287 (2000) 836–839.

[162] T. Mirfakhrai, J.D.W. Madden, R.H. Baughman, Polymer artificial muscles, Mater. Today 10 (2007) 30–38.

[163] P. Brochu, Q. Pei, Advances in dielectric elastomers for actuators and artificial muscles, Macromol. Rapid Commun. 31 (2010) 10–36.

[164] Q.M. Zhang, V. Bharti, X. Zhao, Giant electrostriction and relaxor ferroelectric behavior in electron-irradiated poly(vinylidene fluoride-trifluoroethylene) copolymer, Science 280 (1998) 2101–2104.

[165] Q.M. Zhang, H. Li, M. Poh, F. Xia, Z.-Y. Cheng, H. Xu, C. Huang, An all-organic composite actuator material with a high dielectric constant, Nature 419 (2002) 284–287.

[166] Y. Zhang, C. Ellingford, R. Zhang, J. Roscow, M. Hopkins, P. Keogh, T. McNally, C. Bowen, C. Wan, Electrical and mechanical self-healing in high-performance dielectric elastomer actuator materials, Adv. Funct. Mater. 29 (2019) 1808431.

[167] E. Hajiesmaili, D.R. Clarke, Reconfigurable shape-morphing dielectric elastomers using spatially varying electric fields, Nat. Commun. 10 (2019) 183.

[168] C. Yang, X. Gao, Y. Luo, End-block-curing ABA triblock copolymer towards dielectric elastomers with both high electro-mechanical performance and excellent mechanical properties, Chem. Eng. J. 382 (2020) 123037.

[169] R.H. Baughman, C. Cui, A.A. Zakhidov, Z. Iqbal, J.N. Barisci, G.M. Spinks, G.G. Wallace, A. Mazzoldi, D.D. Rossi, A.G. Rinzler, O. Jaschinski, S. Roth, M. Kertesz, Carbon nanotube Actuators, Science 284 (1999) 1340–1344.

[170] M. Shahinpoor, Y. Bar-Cohen, J.O. Simpson, J. Smith, Ionic polymer-metal composites (IPMCs) as biomimetic sensors, actuators and artificial muscles – a review, Smart Mater. Struct. 7 (1998) R15–R30.

[171] S. Nemat-Nasser, Y. Wu, Comparative experimental study of ionic polymer–metal composites with different backbone ionomers and in various cation forms, J. Appl. Phys. 93 (2003) 5255–5267.

[172] M. Shahinpoor, K.J. Kim, Ionic polymer–metal composites: IV. Industrial and medical applications, Smart Mater. Struct. 14 (2004) 197–214.

[173] P.H. Vargantwar, K.E. Roskov, T.K. Ghosh, R.J. Spontak, Enhanced biomimetic performance of ionic polymer–metal composite actuators prepared with nanostructured block ionomers, Macromol. Rapid Commun. 33 (2012) 61–68.

[174] C. Jo, D. Pugal, I.K. Oh, K.J. Kim, K. Asaka, Recent advances in ionic polymer-metal composite actuators and their modeling and applications, Prog. Polym. Sci. 38 (2013) 1037–1066.

[175] S. Ma, Y. Zhang, Y. Liang, L. Ren, W. Tian, L. Ren, High-performance ionic-polymer–metal composite: toward large-deformation fast-response artificial muscles, Adv. Funct. Mater. 30 (2020) 1908508.

[176] J.D.W. Madden, N.A. Vandesteeg, P.A. Anquetil, P.G.A. Madden, A. Takshi, R.Z. Pytel, S.R. Lafontaine, P.A. Wieringa, I.W. Hunter, Artificial muscle technology: physical principles and naval prospects, IEEE J. Ocean. Eng. 29 (2004) 706–728.

[177] D. Roy, J.N. Cambre, B.S. Sumerlin, Future perspectives and recent advances in stimuli-responsive materials, Prog. Polym. Sci. 35 (2010) 278–301.

[178] R.D. Kornbluh, R. Pelrine, J. Joseph, R. Heydt, Q. Pei, S. Chiba, High-field electrostriction of elastomeric polymer dielectrics for actuation, Smart Struct. Mater. 3669 (1999) 149–161.

[179] D.P. Armstrong, R.J. Spontak, Designing dielectric elastomers over multiple length scales for 21st century soft materials technologies, Rubber Chem. Technol. 90 (2017) 207–224.

[180] P.H. Vargantwar, R. Shankar, A.S. Krishnan, T.K. Ghosh, R.J. Spontak, Exceptional versatility of solvated block copolymer/ionomer networks as electroactive polymers, Soft Matter 7 (2011) 1651.

[181] S. Zhu, J.-H. So, R. Mays, S. Desai, W.R. Barnes, B. Pourdeyhimi, M.D. Dickey, Ultrastretchable fibers with metallic conductivity using a liquid metal alloy core, Adv. Funct. Mater. 23 (2013) 2308–2314.

[182] K.P. Mineart, Y. Lin, S.C. Desai, A.S. Krishnan, R.J. Spontak, M.D. Dickey, Ultrastretchable, cyclable and recyclable 1- and 2-dimensional conductors based on physically cross-linked thermoplastic elastomer gels, Soft Matter 9 (2013) 7695–7700.

[183] E. Bury, S. Chun, A.S. Koh, Recent advances in deformable circuit components with liquid metal, Adv. Electr. Mater. 7 (2021) 2001006.

[184] A. Turgut, M.O. Tuhin, O. Toprakci, M.A. Pasquinelli, R.J. Spontak, H.A.K. Toprakci, Thermoplastic elastomer systems containing carbon nanofibers as soft piezoresistive sensors, ACS Omega 3 (2018) 12648–12657.

[185] H.A.K. Toprakci, S.K. Kalanadhabhatla, R.J. Spontak, T.K. Ghosh, Polymer nanocomposites containing carbon nanofibers as soft printable sensors exhibiting strain-reversible piezoresistivity, Adv. Funct. Mater. 23 (2013) 5536–5542.

[186] H.Y. Jiang, S. Kelch, A. Lendlein, Polymers move in response to light, Adv. Mater. 18 (2006) 1471–1475.

[187] H. Feil, Y.H. Bae, J. Feijen, S.W. Kim, Mutual influence of pH and temperature on the swelling of ionizable and thermosensitive hydrogels, Macromolecules 25 (1992) 5528–5530.

[188] T. Xie, Tunable polymer multi-shape memory effect, Nature 464 (2010) 267–270.

[189] Q. Zhao, H.J. Qi, T. Xie, Recent progress in shape memory polymer: new behavior, enabling materials, and mechanistic understanding, Prog. Polym. Sci. 49–50 (2015) 79–120.

[190] S. Maity, J.R. Bochinski, L.I. Clarke, Metal nanoparticles acting as light-activated heating elements within composite materials, Adv. Funct. Mater. 22 (2012) 5259–5270.

[191] T. Wang, M. Li, H. Zhang, Y. Sun, B. Dong, A multi-responsive bidirectional bending actuator based on polypyrrole and agar nanocomposites, J. Mater. Chem. C 6 (2018) 6416–6422.

[192] F.S. Bates, Polymer-polymer phase-behavior, Science 251 (1991) 898–905.

[193] C.R. López-Barrón, A.H. Tsou, Strain hardening of polyethylene/polypropylene blends via interfacial reinforcement with poly(ethylene-*cb*-propylene) comb block copolymers, Macromolecules 50 (2017) 2986–2995.

[194] Y. Kong, Y. Li, G. Hu, J. Lin, D. Pan, D. Dong, E. Wujick, Q. Shao, M. Wu, J. Zhao, Z. Guo, Preparation of polystyrene-*b*-poly(ethylene/propylene)-*b*-polystyrene grafted glycidyl methacrylate and its compatibility with recycled polypropylene/recycled high impact polystyrene blends, Polymer 145 (2018) 232–241.

[195] S.A. Arvidson, K.E. Roskov, J.J. Patel, R.J. Spontak, S.A. Khan, R.E. Gorga, Modification of melt-spun isotactic polypropylene and poly(lactic acid) bicomponent filaments with a premade block copolymer, Macromolecules 45 (2012) 913–925.

[196] A.J. Oshinski, H. Keskkula, D.R. Paul, Rubber toughening of polyamides with functionalized block copolymers. 1. Nylon-6, Polymer 33 (1992) 268–283.

[197] J. Yan, R.J. Spontak, Toughening poly(lactic acid) with thermoplastic elastomers modified by thiol–ene click chemistry, ACS Sustain. Chem. Eng. 7 (2019) 10830–10839.

[198] C.E. Hoyle, C.N. Bowman, Thiol–ene click chemistry, Angew. Chem. Int. Ed. 49 (2010) 1540–1573.

[199] M.J. Kade, D.J. Burke, C.J. Hawker, The power of thiol-ene chemistry, J. Polym. Sci. A Polym. Chem. 48 (2010) 743–750.

[200] A.B. Lowe, Thiol-ene "click" reactions and recent applications in polymer and materials synthesis, Polym. Chem. 1 (2010) 17–36.

[201] J.J. Ryan, Physico-Chemical Functionalization and Morphological Studies of Nanostructured Block Copolymers and Ionomers, Ph.D. Dissertation, North Carolina State University, Raleigh, NC, 2018.

[202] A. Warshawsky, N. Kahana, A. Deshe, H.E. Gottlieb, R. Arad-Yellin, Halomethylated polysulfone: reactive intermediates to neutral and ionic film-forming polymers, J. Polym. Sci. A Polym. Chem. 28 (1990) 2885–2905.

[203] R.J. Spontak, J.J. Ryan, Polymer blend compatibilization by the addition of block copolymers, in: S. Thomas, A.R. Ajitha (Eds.), Compatibilization of Polymer Blends: Micro and Nano Scale Phase Morphologies, Interphase Characterization and Ultimate Properties, Elsevier, Amsterdam, 2020, pp. 57–102.

[204] Y.A. Elabd, M.A. Hickner, Block copolymers for fuel cells, Macromolecules 44 (2011) 1–11.

[205] C. Huang, Q. Chen, R.A. Weiss, Rheological behavior of partially neutralized oligomeric sulfonated polystyrene ionomers, Macromolecules 50 (2017) 424–431.

[206] R.A. Weiss, A. Sen, C.L. Willis, L.A. Pottick, Block copolymer ionomers: 1. Synthesis and physical properties of sulphonated poly(styrene-ethylene/butylene-styrene), Polymer 32 (1991) 1867–1874.

[207] R.A. Weiss, A. Sen, L.A. Pottick, C.L. Willis, Block copolymer ionomers: 2. Viscoelastic and mechanical-properties of sulfonated poly(styrene-ethylene/butylene-styrene), Polymer 32 (1991) 2785–2792.

[208] P.H. Vargantwar, M.C. Brannock, K. Tauer, R.J. Spontak, Midblock-sulfonated triblock ionomers derived from a long-chain poly[styrene-*b*-butadiene-*b*-styrene] triblock copolymer, J. Mater. Chem. A 1 (2013) 3430–3439.

[209] J. Yan, S. Yan, J.C. Tilly, Y. Ko, B. Lee, R.J. Spontak, Ionic complexation of endblock-sulfonated thermoplastic elastomers and their physical gels for improved thermomechanical performance, J. Colloid Interface Sci. 567 (2020) 419–428.

[210] P.H. Vargantwar, M.C. Brannock, S.D. Smith, R.J. Spontak, Midblock sulfonation of a model long-chain poly(*p*-*tert*-butylstyrene-*b*-styrene-*b*-*p*-*tert*-butylstyrene) triblock copolymer, J. Mater. Chem. 22 (2012) 25262–25271.

[211] K.P. Mineart, J.J. Ryan, M.-S. Appavou, B. Lee, M. Gradzielski, R.J. Spontak, Self-assembly of a midblock-sulfonated pentablock copolymer in mixed organic solvents: a combined SAXS and SANS analysis, Langmuir 35 (2019) 1032–1039.

[212] K.P. Mineart, X. Jiang, H. Jinnai, A. Takahara, R.J. Spontak, Morphological investigation of midblock-sulfonated block ionomers prepared from solvents differing in polarity, Macromol. Rapid Commun. 36 (2015) 432–438.

[213] S.H. Kim, M.J. Misner, T. Xu, M. Kimura, T.P. Russell, Highly oriented and ordered arrays from block copolymers via solvent evaporation, Adv. Mater. 16 (2004) 226–231.

[214] J.N.L. Albert, W.-S. Young, R.L. Lewis, T.D. Bogart, J.R. Smith, T.H. Epps, Systematic study on the effect of solvent removal rate on the morphology of solvent vapor annealed ABA triblock copolymer thin films, ACS Nano 6 (2012) 459–466.

[215] C.O. Osuji, Alignment of self-assembled structures in block copolymer films by solvent vapor permeation, Macromolecules 43 (2010) 3132–3135.

[216] G. Cui, M. Fujikawa, S. Nagano, K. Shimokita, T. Miyazaki, S. Sakurai, K. Yamamoto, Macroscopic alignment of cylinders via directional coalescence of spheres along annealing solvent permeation directions in block copolymer thick films, Macromolecules 47 (2014) 5989–5999.

[217] K.P. Mineart, H.A. Al-Mohsin, B. Lee, R.J. Spontak, Water-induced nanochannel networks in self-assembled block ionomers, Appl. Phys. Lett. 108 (2016) 101907.

[218] K.E. Mineart, J.D. Dickerson, D.M. Love, B. Lee, X. Zuo, R.J. Spontak, Hydrothermal conditioning of physical hydrogels prepared from a midblock-sulfonated multiblock copolymer, Macromol. Rapid Commun. 38 (2017) 1600666.

[219] H.A. Al-Mohsin, K.P. Mineart, R.J. Spontak, Highly flexible aqueous photovoltaic elastomer gels derived from sulfonated block ionomers, Adv. Energy Mater. 5 (2015) 1401941.

[220] H.A. Al-Mohsin, K.P. Mineart, D.P. Armstrong, A. El-Shafei, R.J. Spontak, Quasi-solid-state dye-sensitized solar cells containing a charged thermoplastic elastomeric gel electrolyte and hydrophilic/phobic photosensitizers, Sol. RRL 2 (2018) 1700145.

[221] M. Bui, C.S. Adjiman, A. Bardow, E.J. Anthony, A. Boston, S. Brown, P.S. Fennell, S. Fuss, A. Galindo, L.A. Hackett, J.P. Hallett, H.J. Herzog, G. Jackson, J. Kemper, S. Krevor, G.C. Maitland, M. Matuszewski, I.S. Metcalfe, C. Petit, G. Puxty, J. Reimer, D.M. Reiner, E.S. Rubin, S.A. Scott, N. Shah, B. Smit, J.P.M. Trusler, P. Webley, J. Wilcox, N. Mac Dowell, Carbon capture and storage (CCS): the way forward, Energy Environ. Sci. 11 (2018) 1062–1176.

[222] H. Lin, E.V. Wagner, B.D. Freeman, L.G. Toy, R.P. Gupta, Plasticization-enhanced hydrogen purification using polymeric membranes, Science 311 (2006) 639–642.

[223] J. Deng, Z. Dai, J. Yan, M. Sandru, E. Sandru, R.J. Spontak, L. Deng, Facile and solvent-free fabrication of PEG-based membranes with interpenetrating networks for CO_2 separation, J. Membr. Sci. 570–571 (2019) 455–463.

[224] L. Ansaloni, Z. Dai, J.J. Ryan, K.P. Mineart, Q. Yu, K.T. Saud, M.-B. Hägg, R.J. Spontak, L. Deng, Solvent-templated block ionomers for base- and acid-gas separations: effect of humidity on ammonia and carbon dioxide permeation, Adv. Mater. Interf. 4 (2017) 1700854.

[225] Z. Dai, L. Ansaloni, J.J. Ryan, R.J. Spontak, L. Deng, Incorporation of an ionic liquid into a midblock-sulfonated multiblock polymer for CO_2 capture, J. Membr. Sci. 588 (2019) 117193.

[226] Z. Wu, J.M. McGoogan, Characteristics of and important lessons from the coronavirus disease 2019 (COVID-19) outbreak in China: summary of a report of 72314 cases from the Chinese center for disease control and prevention, J. Am. Med. Assoc. 323 (2020) 1239–1242.

[227] https://coronavirus.jhu.edu/map.html .(Accessed: March 2021).

[228] B.S.T. Peddinti, F. Scholle, M.G. Vargas, S.D. Smith, R.A. Ghiladi, R.J. Spontak, Inherently self-sterilizing charged multiblock polymers that kill drug-resistant microbes in minutes, Mater. Horiz. 6 (2019) 2056–2062.

[229] N. van Doremalen, T. Bushmaker, D.H. Morris, M.G. Holbrook, A. Gamble, B.N. Williamson, A. Tamin, J.L. Harcourt, N.J. Thornburg, S.I. Gerber, J.O. Lloyd-Smith, E. de Wit, V.J. Munster, Aerosol and surface stability of SARS-CoV-2 as compared with SARS-CoV-1, N. Engl. J. Med. 382 (2020) 1564–1567.

[230] J. Cai, W. Sun, J. Huang, M. Gamber, J. Wu, G. He, Indirect virus transmission in cluster of COVID-19 cases, Emerg. Infect. Dis. 26 (2020) 1343–1345.

[231] K. El-Refaie, S.D. Worley, R. Broughton, The chemistry and applications of antimicrobial polymers: a state-of-the-art review, Biomacromolecules 8 (2007) 1359–1384.

[232] S. Li, S. Dong, W. Xu, S. Tu, L. Yan, C. Zhao, J. Ding, X. Chen, Antibacterial hydrogels, Adv. Sci. 5 (2018) 1700527.

[233] B.S.T. Peddinti, R.A. Ghiladi, R.J. Spontak, Next-generation stimuli-responsive antimicrobial coatings with broad-spectrum efficacy, in: P. Zarras, M. Soucek, Y. Wei (Eds.), Handbook of Smart Polymer Coatings and Films, Wiley, New York, 2021 (in press).

[234] R. Laxminarayan, A. Duse, C. Wattal, A.K.M. Zaidi, H.F.L. Wertheim, N. Sumpradit, E. Vlieghe, G.L. Hara, I.M. Gould, H. Goossens, C. Greko, A.D. So, M. Bigdeli, G. Tomson, W. Woodhouse, E. Ombaka, A.Q. Peralta, F.N. Qamar, F. Mir, S. Kariuki, Z.A. Bhutta, A. Coates, R. Bergstrom, G.D. Wright, E.D. Brown, O. Cars, Antibiotic resistance — the need for global solutions, Lancet Infect. Dis. 13 (2013) 1057–1098.

[235] C. Maxouris, A Dangerous Coronavirus Variant Is Wreaking Havoc in Parts of Europe. Experts Fear US Could Be Next, CNN, March 31, 2021. www.cnn.com/2021/03/31/health/us-coronavirus-wednesday/index.html.

Index

'*Note*: Page numbers followed by "f" indicate figures and "t" indicate tables.'

A

Accelerators, 2
Acrylonitrile butadiene rubber (NBR), 129
Additive manufacturing technologies, 25–26
Additives, 11–12
Aerogel materials, 11–12
Agriculture engineering applications
  biodegradability, 72–73
  bio-oil-based elastomeric systems, 73–74
  carbon-based nanomaterials, 72
  crystallization, 72
  polymeric material recycling, 74–75
Aminopropyl terminated poly-dimethylsiloxane (AP-PDMS), 316–317, 316f
Amorphous polyamide (a-PA) and ethylene-1-octene copolymer (EOR) blends, 137, 137t
Anisotropic Newtonian fluids, 88
Artificial heart valves, 66–67, 68f
Atomic force microscopy (AFM), 35–36, 109–112
Atomistic stimulation, 245
Axisymmetric model, 252–254

B

Bamboo charcoal powder (BCP), 117–120
Bio-based engineering elastomer/nano-silica (BEE/SiO_2), 140
Biodegradable elastomers, 6
Biological devulcanization, 283–285
Biological phenolic resin (BPR)
  biomass-based phenolic compounds
    antiseptics, 336
    lignin, 335
    plant biomass, 335
    pyrolysis and liquefaction, 335–336
    tannins, 335
  blended composite
    elastic modulus and compression strength, 343–344
    elongation properties, 344
    hybrid biocomposite, 342
    liquefied oil palm EFB fibers, 342
    mechanical performances, 343–344
    physical attributes, 341–342
    rheological properties, 340–341
    surface hydrophilicity and hydrophobicity, 341–342
    surface morphology and roughness, 341
    tensile strength, 343
    thermal properties, 344–346
    water absorption properties, 342
  glass fiber compatibility, 339
  origin, 334–335
  thermoplastic phenolic resin, 337–339
  thermoset phenolic resin, 337–339
Biomedical engineering applications
  artificial heart valves, 66–67, 68f
  biomedical elastomers, 62
  conductive biopolymer-based materials, 64
  implantable biomedical devices, 65
  nanostructured particles, 65
  poly(glycerol sebacate) (PGS) elastomer, 62–64
  silicone elastomers, 66, 67f
Bio-oil, 332
Biophenols, 333
Biostable elastomer, 18–20
Bis(1-(tert-butylperoxy)-1-methylethyl)-benzene (BIPB), 6
Block random copolymers (BRCs), 358–359, 359f

Boron nitride (BN) content, polyolefin elastomer, 50, 51f
Brabender, 59–60, 60f
Bromo butyl rubber (BIIR)/acrylonitrile-co-butadiene rubber (NBR), 106–109
Burgers model, 172–175, 173t–174t

C

Capillary viscometer, 97
Carbon black, 46
Carbon-fiber-reinforced rubber composites, 252–254
Carbon nanotube (CNT) filler, 11–12, 49–50, 49f
Carbonyl iron (CI)/natural rubber (NR), gamma ferrite additive, 52
Carbonyl-iron particle (CIP), 47
Carreau model, 100
Charge equilibration (QEq), 248–249
Charge transfer interaction mechanism, 36
Chemical devulcanization
  benzoyl peroxide (BPO), 273–274
  carbonaceous adsorbent, 272
  chemical regents, 271–272
  cured rubber powder, 2,6-di-tert-butyl-4-methyl-phenol (BHT), 272
  devulcanized rubber (DVR), 275
  periodic acid, 272
  processing temperature, 274f
  processing time, 274f
  super critical fluids, 271–272
Chemical wet synthesis method, 47
Chlorinated polyethylene rubber(CPR), 106–109
Chlorosulfonated polyethylene (CSM)/isobutylene-co-isoprene (IIR), 106–109
Cholorosulfonated polyethylene (CSM) rubber elastomer, 323, 323f
Classical elasticity, 171–172
Coconut tea leaf sheath (CLS), 344–345
Commodity elastomers, 62
Compatibilization, 34
Compatibilizers, 34, 149
Computer simulations, 243
Condensed-phase optimized molecular potentials for atomistic simulation (COMPASS27) force field, 249–250
Cone-and-plate rheometer, 97
Constitutive law, 91–92
Constitutive rheological models, 98
Conventional fillers, 46
Conventional rubber, 34
Copper nanowires (CuNWs), 47
Cyclic olefin copolymer/polyolefin elastomer (COC/POE) blends, 134, 134t

D

Debye–Scherrer WAXD patterns, 221f
Deviatoric stress tensor, 98
Devulcanization process, 270–271, 270f–271f
  biological methods, 283–285
  chemical method
    benzoyl peroxide (BPO), 273–274
    carbonaceous adsorbent, 272
    chemical regents, 271–272
    cured rubber powder, 2,6-di-tert-butyl-4-methyl-phenol (BHT), 272
    devulcanized rubber (DVR), 275
    periodic acid, 272
    processing temperature, 274f
    processing time, 274f
    super critical fluids, 271–272
  microwave methods
    energy distribution, 276–277
    GTR devulcanization, 277–278
    magnetic rubber composite, 277
    polymer chain growth and sulfur cross-link mechanism, 281f
    solvent extraction and swelling, 279–281
    temperature, 276–277
    waste rubber and barium ferrite, 277
    waste tire rubber (WTR), 278–279
  supercritical methods, 285–286
  thermomechanical methods, 281–283
  ultrasound method
    gel permeation chromatography data, 276

^{1}H transverse relaxation, 276
meshed tire rubbers, 275–276
particle-size, 275–276
roofing membrane elasticity, 275–276
sulfur-cured unfilled butadiene, 276
ultrasonic waves, 275
Devulcanized crumb rubber (DCR), 287–288, 287f
Devulcanized ground tire rubber (DGTR), 278–279
Devulcanized rubber (DVR), 275
3D graphene foam fillers, 50
Dielectric elastomers (D-EAPs), 374–375
Differential scanning calorimetry (DSC) analysis, 152–153
blends, 153–156
elastomeric composites, 161–162
Digital light processing (DLP), 17, 18f
Diglycidyl ether of bisphenol-A (DGEBA) epoxy resin, 141–142
Direct ink writing (DIW) method, 23–25
Direct powder molding, 289
Dissipative particle dynamics (DPD), 245, 361–362
Donor–acceptor Stenhouse adduct (DASA), 314
3D-printed dumbbell-shaped mechanical test specimens, 28
Dynamic mechanical analyzer (DMA), 177
Dynamic mechanical properties (DMPs)
elastomer blends
cyclic olefin copolymer/polyolefin elastomer (COC/POE) blends, 134, 134t
linear low-density polyethylene (LLDPE) and ethylene-co-methyl acrylate (EMA) blends, 133
NR/PP and NR-g-MAH/PP-g-MAH blends, 133, 134t
PA-612/POE-g-MA blends, 136, 136t
polyamide-6 (PA6) and ethylene-butylene (EB) elastomers, 135
PP and ethylene/octene copolymer elastomers (POE) blend, 135, 135t
PP and styrene-(ethylene-butylene)-styrene (SEBS), 135, 135t
silane cross-linked (SC) PP and ethylene propylene diene monomer (EPDM), 134
styrene-ethylene/butylene styrene triblock copolymer (SEBS), 135t
elastomer composites
damping factor reduction, 143
diglycidyl ether of bisphenol-A (DGEBA) epoxy resin, 141–142
epoxy-SAN/glass composites, 142t
natural rubber/HCNF composites, 144, 144t
polycarbonate/ethylene methyl acrylate (PC/EMA) elastomeric composites, 141
PVC and ethylene vinyl acetate (EVA) blend, 142
PVC/EVA/kenaf composites, 142t
PVC/NBR composites, 143, 143t
sisal fibers, 143
thermoplastic copolyester elastomers (TCEs), 141
Dynamic mechanical thermal analysis (DMA), 153
blends
EVA/TPU blends, 156
PCL/NR self-healing blend polymer, 156
storage modulus, 156
elastomeric composites, 162–163
EVA/TPU blends, 156
PCL/NR self-healing blend polymer, 156
storage modulus, 156

E

Elastomer-based blends
applications, 36–41
compatibilization, 34
fabrication methods, 35
food packaging application, 39
immiscible blends, 33–34
mechanical performance, 39–41
nanofillers, 34–35
processing and characterization methods, 35–36

Elastomer-based blends (*Continued*)
properties, 36, 37t
self-healable, 36–39
Elastomer–biological phenolic resin composites
biological phenolic resin. *See* Biological phenolic resin (BPR)
composite material, 334
lignin, 333
phenol, 332
relatively rubber-like materials, 331–332
thermoplastic elastomers, 331–332
Elastomer blends and composites
applications, 2–3
cholorosulfonated polyethylene (CSM), 323
ethylene-propylene-diene monomer (EPDM)-based elastomer, 317–321
fluorocarbon elastomer, 321–323
polyurethane-based elastomer blends, 307–314
silicone-based elastomer blends and composites, 314–317
solid fillers, 305
Brabender, 59–60
dynamic mechanical properties (DMPs). *See* Dynamic mechanical properties (DMPs)
electrical-conductive polymers, 57
engineering applications
agriculture, 71–75
biomedical, 62–67
ocean, 67–71
ethylene propylene diene monomer (EPDM) rubber, 4–5
extrusion, 58–59
internal mixing method, 58
manufacturing methods. *See* Manufacturing methods
material systems, 57
mechanical strength, 57–58
medical applications, 58
modeling study. *See* Modeling study
morphological characteristics. *See* Morphological characteristics
nanocomposites, 2–3
nanotechnology, 2–3
natural rubber (NR)-based elastomers, 3–4
olefin thermoplastic elastomer, 6
processing methods, 58–62
radiation method, 61–62
silicone rubber, 5
simulations. *See* Simulations
two roll mills, 60–61, 61f
viscoelasticity. *See* Viscoelasticity
Elastomers
biodegradable, 6
cross-links, 1
industrial applications, 103–104
natural, 1
properties, 1
synthetic, 1
types, 103–104
vulcanization/cross-linking, 2
Electroactive polymers (EAPs)
dielectric elastomers (D-EAPs), 374–375
electronic, 374–375
styrenic TPEGs, 377–379
vapor-grown carbon nanofiber (CNF), 377–379
Electronegativity equalization method (EEM), 248–249
Empirical and semiempirical models, 86
Engineering applications
agriculture, 71–75
biomedical, 62–67
ocean, 67–71
Environmental scanning microscope (ESEM), 114–117
Epoxidized natural rubber/dodecanedioic acid (ENR/DA), 132, 132t
Epoxidized natural rubber (ENR50), 106–109
Epoxidized soybean oil (ESO), 289
Epoxy resins, 34
Eshelby's equivalent inclusion theory, 252–254
Ethylene-co-vinyl acetate (EVA) and millable polyurethane (MPU) blend, 181–182

Ethylene-co-vinyl acetate (EVA)
elastomers-based blends, 252
Ethylenedioxy diethanethiol (EDDET), 17
Ethylene-octane random copolymer POE elastomers, 130
Ethylene-propylene-diene monomer (EPDM), 106–109
Ethylene-propylene-diene monomer (EPDM)-based elastomer, 4–5
fabrication, 318–319
gel content, 317–318
polyethylene elastomer grafted maleic anhydride (POE-MA), 319–321
polypropylene (PP)/EPDM elastomer blends, 321
and silicone rubber, 317–318
temperature scanning stress relaxation (TSSR) method, 321
Ethylene propylene diene terpolymer and poly(vinyl chloride) (EPDMPVC) blend, 180–181
Excimer, 200
Expanded graphite (EG)
with styrene-butadiene rubber (S-SBR) composites, 50
with styrene isoprene styrene block copolymer, 51
Extrusion, 58–59

F

Fabrication
elastomer-based blends, 35
thermoplastic elastomer gels (TPEGs)
dimensionless temperature–composition, 368f
dynamic rheological analysis, 367–368
electroresponsive media, 366–367
midblock bridging fraction, 369f
shock-absorbing media, 366–367
styrenic TPEs, 367–368
ternary phase diagram, 367f
Field emission scanning electron microscope (FESEM), 114–117
Fillers, 11–12
boron nitride (BN) content, polyolefin elastomer, 50
carbon black, 46
carbon nanotubes and hybrid fillers, 49–50
carbonyl iron (CI)/natural rubber (NR), gamma ferrite additive, 52
copper nanowires (CuNWs), 47
3D graphene foam, 50
glycerol filler, 52
hybrid fillers, 47–48
magnetic, 47
multiwalled carbon nanotubes (MWCNTs), 50
piezoelectric (PZT) and silver-coated glass microsphere fillers, 48
preparation and properties, 46–52
silica, 46
styrene-butadiene-styrene/multiwall (SBS) carbon nanotubes fillers, 49
styrene-isoprene-styrene block copolymer-based conductive polymer composites, 51
Finger prosthesis implant, 66, 67f
Flavonoids, 333
Flow curve, 86
Fluorescence spectroscopy
luminescence studies, 199
mechanoluminescence, 201
nonconjugated and nonfluorescent polymers, 201
nonconjugated polymeric compounds, 199
nonradiative energy transfer (NRAT) technique, 200
poly(9,9-dioctylfluorene) (PFO) and poly(9,9-dioctylfluorene-alt-benzothiadiazole) (F8BT) blend, 200–201
polymer-based LEDs, 200–201
steady-state fluorescence, 199–200
time-resolved measurements, 199–200
Fluorocarbon elastomer
automobile industry, 323
classification, 322
properties, 322–323

Fluorocarbon elastomer (*Continued*)
synthesis monomer, 322f
Fluorosilicone rubber (FSR), 181–182
Food packaging application, 39
Food science and cosmetics, 85
Fourier-transform infrared spectroscopy (FT-IR), 35–36, 196–198
carbon nanotubes, 199f
chitosan/gelatin/thermoplastic polyurethane (TPU) blend nanofibers, 196
grafted poly(styrene-b-ethylene-butylene-b-styrene) (SEBS), 196–198
PVC/TBU blend, 196
styrene-butadiene rubber (SBR) elastomers, 196–198
thermoplastic polyurethane (TPU), 196
thermoplastic polyurethane (TPU)/ ultrathin graphite (UTG) composite, 196–198, 198f
TPU/TPS blends, 196–198
Freeze drying technique, 13–14
Fumed silica, 46

G
Gelation process, 59–60
Generalized Kelvin–Voigt model, 173t–174t
Generalized Maxwell model, 173t–174t
Generalized Newtonian fluid, 89
Geology, 85
Glycerol, 62–63
Glycerol filler, 52
Graphene (MoO_3-GNS) hybrids, 308f
Graphene nanoplatelets (GnPs), 6, 50
Gravimetric capillary rheometers, 97
Ground tire rubber (GTR)/reclaimed tire rubber (RTR), 155
Guinier camera, 224
Guinier law, 224
Guinier–Tennevin method, 224

H
Halloysite nanotube (HNT), 36
Heterocoagulation approah, 15
Heteroinflammation process, 15
High energy chemistry, 61
Hybrid fillers, 47–48
Hybrid magnetic elastomer, 45
Hyperbranched polysiloxane (HPSis), 317

I
In situ polymerization, 15
Internal mixing method, 58
Interpenetrating polymer networks (IPNs), 309
Interphase Monte Carlo (IMC) method, 247, 247f
Isothermal expansion, 28
Isotropic Newtonian fluids, 88

K
Kelvin–Voigt model, 172–175, 173t–174t
Kissinger and Moynihan's model, 250–251
Kissinger method, 152–153

L
Lab on a chip/lab on box (LoC/LoB) devices, 69
Latex stage compounding, 14–15
Lignin, 333
Linear low-density polyethylene and ethylene-co-methyl acrylate (LLDPE/ EMA) blends, 133
Liquid crystal elastomer (LCE), 17–18
Liquid reclaimed rubber (LRR), 282–283, 291
Low-density polyethylene (LDPE), 45
Lower critical solution temperature (LCST), 151, 151f

M
Magnetic fillers, 47
Manufacturing methods
digital light processing (DLP), 17
direct ink writing (DIW) method, 23–25
freeze drying technique, 13–14
heterocoagulation approah, 15
latex stage compounding, 14–15
melt blending/extrusion, 16
patient-specific heart models, 18–20

PCU-Sil creation methodology, 18–20
randomly oriented multimaterial modeling, 25–28, 26f–27f
silicone composites, 28
in situ polymerization, 15
solid-state shear pulverization, 16–17
solvent casting, 13
spray drying technique, 14
Material science, 195
Materials processes and simulations (MAPS), 248–249
Materials science, 85
Mathematical laws, 86
Maxwell model, 172–175, 173t–174t
Mechanical behavior
dynamic mechanical properties (DMPs), 133–136
static mechanical properties, 128–133
Mechanical modeling
axisymmetric model, 252–254
carbon-fiber-reinforced rubber composites, 252–254
2D micromechanical modeling, 252–254
ethylene-co-vinyl acetate (EVA) elastomers-based blends, 252
J-integral, 252–254
parametric study, 252–254
representative volume element (RVE) method, 254, 255f
Rule of Hybrid Mixture (RoHM), 252–254
spherical filler-reinforced rubber matrix, 252–254
Medical applications, 58
Medical-grade elastomers, 62
Melt blending/extrusion, 16
Methyl vinyl silicone rubber (MVQ) composites, 104–106
Microactuator systems, 21
Microwave devulcanization
energy distribution, 276–277
GTR devulcanization, 277–278
magnetic rubber composite, 277
polymer chain growth and sulfur cross-link mechanism, 281f
solvent extraction and swelling, 279–281
temperature, 276–277
waste rubber and barium ferrite, 277
waste tire rubber (WTR), 278–279
Midblock fractions, 363f
Miscibility, blends, 151
Modeling study
mechanical modeling
axisymmetric model, 252–254
carbon-fiber-reinforced rubber composites, 252–254
2D micromechanical modeling, 252–254
ethylene-co-vinyl acetate (EVA) elastomers-based blends, 252
J-integral, 252–254
parametric study, 252–254
representative volume element (RVE) method, 254, 255f
Rule of Hybrid Mixture (RoHM), 252–254
spherical filler-reinforced rubber matrix, 252–254
rheological modeling
magneto rheological elastomers, 256
PEG-grafted graphene oxide, 255
rheological behavior, 254–255
storage and loss moduli, 254–255
thermal modeling
glass transitions kinetics, 250–251
nonequilibrium molecular dynamic (NEMD) simulation methods, 251
polyethylene glycol (PEG)-based epoxy blends, 250–251
thermal conductivity (TC), 251
Molecular dynamics (MD) methods, 244–245
Molecular model system, 249f
Mooney viscometer, 97–98
Mori–Tanaka modified theory, 252–254
Morphological characteristics
atomic force microscopy (AFM), 109–112
field emission scanning electron microscope (FESEM), 114–117
optical microscopy (OM)
chitosan(CS) dispersion, 104–106
2D imaging technology, 104–106

Morphological characteristics (*Continued*)
methyl vinyl silicone rubber (MVQ) composites, 104–106
NR/BIIR blends filled CNTs, 104–106
phase contrast OM(PCOM), 104–106
PLA and PLA/ENR composites, 105f
PS/PANI-loaded rubber blends, 104–106
SR/fluoro rubber(FR), 104–106
plant fiber-reinforced elastomer composites, 117–120
scanning electron microscopy (SEM), 106–109
synthetic fiber-reinforced elastomer composites, 120–121
transmission electron microscopy (TEM), 112–114
Mullins effect, 246
Multimaterial print head design and functionality, 28
Multishape memory effects (multi-SMEs), 312

N

Nanocomposites, 2–3
Nanofillers, 34–35
Nanotechnology, 2–3
Natural rubber/helical carbon nanofibers (HCNF) composites, 144
Natural rubber (NR)-based elastomers
and CNTs composites, 4
composite foams, 3–4
dielectric properties, 4
high-performance, 4
magnetorheological elastomers, 3–4
Natural rubber-polypropylene, 3–4
Natural rubbers, 1
Neo-Hookean model, 176
Newtonian fluids, 86–88
Newton's law, 171–172
Nickel-coated carbon fibers (NiCF), 21
Nonequilibrium molecular dynamic (NEMD) simulation methods, 251
Non-Newtonian fluids, 88–89
Non-petroleum-based sources, 332
Nonradiative energy transfer (NRAT) technique, 200
NR latex (NRL), 14–15
Nuclear magnetic resonance (NMR) spectroscopy
Bisphenol A polycarbonate/poly(methyl methacrylate) (PC/PMMA) blend, 202–203
^{1}H spin-lattice relaxation experiments, 202–203
poly2,5-bis(3-hexadecylthiophene-2-yl) thieno[3,2-b]thiophene (PBTTT), 203
polystyrene (PS)/polyvinyl methyl ether (PVME) blend, 202
spin relaxation times, 202
two-dimensional wide line separation NMR (WISE NMR), 203

O

Ocean engineering applications
automated in-situ measurements, 67–69
filler concentration, 67–69
lab on a chip/lab on box (LoC/LoB) devices, 69
micro plastics, 69–70
physical measurements, 69
polyester degradation, 70t
reactive extrusion, 70–71
seawater degradable polymeric materials, 69–70
Office waste paper (OWP), 117–120
Olefin thermoplastic elastomer, 6
Organomodified layered silicates (OMLS), 189

P

Padding reinforcement, 11–12
Payne effect, 101
PCU-Sil creation methodology, 18–20
Petroleum-based phenol, 332
Petroleum-based PU materials, 73–74
Phase contrast OM(PCOM), 104–106
Phase-modulation fluorometry, 199–200
Phenolic acids, 333

Phosphorus and nitrogen elements (PNPOP) composites ?ame-retardant mechanism, 309f
Physiology, 85
Piezoelectric (PZT) and silver-coated glass microsphere fillers, 48
Pin-hole collimation system, 233–234
Plant fiber-reinforced elastomer composites, 117–120
Plastisol, 353–354
Plunger-type capillary rheometers, 97
Polyamide-6 (PA6) and ethylene-butylene (EB) elastomers, 135
Polyamide-612 (PA-612) and maleic anhydride grafted poly (ethyleneeoctene) (POE-g-MA) elastomer blends, 132, 136
Poly2,5-bis(3-hexadecylthiophene-2-yl) thieno[3,2-b]thiophene (PBTTT), 203
Polycarbonate/ethylene methyl acrylate (PC/EMA) elastomeric composites, 141
Poly-dimethylsiloxane (PDMS), 50
Poly(9,9-dioctylfluorene) (PFO) and poly(9,9-dioctylfluorene-alt-benzothiadiazole) (F8BT) blend, 200–201
Polyethylene elastomer grafted maleic anhydride (POE-MA), 319–321
Poly(glycerol sebacate) (PGS) elastomer, 62–63
  biodegradability rate, 63–64
  carbon nanotubes (CNTs) incorporation, 63–64
  nanobioglass composites, 63–64
  synthesis, 62–63
Polyjet 3D printer, 25–26
Polylactic acid (PLA) blends
  biobased, 39–40
  brittle nature, 40
  supertoughened, 40
Polylactic acid/polyolefin elastomer (PLA/POE) blends, 36
Polyolefin elastomer (POE), boron nitride (BN) content, 50, 51f
Polytetramethylene ether glycol (PTMEG), 310
Polyurea urethane/hollow glass micro spheres (PUUR/HGS) composite, 139, 139t
Polyurethane-based elastomer blends and composites
  flame-retardant elastomer, 306
    aluminum-12 silicon's conductive coatings, 307
    aluminum trihydrate, 307
    graphene (MoO_3-GNS) hybrids, 307, 308f
    phosphorus and nitrogen elements (PNPOP)/PU composites, 309
    Schiff-base polyphosphate ester (SPE), 307, 308f
  self-healing elastomer, 310–312
  sensing elastomer, 313–314
  shape memory polymers (SMPs), 312–313
Polyurethane elastomer, 64
Position sensitive detectors (PSD), 221
Power-law scaling, 226
Precipitated silica, 46
Proanthocyanidins, 333
Pulse fluorometry, 199–200
Pyrolysis, 332

R

Radiation method, 61–62
Radical devulcanization mechanism, tire rubber, 280f
Raman spectroscopy, 196–198
Randomly oriented multimaterial modeling (ROMM) technique, 25–28
Rapid prototyping, 25–26
Reclaiming process, 270–271
Recycled elastomeric materials, value-added products
  acetone extraction, 291f
  cost-effective seismic protection systems, 291–292
  devulcanized crumb rubber (DCR), 287–288, 287f
  devulcanized styrene BR, 286–287
  direct powder molding, 289

Recycled elastomeric materials, value-added products (*Continued*)
  DVR modified epoxy-based composites, 291
  electromagnetic interference shielding (EMI), 292–295
  epoxidized soybean oil (ESO), 289
  ground tire rubber, 286–287
  liquid reclaimed rubber (LRR), 291
  magnetic rubber fabrication, 293f
  mold and molded plates, 290f
  multifunctional reclaiming agent, 289–290
  particle board products, 291
  recycled rubber bearings, 291–292
  thermoplastic polyurethane (TPU), 295–296, 296f
  vitrimers, 289
  wettability crumb rubber, 287f
Recycling methods
  devulcanization process, 270–271, 270f–271f
  reclaiming, 270–271
  scrap rubber disposal, 269–270
Representative volume element (RVE) method, 254, 255f
Rheological modeling
  magneto rheological elastomers, 256
  PEG-grafted graphene oxide, 255
  rheological behavior, 254–255
  storage and loss moduli, 254–255
Rheological tests
  dynamic tests
    complex shear modulus, 95
    complex viscosity, 94–95
    linear viscoelastic region (LVR), 94
    oscillatory regime, 94
    shear stress, 94
  static tests
    deformation, 93
    normal stress components, 93–94
    shear strain rate tensor, 93
    steady shear flow analysis, 92
    velocity gradient, 92
    velocity profile, 93
Rheology
  capillary viscometer, 97
  cone-and-plate rheometer, 97
  constitutive law, 91
  constitutive rheological models, 98
  continuum mechanics domain, 84f
  elastomer blends, 100–101
  elastomer composites, 101
  fluid characteristics, 85
  laboratory tests and instrumentations, 96–97
  material response, 85
  materials, 89–90
  models, 86
  Mooney viscometer, 97–98
  Newtonian fluids, 86–88
  non-Newtonian fluids, 88–89
  production process, 84
  rheological tests, 85, 92–96
  rheometry tests, 86
  system structure effect, micro-/nano-scale, 90–91
  temperature effect, 90
  theoretical rheology, 85
  uncured rubber melts
    Carreau model, 100
    constitutive law, 98
    deviatoric stress tensor, 98
    generalized flow models, 99
    strain rate tensor components, 99
    two-parameter power law, 99
    velocity gradient, 98
Rheometry tests, 86
Rheopectic fluid, 89
Ring-closing metathesis (RCM), 64
Ring-closing polymer (RCP), 64
Rule of Hybrid Mixture (RoHM), 252–254

S

SARS-CoV-2 virus, 390–391
Scanning electron microscopy (SEM), 106–109
Schiff-base polyphosphate ester (SPE), 307
Seawater degradable polymeric materials, 69–70
Self-healable elastomer blends, 36–39

bilayer structure design, 38–39
bromobutyl rubber (BIIR) and terpene resin mixing, 38
charge transfer interaction mechanism, 36
4D printing technology, 38
polyol and zwitterionic polyol blending, 36–37
Self-healing polyurethane elastomer
digital light processing 3D printing, 311
hydroxyl-terminated polybutadiene, 311
networks, 310
polytetramethylene ether glycol (PTMEG), 310
PU/polydopamine composite, 311–312
tensile strength, 311
vacuum-assisted resin transfer molding (VARTM), 310
van der Waals and hydrogen bonds, 310
Shape memory polymers (SMPs), 312–313
Shear strain, 86
Shear strain rate tensor, 93
Shear stress, 86
Shear thickening fluids, 89
Silica, 46
Silicone-based elastomer blends and composites
aminopropyl terminated poly-dimethylsiloxane (AP-PDMS), 316–317
cross-linking, 315f
double-phase glycerol–silicone hybrid elastomer, 317
dynamic disulfide bonds, 316
healable polydimethylsiloxane (HPDMS), 314–315
manufacturing methods, 28
1,4-phthalaldehyde (PTA), 316–317
polydimethylsiloxane (PDMS), 314–315
preparation, 314
thermally reversible hyperbranched polysiloxane (HPSis), 317
Silicone/ethanol composite specimens, 28
Silicone rubber, 5
Simulations
microscopic changes, 243–244
polymer simulation, 244f
thermoplastic elastomers
atomistic molecular dynamics (MD), 246
atomistic stimulation, 245
computational techniques, 244–245
dissipative particle dynamics (DPD), 245, 245f
fuzzy interface, 246
Interphase Monte Carlo (IMC) method, 247
mean square displacement, 246, 246f
molecular dynamics (MD) methods, 244–245
Mullins effect, 246
polyurethanes (TPU), 246
thermosetting elastomers
charge equilibration (QEq), 248–249
computational studies, 249–250
cross-linking technique, 248
electronegativity equalization method (EEM), 248–249
molecular interactions, 249–250
molecular model system, 249f
network polymerization, 248–249
proximity-based MDs, 249–250
thermosetting resins, 248
Single screw extruder, 58–59, 59f
Sisal fibers, 143
Slip-Tube Network (STN) model, 368–372
Slit collimation system, 232–233
Small-angle X-ray scattering (SAXS), 35–36
average interdomain distance, 234
configurations, 233f
diffuse small-angle scattering
dilute system, 225
fractal structure, 224–228, 226f
Guinier law, 224
scattering equivalents, 225–226
discrete small-angle scattering
crystallizable and nanostructured polymers, 228–229
extruded polyethylene (PE) film, 228–229, 229f
invariant and radial correlation function, 228–232

Small-angle X-ray scattering (SAXS) (*Continued*)
meridional reflections, 228
spatially periodic systems, 228
synthetic and natural materials, 229f
synthetic spherical latex particles, 228
two-phase model and Lorentz correction, 227–228
elastomer/nanoclay composite, 225f
hybrid nanocomposite PI-b-PEO, 235, 236f
instrumentation, 224, 230
intensity maxima position, 234–235, 235t–236t
pin-hole collimation system, 233–234
and polymers, 224–234
slit collimation system, 232–233
and WAXD, 235–237
Soak and irradiate method, 13–14
Soft and biostable elastomer, 18–20
Soft elastomer, 18–20
Soft Robotics projects, 28
Solid mechanics approach, 85
Solid-state shear pulverization, 16–17
Solvent casting method, 13
Solvent-vapor annealing (SVA), 386–388
Spectroscopic techniques
fluorescence spectroscopy, 199–201
Fourier-transform infrared spectroscopy (FT-IR), 196–198
nuclear magnetic resonance (NMR) spectroscopy, 202–203
Raman spectroscopy, 196–198
Spray drying technique, 14
Standard linear solid model, 172–175, 173t–174t
Static mechanical properties (SMP)
elastomer blends
acrylonitrile butadiene rubber (NBR) and polyvinyl chloride (PVC) blends, 129–130, 130t
brittle characteristics, 129
elastic modulus, 130
epoxidized natural rubber/dodecanedioic acid (ENR/DA), 132, 132t
ethylene-octane random copolymer POE elastomers, 130
polyamide-612 (PA-612) and maleic anhydride grafted poly (ethyleneeoctene) elastomer (POE-g-MA) elastomer blends, 132, 133t
poly-lactic acid and NBR blend, 130–131, 131t
poly-lactic acid/thermoplastic polyurethane (PLA/TPU) blends, 128–129, 129t
polyolefin elastomer blends, 131t
polyolefin (POE) and propylene-based elastomer (PBE), 130
polyoxymethylene/thermoplastic polyurethane (POM/TPU) blends, 129, 129t
thermoplastic polyester elastomer/poly (butylene terephthalate) (TPEE/PBT) elastomer blends, 132, 132t
TPU/NBR34 blends, 132, 133t
elastomer composites
a-PA/EOR/MMT composites, 137, 137t
bio-based engineering elastomer/nano-silica (BEE/SiO_2), 140, 140t
PLA/TPU blends and composites, 136, 136t
polyurea urethane/hollow glass micro spheres (PUUR/HGS) composite, 139, 139t
TPEE and composites, 138, 138t
TPNR-KF and PP/EPDM-KF, 137, 137t
TPU/CNS and TPU/CNT composites, 138, 138t
TPU/WF composites, 139, 139t
XNBR/BN-PCPA composites, 140, 140t
Statistical associating fluid theory (SAFT), 249–250
Steady shear flow analysis, 92
Stereo complex-type polylactide (SC-PLA), 38–39
Strain tensor, 87
Stretchable piezoelectric nanogenerator (SPENG), 48, 48f

Styrene-butadiene-styrene/multiwall (SBS)
carbon nanotubes fillers, 49
Styrene-ethylene/butylene styrene triblock
copolymer (SEBS), 135, 135t
Styrene-isoprene-styrene block copolymer-
based conductive polymer
composites, 51
Super critical fluids (SCFs), 271–272
Supercritical methods, 285–286
Surface-modified flax elastomer
composites, 23–25
Synchrotron X-ray scattering
high flux and modern detectors, 237–238
intensity maps, 238f
tensile deformation, 239
uniaxially stretched linear low-density
polyethylene (LLDPE), 237–238, 238f
Synthetic elastomers, 1
Synthetic fiber-reinforced elastomer
composites, 120–121

T

Tannins, 333
Temperature effect, 90
Temperature scanning stress relaxation
(TSSR) method, 321
Theoretical rheology, 85
Thermal behavior
compatibilizer, 154
differential scanning calorimetry (DSC)
analysis, 152–153
blends, 153–156
elastomeric composites, 161–162
di- or multiphase elastomer blending,
152–153
dynamic mechanical thermal analysis
(DMA), 153
elastomeric composites, 162–163
EVA/TPU blends, 156
PCL/NR self-healing blend polymer, 156
storage modulus, 156
electrospinning method, 154, 155f
marble waste (MW)-filled polypropylene
(PP) and acrylonitrile butadiene (NBR)
rubber blend, 149–150
melt-blended polycaprolactone (PCL) and
natural rubber (NR), 154
miscibility, 151
MWCNT-filled fluoroelastomer
nanocomposite, 150
rubber–rubber and rubber–polymer
blend, 152
short natural fiber-reinforced polymer
composites, 150
thermodynamics, 150–152
thermogravimetric analysis (TGA)
blends, 157–161
elastomeric composites, 163–164
waste tire rubber (WTR), 155
Thermal modeling
glass transitions kinetics, 250–251
nonequilibrium molecular dynamic
(NEMD) simulation methods, 251
polyethylene glycol (PEG)-based epoxy
blends, 250–251
thermal conductivity (TC), 251
Thermogravimetric analysis (TGA)
blends, 157–161
amine modified GTR and MA grafted
polymer, 159, 159f
compatibilizers, 160
ground tire rubber (GTR) and reclaimed
tire rubber (RTR), 158–159
interfacial area per unit volume, 157
isotactic polypropylene (iPP) and nitrile
rubber (NBR) blend, 157
phenol modified PP, 158f
TPU and EVA blend, 157
elastomeric composites, 163–164
Thermoplastic copolyester elastomers
(TCEs), 141
Thermoplastic elastomer gels (TPEGs)
electroactive polymers (EAPs), 374–381
fabrication and properties
dimensionless temperature–
composition, 368f
dynamic rheological analysis, 367–368
electroresponsive media, 366–367
midblock bridging fraction, 369f
shock-absorbing media, 366–367

Thermoplastic elastomer gels (TPEGs) (*Continued*)
styrenic TPEs, 367–368
ternary phase diagram, 367f
low-dielectric, 375–377
nonpolar, 366–381
chemical modification, 381–385
physical modification, 366–381
shape-memory mechanism, 380f
stimuli-responsive, 374–381
Thermoplastic elastomers (TPEs)
atomistic molecular dynamics (MD), 246
atomistic stimulation, 245
biomass to bio-oil conversion, 332
block copolymers
ABA triblock copolymers, 357f
block purity, 358–359
block random copolymers (BRCs), 358–359
cutting-edge synthetic protocols, 362–364
designer molecules synthesis, 364–366
dissipative particle dynamics (DPD) simulations, 361–362, 370f
field phase diagram, 356–358
hydrogenation, 364–366
microphase-separated ABA triblock copolymer, 362f
midblock fractions, 363f
molecular asymmetry, 362–364
molecular transition, 361–362
morphological development, 356–358
nanostructure assembly, 356–358
olefinic TPEGs (OTPEGs), 372–374
Slip-Tube Network (STN) model, 368–372
time–composition superpositioning (tCS), 372
chemically incompatible hard and soft segments, 354–355
chemical modification
block copolymers, 381–382
poly(lactic acid) (PLA), 381–382
styrenic endblocks, 383–385
sulfonation, 385
thiol–ene click reaction, 383f
commercial uses, 353–354
composite material, 334
computational techniques, 244–245
dissipative particle dynamics (DPD), 245, 245f
elastomeric behavior, 331–332
fuzzy interface, 246
Interphase Monte Carlo (IMC) method, 247
liquefaction, 332
mean square displacement, 246, 246f
microphase-separated styrenic TPE, 355–356
microphase-separated TPU, 354–355
molecular architectures, 364f
molecular dynamics (MD) methods, 244–245
morphological characteristics
energy, 388–389
environment, 389–390
healthcare, 390–391
solvent-vapor annealing (SVA), 386–388
tetrahydrofuran (THF), 385–386
Mullins effect, 246
phenol, 332–333
polymeric materials and subclassification, 353–354, 353f
polyurethanes (TPU), 246
pyrolysis, 332
semicrystalline, 355f
vitrifiable hard segment, 355–356
Thermoplastic polyester elastomer/poly (butylene terephthalate) (TPEE/PBT) elastomer blends, 132, 132t
Thermoplastic polyurethane elastomer (TPU), 36
Thermosetting elastomers
charge equilibration (QEq), 248–249
computational studies, 249–250
cross-linking technique, 248
electronegativity equalization method (EEM), 248–249
molecular interactions, 249–250
molecular model system, 249f

network polymerization, 248–249
proximity-based MDs, 249–250
thermosetting resins, 248
Thiol–ene click reaction, 383f
Thixotropic fluid, 89
Time-temperature superposition principle, 4–5
Titanium dioxide functionalized graphene, 47–48, 47f
Titanium dioxide-reduced graphene oxide (TiO_2-RGO), 47f
Transmission electron microscopy (TEM), 112–114
Twin screw extruder, 58–59, 59f
Two-dimensional wide line separation NMR (WISE NMR), 203
Two-parameter power law, 99
Two roll mills, 60–61, 61f

U

Ultrafine full-vulcanized powdered styrene-butadiene rubber (UFPSBR) blends, 106–109
Ultrasound devulcanization
gel permeation chromatography data, 276
^{1}H transverse relaxation, 276
meshed tire rubbers, 275–276
particle-size, 275–276
roofing membrane elasticity, 275–276
sulfur-cured unfilled butadiene, 276
ultrasonic waves, 275
Universal testing machine (UTM), 35–36
Upper critical solution temperature (UCST), 151, 151f

V

Vacuum-assisted resin transfer molding (VARTM), 310
Vegetable oil-based PU materials, 73–74
Vinyl trimethoxysilane (VTMS), 114–117
Viscoelasticity
classical elasticity, 171–172
constitutive models, 172–176, 173t–174t
dynamic loading and responses
creep and relaxation test, 176
dynamic mechanical analyzer (DMA), 177
input and output wave pattern, 176f
linear viscoelastic regime, 177
loss tangent, 177
oscillatory load, 176
polymer volume, 178–179
stress response, 177
transition temperature, 179–180, 179f
elastomer blends
compatibilizers, 180
ethylene-co-vinyl acetate (EVA) and millable polyurethane (MPU), 181–182
ethylene propylene diene terpolymer and poly(vinyl chloride) (EPDMPVC) blend, 180–181
ground tire rubber (GTR) recycling, 184
natural rubber (NR)/polycaprolactone (NR/PCL) blends, 182
reclaimed rubber (RR) and poly(-ε caprolactone) (PCL) blends, 184
silane cross-linked polypropylene (PP)/EPDM elastomer blends, 183–184
ternary blends, 182, 183f
vulcanized PVC-epoxidized natural-rubber (ENR) thermoplastic elastomers blend, 181–182
elastomer composites
bromobutyl rubber (BIIR)/polyepichlorohydrin rubber, 187–188
carbon black and nanoclay fillers, 186
chain mobility, 185–186
damping, 185–186
filler geometry, 185–186
graphene content, carbon-black-filled nitrile butadiene rubber (NBR), 188–189
multifunctional thermoplastic polyurethane (TPU) elastomer composites, 190
nanoclay filled composites without carbon black, 186
organomodified layered silicates (OMLS), 189

Viscoelasticity (*Continued*)
silica-particle-filled bromo butyl rubber (BIIR), 189
physical and mathematical representation, 173t–174t
Viscoplastic fluids, 88
Viscosity curve, 86
Vitrimers, 289
Vulcanization, 2

W
Waste egg-shell derived calcium carbonate nanoparticles (WESNCC), 114–117
Waste tire rubber (WTR), 155, 278–279
Weissenberg effect, 172f
Wide-angle X-ray diffraction (WAXD), 35–36
amorphous state, 217–218
asymmetrical reflection, 217
configurations, 215–217, 217f
1D detectors, 221
2D detectors, 221, 223
Debye–Scherrer WAXD patterns, 221f
diffractometers, 222f
Eulerian cradle and transmission mode, 222f
film-to-specimen distance, 218
random microcrystallinity, 217–218
symmetrical fiber diffraction, 231
symmetrical reflection, 231–232
symmetrical transmission mode, 231
X-ray patterns and orientation, 217–224
Wide-angle X-ray scattering (WAXS)
academic research, 213
angular range and distance, diffractometers, 216, 216t
atomic theory, 212–213
Bragg's relationship, 211–212
circular halo and rings, 217
experimental setup, 212, 212f
filter, 213t
filters *versus* monochromators, 213
history, 211
instrumentation, 215
oriented amorphous polymer, 214f
radially averaged intensity trace, 216f
scattering angle, 216
theoretical calculations, 211
wavelength, 210–213, 213t
X-ray properties, 210–214

X
X-ray scattering
advantage, 209–210
mathematical treatment, 210
small-angle X-ray scattering (SAXS). *See* Small-angle X-ray scattering (SAXS)
wide-angle X-ray diffraction (WAXD). *See* Wide-angle X-ray diffraction (WAXD)
wide-angle X-ray scattering (WAXS). *See* Wide-angle X-ray scattering (WAXS)

Printed in the United States
by Baker & Taylor Publisher Services